MATHEMATICAL ELASTICITY

VOLUME I: THREE-DIMENSIONAL ELASTICITY

STUDIES IN MATHEMATICS AND ITS APPLICATIONS

VOLUME 20

NORTH-HOLLAND
AMSTERDAM · LONDON · NEW YORK · TOKYO

MATHEMATICAL ELASTICITY

VOLUME I:
THREE-DIMENSIONAL ELASTICITY

PHILIPPE G. CIARLET

Université Pierre et Marie Curie, Paris

and

City University of Hong Kong

With 44 figures

NORTH-HOLLAND
AMSTERDAM • LONDON • NEW YORK • TOKYO

ELSEVIER B.V.
Sara Burgerhartstraat 25
P.O. Box 211, 1000 AE Amsterdam
The Netherlands

ELSEVIER Inc.
525 B Street, Suite 1900
San Diego, CA 92101-4495
USA

ELSEVIER Ltd
The Boulevard, Langford Lane
Kidlington, Oxford OX5 1GB
UK

ELSEVIER Ltd
84 Theobalds Road
London WC1X 8RR
UK

First impression hardbound: 1988
First impression paperback: 1993
Second impression paperback: 2004

Library of Congress Cataloging in Publication Data
A catalog record is available from the Library of Congress.

British Library Cataloguing in Publication Data
A catalogue record is available from the British Library.

ISBN hardbound: 0 444 70259 8
ISBN paperback: 0 444 81776 x

Printed and bound by Antony Rowe Ltd, Eastbourne
Transferred to digital print on demand, 2005

To Jacques-Louis Lions

MATHEMATICAL ELASTICITY: GENERAL PLAN

PREFACE

This book is intended to be both a *thorough introduction to contemporary research in elasticity*, and a *working textbook at the graduate level for courses in pure or applied mathematics or in continuum mechanics*.

During the past decades, elasticity has become the object of a considerable renewed interest, both in its physical foundations and in its mathematical theory. One reason behind this recent attention is that it has been increasingly acknowledged that the classical *linear models* of elasticity, whose mathematical theory is now firmly established, have a limited range of applicability, outside of which they should be replaced by the genuine *nonlinear models* that they in effect approximate. Another reason, similar in its principle, is that the validity of the classical *lower-dimensional models*, such as the two-dimensional von Kármán equations for nonlinearly elastic plates, is no longer left unquestioned. A need has been felt for a better assessment of their relation to the corresponding three-dimensional models that they are supposed to approximate.

This book illustrates at length these recent trends, as shown by the *main topics covered*:

– A thorough *description*, with a pervading emphasis on the *nonlinear aspects*, of the two competing mathematical models of *three-dimensional elasticity*, either as a *boundary value problem consisting of a system of three quasilinear partial differential equations of the second order* together with specific boundary conditions, or as a *minimization problem for the associated energy* (Chapters 1 to 5);

– A *mathematical analysis* of these models, comprising in particular *complete proofs of the most recent existence results* (Chapters 6 and 7);

– A systematic *derivation of two-dimensional plate models from three-dimensional elasticity* by means of the asymptotic expansion method, which includes in particular a rigorous *convergence analysis* in the linear case and a justification of well known plate models, such as the *von Kármán equations* (Volume II);

– A *mathematical analysis of the two-dimensional plate models*, which includes in particular a review of the *existence theorems* in the nonlinear case, and an introduction to *bifurcation theory* (Volume II).

– A systematic *derivation of one-dimensional rod models from three-dimensional elasticity* (Volume II).

– A systematic derivation of mathematical models of *junctions between plates and three-dimensional structures*, or *between plates and rods*, and of *folded plates* (Volume II).

While substantial progress has been made in the *study of statics* (which is all that we consider here), the analysis of time-dependent elasticity is still at an early stage. Deep results have been recently obtained for one space variable, but formidable difficulties stand in the way of further progress in this area. It is therefore probable that a substantial time will elapse before a "dynamic" counterpart of this book can be written.

Although the emphasis is definitely on the mathematical side, every effort has been made to keep the prerequisites, whether from mathematics or continuum mechanics, to a minimum, notably by making the book as largely *self-contained* as possible. The reading of the book only presupposes some familiarity with basic topics from analysis and functional analysis.

A fascinating aspect of elasticity is that, in the course of its study, one *naturally* feels the need for studying *basic mathematical techniques of analysis and functional analysis*; how could one possibly find a better motivation? For instance:

– Both common and uncommon results from *matrix theory* are constantly needed, such as the polar factorization theorem (Theorem 3.2-2) or the celebrated Rivlin–Ericksen representation theorem (Theorem 3.6-1). In the same vein, who would think that the inequality $|\mathrm{tr}\, \boldsymbol{AB}| \leq \Sigma_i\, v_i(\boldsymbol{A})v_i(\boldsymbol{B})$, where the v_i's denote the singular values in increasing order of the matrices $\boldsymbol{A}$ and $\boldsymbol{B}$, comes up naturally in the analysis of a wide class of actual stored energy functions? Incidentally, this seemingly innocuous inequality is not easy to prove (Theorem 3.2-4)!

– The understanding of the "geometry of deformations" (Chapter 1) relies on a perhaps elementary, but "applicable", knowledge of *differential geometry*. For instance, my experience is that, among those of my students who had previously been exposed to modern differential geometry, very few could compute effectively the formula $\mathrm{d}a^{\varphi} = |\mathbf{Cof}\, \nabla\varphi\, \boldsymbol{n}|\, \mathrm{d}a$ relating reference and deformed area elements (Theorem 1.7-1).

– The study of the geometrical properties (orientation-preserving character, injectivity) of mappings in $\mathbb{R}^3$ naturally leads to using such basic tools as the *invariance of domain theorem* (Theorems 1.2-5 and 1.2-6) or the *topological degree* (Sect. 5.4); yet these are unfortunately all too often left out from standard analysis courses.

– *Differential calculus in Banach spaces* is an indispensable tool which is used throughout the book, and the unaccustomed reader should quickly become convinced of the many merits of the *Fréchet derivative* and of the *implicit function theorem*, which are the keystones to the existence theory developed in Chapter 6.

– The fundamental existence theorem for *ordinary differential equations in Banach spaces*, as well as the convergence of its *approximation by Euler's method*, are needed in the analysis of incremental methods, often used in the numerical approximation of the equations for nonlinearly elastic structures (Chapter 6).

– Basic topics from *functional analysis* and the *calculus of variations*, such as *Sobolev spaces* (which in elasticity are simply the "spaces of finite energy"), *weak convergence*, *existence of minimizers for weakly lower semi-continuous functionals*, pervade the treatment of existence results in three-dimensional elasticity (Chapters 6 and 7) and in two-dimensional plate theory (Volume II).

– Key results about *elliptic linear systems of partial differential equations*, notably sufficient conditions for the $\boldsymbol{W}^{2,p}(\Omega)$-regularity of their solutions (Theorem 6.3-6), are needed preliminaries for the existence theory of Chapter 6.

– Mostly as the result of John Ball's pioneering works in three-dimensional elasticity, *convexity* plays a particularly important rôle throughout Volume I. In particular, we shall naturally be led to finding nontrivial examples of *convex hulls*, such as that of the set of all square matrices whose determinant is >0 (Theorem 4.7-4), and of *convex functions of matrices*. For instance, functions such as $\boldsymbol{F} \rightarrow \Sigma_i \{\lambda_i(\boldsymbol{F}^{\mathrm{T}}\boldsymbol{F})\}^{\alpha/2}$ with $\alpha \geqslant 1$ naturally come up in the study of *Ogden's materials* in Chapter 4; while proving that such functions are convex is elementary for $\alpha = 2$, it becomes surprisingly difficult for the other values of $\alpha \geqslant 1$ (Sect. 4.9). Such functions are examples of John Ball's *polyconvex stored energy functions*, a concept of major importance in elasticity (Chapters 4 and 7).

– In Chapter 7, we shall come across the notion of *compensated compactness*. This technique, discovered and studied by François Murat and Luc Tartar is now recognized as a powerful tool for studying nonlinear partial differential equations.

– In Volume II, it will be shown that the derivation of two-dimensional plate models and one-dimensional rod models from three-dimensional elasticity makes constant use of the techniques of *asymptotic expansions* (formal expansions, error analysis, correctors, boundary layers, etc. . .), as developed by Jacques-Louis Lions for problems in variational form.

– *Bifurcation theory* naturally arises in the analysis of *nonlinear plate models* (Volume II). Its problems constitute a wealth of remarkable "real-life" applications of that theory (buckling, turning points, multiplicity of solutions, perturbation theory, etc.).

Another fascinating aspect of mathematical elasticity, even in the static case considered here, is that it gives rise to a number of *open problems*, e.g.,

– The extension of the "local" analysis of Chapter 6 (existence theory, continuation of the solution as the forces increase, analysis of incremental methods) to genuine mixed displacement–traction problems;

– "Filling the gap" between the existence results based on the implicit function theorem (Chapter 6) and the existence results based on the minimization of the energy (Chapter 7);

– An analysis of the nonuniqueness of solutions (cf. the examples given in Sect. 5.8);

– A mathematical analysis of contact with friction (contact, or self-contact, without friction is studied in Chapters 5 and 7);

– Finding reasonable conditions under which the minimizers of the energy (Chapter 7) are solutions of the associated Euler–Lagrange equations;

– Existence of solutions of three-dimensional nonlinear plate problems obtained through a proper extension of the two-dimensional solutions that are known to exist (Volume II);

– A numerical comparison between two- and three-dimensional plate problems which, surprisingly, is still lacking at the present time (even in the linear case, where the theoretical analysis is otherwise at a satisfactory state);

– A mathematical study of elasto-plasticity, which has been so far considered only in the framework of linearized elasticity.

This book will have therefore fulfilled its purposes if the above messages have been conveyed to its readers, that is,

– if it has convinced its more application-minded readers, such as continuum mechanicists, engineers, "applied" mathematicians, that mathematical analysis is an indispensable tool for a genuine understanding of elasticity, whether it be for its modeling or for its analysis, essentially because more and more emphasis is put on the nonlinearities (e.g., polyconvexity, bifurcation, etc. . .) whose consideration requires, even at the onset, some degree of mathematical sophistication;

– if it has convinced its more mathematically oriented readers that elasticity, far from being a dusty classical field, is on the contrary a prodigious source of challenging open problems.

The book, which comprises two volumes, is divided into consecutively numbered *chapters*. Chapter m contains an introduction, several sections numbered Sect. $m.1$, Sect. $m.2$, etc., and is concluded by a set of exercises. Within Sect. $m.n$, theorems are consecutively numbered, as Theorem $m.n$-1, Theorem $m.n$-2, etc., and figures are likewise consecutively numbered, as Figure $m.n$-1, Figure $m.n$-2, etc. Remarks and formulas are not numbered. The end of a theorem or of a remark is indicated by the symbol ■ in the right margin. In Chapter m, exercises are numbered as Exercise $m.1$, Exercise $m.2$, etc.

All the important results are stated in the form of *theorems* (there are no lemmas, propositions, or corollaries), which therefore represent the core of the text. At the other extreme, the *remarks* are intended to point out some interpretations, extensions, counter-examples, relations with other results, that in principle can be skipped during a first reading; yet, they could be helpful for a better understanding of the material. When a term is rigorously defined for the first time, it is set in boldface if it is deemed important, or in italics otherwise. Terms that are only given a loose or intuitive meaning are put between quotation marks.

Special attention has been given to the *notation*, which so often has a distractive and depressing effect in a first encounter with elasticity. In particular, the book begins with a special section, which the reader is urged to read first, about the rules that have guided the choices of notation here. This section also reviews the main *definitions* and *formulas* that will be used throughout the text.

Complete proofs are generally given. In particular, whenever a mathematical result is of particular significance in elasticity, its proof has been included: This is the case for instance of the polar factorization theorem, of the Rivlin–Ericksen representation theorem (which is seldom proved in books on matrices), or of the convexity of the function $\boldsymbol{F} \rightarrow \Sigma_i \{\lambda_i(\boldsymbol{F}^{\mathrm{T}}\boldsymbol{F})\}^{\alpha/2}$ for $\alpha \geq 1$ (which is seldom mentioned as a nontrivial example of convex function), etc. More standard mathematical prerequisites are presented (usually without proofs) in special *starred sections*, scattered throughout the book according to the local needs.

Exercises of varying difficulty are included at the end of each chapter. Some are straightforward applications of, or complements to, the text; others, which are more challenging, are usually provided with hints or references.

Although more than 570 items are listed in the bibliography, there has been no attempt to compile an exhaustive list of references. The interested readers should look at the extensive bibliography covering the years 1678–1965 in the treatise of Truesdell & Noll [1965], at the additional references found in the recent books by Marsden & Hughes [1983], Hanyga [1985], and Oden [1986], and at the papers of Antman [1983] and Truesdell [1983], which give short and illuminating historical perspectives on the interplay between elasticity and analysis.

The readers of this book are strongly advised to complement the material given here by consulting a few other books, and in this respect, we particularly recommend the following *general references on three-dimensional elasticity* (general references on lower-dimensional theories of plates and rods are given in Volume II):

– In-depth perspectives in continuum mechanics in general, and in elasticity in particular: the treatises of Truesdell & Toupin [1960] and Truesdell & Noll [1965], and the books by Germain [1972], Truesdell [1977], Gurtin [1981b].

– Classical and modern expositions of elasticity: Love [1927], Murnaghan [1951], Timoshenko [1951], Novozhilov [1953], Sokolnikoff [1956], Novozhilov [1961], Eringen [1962], Landau & Lifchitz [1967], Green & Zerna [1968], Stoker [1968], Green & Adkins [1970], Knops & Payne [1971], Duvaut & Lions [1972], Fichera [1972a, 1972b], Gurtin [1972], Wang & Truesdell [1973], Villagio [1977], Gurtin [1981a], Nečas & Hlaváček [1981], Ogden [1984].

– Mathematically oriented treatments in nonlinear elasticity: Marsden & Hughes [1983], Hanyga [1985], Oden [1986], and the forthcoming book of Antman [1988].

In our description of continuum mechanics and elasticity, we have only singled out *two axioms*: the stress principle of Euler and Cauchy (Sect. 2.2) and the axiom of material frame-indifference (Sect. 3.3), thus considering that all the other notions are *a priori* given. The reader interested in a more axiomatic treatment of the basic concepts, such as frame of reference, body, reference configuration, mass, forces, material frame-indifference, isotropy, should consult the treatise of Truesdell & Noll [1965], the book of Wang & Truesdell [1973], and the fundamental contributions of Noll [1959, 1966, 1972, 1973, 1978].

At the risk of raising the eyebrows of some of our readers, and at the expense of various *abus de langage*, we have also ignored the difference between second-order tensors and matrices. The readers disturbed by this approach should look at the books of Abraham, Marsden & Ratiu [1983] and, especially, of Marsden & Hughes [1983], where they will find all the tensorial and differential geometric aspects of elasticity explained in depth and put in their proper perspective.

This book is an outgrowth of lectures that I have given over the past years at the Tata Institute of Fundamental Research, at the University of Stuttgart, at the Université Pierre et Marie Curie, and at the Ecole Normale Supérieure. During this period, I had the fortune of working with colleagues and students, to all of whom I am particularly indebted for their cooperation, in particular: Michel Bernadou, Dominique Blanchard, Jean-Louis Davet, Philippe Destuynder, Giuseppe Geymonat, Hervé Le Dret, Hu Jian-Wei, Srinivasan Kesavan, Klaus Kirchgässner, Florian Laurent, Jindřich Nečas, Robert Nzengwa, Jean-Claude Paumier, Peregrina Quintela-Estevez, Patrick Rabier, and Annie Raoult. Special thanks are also due to Stuart Antman, Irene Fonseca, Morton Gurtin, Patrick Le Tallec, Bernadette Miara, François Murat, Tinsley Oden, and Gérard Tronel, who were kind enough to read early drafts of Volume I and to suggest significant improvements. For their especially expert and diligent assistance as regards the material realization of the book, I very sincerely thank Ms. Bugler, Ms. Dampérat, and Ms. Ruprecht.

Last but not least, this book is dedicated to Jacques-Louis Lions, as an expression of my deep appreciation and gratitude.

August, 1986 Philippe G. Ciarlet

TABLE OF CONTENTS

* Indicates a section dealing exclusively with mathematical preliminaries.

MAIN NOTATION, DEFINITIONS, AND FORMULAS

1. GENERAL

Preliminary remarks

The variety of notations used in books and papers devoted to Elasticity is often a confusing and dismaying fact, especially for the beginner. To ease the reader's pain, we have systematically tried to strive towards the greatest "transparency", notably by:

– minimizing the number of different symbols and alphabets that are introduced (sometimes at the expense of longer formulas);

– consistently following simple rules governing the usage of the various types of characters found in the book (these rules are explained below);

– "interpolating at best" between the various notation found in the literature, with a definite preference for the notation commonly used in the "partial differential equation literature" (for instance the generic point in the reference configuration is denoted x and the reference configuration is denoted $\bar{\Omega}$, while these are usually denoted X and B, respectively, in books of mechanics, etc.).

General conventions

Unless otherwise indicated, *all numbers, vectors, matrices, functions, etc., considered in this book are real.*

If a colon is set before an equality sign, as in $:=$, the right-hand side defines the left-hand side.

If a group of words is set between quotation marks, it is to be understood in a naive or intuitive sense. Quotation marks are also used to delineate mathematical expressions that are not displayed.

The symbol $\Rightarrow$ means "implies".

The symbol $\Leftrightarrow$ means "is equivalent to".

The symbols c, c_1, c_2, etc. (or $c(a)$, $c_1(a)$, $c_2(a)$, etc., if a dependence on some variable a must be indicated) denote constants, usually found in inequalities, which are not necessarily the same at their various occurrences.

The *repeated index convention for summation* is systematically used in any formula where the same Latin index (i, j, etc.) appears twice, unless the formula is followed by the indication "(no summation)". The range of a Latin index is either $\{1, 2, \ldots, n\}$ where n is an arbitrary integer, or $\{1, 2, 3\}$ in any application to three-dimensional elasticity (unless of course it is used for indexing an infinite sequence or a series); whichever case applies should be always clear from the context. For instance,

$$-\partial_j(\sigma_{ij} + \sigma_{kj}\partial_k u_i) = f_i \text{ means } -\sum_{j=1}^{3} \partial_j\Big(\sigma_{ij} + \sum_{k=1}^{3} \sigma_{kj}\partial_k u_i\Big) = f_i$$

for $i = 1, 2, 3$,

$$\det A = \tfrac{1}{6}\varepsilon_{ijk}\varepsilon_{pqr}a_{ip}a_{jq}a_{kr} \text{ means } \det A = \tfrac{1}{6}\sum_{i,j,k,p,q,r=1}^{3} \varepsilon_{ijk}\varepsilon_{pqr}a_{ip}a_{jq}a_{kr}$$

for a matrix $A = (a_{ij})$ of order three, etc.

In the following list of notations, definitions, and formulas, we have indicated the sections where more information can be found on a particular notion.

A *box* indicates a definition, or a relation, that is particularly important in elasticity.

Sets, topological spaces, mappings

$\emptyset$: empty set.
$\mathbb{N} = \{0, 1, 2, \ldots\}$: set of ≥ 0 integers.
$\mathbb{Z} = \{\ldots, -2, -1, 0, 1, 2, \ldots\}$: set of integers.
$\mathbb{R}$: set of real numbers.
$\{-\infty\} \cup \mathbb{R} \cup \{+\infty\}$: set of extended real numbers (Sect. 4.7).

Remarks: We do *not* use the often employed symbols $\mathbb{R}_+$ and $\mathbb{R}_+^*$ to denote the sets $[0, +\infty[$ and $]0, +\infty[$ respectively, for these would be inconsistent with the notation $\mathbb{M}_+^n$ used for the set of matrices with determinant >0. In the set of extended real numbers, the values $a = -\infty$ and $b = +\infty$ are allowed in the definition of closed intervals $[a, b]$; thus for instance,

$$[a, +\infty] = [a, +\infty[\cup \{+\infty\}, [-\infty, +\infty] = \{-\infty\} \cup \mathbb{R} \cup \{+\infty\}, \text{ etc.}$$

$A \subsetneqq B$: A is strictly contained in B.

$B - A = B \cap (X - A)$: complement of a subset $A \subset X$ with respect to a subset $B \subset X$.

$X - A = \{y \in X;\ y \notin A\}$: complement of a subset $A \subset X$.

Points in $\mathbb{R}^n$ or in a general set are denoted by lightface minuscules (examples: $x, y, \ldots$).

Subsets in $\mathbb{R}^n$ or in a general set are denoted by lightface majuscules (examples: $\bar{\Omega}, A, \ldots$).

$\mathfrak{S}_n$: set of all permutations of $\{1, 2, \ldots, n\}$.

$\bar{A}$ or $\{A\}^-$ or $\mathrm{cl}\,A$: closure of a set A.

$\mathring{A}$ or $\mathrm{int}\,A$: interior of a set A.

∂A: boundary of a set A.

$\mathrm{card}\,A$: number of elements of a set A.

$f: X \to Y$, or $f: x \in X \to f(x) \in Y$: mapping, or function, from X into Y.

$f: A \subset X \to Y$: mapping f from a subset A of X into Y.

$f \circ g$: composite mapping.

$f|_A$: restriction of the mapping f to the set A.

$f(\cdot, b)$: partial mapping $x \to f(x, b)$.

$f(A) = \{y \in Y;\ y = f(x) \text{ for some } x \in X\}$: image of a subset $A \subset X$ by the mapping $f: X \to Y$ (also denoted $\mathrm{Im}(A)$ if A is a linear mapping).

$f^{-1}(B) = \{x \in X;\ f(x) \in B\}$: inverse image of a subset $B \subset Y$ by the mapping $f: X \to Y$.

A mapping $f: X \to Y$ is *surjective*, or onto, if $f(X) = Y$; *injective*, or one-to-one, if $\mathrm{card}\, f^{-1}(\{y\}) = 0$ or 1 for all $y \in Y$; *bijective* if it is both injective and surjective.

f^{-1}: inverse mapping of a bijective mapping.

$\mathrm{supp}\, f = \{x \in X;\ f(x) \neq 0\}^-$: support of a function $f: X \to \mathbb{R}$.

Numbers and real-valued functions are represented by symbols beginning by a lightface minuscule or majuscule (examples: c, r, ρ, $\iota_1(A)$, $\det A$, u_i, E_{ij}, A_{ii}, etc.) or, in two instances only, by special Roman characters ($\mathbb{L}(\boldsymbol{F}, \mathbf{Cof}\,\boldsymbol{F}, \det \boldsymbol{F})$ and $\mathbb{W}(\boldsymbol{F}, \mathbf{Cof}\,\boldsymbol{F}, \det \boldsymbol{F})$).

A functional, i.e., a mapping from a space of real-valued, or vector-valued, functions into $\mathbb{R}$ or $\mathbb{R} \cup \{+\infty\}$ is denoted by a light-face majuscule; examples:

$$F(\boldsymbol{\psi}) = \int_\Omega \hat{f}(x, \boldsymbol{\psi}(x))\,\mathrm{d}x, \quad L(\boldsymbol{v}) = \int_\Omega \boldsymbol{f} \cdot \boldsymbol{v}\,\mathrm{d}x\,.$$

Whenever some confusion could arise (in particular when functions are differentiated), we have chosen different symbols for representing the same "function" in terms of different variables, and its values (e.g. $\Sigma = \hat{\Sigma}(F) = \tilde{\Sigma}(C)$).

id, or id_X: identity mapping in a set X ($f = id_X$ means $f(x) = x$ for all $x \in X$).

$\operatorname{sgn} \alpha = +1$ if $\alpha > 0$, -1 if $\alpha < 0$.

$\deg(\varphi, \Omega, b)$: topological degree of a mapping $\varphi \in \mathscr{C}^0(\bar{\Omega}; \mathbb{R}^n)$ at a point $b \notin \varphi(\partial\Omega)$ with respect to a bounded open subset Ω of $\mathbb{R}^n$ (Sect. 5.4).

$(\varphi_k)_{k=l}^{\infty}$, or (φ_k): sequence of elements $\varphi_l, \varphi_{l+1}, \ldots, \varphi_k, \ldots$.

$\varphi = \lim_{k\to\infty} \varphi_k$, or $\varphi_k \xrightarrow[k\to\infty]{} \varphi$: the sequence (φ_k) converges, and its limit is φ.

$x \to a^+$: the real numbers $x > a$ converge to $a \in \mathbb{R}$.

$x \to a^-$: the real numbers $x < a$ converge to $a \in \mathbb{R}$.

$\liminf_{k\to\infty} \varphi_k$, $\limsup_{k\to\infty} \varphi_k$: limit inferior, limit superior, of a sequence (φ_k) of numbers, or of functions with values, in the set $\{-\infty\} \cup \mathbb{R} \cup \{+\infty\}$ (Sect. 7.2).

When no confusion should arise, the symbol "$k \to \infty$" is omitted for notational brevity (e.g. $\varphi = \lim \varphi_k$, $\varphi = \limsup \varphi_k$, $\varphi_k \to \varphi$, etc.).

Vector spaces

$[a, b] = \{ta + (1-t)b; 0 \leq t \leq 1\}$: closed segment with end-points a and b.

$]a, b[= \{ta + (1-t)b;\ 0 < t < 1\}$: open segment with end-points a and b.

$\operatorname{co} A$: convex hull of a set A (smallest convex set containing the set A; Sect. 4.7).

$\operatorname{Ker} L = \{x \in X; Lx = 0\}$: kernel of the linear mapping $L : X \to Y$.

$\operatorname{Im} L = \{y \in Y; y = Lx \text{ for some } x \in X\}$: image of the space X by the linear mapping $L : X \to Y$ (also denoted $L(X)$).

$\operatorname{Coker} L = Y/\operatorname{Im} L$: cokernel of the linear mapping $L : X \to Y$.

$\|\cdot\|$, or $\|\cdot\|_X$: norm in a vector space X.

$B_r(a) = \{x \in X;\ \|x - a\| < r\}$: open ball of radius r centered at x.

$B_r = B_r(0) = \{x \in X;\ \|x\| < r\}$: open ball of radius r centered at the origin.

$S_r = \{x \in X;\ \|x\| = r\}$: sphere of radius r centered at the origin.

$|\cdot|$: semi-norm (which may be a norm).

X': (topological) dual of a normed vector space X (Sect. 1.2).

$\|\cdot\|'$: norm in the dual space.
X/Y: quotient of a vector space X by a vector subspace Y of X.
$X \hookrightarrow Y$: X is contained in Y with a continuous injection.
$X \Subset Y$: X is contained in Y with a compact injection.

Let $f: A \subset X \to Y$ and $\alpha \geqslant 0$. Then

$$\left.\begin{array}{l} f(x) = O(\|x\|^\alpha) \\ \text{or simply } f(x) = O(x) \text{ if } \alpha = 1 \end{array}\right\} \Leftrightarrow \left\{\begin{array}{l} \text{there exist a constant } c \text{ and a} \\ \text{neighborhood } V \text{ of the origin} \\ \text{in } X \text{ such that} \\ \|f(x)\|_Y \leqslant c(\|x\|_X)^\alpha \text{ for all} \\ x \in A \cap V. \end{array}\right.$$

$$\left.\begin{array}{l} f(x) = o(\|x\|^\alpha) \\ \text{or simply } f(x) = o(x) \text{ if } \alpha = 1 \end{array}\right\} \Leftrightarrow \left\{ \lim_{x \to 0} \frac{\|f(x)\|_Y}{(\|x\|_X)^\alpha} = 0 \right. .$$

$$f(x, y) = o_y(x) = o(x; y) \Leftrightarrow \text{for each } y, \lim_{x \to 0} \frac{\|f(x, y)\|}{\|x\|} = 0 .$$

$\varphi_k \to \varphi$, or $\varphi = \lim \varphi_k$: strong convergence in $X \Leftrightarrow \lim \|\varphi_k - \varphi\|_X = 0$.
$\varphi_k \rightharpoonup \varphi$: weak convergence in $X \Leftrightarrow \lim L(\varphi_k) = L(\varphi)$ for all $L \in X'$ (Sect. 7.1).

Some function spaces

$\mathscr{L}(X; Y)$: space of all continuous linear mappings from a normed vector space X into a normed vector space Y.
$\mathscr{L}(X) = \mathscr{L}(X; X)$ (Sect. 1.2).
$X' = \mathscr{L}(X; \mathbb{R})$ (Sect. 1.2).
$\mathscr{I}som(X; Y) = \{A \in \mathscr{L}(X; Y);\ A$ is bijective, and $A^{-1} \in \mathscr{L}(Y; X)\}$. (Sect. 1.2).
$\mathscr{I}som(X) = \mathscr{I}som(X; X)$ (Sect. 1.2).

Remark. $\mathscr{I}som(X; Y)$ is *not* a vector space.

$\mathscr{L}_k(X; Y)$: space of all continuous k-linear mappings from a normed vector space X into a normed vector space Y, $k \geqslant 2$ (Sect. 1.3).
$\mathscr{C}^0(E; F)$: set of all continuous mappings from a topological space E into a topological space F.
$\mathscr{C}^0(E) = \mathscr{C}^0(E; \mathbb{R})$.
$\mathscr{C}^m(\Omega; Y)$: space of all m times continuously differentiable mappings

from an open subset Ω of a normed vector space X into a normed vector space $Y, 1 \leq m \leq \infty$ (Sects. 1.2 and 1.3).

$\mathscr{C}^m(\Omega) = \mathscr{C}^m(\Omega; \mathbb{R})$.

$\mathscr{C}^m(\bar{\Omega})$, where Ω is a bounded open subset of $\mathbb{R}^n$, and $1 \leq m \leq \infty$: space of all functions $v \in \mathscr{C}^m(\Omega)$ such that, for each multi-index α with $|\alpha| \leq m$, there exists a function $v^\alpha \in \mathscr{C}^0(\bar{\Omega})$ such that $v^\alpha|_\Omega = \partial^\alpha v$ (Sect. 1.3).

$\|v\|_{\mathscr{C}^m(\bar{\Omega})} = \max_{|\alpha| \leq m} \sup_{x \in \bar{\Omega}} |v^\alpha(x)|$ (Sect. 1.3).

$\mathscr{C}^{m,\lambda}(\bar{\Omega})$, where Ω is a bounded open subset of $\mathbb{R}^n$, $1 \leq m < \infty$, $0 < \alpha \leq 1$: space of all functions $v \in \mathscr{C}^m(\bar{\Omega})$ whose mth partial derivatives satisfy on $\bar{\Omega}$ a Hölder condition with exponent λ (Sect. 1.3).

$$\|v\|_{\mathscr{C}^{m,\lambda}(\bar{\Omega})} = \|v\|_{\mathscr{C}^m(\bar{\Omega})} + \max_{|\alpha|=m} \sup_{\left\{\substack{x, y \in \bar{\Omega} \\ x \neq y}\right.} \frac{|\partial^\alpha v(x) - \partial^\alpha v(y)|}{|x-y|^\lambda} \quad \text{(Sect. 1.3).}$$

In what follows, Ω is an open subset of $\mathbb{R}^n$.

$\mathscr{D}(\Omega) = \{v \in \mathscr{C}^\infty(\Omega),\ \operatorname{supp} v \text{ is compact}\}$ (Sect. 6.1).

$\mathscr{D}'(\Omega)$: space of distributions on Ω (Sect. 6.1).

$L^p(\Omega)$: space of equivalence classes of dx – almost everywhere equal functions v that satisfy

$$|v|_{0,p,\Omega} = \left\{\int_\Omega |v(x)|^p \, dx\right\}^{1/p} < +\infty \quad \text{if } 1 \leq p < \infty,$$

$$|v|_{0,\infty,\Omega} = \inf\{\alpha \geq 0;\ dx - \operatorname{meas}\{x \in \Omega;\ |v(x)| \geq \alpha\} = 0\} < +\infty, \quad \text{if } p = \infty.$$

$L^p(\partial\Omega)$: space of equivalence classes of da – almost everywhere equal functions that satisfy

$$\|v\|_{L^p(\partial\Omega)} = \left\{\int_{\partial\Omega} |v|^p \, da\right\}^{1/p} < +\infty,\ 1 \leq p < \infty \text{ (Sect. 1.6)}.$$

$W^{m,p}(\Omega) = \{v \in L^p(\Omega);\ \partial^\alpha v \in L^p(\Omega) \text{ for all } |\alpha| \leq m\}, 1 \leq p \leq \infty$ (Sect. 6.1).

$W_0^{m,p}(\Omega) =$ closure of $\mathscr{D}(\Omega)$ in $W^{m,p}(\Omega)$, $1 \leq p < \infty$ (Sect. 6.1).

$$\|v\|_{m,p,\Omega} = \begin{cases} \left\{\int_\Omega \sum_{|\alpha|\le m} |\partial^\alpha v|^p \,\mathrm{d}x\right\}^{1/p}, & 1 \le p < \infty, \\ \max_{|\alpha|\le m} |\partial^\alpha v|_{0,\infty,\Omega}, & p = \infty \text{ (Sect. 6.1)}. \end{cases}$$

$$|v|_{m,p,\Omega} = \begin{cases} \left\{\int_\Omega \sum_{|\alpha|= m} |\partial^\alpha v|^p \,\mathrm{d}x\right\}^{1/p}, & 1 \le p < \infty, \\ \max_{|\alpha|= m} |\partial^\alpha v|_{0,\infty,\Omega}, & p = \infty \text{ (Sect. 6.1)}. \end{cases}$$

$H^m(\Omega) = W^{m,2}(\Omega)$ (Sect. 6.1).
$H_0^m(\Omega) = W_0^{m,2}(\Omega)$ (Sect. 6.1).
$\|v\|_{m,\Omega} = \|v\|_{m,2,\Omega}$ (Sect. 6.1).
$|v|_{m,\Omega} = |v|_{m,2,\Omega}$ (Sect. 6.1).

Remark. Notice that $|v|_{0,\Omega} = \|v\|_{0,\Omega}$; hence $|\cdot|_{0,\Omega}$ also denotes the norm in the space $L^2(\Omega)$.

If $X(\Omega)$ denotes a space of real-valued functions defined over Ω, $\boldsymbol{X}(\Omega)$ denotes any space of vector-valued, or tensor-valued, mappings whose components, or elements, are in $X(\Omega)$, e.g.,

$$\boldsymbol{\mathcal{D}}(\Omega) = \{\boldsymbol{v} = (v_i);\ v_i \in \mathcal{D}(\Omega)\},$$

$$\boldsymbol{W}^{1,p}(\Omega) = \{\boldsymbol{v} = (v_i);\ v_i \in W^{1,p}(\Omega)\},$$

$$\boldsymbol{L}^2(\Omega) = \{\boldsymbol{\Sigma} = (\sigma_{ij});\ \sigma_{ij} \in L^2(\Omega)\}, \text{ etc.},$$

and the associated norms or semi-norms are denoted by the same symbols, e.g.

$$\|\boldsymbol{v}\|_{1,p,\Omega} = \left(\sum_{i=1}^{3} |v_i|_{1,p,\Omega}^p\right)^{1/p} \quad \text{for } \boldsymbol{v} \in \boldsymbol{W}^{1,p}(\Omega),$$

$$|\boldsymbol{\Sigma}|_{0,\Omega} = \left(\sum_{i,j=1}^{3} |\sigma_{ij}|_{0,\Omega}^2\right)^{1/2} \quad \text{for } \boldsymbol{\Sigma} \in \boldsymbol{L}^2(\Omega), \text{ etc.}$$

Differential calculus

Let X and Y be normed vector spaces, and let $f:\Omega\subset X\to Y$ be given, Ω: open subset of X.

$f'(a)\in\mathscr{L}(X;Y)$: derivative of f at $a\in\Omega$ (Sect. 1.2).

$\dfrac{\mathrm{d}f}{\mathrm{d}x}(a)=f'(a)$ if $X=\mathbb{R}$.

$\partial_j f(a)\in\mathscr{L}(X_j;Y)$, or $\dfrac{\partial f}{\partial x_j}(a)$: jth partial derivative of f at a, when $X=\prod_j X_j$ (Sect. 1.2).

$\mathbf{grad}\, f(a)=\left(\dfrac{\partial f}{\partial x_i}(a)\right)\in\mathbb{R}^n$: gradient of a function $f:\Omega\subset\mathbb{R}^n\to\mathbb{R}$ at $a\in\Omega$ (Sect. 1.2).

$\mathbf{grad}_x\, f(a,y)=\left(\dfrac{\partial f}{\partial x_i}(a,y)\right)\in\mathbb{R}^n$ (Sect. 1.2).

$\dfrac{\partial W}{\partial F}(F)=\left(\dfrac{\partial W}{\partial F_{ij}}(F)\right)\in\mathbb{M}^n$: gradient of a function $W:\Omega\subset\mathbb{M}^n\to\mathbb{R}$ at $F\in\mathbb{M}^n$ (Sect. 1.2).

$\nabla\varphi(a)=(\partial_j\varphi_i(a))\in\mathbb{M}^3$: deformation gradient at $a\in\Omega$ of a mapping $\varphi:\Omega\subset\mathbb{R}^3\to\mathbb{R}^3$ (Sect. 1.4).

$\operatorname{div} u(a)=\partial_i u_i(a)\in\mathbb{R}$: divergence of a vector field $u=(u_i)$: $\Omega\subset\mathbb{R}^n\to\mathbb{R}^n$ at $a\in\Omega$ (Sect. 1.6).

$\mathbf{div}\, T(a)=\partial_j T_{ij}(a)e_i\in\mathbb{R}^n$: divergence of a tensor field $T=(T_{ij}):\Omega\subset\mathbb{R}^n\to\mathbb{M}^n$ at $a\in\Omega$ (Sect. 1.7).

Green's formulas ($n=(n_i)$ is the unit outward normal vector along $\partial\Omega$):

$$\int_\Omega \partial_i u \, dx = \int_{\partial\Omega} u n_i \, da, \text{ with } u : \bar{\Omega} \to \mathbb{R} \text{ (Sects. 1.6 and 6.1)},$$

$$\int_\Omega \operatorname{div} \boldsymbol{u} \, dx = \int_{\partial\Omega} \boldsymbol{u} \cdot \boldsymbol{n} \, da, \text{ with } \boldsymbol{u} : \bar{\Omega} \to \mathbb{R}^n \text{ (Sect. 1.6)},$$

$$\int_\Omega \operatorname{div} \boldsymbol{T} \, dx = \int_{\partial\Omega} \boldsymbol{T}\boldsymbol{n} \, da, \text{ with } \boldsymbol{T} : \bar{\Omega} \to \mathbb{M}^n \text{ (Sect. 1.7)},$$

$$\int_\Omega \operatorname{div} \boldsymbol{T} \cdot \boldsymbol{\theta} \, dx = -\int_\Omega \boldsymbol{T} : \boldsymbol{\nabla\theta} \, dx + \int_{\partial\Omega} \boldsymbol{T}\boldsymbol{n} \cdot \boldsymbol{\theta} \, da, \text{ with}$$
$$\boldsymbol{T} : \bar{\Omega} \to \mathbb{M}^n \text{ and } \boldsymbol{\theta} : \bar{\Omega} \to \mathbb{R}^n \text{ (Sect. 2.4)},$$

$$\int_\Omega \operatorname{div} \boldsymbol{S} \cdot \boldsymbol{v} \, dx = -\int_\Omega \boldsymbol{S} : \boldsymbol{e}(\boldsymbol{v}) \, dx + \int_{\partial\Omega} \boldsymbol{S}\boldsymbol{n} \cdot \boldsymbol{v} \, da, \text{ with}$$
$$\boldsymbol{S} : \bar{\Omega} \to \mathbb{S}^n \text{ and } \boldsymbol{v} : \bar{\Omega} \to \mathbb{R}^n \text{ (Sect. 6.3)}.$$

$f''(a) \in \mathscr{L}_2(X; Y)$: second derivative of f at $a \in \Omega$ (Sect. 1.3).

$\partial_{ij} f(a) = \dfrac{\partial^2 f}{\partial x_i \partial x_j}(a) \in \mathbb{R}$: second order partial derivative of $f : \Omega \subset \mathbb{R}^n \to \mathbb{R}$ at $a \in \Omega$ (Sect. 1.3).

$f^{(m)}(a) \in \mathscr{L}_m(X; Y)$: mth derivative of a mapping f at $a \in \Omega$ (Sect. 1.3).

$f^{(m)}(a)h^m = f^{(m)}(a)(h_1, h_2, \ldots, h_m) \in Y$ when $h_i = h$, $1 \le i \le m$ (Sect. 1.3).

$\partial^\alpha v(a) = \dfrac{\partial^{|\alpha|} v}{\partial x_1^{\alpha_1} \cdots \partial x_n^{\alpha_n}}$, $|\alpha| = \alpha_1 + \cdots + \alpha_n$: multi-index notation for partial derivatives of functions $v : \Omega \subset \mathbb{R}^n \to \mathbb{R}$, with $\alpha = (\alpha_1, \ldots, \alpha_n) \in \mathbb{N}^n$ (Sect. 1.3).

Remark: The notations $\partial_j f$, $\partial f / \partial x_j$, $\partial^2 f / \partial x_i \partial x_j$, $\partial^\alpha v$, are also used for denoting partial derivatives in the sense of distribution.

Vectors, matrices, tensors

Vectors and vector-valued functions are represented by symbols beginning by a boldface minuscule (examples: $\boldsymbol{o}$, $\boldsymbol{n}^{\varphi}$, $\boldsymbol{t}(x, \boldsymbol{n})$, $\boldsymbol{\varphi}(x)$, $\mathbf{div}\, \boldsymbol{T}$, $\boldsymbol{id}$, $\mathbf{grad}_x\, f(a, y)$, . . .).

Remark. Once a basis is specified, we identify a point $x \in \mathbb{R}^n$ with the vector $\boldsymbol{ox}$, and we identify the difference $(y - x)$ with the vector $\boldsymbol{xy}$.

Matrices and matrix-valued functions are represented by symbols beginning by a boldface majuscule (examples: $\boldsymbol{T}$, $\boldsymbol{E}(\boldsymbol{u})$, $\boldsymbol{I}$, $\boldsymbol{0}$, $\mathbf{Diag}\, \mu_i$, $\mathbf{Cof}\, \boldsymbol{A}$, $\check{\boldsymbol{\Sigma}}(\boldsymbol{E})$, etc.), with the following exceptions: $\nabla\boldsymbol{\varphi}$, $\nabla\boldsymbol{u}$ (deformation and displacement gradients), $\partial W / \partial \boldsymbol{F}$ (gradient of a function $W : \Omega \subset \mathbb{M}^n \to \mathbb{R}$), $\boldsymbol{e}(\boldsymbol{u})$ (linearized strain tensor).

Only two higher-order tensors are found in this book, the *orientation tensor* (ε_{ijk}), a tensor of order 3 defined by (Sect. 1.1)

$$\varepsilon_{ijk} = \begin{cases} +1 \text{ if } \{i, j, k\} \text{ is an even permutation of } \{1, 2, 3\}, \\ -1 \text{ if } \{i, j, k\} \text{ is an odd permutation of } \{1, 2, 3\}, \\ \ \ 0 \text{ if at least two indices are equal,} \end{cases}$$

and the *elasticity tensor*, a tensor of order 4 denoted by $\mathsf{A}(x, \boldsymbol{F}) = (a_{ijkl}(x, \boldsymbol{F}))$ (Sect. 5.9).

Sets of matrices are denoted by special Roman characters (examples: $\mathbb{M}^3$, $\mathbb{M}^3_+$, $\mathbb{O}^3$, $\mathbb{S}^3_>$). Special Roman characters are also used for polyconvex representations of functions of $\boldsymbol{F} \in \mathbb{M}^3_+$ (example: $W(\boldsymbol{F}) = \mathbb{W}(\boldsymbol{F}, \mathbf{Cof}\, \boldsymbol{F}, \det \boldsymbol{F})$).

Let $\{\boldsymbol{e}_1, \boldsymbol{e}_2, \ldots, \boldsymbol{e}_n\}$ denote the canonical basis of the space $\mathbb{R}^n$. If $\boldsymbol{u}$ is a vector in $\mathbb{R}^n$, we denote by $(\boldsymbol{u})_i$, or u_i, its components, and we write

$$\boldsymbol{u} = (u_i) = (u_1, u_2, \ldots, u_n) = \sum_{i=1}^{n} u_i \boldsymbol{e}_i \,.$$

Remark: This notation differs (by the commas) from that used to denote the transpose of a column vector (see below).

If $\boldsymbol{A}$ is a matrix of type (m, n) (m rows, n columns), we denote by $(\boldsymbol{A})_{ij}$, or A_{ij}, or a_{ij}, its element at the ith row and jth column, and we write

$$\boldsymbol{A} = (A_{ij}) = (a_{ij}) = \begin{pmatrix} a_{11} & a_{12} \cdots a_{1n} \\ a_{21} & a_{22} \cdots a_{2n} \\ \vdots & \vdots \quad\quad \vdots \\ a_{m1} & a_{m2} \cdots a_{mn} \end{pmatrix}.$$

When viewed as a matrix, a vector in $\mathbb{R}^n$ is *always* identified with a column vector, i.e., a matrix of type $(n, 1)$. We use the following notation for *vectors*:

$u^T = (u_1 u_2 \cdots u_n)$: transpose of the vector u (a row vector, i.e., a matrix of type $(1, n)$).

$u \cdot v = u^T v$: Euclidean inner product in $\mathbb{R}^n$.

$|u| = \sqrt{u \cdot u}$: Euclidean norm in $\mathbb{R}^n$.

$u \otimes v = uv^T = (u_i v_j)$: tensor product in $\mathbb{R}^n$.

$u \wedge v = \varepsilon_{ijk} u_j v_k e_i$: exterior product in $\mathbb{R}^3$.

We use the following notation for *matrices*:

A^T: transpose of the matrix A.

A^{-1}: inverse of the matrix A.

$A^{-T} = (A^{-1})^T = (A^T)^{-1}$.

$A^{1/2}$ = square root of a symmetric positive definite matrix A (Sect. 3.2).

A^α, $\alpha > 0$: αth power of a symmetric positive definite matrix A (Sect. 4.9).

$I = (\delta_{ij})$: unit matrix.

Diag μ_i, or **Diag**$(\mu_1, \mu_2, \ldots, \mu_n)$: diagonal matrix whose diagonal elements are $\mu_1, \mu_2, \ldots, \mu_n$ (in this order).

tr A: trace of the matrix A (tr also denotes the trace operator in Sobolev spaces).

det A: determination of the matrix A.

$\lambda_i = \lambda_i(A)$: eigenvalues of the matrix A.

$v_i = v_i(A) = \{\lambda_i(A^T A)\}^{1/2}$: singular values of the matrix A (Sect. 3.2).

$|A| = \sup_{v \neq 0}(|Av|/|v|) = \max_i \{\lambda_i(A^T A)\}^{1/2}$: spectral norm of the matrix A.

$A : B = \operatorname{tr} A^T B$: matrix inner product in $\mathbb{M}^n$.

$\|A\| = \{A : A\}^{1/2}$: matrix norm in $\mathbb{M}^n$.

Cof A: cofactor matrix of the matrix A (**Cof** $A = (\det A)A^{-T}$ if A is invertible) (Sect. 1.1).

$\iota_1(A) = \operatorname{tr} A$ (Sects. 1.2, 3.5).

$\iota_{n-1}(A) = \operatorname{tr} \mathbf{Cof}\, A$ (=det A tr A^{-1} if A is invertible) (Sects. 1.2, 3.5).

$\iota_n(A) = \det A$ (Sects. 1.2, 3.5).

$\iota_k = \iota_k(A)$, $k = 1, 2, 3$, for a matrix A of order 3 (Sect. 3.5).

$\iota_A = (\iota_1(A), \iota_2(A), \iota_3(A))$: triple formed by the principal invariants of a matrix A of order 3 (Sect. 3.5).

The following sets of matrices will be of particular interest in this book. Other common notations for the same sets, or for the corresponding sets of linear transformations, are indicated between parentheses.

$\mathbb{M}^n$: set of all real square matrices of order n (GL(n), Lin).

$\mathbb{M}^n_+ = \{F \in \mathbb{M}^n;\ \det F > 0\}(\mathrm{GL}^+(n), \mathrm{Lin}^+)$.

$\mathbb{O}^n = \{P \in \mathbb{M}^n;\ PP^{\mathrm{T}} = P^{\mathrm{T}}P = I\}$: set of all orthogonal matrices of order n ($\mathrm{O}(n)$, Orth).

$\mathbb{O}^n_+ = \{P \in \mathbb{O}^n;\ \det P > 0\} = \{P \in \mathbb{O}^n;\ \det P = 1\}$: set of all rotations in $\mathbb{R}^n$, also called proper orthogonal matrices ($SO(n)$; Orth^+).

$\mathbb{S}^n = \{B \in \mathbb{M}^n:\ B = B^{\mathrm{T}}\}$: set of all symmetric matrices of order n(Sym).

$\mathbb{S}^n_>$: set of all symmetric, positive definite, matrices of order n(Psym, Sym^+).

$\mathbb{S}^n_\geqslant$: set of all symmetric, positive-semidefinite, matrices of order n.

2. ELASTICITY

Geometry of the reference configuration; deformations; strain tensors

dx: volume element in $\mathbb{R}^n$.
dx-meas A, or vol A if $n = 3$: volume of a subset $A \subset \mathbb{R}^n$ (Sect. 1.5).
da: area element in $\mathbb{R}^n$ (Sect. 1.6).
da-meas Δ, or area Δ if $n = 3$: area of a da-measurable subset Δ of the boundary of a domain in $\mathbb{R}^n$ (Sect. 1.6).
length γ: length of a curve γ in $\mathbb{R}^3$ (Sect. 1.8).
$\bar{\Omega} \subset \mathbb{R}^3$: reference configuration (Ω is a domain in $\mathbb{R}^3$) (Sect. 1.4).
$\Gamma = \partial\Omega$: boundary of the set Ω.
Γ_a: da-measurable subset of Γ.
$x \in \bar{\Omega}$: generic point of the reference configuration, with coordinates x_i.
$\boldsymbol{n}$: unit ($|\boldsymbol{n}| = 1$) outer normal vector along $\partial\Omega$, or along the boundary of any subdomain of $\bar{\Omega}$ (Sect. 1.6).
$\boldsymbol{\varphi}, \boldsymbol{\psi}: \bar{\Omega} \to \mathbb{R}^3$: usual notations for deformations of the reference configuration; a deformation is a smooth enough mapping

that is orientation-preserving and injective, except possibly on Γ (Sect. 1.4).

$\boldsymbol{u}, \boldsymbol{v}: \bar{\Omega} \to \mathbb{R}^3$: usual notations for displacements, with $\boldsymbol{\varphi} = \boldsymbol{id} + \boldsymbol{u}$, $\boldsymbol{\psi} = \boldsymbol{id} + \boldsymbol{v}$ (Sect. 1.4).

$\nabla\boldsymbol{\varphi} = (\partial_j \varphi_i) = \boldsymbol{I} + \nabla\boldsymbol{u}: \bar{\Omega} \to \mathbb{M}^3$: deformation gradient (Sect. 1.4).

$\nabla\boldsymbol{u} = (\partial_j u_i): \bar{\Omega} \to \mathbb{M}^3$: displacement gradient (Sect. 1.4).

$\partial_j \boldsymbol{\varphi} = \partial_j \varphi_i \boldsymbol{e}_i : \bar{\Omega} \to \mathbb{R}^3$: tangent vectors to the coordinate lines (Sect. 1.4).

$\boldsymbol{C} = \nabla\boldsymbol{\varphi}^{\mathrm{T}}\nabla\boldsymbol{\varphi} \in \mathbb{S}^3_>$: right Cauchy–Green strain tensor (Sect. 1.8).

$\boldsymbol{B} = \nabla\boldsymbol{\varphi}\nabla\boldsymbol{\varphi}^{\mathrm{T}} \in \mathbb{S}^3_>$: left Cauchy–Green strain tensor (Sect. 1.8).

$\boldsymbol{E} = \frac{1}{2}(\boldsymbol{C} - \boldsymbol{I}) = \frac{1}{2}(\nabla\boldsymbol{u}^{\mathrm{T}} + \nabla\boldsymbol{u} + \nabla\boldsymbol{u}^{\mathrm{T}}\nabla\boldsymbol{u}) \in \mathbb{S}^3$: Green–St Venant strain tensor (Sect. 1.8).

Geometry of the deformed configuration

$\bar{\Omega}^{\varphi} = \boldsymbol{\varphi}(\bar{\Omega})$: deformed configuration (Sect. 1.6).

Ω^{φ}: interior of the deformed configuration ($\Omega^{\varphi} = \operatorname{int} \boldsymbol{\varphi}(\bar{\Omega})$ when $\boldsymbol{\varphi}$ is injective on $\bar{\Omega}$; cf. Sect. 1.6).

Γ^{φ}: boundary of the deformed configuration ($\Gamma^{\varphi} = \partial\{\Omega^{\varphi}\}$ when $\boldsymbol{\varphi}$ is injective on $\bar{\Omega}$; cf. Sect. 1.6).

$\Gamma^{\varphi}_{\alpha}$: da^{φ}-measurable subset of Γ^{φ}.

$x^{\varphi} = \boldsymbol{\varphi}(x) \in \bar{\Omega}^{\varphi}$: generic point of the deformed configuration, with coordinates x^{φ}_i (Sect. 1.4).

$\partial^{\varphi}_j = \partial/\partial x^{\varphi}_j$.

$\boldsymbol{n}^{\varphi}$: unit ($|\boldsymbol{n}^{\varphi}| = 1$) outer normal vector along Γ^{φ}, or along the boundary of any subdomain of $\bar{\Omega}^{\varphi}$.

$\mathrm{d}x^{\varphi} = \det \nabla\varphi \, \mathrm{d}x$: volume element in the deformed configuration (Sect. 1.5).

$\mathrm{d}a^{\varphi} = |\mathbf{Cof}\, \nabla\varphi \, \boldsymbol{n}| \, \mathrm{d}a = \det \nabla\varphi |\nabla\varphi^{-\mathrm{T}} \boldsymbol{n}| \, \mathrm{d}a$: area element in the deformed configuration (Sect. 1.7).

$\mathrm{d}l^{\varphi} = \{\mathrm{d}x^{\mathrm{T}} \nabla\varphi^{\mathrm{T}} \nabla\varphi \, \mathrm{d}x\}^{1/2}$: length element in the deformed configuration (Sect. 1.8).

$\mathbf{div}^{\varphi} \boldsymbol{T}^{\varphi} = \partial_j^{\varphi} T_{ij}^{\varphi} \boldsymbol{e}_i : \bar{\Omega}^{\varphi} \to \mathbb{R}^3$: divergence of a tensor field $\boldsymbol{T}^{\varphi} : \bar{\Omega}^{\varphi} \to \mathbb{M}^3$ with respect to the variable x^{φ}.

Applied forces

$\boldsymbol{f}^{\varphi} : \Omega^{\varphi} \to \mathbb{R}^3$: density of the applied body force per unit volume in the deformed configuration (Sect. 2.1).
$\boldsymbol{f} : \Omega \to \mathbb{R}^3$: density of the applied body force per unit volume in the reference configuration (Sect. 2.6).
These densities are related by:

$$\boldsymbol{f}^{\varphi}(x^{\varphi}) \, \mathrm{d}x^{\varphi} = \boldsymbol{f}(x) \, \mathrm{d}x \, .$$

Example (gravity field; cf. Sects. 2.1, 2.7):

$$\boldsymbol{f}^{\varphi}(x^{\varphi}) = -g\rho^{\varphi}(x^{\varphi})\boldsymbol{e}_3 \, , \quad \boldsymbol{f}(x) = -g\rho(x)\boldsymbol{e}_3 \, ,$$

where ρ^{φ} and ρ are the mass densities in the deformed and reference configurations, respectively.

Conservative applied body force (Sect. 2.7): $\boldsymbol{f}(x) = \hat{\boldsymbol{f}}(x, \varphi(x))$, $x \in \Omega$, and

$$\hat{\boldsymbol{f}}(x, \boldsymbol{\eta}) = \mathbf{grad}_{\eta} \hat{F}(x, \boldsymbol{\eta}) \quad \text{for all } x \in \Omega, \; \boldsymbol{\eta} \in \mathbb{R}^3 \, .$$

Then

$$\int_\Omega \hat{f}(x, \varphi(x)) \cdot \theta(x)\, dx = F'(\varphi)\theta \quad \text{for all } \varphi, \theta : \bar{\Omega} \to \mathbb{R}^3 ,$$

$$\text{where } F(\psi) = \int_\Omega \hat{F}(x, \psi(x))\, dx .$$

$\hat{F} : \Omega \times \mathbb{R}^3 \to \mathbb{R}$: potential of the applied body force.

Example (dead load; cf. Sect. 2.7):

$$\hat{f}(x, \eta) = f(x) \Rightarrow \hat{F}(x, \eta) = f(x) \cdot \eta .$$

$g^\varphi : \Gamma_1^\varphi \to \mathbb{R}^3$: density of the applied surface force per unit area in the deformed configuration (Sect. 2.1).

$g : \Gamma_1 \to \mathbb{R}^3$: density of the applied surface force per unit area in the reference configuration (Sect. 2.6).

These densities are related by:

$$g^\varphi(x^\varphi)\, da^\varphi = g(x)\, da .$$

Example (pressure load; cf. Sect. 2.7):

$$g^\varphi(x^\varphi) = -\pi n^\varphi(x^\varphi),$$

$$g(x) = -\pi (\det \nabla\varphi(x)) \nabla\varphi(x)^{-\mathrm{T}} n(x) = \hat{g}(x, \nabla\varphi(x)), \quad \pi \in \mathbb{R} .$$

Conservative applied surface force (Sect. 2.7): $g(x) = \hat{g}(x, \nabla\varphi(x))$, $x \in \Gamma_1$, and

$$\int_{\Gamma_1} \hat{g}(x, \nabla\varphi(x)) \cdot \theta(x)\, da = G'(\varphi)\theta \quad \text{for all } \varphi, \theta : \bar{\Omega} \to \mathbb{R}^3 \text{ with } \theta = o \text{ on } \Gamma - \Gamma_1,$$

$$\text{where } G(\psi) = \int_{\Gamma_1} \hat{G}(x, \psi(x), \nabla\psi(x))\, da .$$

$\hat{G} : \Gamma \times \mathbb{R}^3 \times \mathbb{M}^3_+ \to \mathbb{R}$: potential of the applied surface force.

Example (dead load; cf. Sect. 2.7):

$$\hat{\boldsymbol{g}}(x, \boldsymbol{\eta}) = \boldsymbol{g}(x) \Rightarrow \hat{G}(x, \boldsymbol{\eta}) = \boldsymbol{g}(x) \cdot \boldsymbol{\eta}$$

Stress tensors, constitutive equations

$\boldsymbol{T}^{\varphi}(x^{\varphi}) \in \mathbb{S}^3$: Cauchy stress tensor at the point $x^{\varphi} \in \bar{\Omega}^{\varphi}$ (Sect. 2.3).

$\boldsymbol{t}^{\varphi}(x^{\varphi}, \boldsymbol{n}^{\varphi}) = \boldsymbol{T}^{\varphi}(x^{\varphi})\boldsymbol{n}^{\varphi}$, $x^{\varphi} \in \bar{\Omega}^{\varphi}$, $|\boldsymbol{n}^{\varphi}| = 1$: Cauchy stress vector (Sects. 2.2, 2.3).

$\boldsymbol{T}(x) = \boldsymbol{T}^{\varphi}(x^{\varphi})\, \mathbf{Cof}\, \boldsymbol{\nabla\varphi}(x) = (\det \boldsymbol{\nabla\varphi}(x))\boldsymbol{T}^{\varphi}(x^{\varphi})\boldsymbol{\nabla\varphi}(x)^{-\mathrm{T}}$, $x \in \bar{\Omega}$: first Piola–Kirchhoff stress tensor (Sect. 2.5).

The Cauchy and first Piola–Kirchhoff stress tensors are related by (Sect. 1.7):

$$\boldsymbol{T}(x)\boldsymbol{n}\, \mathrm{d}a = \boldsymbol{T}^{\varphi}(x^{\varphi})\boldsymbol{n}^{\varphi}\, \mathrm{d}a^{\varphi}, \quad \mathbf{div}\, \boldsymbol{T}(x) = (\det \boldsymbol{\nabla\varphi}(x))\, \mathbf{div}\, \boldsymbol{T}^{\varphi}(x^{\varphi}).$$

$\boldsymbol{\Sigma}(x) = \boldsymbol{\nabla\varphi}(x)^{-1}\boldsymbol{T}(x) = (\det \boldsymbol{\nabla\varphi}(x))\boldsymbol{\nabla\varphi}(x)^{-1}\boldsymbol{T}^{\varphi}(x^{\varphi})\boldsymbol{\nabla\varphi}(x)^{-\mathrm{T}} \in \mathbb{S}^3$, $x \in \bar{\Omega}$: second Piola–Kirchhoff stress tensor (Sect. 2.5).

Constitutive equations for an isotropic elastic material (Sect. 3.6):

$$\boldsymbol{T}^{\varphi}(x^{\varphi}) = \hat{\boldsymbol{T}}^{\mathrm{D}}(x, \boldsymbol{\nabla\varphi}(x)) = \bar{\boldsymbol{T}}^{\mathrm{D}}(x, \boldsymbol{\nabla\varphi}(x)\boldsymbol{\nabla\varphi}(x)^{\mathrm{T}}), \text{ with}$$
$$\bar{\boldsymbol{T}}^{\mathrm{D}}(x, \boldsymbol{B}) = \beta_0(x, \boldsymbol{\iota}_B)\boldsymbol{I} + \beta_1(x, \boldsymbol{\iota}_B)\boldsymbol{B} + \beta_2(x, \boldsymbol{\iota}_B)\boldsymbol{B}^2, \boldsymbol{B} \in \mathbb{S}^3_>,$$

$$\boldsymbol{\Sigma}(x) = \hat{\boldsymbol{\Sigma}}(x, \boldsymbol{\nabla\varphi}(x)) = \tilde{\boldsymbol{\Sigma}}(x, \boldsymbol{\nabla\varphi}(x)^{\mathrm{T}}\boldsymbol{\nabla\varphi}(x)), \text{ with}$$
$$\tilde{\boldsymbol{\Sigma}}(x, \boldsymbol{C}) = \gamma_0(x, \boldsymbol{\iota}_C)\boldsymbol{I} + \gamma_1(x, \boldsymbol{\iota}_C)\boldsymbol{C} + \gamma_2(x, \boldsymbol{\iota}_C)\boldsymbol{C}^2, \boldsymbol{C} \in \mathbb{S}^3_>.$$

$T_R(x) = \hat{T}^D(x, I) = \hat{\Sigma}(x, I)$: residual stress tensor (Sect. 3.6).

If the material is isotropic, homogeneous, and if the reference configuration is a natural state (Sect. 3.7),

$$\boxed{\tilde{\Sigma}(C) = \check{\Sigma}(E) = \lambda(\operatorname{tr} E)I + 2\mu E + o(E),\ C = I + 2E\ .}$$

Experimental evidence shows that the *Lamé constants* λ and μ are both >0 (Sect. 3.8).

Constitutive equation of a *St Venant–Kirchhoff material* (Sect. 3.9):

$$\boxed{\check{\Sigma}(E) = \lambda(\operatorname{tr} E)I + 2\mu E\ .}$$

$$\nu = \frac{\lambda}{2(\lambda + \mu)}: \text{ Poisson's ratio (Sect. 3.8)},$$

$$E = \frac{\mu(3\lambda + 2\mu)}{\lambda + \mu}: \text{ Young's modulus (Sect. 3.8)}.$$

$$\boxed{\mathsf{A}(x, F) = \left(\frac{\partial \hat{T}_{ij}}{\partial F_{kl}}(x, F)\right): \text{ elasticity tensor (Sect. 5.9).}}$$

Rule: If a function $\hat{S} : F \in \mathbb{M}^3_+ \to \hat{S}(F)$ happens to be also a function of particular functions of F, such as FF^T, F^TF, etc., the following self-explanatory notations are used:

$$S = \hat{S}(F) = \bar{S}(B) = \tilde{S}(C) = \check{S}(E) = \dot{S}(\iota_C) = \mathbb{S}(F, \operatorname{Cof} F, \det F),$$

with $B = FF^T$, $C = F^TF = I + 2E$.

Hyperelasticity

Hyperelastic material (Sect. 4.1): $T(x) = \hat{T}(x, \nabla\varphi(x))$, $x \in \Omega$, and

$$\boxed{\hat{T}(x, F) = \frac{\partial \hat{W}}{\partial F}(x, F) \quad \text{for all } x \in \bar{\Omega},\ F \in \mathbb{M}^3_+\ .}$$

$$\int_{\Omega} \hat{\boldsymbol{T}}(x, \boldsymbol{\nabla\varphi}(x)) : \boldsymbol{\nabla\theta}(x)\,\mathrm{d}x = W'(\boldsymbol{\varphi})\boldsymbol{\theta} \quad \text{for all } \boldsymbol{\varphi}, \boldsymbol{\theta} : \Omega \to \mathbb{R}^3,$$

$$\text{where } W(\boldsymbol{\psi}) = \int_{\Omega} \hat{W}(x, \boldsymbol{\nabla\psi}(x))\,\mathrm{d}x\,.$$

$\hat{W} : \bar{\Omega} \times \mathbb{M}^3_+ \to \mathbb{R}$: stored energy function (Sect. 4.1):
$W(\cdot)$: strain energy (Sect. 4.1).
Let $\hat{W}(x, \boldsymbol{F}) = \tilde{W}(x, \boldsymbol{C}) = \check{W}(x, \boldsymbol{E})$, $\boldsymbol{C} = \boldsymbol{F}^{\mathrm{T}}\boldsymbol{F} = \boldsymbol{I} + 2\boldsymbol{E}$. Then

$$\tilde{\boldsymbol{\Sigma}}(x, \boldsymbol{C}) = 2\,\frac{\partial \tilde{W}}{\partial \boldsymbol{C}}(x, \boldsymbol{C}), \quad \check{\boldsymbol{\Sigma}}(x, \boldsymbol{E}) = \frac{\partial \check{W}}{\partial \boldsymbol{E}}(x, \boldsymbol{E})\,.$$

Stored energy function of an isotropic hyperelastic material (Sect. 4.4):

$$\hat{W}(x, \boldsymbol{F}) = \dot{W}(x, \boldsymbol{\iota}_{\boldsymbol{F}^{\mathrm{T}}\boldsymbol{F}})\,.$$

If the material is homogeneous and the reference configuration is a natural state (Sect. 4.5):

$$\hat{W}(\boldsymbol{F}) = \frac{\lambda}{2}(\operatorname{tr}\boldsymbol{E})^2 + \mu \operatorname{tr}\boldsymbol{E}^2 + o(\|\boldsymbol{E}\|^2), \; \boldsymbol{F}^{\mathrm{T}}\boldsymbol{F} = \boldsymbol{I} + 2\boldsymbol{E}\,.$$

Stored energy function of a *St Venant–Kirchhoff material* (Sect. 4.4):

$$\hat{W}(\boldsymbol{F}) = \check{W}(\boldsymbol{E}) = \frac{\lambda}{2}(\operatorname{tr}\boldsymbol{E})^2 + \mu \operatorname{tr}\boldsymbol{E}^2, \; \boldsymbol{F}^{\mathrm{T}}\boldsymbol{F} = \boldsymbol{I} + 2\boldsymbol{E}\,.$$

Stored energy function of an *Ogden's material* (Sect. 4.10):

$$\hat{W}(\boldsymbol{F}) = \sum_{i=1}^{M} a_i \operatorname{tr}(\boldsymbol{F}^{\mathrm{T}}\boldsymbol{F})^{\gamma_i/2} + \sum_{j=1}^{N} b_j \operatorname{tr}(\mathbf{Cof}\boldsymbol{F}^{\mathrm{T}}\boldsymbol{F})^{\delta_j/2} + \Gamma(\det \boldsymbol{F})\,,$$

$a_i > 0$, $\gamma_i \geqslant 1$, $b_j > 0$, $\delta_j \geqslant 1$, $\Gamma :]0, +\infty[\to \mathbb{R}$ is convex,
$\lim_{\delta \to 0^+} \Gamma(\delta) = +\infty$.

$\mathsf{A}(x, \boldsymbol{F}) = \left(\dfrac{\partial^2 \hat{W}}{\partial F_{ij} \partial F_{kl}}(x, \boldsymbol{F})\right)$: elasticity tensor of a hyperelastic material (Sect. 5.9).

The basic equations

For definiteness, we consider a displacement–traction problem with dead loads.

Equilibrium equations in the reference configuration (Sect. 2.6 and 5.1):

$$
\begin{aligned}
&-\operatorname{div} \hat{T}(x, \nabla\varphi(x)) = f(x),\ x \in \Omega, \\
&\varphi(x) = \varphi_0(x),\ x \in \Gamma_0, \\
&\hat{T}(x, \nabla\varphi(x))n = g(x),\ x \in \Gamma_1 .
\end{aligned}
$$

Principle of virtual work in the reference configuration (Sect. 2.6): The equilibrium equations are formally equivalent to finding $\varphi : \bar{\Omega} \to \mathbb{R}^3$ such that $\varphi = \varphi_0$ on Γ_0 and

$$
\int_\Omega \hat{T}(x, \nabla\varphi(x)) : \nabla\theta(x)\, dx = \int_\Omega f(x) \cdot \theta(x)\, dx + \int_{\Gamma_1} g(x) \cdot \theta(x)\, da ,
$$

for all $\theta : \bar{\Omega} \to \mathbb{R}^3$ that vanish on Γ_0.

In *hyperelasticity*, where

$$
\hat{T}(x, F) = \frac{\partial \hat{W}}{\partial F}(x, F) ,
$$

the equilibrium equations are formally equivalent to finding

$$
\varphi \in \Phi = \{\psi : \bar{\Omega} \to \mathbb{R}^3 \text{ smooth enough; } \det \nabla\psi > 0 \text{ in } \Omega,\ \psi = \varphi_0 \text{ on } \Gamma_0\} ,
$$

such that

$$
I'(\varphi)\theta = 0 \quad \text{for all } \theta : \Omega \to \mathbb{R}^3 \text{ that vanish on } \Gamma_0 ,
$$

where the *total energy* I is given by

$$I(\boldsymbol{\psi}) = \int_\Omega \hat{W}(x, \nabla\boldsymbol{\psi}(x))\,dx - \left\{ \int_\Omega \boldsymbol{f}(x)\cdot\boldsymbol{\psi}(x)\,dx + \int_{\Gamma_1} \boldsymbol{g}(x)\cdot\boldsymbol{\psi}(x)\,da \right\}.$$

Unilateral boundary condition of place (Sect. 5.3):

$$\boldsymbol{\varphi}(\Gamma_2) \subset C, \quad \text{where } \Gamma_2 \subset \Gamma \text{ and } C \text{ is a closed subset of } \mathbb{R}^3.$$

Confinement condition (Sect. 5.6):

$$\boldsymbol{\varphi}(\bar{\Omega}) \subset B, \quad \text{where } B \text{ is a closed subset of } \mathbb{R}^3.$$

Injectivity condition (Sect. 5.6):

$$\int_\Omega \det \nabla\boldsymbol{\varphi}\,dx \leqslant \operatorname{vol} \boldsymbol{\varphi}(\Omega).$$

Equilibrium equations in terms of the displacement:

$$\boldsymbol{A}(\boldsymbol{u}) := -\operatorname{\mathbf{div}}\{(\boldsymbol{I} + \nabla\boldsymbol{u})\check{\boldsymbol{\Sigma}}(\boldsymbol{E}(\boldsymbol{u}))\} = \boldsymbol{f} \text{ in } \Omega, \quad \boldsymbol{u} = \boldsymbol{u}_0 \text{ on } \Gamma_0,$$

$$\boldsymbol{B}(\boldsymbol{u}) := (\boldsymbol{I} + \nabla\boldsymbol{u})\check{\boldsymbol{\Sigma}}(\boldsymbol{E}(\boldsymbol{u}))\boldsymbol{n} = \boldsymbol{g} \text{ on } \Gamma_1.$$

Equations of linearized elasticity for a homogeneous isotropic material whose reference configuration is a *natural state*; we also assume that $\boldsymbol{u}_0 = \boldsymbol{o}$ (Sect. 6.2):

$$\boldsymbol{A}'(\boldsymbol{o})\boldsymbol{u} = -\operatorname{\mathbf{div}}\{\lambda(\operatorname{tr} \boldsymbol{e}(\boldsymbol{u}))\boldsymbol{I} + 2\mu\boldsymbol{e}(\boldsymbol{u})\} = \boldsymbol{f} \text{ in } \Omega, \quad \boldsymbol{u} = \boldsymbol{o} \text{ on } \Gamma_0,$$

$$\boldsymbol{B}'(\boldsymbol{o})\boldsymbol{u} = \{\lambda(\operatorname{tr} \boldsymbol{e}(\boldsymbol{u}))\boldsymbol{I} + 2\mu\boldsymbol{e}(\boldsymbol{u})\}\boldsymbol{n} = \boldsymbol{g} \text{ on } \Gamma_1,$$

$$\boldsymbol{e}(\boldsymbol{u}) = \tfrac{1}{2}(\nabla\boldsymbol{u}^{\mathrm{T}} + \nabla\boldsymbol{u}): \text{ linearized strain tensor.}$$

DE LA PRESSION OU TENSION

DANS UN CORPS SOLIDE.

Les géomètres qui ont recherché les équations d'équilibre ou de mouvement des lames ou des surfaces élastiques ou non élastiques ont distingué deux espèces de forces produites, les unes par la dilatation ou la contraction, les autres par la flexion de ces mêmes surfaces. De plus, ils ont généralement supposé, dans leurs calculs, que les forces de la première espèce, nommées *tensions*, restent perpendiculares aux lignes contre lesquelles elles s'exercent. Il m'a semblé que ces deux espèces de forces pouvaient être réduites à une seule, qui doit constamment s'appeler *tension* ou *pression*, qui agit sur chaque élément d'une section faite à volonté, non seulement dans une surface flexible, mais encore dans un solide élastique ou non élastique, et qui est de la même nature que la pression hydrostatique exercée par un fluide en repos contre la surface extérieure d'un corps. Seulement la nouvelle pression ne demeure pas toujours perpendiculaire aux faces qui lui sont soumises, ni la même dans tous les sens en un point donné. En développant cette idée, je suis parvenu à reconnaître que la pression ou tension exercée contre un plan quelconque en un point donné d'un corps solide se déduit très aisément, tant en grandeur qu'en direction, des pressions ou tensions exercées contre trois plans rectangulaires menés par le même point. Cette proposition, que j'ai déjà indiquée dans le *Bulletin de la Société philomathique* de janvier 1823 ([1]), peut être établie à l'aide des considérations suivantes.

([1]) *OEuvres de Cauchy*, S. II, T. II.

Reprinted from CAUCHY, A.-L.: De la pression ou tension dans un corps solide, *Exercises de Mathématiques* **2** (1827), 42–56.

PART A

DESCRIPTION OF THREE-DIMENSIONAL ELASTICITY

CHAPTER 1

GEOMETRICAL AND OTHER PRELIMINARIES

INTRODUCTION

A central problem in nonlinear, three-dimensional, elasticity consists in finding the equilibrium position of an elastic body that occupies a *reference configuration* $\bar{\Omega}$ in the absence of applied forces, where Ω is a bounded open connected subset of $\mathbb{R}^3$ with a Lipschitz-continuous boundary. When subjected to applied forces, the body occupies a *deformed configuration* $\boldsymbol{\varphi}(\bar{\Omega})$, characterized by a mapping $\boldsymbol{\varphi}: \bar{\Omega} \to \mathbb{R}^3$ that must be in particular *orientation-preserving* in the set $\bar{\Omega}$ and *injective on the set* Ω, in order to be physically acceptable (Sect. 1.4).

Such mappings $\boldsymbol{\varphi}$ are called *deformations*, and the object in this chapter is to study their *geometrical properties*. It is shown in particular that the changes of *volumes*, *surfaces*, and *lengths* associated with a deformation $\boldsymbol{\varphi}$, are respectively governed by the *scalar* det $\boldsymbol{\nabla\varphi}$ (Sect. 1.5), the *matrix* $\mathbf{Cof}\, \boldsymbol{\nabla\varphi}$ (Sect. 1.7; Theorem 1.7-1), and the *right Cauchy–Green strain tensor* $\boldsymbol{C} = \boldsymbol{\nabla\varphi}^{\mathrm{T}} \boldsymbol{\nabla\varphi}$ (Sect. 1.8). It is further shown (Theorems 1.8-1 and 1.8-2) that the *Green–St Venant strain tensor* $\boldsymbol{E} = \frac{1}{2}(\boldsymbol{C} - \boldsymbol{I})$ associated with a deformation $\boldsymbol{\varphi}$ measures the deviation between $\boldsymbol{\varphi}$ and a rigid deformation (which corresponds to $\boldsymbol{C} = \boldsymbol{I}$). Both strain tensors $\boldsymbol{C}$ and $\boldsymbol{E}$ will play a fundamental role in the subsequent chapters.

*1.1. THE COFACTOR MATRIX

The cofactor matrix of the deformation gradient (Sect. 1.4) appears in the Piola transform of tensor fields and in the formula relating area elements in a deformation in $\mathbb{R}^3$ (Sect. 1.7); it also appears as a natural variable in the stored energy functions of hyperelastic materials (Chapter 4).

Let $\boldsymbol{A} = (a_{ij})$ be a matrix of order n. For each pair (i, j) of indices, let $\boldsymbol{A}'_{ij}$ be the matrix of order $(n-1)$ obtained by deleting the ith row and the jth column in the matrix $\boldsymbol{A}$. The scalar

$$d_{ij} := (-1)^{i+j} \det A'_{ij}$$

is called the (i, j)-*cofactor* of the matrix A, and the matrix

$$\mathrm{Cof}\, A := (d_{ij})$$

is called the **cofactor matrix** of the matrix A.

Remark. Some authors prefer to introduce the *transpose* $(\mathrm{Cof}\, A)^{\mathrm{T}}$ of the cofactor matrix, which they call the *adjugate* of the matrix A. ■

The well known formulas

$$\det A = \sum_{j=1}^{n} a_{ij} d_{ij},\ 1 \le i \le n, \quad \text{and} \det A = \sum_{i=1}^{n} a_{ij} d_{ij}\,, \quad 1 \le j \le n\,,$$

are equivalent to the relations

$$A(\mathrm{Cof}\, A)^{\mathrm{T}} = (\mathrm{Cof}\, A)^{\mathrm{T}} A = (\det A) I\,.$$

If the matrix A is invertible,

$$\boxed{\mathrm{Cof}\, A = (\det A) A^{-\mathrm{T}}\,,}$$

with $A^{-\mathrm{T}} = (A^{-1})^{\mathrm{T}}$, and in this case, $(\mathrm{Cof}\, A)^{\mathrm{T}}$ is the only matrix B that satisfies $AB = BA = (\det A) I$. The following relations are easily established (they clearly hold for invertible matrices and hence can be extended to arbitrary matrices, since the mapping $A \in \mathbb{M}^n \rightarrow \mathrm{Cof}\, A \in \mathbb{M}^n$ is continuous):

$$\mathrm{Cof}\,(A^{\mathrm{T}}) = (\mathrm{Cof}\, A)^{\mathrm{T}},\ \mathrm{Cof}\, AB = (\mathrm{Cof}\, A)(\mathrm{Cof}\, B)\,.$$

In the special case $n = 3$, which will be of central interest in this book, one has

$$\mathrm{Cof}\, A = \begin{pmatrix} a_{22}a_{33} - a_{23}a_{32} & a_{23}a_{31} - a_{21}a_{33} & a_{21}a_{32} - a_{22}a_{31} \\ a_{32}a_{13} - a_{33}a_{12} & a_{33}a_{11} - a_{31}a_{13} & a_{31}a_{12} - a_{32}a_{11} \\ a_{12}a_{23} - a_{13}a_{22} & a_{13}a_{21} - a_{11}a_{23} & a_{11}a_{22} - a_{12}a_{21} \end{pmatrix},$$

or

$$(\mathbf{Cof}\, A)_{ij} = a_{i+1,j+1}a_{i+2,j+2} - a_{i+1,j+2}a_{i+2,j+1}\,, \quad 1 \leq i, j \leq 3 \quad \text{(no summation)}\,,$$

counting the indices modulo 3. Introducing the third-order *orientation tensor* (ε_{ijk}) defined by

$$\varepsilon_{ijk} = \begin{cases} +1 \text{ if } \{i, j, k\} \text{ is an even permutation of } \{1, 2, 3\}\,, \\ -1 \text{ if } \{i, j, k\} \text{ is an odd permutation of } \{1, 2, 3\}\,, \\ \ \ 0 \text{ if at least two indices are equal}\,, \end{cases}$$

one also has (using the repeated index summation convention)

$$(\mathbf{Cof}\, A)_{ij} = \tfrac{1}{2}\varepsilon_{mni}\varepsilon_{pqj}a_{mp}a_{nq}\,.$$

Notice that the orientation tensor is also useful for computing the determinant of a matrix of order three:

$$\det A = \tfrac{1}{6}\varepsilon_{ijk}\varepsilon_{pqr}a_{ip}a_{jq}a_{kr}\,.$$

Finally, we prove a useful property of the cofactor matrix (the extension to matrices of arbitrary order is left to the reader).

Theorem 1.1-1. *Let $\lambda_1, \lambda_2, \lambda_3$ denote the eigenvalues of a matrix A of order three. Then the eigenvalues of the matrix* **Cof** A *are $\lambda_2\lambda_3, \lambda_3\lambda_1, \lambda_1\lambda_2$.*

Proof. By a well known result about matrices (see e.g. Ciarlet [1982, Th. 1.2-1] or Strang [1976, p. 223]), there exists an invertible matrix P such that the matrix $P^{-1}AP$ is upper triangular:

$$P^{-1}AP = \begin{pmatrix} \lambda_1 & t_{12} & t_{13} \\ 0 & \lambda_2 & t_{23} \\ 0 & 0 & \lambda_3 \end{pmatrix}.$$

Using this relation, we obtain

$$(\mathbf{Cof}\, P)^{-1}\, \mathbf{Cof}\, A(\mathbf{Cof}\, P) = (\mathbf{Cof}\, P^{-1})\, \mathbf{Cof}\, A(\mathbf{Cof}\, P) = \mathbf{Cof}(P^{-1}AP)\,,$$

and since

$$\mathbf{Cof}(P^{-1}AP) = \begin{pmatrix} \lambda_2\lambda_3 & 0 & 0 \\ -\lambda_3 t_{12} & \lambda_3\lambda_1 & 0 \\ (t_{12}t_{23} - \lambda_2 t_{13}) & -\lambda_1 t_{23} & \lambda_1\lambda_2 \end{pmatrix},$$

the conclusion follows. ∎

Remark. Since the result of Theorem 1.1-1 clearly holds for invertible matrices (by virtue of the relation $\mathbf{Cof}\, A = (\mathbf{det}\, A)A^{-T}$), one can think of an alternative proof "by continuity". Such a proof turns out however to be more delicate; cf. Exercise 1.2. ∎

*1.2. THE FRÉCHET DERIVATIVE

All vector spaces considered are real.

Since differential calculus in normed vector spaces will be widely used in this book, and notably in this chapter, we first review some basic results concerning differential mappings. For more detailed accounts, see notably Avez [1983], Cartan [1967], Dieudonné [1968], Schwartz [1967].

Given two normed vector spaces X and Y, let

$$\mathcal{L}(X; Y)\,, \quad \text{or simply } \mathcal{L}(X) \text{ if } X = Y\,,$$

denote the vector space formed by all continuous linear mappings $A: X \to Y$. Equipped with the norm

$$\|A\| := \sup_{\substack{x \in X \\ x \neq 0}} \frac{\|Ax\|_Y}{\|x\|_X}\,,$$

the space $\mathcal{L}(X; Y)$ becomes itself a normed vector space, which is complete if the space Y is complete. If $X = Y = \mathbb{R}^n$, an element $A \in \mathcal{L}(\mathbb{R}^n; \mathbb{R}^n)$ is identified with the matrix that represents it, and if both spaces are equipped with the Euclidean vector norm $|\cdot|$, the associated norm of the matrix A is the *spectral norm*, also denoted $|\cdot|$. When $Y = \mathbb{R}$, the space

$$X' := \mathcal{L}(X; \mathbb{R})$$

is called the (topological) **dual space** of the space X.

For notational brevity, let us agree that whenever the notation

$$f : \Omega \subset X \to Y$$

is used in this section and in the next one, it means that X and Y are normed vector spaces (whose norms are denoted by the same symbol $\|\cdot\|$ whenever no confusion should arise), that Ω is an open subset of the space X, and that f is a mapping defined on the set Ω, with values in the space Y.

A mapping $f : \Omega \subset X \to Y$ is **differentiable** *at a point* $a \in \Omega$ if there exists an element $f'(a)$ of the space $\mathscr{L}(X; Y)$ such that

$$\boxed{f(a+h) = f(a) + f'(a)h + o(h)\,,}$$

where the notation $o(h)$ means that

$$o(h) = \|h\|\,\varepsilon(h) \quad \text{with } \lim_{h \to 0} \varepsilon(h) = 0 \text{ in } Y\,.$$

Of course, only points $(a+h)$ that belong to the set Ω should be considered in the above relation; since the set Ω is open by assumption, the set of admissible vectors h contains a ball centered at the origin in the space X. If a mapping f is differentiable at $a \in \Omega$, it is easily seen that f is continuous at a and that the element $f'(a) \in \mathscr{L}(X; Y)$ appearing in the definition is necessarily unique. The element $f'(a) \in \mathscr{L}(X; Y)$ is called the **Fréchet derivative**, or simply the **derivative**, of the mapping f at the point a. If $X = \mathbb{R}$ and x denotes the generic point of $\mathbb{R}$, the derivative is also noted:

$$f'(a) = \frac{\mathrm{d}f}{\mathrm{d}x}(a)\,.$$

If a mapping $f : \Omega \subset X \to Y$ is differentiable at all points of the open set Ω, it is said to be *differentiable in* Ω. If the mapping

$$f' : x \in \Omega \subset X \to f'(x) \in \mathscr{L}(X; Y)\,,$$

which is well defined in this case, is continuous, the mapping f is said to be *continuously differentiable in* Ω, or simply *of class* $\mathscr{C}^1$. We denote by

$$\mathscr{C}^1(\Omega; Y), \text{ or simply } \mathscr{C}^1(\Omega) \text{ if } Y = \mathbb{R}\,,$$

the space of all continuously differentiable mappings from Ω into Y.

Consider for example an *affine continuous mapping*

$$f: x \in X \to f(x) = Ax + b \quad \text{with } A \in \mathscr{L}(X; Y) \text{ and } b \in Y .$$

Since $f(a+h) = f(a) + Ah$ for all $a, h \in X$, such a mapping is continuously differentiable in X, with

$$f'(x) = A \quad \text{for all } x \in \Omega ,$$

i.e., the mapping f' is constant in this case. Conversely it can be shown (using the mean value theorem; cf. Theorem 1.2-2 below) that, if $f'(x) = A \in \mathscr{L}(X; Y)$ for all $x \in \Omega$ and if in addition the open set Ω is *connected*, there exists a vector $b \in Y$ such that $f(x) = Ax + b$ for all $x \in \Omega$.

If the space Y is a *product* $Y = Y_1 \times Y_2 \times \cdots \times Y_m$ of normed vector spaces Y_i, a mapping $f : \Omega \subset X \to Y$ is defined by m *component mappings* $f_i : \Omega \subset X \to Y_i$, and it is easily seen that the mapping f is differentiable at a point $a \in \Omega$ if and only if each mapping f_i is differentiable at the same point a. In this case, the derivative $f'(a) \in \mathscr{L}(X; Y)$ can be identified with the element $(f_1'(a), f_2'(a), \ldots, f_m'(a))$ of the product space $\mathscr{L}(X; Y_1) \times \mathscr{L}(X; Y_2) \times \cdots \times \mathscr{L}(X; Y_m)$.

Consider next the case where the space X is a product $X = X_1 \times X_2 \times \cdots \times X_n$ of normed vector spaces X_j. Given a point $a = (a_1, a_2, \ldots, a_n)$ of an open subset Ω of the space X, there exists for each index j an open subjet Ω_j of the space X_j containing the point a_j such that the open set $\Omega_1 \times \Omega_2 \times \cdots \times \Omega_n$ is contained in Ω. If for some index j the *partial mapping*:

$$f(a_1, \ldots, a_{j-1}, \cdot, a_{j+1}, \ldots, a_n) : \Omega_j \subset X_j \to Y$$

is differentiable at the point $a_j \in \Omega_j$, its derivative

$$\partial_j f(a) \in \mathscr{L}(X_j; Y)$$

is called the jth **partial derivative** of the mapping f at the point a. If x_j denotes a generic point of the space X_j, the partial derivatives are also noted:

$$\partial_j f(a) = \frac{\partial f}{\partial x_j}(a) .$$

Remark. The notation $\partial f/\partial A$ is also used in a different setting, to denote the *gradient* of a function $f:\Omega\subset\mathbb{M}^n\to\mathbb{R}$ (cf. the end of this section). ■

If a mapping $f:\Omega\subset X=X_1\times X_2\times\cdots X_n$ is differentiable at a point $a\in\Omega$, it is easy to see that the n partial derivatives $\partial_j f(a)$ exist and that

$$f'(a)h=\sum_{j=1}^{n}\partial_j f(a)h_j \quad \text{for all } h=(h_1,h_2,\ldots,h_n)\in X_1\times X_2\times\cdots X_n\,.$$

Using the mean value theorem, it can be shown that *conversely, if the partial derivatives are defined and continuous on Ω, the equivalence*

$$f\in\mathscr{C}^1(\Omega;Y)\Leftrightarrow\partial_j f\in\mathscr{C}^0(\Omega;\mathscr{L}(X_j;Y))\,,\ 1\leq j\leq n\,,$$

holds, where $\mathscr{C}^0(E;F)$ denotes in general the set of all continuous mappings from a topological space E into a topological space F.

Let X_1, X_2, Y be normed vector spaces. A mapping B: $X_1\times X_2\to Y$ is *bilinear* if it satisfies

$$B(\alpha_1x_1+\alpha_1'x_1',x_2)=\alpha_1B(x_1,x_2)+\alpha_1'B(x_1',x_2)\,,$$
$$B(x_1,\alpha_2x_2+\alpha_2'x_2')=\alpha_2B(x_1,x_2)+\alpha_2'B(x_1,x_2')\,,$$

for all $x_1,x_1'\in X_1$, $x_2,x_2'\in X_2$, $\alpha_1,\alpha_1',\alpha_2,\alpha_2'\in\mathbb{R}$. If it is in addition *continuous*, i.e., if and only if

$$\|B\|:=\sup_{\substack{x_1\in X_1,x_2\in X_2\\ x_1\neq 0,x_2\neq 0}}\frac{\|B(x_1,x_2)\|_Y}{\|x_1\|_{X_1}\|x_2\|_{X_2}}<+\infty\,,$$

it is differentiable in the space $X_1\times X_2$, since (by the bilinearity):

$$B(a_1+h_1,a_2+h_2)=B(a_1,a_2)+B(h_1,a_2)+B(a_1,h_2)+B(h_1,h_2)$$

for all $(a_1,a_2)\in X_1\times X_2$ and all $(h_1,h_2)\in X_1\times X_2$, and since (by the continuity)

$$\|B(h_1,h_2)\|\leq\|B\|\,\|h_1\|_{X_1}\|h_2\|_{X_2}\leq\|B\|\max\{\|h_1\|_{X_1},\|h_2\|_{X_2}\}^2\,.$$

The derivative and the partial derivatives are thus respectively given by

$$B'(a_1, a_2)(h_1, h_2) = B(h_1, a_2) + B(a_1, h_2) ,$$

$$\partial_1 B(a_1, a_2)h_1 = B(h_1, a_2), \quad \partial_2 B(a_1, a_2)h_2 = B(a_1, h_2) .$$

If $X_1 = X_2 = X$, a similar computation shows that the mapping $f: x \in X \rightarrow B(x, x) \in Y$ is also differentiable, with $f'(a)h = B(a, h) + B(h, a)$ for all $a, h \in X$. If in addition the bilinear mapping $B: X \times X \rightarrow Y$ is *symmetric*, i.e., if $B(x, x') = B(x', x)$ for all $x, x' \in X$, the above formula reduces to $f'(a)h = 2B(a, h)$.

The derivative $f'(a) \in \mathscr{L}(X; Y)$ *is usually computed in terms of its action on vectors of* X, i.e., the vectors

$$f'(a)h = \lim_{\theta \to 0} \frac{f(a + \theta h) - f(a)}{\theta} = \frac{\mathrm{d}}{\mathrm{d}\theta} f(a + \theta h)\bigg|_{\theta = 0} \in Y$$

are computed for arbitrary vectors h of the space X. Such a vector $f'(a)h \in Y$ is called a **directional derivative**, or a **Gâteaux derivative**, *in the direction of the vector h*.

As an illustration, let us compute the derivatives of the mappings

$$\iota_1 = A \in \mathbb{M}^n \rightarrow \iota_1(A) := \operatorname{tr} A \text{ and } \iota_n : A \in \mathbb{M}^n \rightarrow \iota_n(A) := \det A .$$

Since the mapping ι_1 is linear and continuous, it is differentiable over the space $\mathbb{M}^n$ with

$$\iota_1'(A)H = \iota_1(H) = \operatorname{tr} H .$$

The mapping ι_n being a polynomial of degree n with respect to the n^2 elements of the matrix, it is continuously differentiable over the space $\mathbb{M}^n$. If the matrix A is invertible, we can write

$$\begin{aligned} \iota_n(A + H) &= \det(A + H) = \det A \det(I + A^{-1}H) \\ &= (\det A)(1 + \operatorname{tr}(A^{-1}H) + o(H)) ; \end{aligned}$$

the last equality is deduced from the relation

$$\det(I + E) = 1 + \operatorname{tr} E + \{\text{monomials of degree} \geq 2\} ,$$

which itself follows from the definition of the determinant. We have thus proved that (cf. also Exercise 1.3):

$$\iota_n'(A)H = \det A \operatorname{tr}(A^{-1}H) = \operatorname{tr}\{(\mathbf{Cof}\,A)^{\mathrm{T}}H\}$$

when the matrix A is invertible. Since the mapping $A \in \mathbb{M}^n \to \mathbf{Cof}\,A \in \mathbb{M}^n$ is continuous, we conclude that the second equality $\iota_n'(A)H = \operatorname{tr}\{(\mathbf{Cof}\,A)^{\mathrm{T}}H\}$ holds even if the matrix A is singular.

In various instances, the mapping to be differentiated is itself composed of simpler mappings whose derivatives are known. In this case, the following result is particularly useful:

Theorem 1.2-1 (chain rule). *Let X, Y, Z be normed vector spaces, let U and V be open subsets of the spaces X and Y respectively, let $f : U \subset X \to V \subset Y$ be a mapping differentiable at a point $a \in U$ and let $g : V \subset Y \to Z$ be a mapping differentiable at the point $f(a)$. Then the composite mapping $g \circ f : U \subset X \to Z$ is differentiable at the point $a \in U$ and*

$$\boxed{(g \circ f)'(a) = g'(f(a))f'(a)\,.}$$ ■

As an application of the chain rule, we compute the derivative of the mapping

$$\iota_{n-1} : A \in \mathbb{U}^n \subset \mathbb{M}^n \to \iota_{n-1}(A) := \det A \operatorname{tr} A^{-1} = \operatorname{tr} \mathbf{Cof}\,\mathbf{A}\,,$$

where $\mathbb{U}^n$ denotes the open subset of the set $\mathbb{M}^n$ formed by all invertible matrices of order n. We can write

$$\iota_{n-1}(A) = \iota_n(A)h(A) \text{ with } h(A) = (\iota_1 \circ f)(A),\ f(A) = A^{-1}\,,$$

so that combining the formula for the derivative of a bilinear mapping with the chain rule, we obtain

$$\iota_{n-1}'(A)H = h(A)\iota_n'(A)H + \iota_n(A)\iota_1'(f(A))f'(A)H\,.$$

Since the matrix $(I + A^{-1}H)$ is invertible for $|H| < |A^{-1}|^{-1}$, with

$$(I + A^{-1}H)^{-1} = I - A^{-1}H + o(H)\,,$$

we can write

$$f(A+H)=(A+H)^{-1}=(I+A^{-1}H)^{-1}A^{-1}$$
$$=A^{-1}-A^{-1}HA^{-1}+o(H),$$

and whence

$$f'(A)H=-A^{-1}HA^{-1}.$$

Using the expressions found for the derivatives ι_1' and ι_n', we thus obtain

$$\iota_{n-1}'(A)H=\det A\{\operatorname{tr}(A^{-1}H)\operatorname{tr}A^{-1}-\operatorname{tr}(A^{-1}HA^{-1})\},$$
$$=\operatorname{tr}\{(\operatorname{Cof}A)^{\mathrm{T}}((\operatorname{tr}A^{-1})I-A^{-1})H\}.$$

The mappings ι_1, ι_{n-1}, ι_n are instances of the *principal invariants of a matrix*, which will be mostly used in the sequel for matrices or order three (cf. in particular Sect. 3.6). Further results concerning derivatives of the principal invariants can be found in Exercises 1.4 and 1.5.

Let us now state some basic results from differential calculus that will be needed in the sequel. The first is a generalization of the mean value theorem for a real-valued function f, continuous on a compact interval $[a,b]\subset\mathbb{R}$, and differentiable on the open interval $]a,b[$: There exists a point $c\in\,]a,b[$ such that $f(b)-f(a)=f'(c)(b-a)$. This formula cannot be generalized for *vector-valued functions*: For instance the mapping $f:t\in[0,2\pi]\to f(t)=(\cos t,\sin t)\in\mathbb{R}^2$ satisfies $f(2\pi)-f(0)=0$, yet its derivative $f'(t)=(-\sin t,\cos t)$ never vanishes. What can be generalized however (Theorem 1.2-2) is the *inequality*

$$|f(b)-f(a)|\leq\sup_{t\in]a,b[}|f'(t)||b-a|,$$

which is a consequence of the relation $f(b)-f(a)=f'(c)(b-a)$.

If a and b are two points in a vector space, let

$$[a,b]=\{x=ta+(1-t)b;\,t\in[0,1]\},$$
$$]a,b[=\{x=ta+(1-t)b;\,t\in]0,1[\},$$

denote respectively the *closed segment*, and the *open segment*, with *end-points* a and b.

Theorem 1.2-2 (mean value theorem). *Let there be given two normed vector spaces X and Y, an open subset Ω of X containing a closed segment*

$[a, b]$, and a mapping $f : \Omega \subset X \to Y$, continuous on the closed segment $[a, b]$ and differentiable on the open segment $]a, b[$. Then

$$\|f(b) - f(a)\|_Y \leqslant \sup_{x \in]a,b[} \|f'(x)\|_{\mathscr{L}(X;Y)} \|b - a\|_X . \qquad \blacksquare$$

With the mean value theorem and the contraction mapping theorem, one can prove a result of paramount importance, which gives sufficient conditions under which an equation of the form $\varphi(x_1, x_2) = b$ is *locally* equivalent to an equation of the form $x_2 = f(x_1)$("locally" means in a neighborhood of a particular solution of the equation $\varphi(x_1, x_2) = b$). Such a function f is called an **implicit function**. If X and Y are two normed vector spaces, let

$$\mathscr{I}som(X; Y) , \quad \text{or simply } \mathscr{I}som(X) \text{ if } X = Y ,$$

denote the set of all continuous linear mappings $A : X \to Y$ that are bijective (one-to-one and onto) and whose inverse $A^{-1} : Y \to X$ is also continuous. Observe that $\mathscr{I}som(X; Y)$ is a subset, but *not* a subspace, of the space $\mathscr{L}(X; Y)$. Mappings of class $\mathscr{C}^m$, $m \geqslant 2$, are defined in Sect. 1.3.

Theorem 1.2-3 (implicit function theorem). *Let there be given two normed vector spaces X_1 and Y, a Banach space X_2, an open subset Ω of the space $X_1 \times X_2$ containing a point (a_1, a_2), a mapping $\varphi : \Omega \subset X_1 \times X_2 \to Y$ satisfying*

$$\varphi \in \mathscr{C}^1(\Omega; Y), \ \partial_2\varphi(a_1, a_2) \in \mathscr{I}som(X_2; Y) ,$$

and let $\varphi(a_1, a_2) = b \in Y$. Then there exist open subsets O_1 and O_2 of the spaces X_1 and X_2 respectively with $(a_1, a_2) \in O_1 \times O_2 \subset \Omega$, and an implicit function (Fig. 1.2-1) *$f : O_1 \subset X_1 \to O_2 \subset X_2$ such that*

$$\begin{aligned}
&\{(x_1, x_2) \in O_1 \times O_2; \varphi(x_1, x_2) = b\} \\
&\qquad\qquad = \{(x_1, x_2) \in O_1 \times O_2; x_2 = f(x_1)\} , \\
&f \in \mathscr{C}^1(O_1; X_2) , \\
&f'(x_1) = -\{\partial_2\varphi(x_1, f(x_1))\}^{-1}\partial_1\varphi(x_1, f(x_1)) \quad \text{for all } x_1 \in O_1 .
\end{aligned}$$

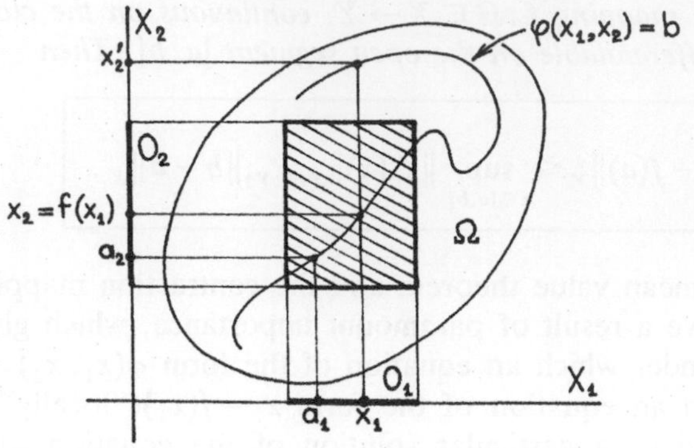

Fig. 1.2-1. If the equation $\varphi(x_1, x_2) = b$ is locally equivalent to the equation $x_2 = f(x_1)$, the function f is called an implicit function.

If the mapping $\varphi : \Omega \subset X_1 \times X_2 \to Y$ *is of class* $\mathscr{C}^m$, $m \geq 2$, *the implicit function* $f : O_1 \to X_2$ *is also of class* $\mathscr{C}^m$. ■

Remark. This result is essentially *local*: It may happen that there exist elements $x_2' \in (X_2 - O_2)$ such that $\varphi(x_1, x_2') = b$ (Fig. 1.2-1). ■

A mapping $f : \Omega \subset X \to f(\Omega) \subset Y$ is a $\mathscr{C}^m$*-diffeomorphism*, $m \geq 1$, if it is of class $\mathscr{C}^m$, if it is injective, and if the inverse mapping $f^{-1} : f(\Omega) \subset Y \to \Omega \subset X$ is also of class $\mathscr{C}^m$. In the special case where $Y = X_1$ and the mapping φ is of the form $\varphi(x_1, x_2) = x_1 - g(x_2)$, the implicit function found in Theorem 1.2-3 can be shown to be a $\mathscr{C}^m$-diffeomorphism, according to the following result.

Theorem 1.2-4 (local inversion theorem). *Let there be given two Banach spaces* X_1 *and* X_2, *an open subset* Ω_2 *of the space* X_2 *containing a point* a_2, *a mapping* $g : \Omega_2 \subset X_2 \to X_1$ *satisfying*

$$g \in \mathscr{C}^1(\Omega_2; X_1), \quad g'(a_2) \in \mathscr{I}som(X_2; X_1),$$

and let $a_1 = g(a_2)$. *Then there exist open subsets* O_1 *and* O_2 *of the spaces* X_1 *and* X_2 *respectively, with* $(a_1, a_2) \in O_1 \times O_2$, $O_2 \subset \Omega_2$, *and an implicit function* $f : O_1 \subset X_1 \to O_2 \subset X_2$ *such that*

$$\begin{cases} \{(x_1, x_2) \in O_1 \times O_2; x_1 = g(x_2)\} \\ \qquad\qquad = \{(x_1, x_2) \in O_1 \times O_2; x_2 = f(x_1)\}\,, \\ O_2 = f(O_1) \text{ and } f : O_1 \subset X_1 \to O_2 \subset Y \text{ is a } \mathscr{C}^1\text{-diffeomorphism}\,, \\ f'(x_1) = \{g'(f(x_1))\}^{-1} \text{ for all } x_1 \in O_1\,. \end{cases}$$

If in addition the mapping $g : \Omega_2 \subset X_2 \to X_1$ *is of class* $\mathscr{C}^m$, $m \geq 2$, *the implicit function* $f : O_1 \subset X_1 \to O_2 \subset Y$ *is a* $\mathscr{C}^m$*-diffeomorphism.* ■

Another important consequence of the implicit function theorem is the following result, due to L.E.J. Brouwer (for a proof, see e.g. Schwartz [1967, p. 294], Zeidler [1986]).

Theorem 1.2-5 (invariance of domain in Banach spaces). *Let there be given two Banach spaces X and Y, an open subset* Ω *of X, and a mapping* $f : \Omega \subset X \to Y$ *satisfying*

$$f \in \mathscr{C}^1(\Omega; Y),\ f'(x) \in \mathscr{I}som(X; Y) \quad \text{for all } x \in \Omega\,.$$

Then the set

$$f(\Omega) := \{y \in Y;\ y = f(x) \quad \text{for some } x \in \Omega\}$$

is open in Y.

If in addition the mapping $f : \Omega \to Y$ *is injective, the mapping* $f : \Omega \to f(\Omega)$ *is a* $\mathscr{C}^1$*-diffeomorphism.* ■

If we drop the differentiability assumptions, a similar result still holds for *injective* mappings in *finite-dimensional spaces*; for a proof, see e.g. Nirenberg [1974, Corollary 2, p. 17], Hurewicz & Wallman [1948, pp. 95–97], Rado & Reichelderfer [1955, p. 135], Dieudonné [1982, Th. 24.8.7].

Theorem 1.2-6 (invariance of domain in $\mathbb{R}^n$**).** *Let* Ω *be an open subset in* $\mathbb{R}^n$, *and let* $f \in \mathscr{C}^0(\Omega; \mathbb{R}^n)$ *be an injective mapping. Then the set* $f(\Omega)$ *is open.* ■

The following consequences of the invariance of domain theorem will be of particular interest in elasticity.

Theorem 1.2-7. *Let Ω be a bounded open subset of $\mathbb{R}^n$ and let $f \in \mathscr{C}^0(\bar{\Omega}; \mathbb{R}^n)$ be a mapping whose restriction to the set Ω is injective. Then*

$$f(\bar{\Omega}) = \{f(\Omega)\}^-, \; f(\Omega) \subset \text{int}\, f(\bar{\Omega}), \; f(\partial\Omega) \supset \partial f(\bar{\Omega}).$$

Proof. Let $y \in f(\bar{\Omega})$. By definition, there exists $x \in \bar{\Omega}$ such that $f(x) = y$. Let $x_k \in \Omega$ be such that $\lim_{k\to\infty} x_k = x$. Since f is continuous, $f(x) = \lim_{k\to\infty} f(x_k) \in \{f(\Omega)\}^-$. Hence $f(\bar{\Omega}) \subset \{f(\Omega)\}^-$. Since $\bar{\Omega}$ is compact and f is continuous, $f(\bar{\Omega})$ is compact, hence closed. Thus

$$f(\Omega) \subset f(\bar{\Omega}) \Rightarrow \{f(\Omega)\}^- \subset \{f(\bar{\Omega})\}^- = f(\bar{\Omega}),$$

which shows that $f(\bar{\Omega}) = \{f(\Omega)\}^-$.

Since the set $f(\Omega)$ is open by the invariance of domain theorem and since it is contained in $f(\bar{\Omega})$, we conclude that $f(\Omega) \subset \text{int}\, f(\bar{\Omega})$. If A is an arbitrary subset of a topological space, we always have $\bar{A} = (\text{int}\, A) \cup \partial A$ and $(\text{int}\, A) \cap \partial A = \emptyset$. Hence

$$f(\bar{\Omega}) = (\text{int}\, f(\bar{\Omega})) \cup \partial f(\bar{\Omega}) \text{ and } (\text{int}\, f(\bar{\Omega})) \cap \partial f(\bar{\Omega}) = \emptyset$$

on the one hand. Since

$$f(\bar{\Omega}) = f(\Omega \cup \partial\Omega) = f(\Omega) \cup f(\partial\Omega) \text{ and } f(\Omega) \subset \text{int}\, f(\bar{\Omega})$$

on the other, we conclude from these relations that $f(\partial\Omega) \supset \partial f(\bar{\Omega})$. ∎

With the additional assumptions that the mapping f is injective "up to the boundary" and that int $\bar{\Omega} = \Omega$, we obtain the stronger result that not only the closure, but also the interior and the boundary, are "preserved by f":

Theorem 1.2-8. *Let Ω be a bounded open subset of $\mathbb{R}^n$ that satisfies*

$$\text{int}\, \bar{\Omega} = \Omega,$$

and let $f \in \mathscr{C}^0(\bar{\Omega}; \mathbb{R}^n)$ be an injective mapping. Then

$$f(\bar{\Omega}) = \{f(\Omega)\}^-, \; f(\Omega) = \text{int}\, f(\bar{\Omega}), \; f(\partial\Omega) = \partial f(\Omega) = \partial f(\bar{\Omega}).$$

Proof. From Theorem 1.2-7, we already infer that $f(\bar{\Omega}) = \{f(\Omega)\}^-$

and $f(\Omega) \subset \text{int } f(\bar{\Omega})$. To prove that $f(\Omega) = \text{int } f(\bar{\Omega})$, let y be such that $y \in \text{int } f(\bar{\Omega})$ and $y \not\in f(\Omega)$. Since the continuous mapping $f : \bar{\Omega} \to f(\bar{\Omega})$ is bijective and $\bar{\Omega}$ is compact, the mapping $f^{-1} : f(\bar{\Omega}) \to \bar{\Omega}$ is also continuous. Hence by the invariance of domain theorem, $f^{-1}(\text{int } f(\bar{\Omega}))$ is an open subset of $\bar{\Omega}$ that contains the point $f^{-1}(y)$. Since $f^{-1}(y) \not\in \Omega$ (f is a bijection), we have found an open subset of $\bar{\Omega}$ that strictly contains Ω, in contradiction with the assumption $\text{int } \bar{\Omega} = \Omega$. Hence $f(\Omega) = \text{int } f(\bar{\Omega})$.

If A is an arbitrary open set, we have $\bar{A} = A \cup \partial A$ and $A \cap \partial A = \emptyset$. Hence

$$\bar{\Omega} = \Omega \cup \partial\Omega \text{ and } \Omega \cap \partial\Omega = \emptyset ,$$

$$\{f(\Omega)\}^- = f(\Omega) \cup \partial f(\Omega) \text{ and } f(\Omega) \cap \partial f(\Omega) = \emptyset .$$

Since $f : \bar{\Omega} \to f(\bar{\Omega}) = \{f(\Omega)\}^-$ is a bijection, we conclude that $f(\partial\Omega) = \partial f(\Omega)$. Since $f(\Omega) = \text{int } f(\bar{\Omega})$, we also have

$$\{f(\Omega)\}^- = f(\Omega) \cup \partial f(\bar{\Omega}) \text{ and } f(\Omega) \cap \partial f(\bar{\Omega}) = \emptyset ,$$

and thus $\partial f(\Omega) = \partial f(\bar{\Omega})$. ∎

To illustrate these results, consider the open set

$$U = \{x \in \mathbb{R}^2;\ \tfrac{1}{2} < x_1 < 1,\ -1 < x_2 < 1\} ,$$

and the mapping

$$f_\theta : x \in \bar{U} \to f_\theta(x) = (x_1 \cos(\theta x_2),\ x_1 \sin(\theta x_2)) ,$$

where $\theta > 0$ is a parameter: If $0 < \theta < \pi$, Theorem 1.2-8 applies, while only Theorem 1.2-7 applies if $\theta = \pi$ (Fig. 1.2-2). The special case $\theta = \pi$ also shows why the assumption $\text{int } \bar{\Omega} = \Omega$ is essential in the second theorem: The open set $\Omega := f_\pi(U)$ does not satisfy $\text{int } \bar{\Omega} = \Omega$ and the injective mapping $f := id$ does not satisfy the last two equalities of Theorem 1.2-8.

Remarks. (1) Sufficient conditions for a mapping $f : \bar{\Omega} \subset \mathbb{R}^n \to \mathbb{R}^n$ to be injective will be given in Theorems 5.5-1, 5.5-2, 5.5-3, and 5.6-1.

(2) As exemplified by the open set $\Omega = f_\pi(U)$ of Fig. 1.2-2, the assumption $\text{int } \bar{\Omega} = \Omega$ precludes in a sense open sets with "too wild"

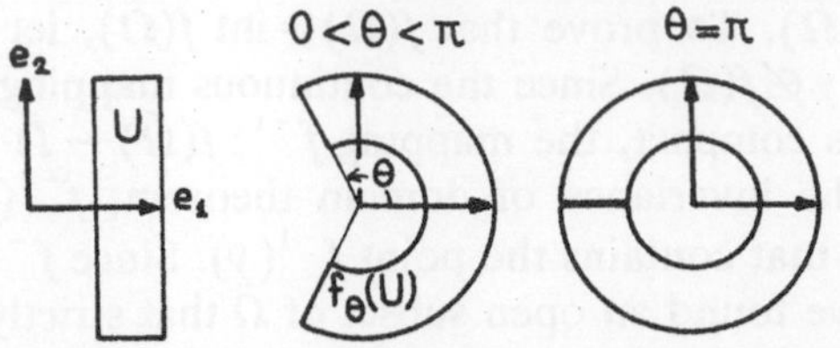

Fig. 1.2-2. The mapping $f_\theta : \bar{U} \to \mathbb{R}^2$ is injective for $0 < \theta < \pi$; the mapping $f_\pi : \bar{U} \to \mathbb{R}^2$ is injective on U but not on $\bar{U}$, and $f_\pi(U) \subsetneqq \text{int } f_\pi(\bar{U})$, $f_\pi(\partial U) \supsetneqq \partial f_\pi(\bar{U})$.

boundaries. This assumption is however satisfied by a reasonably wide class of open subsets of $\mathbb{R}^n$, which are for $n = 3$ particularly well adapted to elasticity (Sect. 1.6).

(3) The book of Rado & Reichelderfer [1955] contains a wealth of results similar in spirit to that of Theorem 1.2-8. ∎

To conclude this review, we consider properties of the derivative that are particular to *real-valued functions*. A point $a \in \Omega$ is a *local extremum* of a function $f : \Omega \subset X \to \mathbb{R}$ if there exists a neighborhood V of the point a such that either $f(a) \leq f(x)$ for all $x \in V$ **(local minimum)** or $f(x) \leq f(a)$ for all $x \in V$ **(local maximum)**. If the function f is derivable at a local extremum a, then necessarily $f'(a) = 0$. If conversely $f'(a) = 0$, additional conditions (involving the second derivative as in Theorem 1.3-1, or the convexity, as in Theorem 4.7-8) are needed in order to conclude that it is a local extremum. A point $a \in \Omega$ at which $f'(a) = 0$ is called a **stationary point** of the function f.

When X is a *Hilbert space*, the derivative of a real-valued function $f : \Omega \subset X \to \mathbb{R}$ can be identified with an element of the space X: If the mapping f is differentiable at a point $a \in \Omega$, its derivative $f'(a)$ is by definition an element of the dual space $X' = \mathscr{L}(X; \mathbb{R})$ and thus, since the space X is a Hilbert space, there exists by the *Riesz representation theorem* a unique element grad $f(a)$ in the space X that satisfies

$$f'(a)h = (\text{grad } f(a), h) \quad \text{for all } h \in X ,$$

where $(\cdot, \cdot)$ denotes the inner product of the space X. The element grad $f(a) \in X$ is called the **gradient** of the function f at the point a. While the derivative $f'(a)$ is intrinsically defined as an element of the dual space X', the gradient is on the contrary an element of the space X that depends on the inner product.

As a first example, consider the case where X is the space $\mathbb{R}^n$ equipped with the Euclidean inner product $\boldsymbol{u}\cdot\boldsymbol{v} = \boldsymbol{u}^{\mathrm{T}}\boldsymbol{v}$. If a real-valued function $f:\Omega \subset \mathbb{R}^n \to \mathbb{R}$ is differentiable at $a \in \Omega$, its gradient is the *vector*

$$\mathbf{grad}\, f(a) := \begin{pmatrix} \dfrac{\partial f}{\partial x_1}(a) \\ \vdots \\ \dfrac{\partial f}{\partial x_n}(a) \end{pmatrix} \in \mathbb{R}^n ,$$

whose components are the partial derivatives of the mapping f. By definition, one has

$$f'(a)\boldsymbol{h} = \mathbf{grad}\, f(a)\cdot \boldsymbol{h} = \sum_{i=1}^{n} \frac{\partial f}{\partial x_i}(a) h_i \quad \text{for all vectors } \boldsymbol{h} \in \mathbb{R}^n .$$

In the case of a function of two variables:

$$f:(t, x) \in \Omega \subset (T \times \mathbb{R}^n) \to f(t, x) \in \mathbb{R} ,$$

we likewise define the vector

$$\mathbf{grad}_x\, f(t, a) = \begin{pmatrix} \dfrac{\partial f}{\partial x_1}(t, a) \\ \vdots \\ \dfrac{\partial f}{\partial x_n}(t, a) \end{pmatrix} \in \mathbb{R}^n ,$$

at each point $(t, a) \in \Omega$ where the function f has a partial derivative with respect to the second variable x.

As a second example, consider the case where X is the space $\mathbb{M}^n$ equipped with the *matrix inner product* $\boldsymbol{A}:\boldsymbol{B} = \operatorname{tr} \boldsymbol{A}^{\mathrm{T}}\boldsymbol{B}$. If a real-valued function $f:\Omega \subset \mathbb{M}^n \to \mathbb{R}$ is differentiable at $\boldsymbol{A} \in \Omega$, its gradient is the *matrix*

$$\frac{\partial f}{\partial A}(A) := \begin{pmatrix} \frac{\partial f}{\partial A_{11}}(A) & \cdots & \frac{\partial f}{\partial A_{1n}}(A) \\ \vdots & & \vdots \\ \frac{\partial f}{\partial A_{n1}}(A) & \cdots & \frac{\partial f}{\partial A_{nn}}(A) \end{pmatrix} \in \mathbb{M}^n ,$$

whose elements are the partial derivatives of the mapping f. By definition, one has

$$f'(A)H = \frac{\partial f}{\partial A}(A) : H = \sum_{i,j=1}^{n} \frac{\partial f}{\partial A_{ij}}(A) H_{ij} \text{ for all matrices } H \in \mathbb{M}^n .$$

For instance the gradients of the three mappings $\iota_1, \iota_{n-1}, \iota_n$, whose derivatives have been computed earlier in this section, are given by

$$\frac{\partial \iota_1}{\partial A}(A) = I ,$$

$$\frac{\partial \iota_{n-1}}{\partial A}(A) = (\text{tr}(A^{-1})I - A^{-T}) \, \text{Cof}\, A ,$$

$$\frac{\partial \iota_n}{\partial A}(A) = (\det A)A^{-T} = \text{Cof}\, A .$$

In the case of a function of two variables:

$$f : (y, A) \in \Omega \subset (Y \times \mathbb{M}^n) \to f(y, A) \in \mathbb{R} ,$$

we likewise define the matrix

$$\frac{\partial f}{\partial A}(y, A) := \left(\frac{\partial f}{\partial A_{ij}}(y, A)\right) \in \mathbb{M}^n$$

at each point $(y, A) \in \Omega$ where the function f has a partial derivative with respect to the second variable A.

Remarks. (1) We avoid the notation ∇f for denoting the gradient of a real-valued function f because the symbol ∇ is commonly, albeit improperly, used in elasticity for denoting the matrix $\nabla\boldsymbol{\varphi} = (\partial_j \varphi_i)$ representing the derivative of a mapping $\boldsymbol{\varphi} \subset \mathbb{R}^3 \rightarrow \mathbb{R}^3$ (Sect. 1.4). To make the matter even more confusing, another misleading habit consists in calling "deformation gradient" the matrix $\nabla\boldsymbol{\varphi}$, although it has nothing to do with the representation of a linear functional by an inner product.

(2) The notations $(\partial f/\partial A)(A)$ or $(\partial f/\partial A)(y, A)$ are equally misleading, since they are not partial derivatives in the sense heretofore understood. ∎

*1.3. HIGHER-ORDER DERIVATIVES

Let there be given two normed vector spaces X and Y, an open subset Ω of X, and a mapping $f : \Omega \subset X \rightarrow Y$ differentiable in Ω. If the derivative mapping

$$f' : x \in \Omega \subset X \rightarrow f'(x) \in \mathcal{L}(X; Y)$$

is differentiable at a point $a \in \Omega$, its derivative

$$f''(a) := (f')'(a) \in \mathcal{L}(X; \mathcal{L}(X; Y)),$$

is called the **second derivative** of f at the point a, and the mapping f is said to be **twice differentiable** *at the point* a. If the mapping f is twice differentiable at all the points of the open set Ω, it is said to be *twice differentiable in* Ω. If the mapping

$$f'' : x \in \Omega \subset X \rightarrow f''(x) \in \mathcal{L}(X; \mathcal{L}(X; Y)),$$

which is well defined in this case, is continuous, the mapping f is said to be *twice continuously differentiable in* Ω, or simply *of class* $\mathscr{C}^2$. We denote by

$$\mathscr{C}^2(\Omega; Y), \quad \text{or simply } \mathscr{C}^2(\Omega) \text{ if } Y = \mathbb{R},$$

the space of all twice continuously differentiable mappings from Ω into Y.

The second derivative at a point can be identified with a continuous bilinear mapping from X into Y by letting

$$(f''(a)h)k = f''(a)(h, k) \quad \text{for all } h, k \in X,$$

and using the mean value theorem, it can be shown that it is a *symmetric* bilinear mapping, in the sense that

$$f''(a)(h, k) = f''(a)(k, h) \quad \text{for all } h, k \in X .$$

The actual computations of second derivatives often rely on the following observation, which reduces them to the computation of first derivatives: *Given two vectors $h, k \in X$, the vector $f''(a)(h, k) \in Y$ is the Gâteaux derivative of the mapping $x \in \Omega \to f'(x)k \in Y$ at the point $x = a \in \Omega$ in the direction of the vector h.* To illustrate this point, we compute the second derivative of a mapping of the form $f(x) = B(x, x)$, where $B : X \times X \to Y$ is a continuous bilinear mapping. We have seen in Sect. 1.2 that the mapping $x \in X \to f'(x)k \in Y$ is given in this case by

$$x \in X \to f'(x)k = B(x, k) + B(k, x) .$$

Since for a fixed vector $k \in X$, the above mapping is affine and continuous, we obtain

$$f''(a)(h, k) = B(h, k) + B(k, h).$$

Note that $f''(a)(h, k) = 2B(h, k)$ if the mapping B is symmetric.

With the knowledge of the second derivative, one can state necessary and sufficient conditions for a point to be a local extremum of a real-valued function. For definiteness, we consider the case of a local minimum.

Theorem 1.3-1 (sufficient conditions for a local minimum). *Let Ω be an open subset of a normed vector space X, let $f : \Omega \subset X \to \mathbb{R}$ be a function differentiable in Ω, and let $a \in \Omega$ be such that $f'(a) = 0$.*

(a) *If the function f is twice differentiable at the point a and if there exists a number α such that*

$$\alpha > 0 \text{ and } f''(a)(h, h) \geq \alpha \|h\|^2 \quad \text{for all } h \in X ,$$

then the point a is a local minimum of the function f.

(b) *If the function f is twice differentiable in Ω and if there exists a neighbourhood $V \subset \Omega$ of the point a such that*

$$f''(x)(h, h) \geq 0 \quad \text{for all } x \in V \text{ and all } h \in X ,$$

then the point a is a local minimum of the function f. ∎

Notice that the local minimum found under assumptions (a) is in fact *strict*, in the sense that there exists a neighborhood V of the point a such that $f(a) < f(x)$ for all $x \in V$, $x \neq a$. The local minimum found in (b) is strict under the stronger assumption:

$$f''(x)(h, h) > 0 \quad \text{for all } x \in V \text{ and all } h \in X, h \neq 0 .$$

Theorem 1.3-2 (necessary conditions for a local minimum). *Let Ω be an open subset of a normed vector space X and let $f : \Omega \subset X \to Y$ be a function differentiable in Ω, and twice differentiable at a point $a \in \Omega$. If a is a local minimum of the function f, then*

$$f'(a) = 0 \text{ and } f''(a)(h, h) \geq 0 \quad \text{for all } h \in X .$$ ■

Notice that *neither theorem is the converse of the other*. Their proofs rely on the Taylor–Young and Taylor–MacLaurin formulas for twice differentiable functions (Theorem 1.3-3).

In the special case where $X = \mathbb{R}^n$ and $Y = \mathbb{R}$, we let

$$\partial_{ij} f(a) := \partial_i(\partial_j f)(a) ,$$

or $(\partial^2 f / \partial x_i \partial x_j)(a)$ if (x_i) denotes a generic point in $\mathbb{R}^n$, denote the usual *second-order partial derivatives*, so that

$$f''(a)(h, k) = \sum_{i,j=1}^{n} \partial_{ij} f(a) h_i k_j \quad \text{for all } h = (h_i), k = (k_i) \in \mathbb{R}^n .$$

Higher-order derivatives are similarly defined. For each integer $k \geq 2$, let $\mathscr{L}_k(X; Y)$ denote the space of all continuous k-linear mappings from X into Y; the space $\mathscr{L}_k(X; Y)$ is isomorphic to, and thus can be identified with, the space $\mathscr{L}(X; \mathscr{L}_{k-1}(X; Y))$, with $\mathscr{L}_1(X; Y) = \mathscr{L}(X; Y)$. Then the ***m*th derivative**

$$f^{(m)}(a) \in \mathscr{L}(X; \mathscr{L}_{m-1}(X; Y)) = \mathscr{L}_m(X; Y)$$

at a point $a \in \Omega$ of a mapping $f : \Omega \subset X \to Y$ is the derivative at the point a of the $(m-1)$th derivative mapping

$$f^{(m-1)} : x \in \Omega \to f^{(m-1)}(x) \in \mathscr{L}_{m-1}(X; Y) .$$

If the mth derivative $f^{(m)}(a)$ exists, the mapping f is said to be ***m* times**

differentiable *at the point a*. The mapping f is *m times differentiable in* Ω if it is m times differentiable at all points in Ω. If the mth derivative mapping $f^{(m)}: \Omega \to \mathscr{L}_m(X; Y)$ is continuous, the mapping f is said to be *m times continuously differentiable*, or simply *of class* $\mathscr{C}^m$; we let

$$\mathscr{C}^m(\Omega; Y)\,, \quad \text{or simply } \mathscr{C}^m(\Omega) \text{ if } Y = \mathbb{R}\,,$$

denote the space of all m times continuously differentiable mappings from X into Y.

Finally, we let

$$\mathscr{C}^\infty(\Omega; Y) := \bigcap_{m=0}^{\infty} \mathscr{C}^m(\Omega; Y)\,, \quad \text{or simply } \mathscr{C}^\infty(\Omega) \text{ if } Y = \mathbb{R}\,,$$

denote the space of *infinitely differentiable mappings* from Ω into Y.

Using the abbreviated notation

$$f^{(m)}(a)h^m = f^m(a)(h_1, h_2, \ldots, h_m) \quad \text{when } h_i = h, 1 \leq i \leq m\,,$$

we now state several useful *Taylor formulas*. The first generalizes the definition of the derivative; the second generalizes the mean value theorem; the third and fourth give explicit forms of the remainder; the third is a generalization of the classical mean value theorem for real-valued functions and the fourth generalizes the well known formula $f(a+h) - f(a) = \int_a^{a+h} f'(\eta)\, d\eta$ for real-valued functions.

Theorem 1.3-3 (Taylor formulas). *Let X and Y be two normed vector spaces, let* Ω *be an open subset of X, let* $[a, a+h]$ *be a closed segment contained in* Ω, *let* $f: \Omega \subset X \to Y$ *be a given mapping, and let m be an integer* ≥ 1.

(a) (*Taylor–Young formula*). *If f is* $(m-1)$ *times differentiable in* Ω *and m times differentiable at the point a, then*

$$f(a+h) = f(a) + f'(a)h + \cdots + \frac{1}{m!} f^{(m)}(a)h^m + \|h\|^m \varepsilon(h)\,,$$

$$\lim_{h \to 0} \varepsilon(h) = 0\,.$$

(b) (*generalized mean value theorem*). *If f is* $(m-1)$ *times continuously differentiable in* Ω *and m times differentiable on the open segment* $]a, a+h[$, *then*

$$\left\| f(a+h) - \left\{ f(a) + f'(a)h + \cdots + \frac{1}{(m-1)!} f^{(m-1)}(a)h^{m-1} \right\} \right\|$$

$$\leq \frac{1}{m!} \sup_{x \in]a, a+h[} \|f^{(m)}(x)\| \, \|h\|^m .$$

(c) (*Taylor–MacLaurin formula*). *If $Y = \mathbb{R}$ and f is $(m-1)$ times continuously differentiable in Ω and m times differentiable on the open segment $]a, a+h[$, there exists a number $\theta \in]0, 1[$ such that*

$$f(a+h) = f(a) + f'(a)h + \cdots + \frac{1}{(m-1)!} f^{(m-1)}(a)h^{m-1}$$

$$+ \frac{1}{m!} f^{(m)}(a + \theta h)h^m .$$

(d) (*Taylor formula with integral remainder*). *If f is m times continuously differentiable in Ω and if Y is complete, then*

$$f(a+h) = f(a) + f'(a)h + \cdots + \frac{1}{(m-1)!} f^{(m-1)}(a)h^{m-1}$$

$$+ \int_0^1 \frac{(1-t)^{m-1}}{(m-1)!} \{ f^{(m)}(a + th)h^m \} \, dt . \quad \blacksquare$$

We shall also use the *multi-index notation* for higher-order partial derivatives of mappings $f : \Omega \subset \mathbb{R}^n \to \mathbb{R}$: Given a multi-index $\alpha = (\alpha_1, \alpha_2, \ldots, \alpha_n) \in \mathbb{N}^n$, we let $|\alpha| = \Sigma_{i=1}^n \alpha_i$. Then

$$\partial^\alpha f(a) := f^{(|\alpha|)}(a)(e_1, \ldots, e_1, e_2, \ldots, e_2, \ldots, e_n, \ldots, e_n) ,$$

where each basis vector e_i of $\mathbb{R}^n$ occurs α_i times, $1 \leq i \leq n$. For instance, if $n = 3$,

$$\frac{\partial^2 f}{\partial x_1^2}(a) = \partial^{(2,0,0)} f(a), \quad \frac{\partial^3 f}{\partial x_1 \partial x_3^2}(a) = \partial^{(1,0,2)} f(a), \text{ etc} .$$

This notational device allows for a simple representation of the sets formed by all partial derivatives of a given order m, or of order less than or equal to m, *viz.*,

$$\{ \partial^\alpha f(a); |\alpha| = m \} , \quad \text{or } \{ \partial^\alpha f(a); |\alpha| \leq m \} .$$

In the special case where Ω is a *bounded open subset of* $\mathbb{R}^n$, we let $\mathscr{C}^m(\bar{\Omega})$ denote for each integer $m \geq 1$ the subspace of $\mathscr{C}^m(\Omega)$ consisting of those functions v for which there exist functions $v^\alpha \in \mathscr{C}^0(\bar{\Omega})$ such that $v^\alpha|_\Omega = \partial^\alpha v$ for all multi-index α satisfying $|\alpha| \leq m$ (in particular, $\mathscr{C}^m(\bar{\Omega})$ is a subspace of $\mathscr{C}^0(\bar{\Omega})$). *Equipped with the norm*

$$\|v\|_{\mathscr{C}^m(\bar{\Omega})} = \max_{|\alpha| \leq m} \sup_{x \in \Omega} |v^\alpha(x)|\,,$$

the space $\mathscr{C}^m(\bar{\Omega})$ *is a Banach space*. We also let

$$\mathscr{C}^\infty(\bar{\Omega}) := \bigcap_{m=0}^{\infty} \mathscr{C}^m(\bar{\Omega})\,.$$

Assuming again that Ω is a bounded open subset of $\mathbb{R}^n$, we let $\mathscr{C}^{m,\lambda}(\bar{\Omega})$ denote for each integer $m \geq 0$ and each number $\lambda \in \,]0, 1]$ the subspace of $\mathscr{C}^m(\bar{\Omega})$ consisting of those functions v that satisfy

$$\|v\|_{\mathscr{C}^{m,\lambda}(\bar{\Omega})} := \|v\|_{\mathscr{C}^m(\bar{\Omega})} + \max_{|\alpha|=m} \sup_{\substack{x, y \in \Omega \\ x \neq y}} \frac{|\partial^\alpha v(x) - \partial^\alpha v(y)|}{|x-y|^\lambda} < +\infty\,.$$

The functions in the space $\mathscr{C}^{0,\lambda}(\bar{\Omega})$ are said to satisfy a *Hölder condition with exponent* λ if $\lambda < 1$, and to be **Lipschitz-continuous** if $\lambda = 1$. *Equipped with the norm* $\|\cdot\|_{\mathscr{C}^{m,\lambda}(\bar{\Omega})}$, *the space* $\mathscr{C}^{m,\lambda}(\bar{\Omega})$ *is a Banach space.*

If $X(\Omega)$, or $Y(\bar{\Omega})$, denotes any one of the vector spaces encountered in this section, we shall denote by $\boldsymbol{X}(\Omega)$, or $\boldsymbol{Y}(\bar{\Omega})$, any space of vector-valued, or tensor-valued, functions with components in $X(\Omega)$, or $Y(\bar{\Omega})$. If $\|\cdot\|$ denotes a norm over the space $Y(\bar{\Omega})$, the same symbol denotes the corresponding product norm in the space $\boldsymbol{Y}(\bar{\Omega})$.

1.4. DEFORMATIONS IN $\mathbb{R}^3$

We assume *once and for all* that an origin $\boldsymbol{o}$ and an orthonormal basis $\{\boldsymbol{e}_1, \boldsymbol{e}_2, \boldsymbol{e}_3\}$ have been chosen in three-dimensional Euclidean space, which will therefore be identified with the space $\mathbb{R}^3$: From the notational viewpoint, we identify the point x with the vector $\boldsymbol{ox}$. Whenever we consider components of vectors in $\mathbb{R}^3$, or elements of matrices in $\mathbb{M}^3$, we make the convention that *Latin indices* $(i, j, p, \ldots)$ *always take their values in the set* $\{1, 2, 3\}$, and we combine this rule with the standard *summation convention*.

Let there be given a bounded, open, connected, subset Ω of $\mathbb{R}^3$ with a sufficiently smooth boundary (specific smoothness assumptions will be made subsequently). We shall think of the closure $\bar{\Omega}$ of the set Ω as representing the volume occupied by a body "before it is deformed"; for this reason, the set $\bar{\Omega}$ is called the **reference configuration**.

A **deformation** of the reference configuration $\bar{\Omega}$ is a vector field

$$\varphi : \bar{\Omega} \to \mathbb{R}^3$$

that is *smooth enough*, *injective except possibly on the boundary of the set* Ω, and *orientation-preserving*.

Remarks. (1) The reason a deformation may loose its injectivity on the boundary of Ω is that "self-contact" must be allowed. We will discuss this aspect at length in Chapter 5.

(2) The expression "smooth enough" is simply a convenient way of saying that in a given definition, theorem, proof, etc., the smoothness of the deformations involved is such that all arguments make sense. As a consequence, the underlying degree of smoothness varies from place to place. For instance, the existence of the deformation gradient (to be next introduced) implies that a deformation is *differentiable at all points* of the reference configuration; Theorem 1.7-1 relies on the Piola identity, which makes sense, at least in a classical setting, only for *twice differentiable* deformations; the characterization of rigid deformations (Theorem 1.8-1) is established for deformations that are *continuously differentiable*, etc. By contrast, even the assumption of everywhere differentiability will be relaxed in Chapter 7, where the partial derivatives of the "deformations", then understood *in the sense of distributions*, only need to be defined almost everywhere (typically, they lie in some $L^p(\Omega)$ spaces).

(3) Deformations are synonymously called *configurations*, or *placements*, by some authors. ■

We denote by x a generic point in the set $\bar{\Omega}$, by x_i its components with respect to the basis $\{e_i\}$, and by

$$\partial_i = \partial / \partial x_i$$

the partial derivative with respect to the variable x_i. Given a deformation

$$\boldsymbol{\varphi} = \varphi_i \boldsymbol{e}_i ,$$

we define at each point of the set Ω the matrix

$$\boxed{\nabla\boldsymbol{\varphi} := \begin{pmatrix} \partial_1\varphi_1 & \partial_2\varphi_1 & \partial_3\varphi_1 \\ \partial_1\varphi_2 & \partial_2\varphi_2 & \partial_3\varphi_2 \\ \partial_1\varphi_3 & \partial_2\varphi_3 & \partial_3\varphi_3 \end{pmatrix} .}$$

The matrix $\nabla\boldsymbol{\varphi}$ is called the **deformation gradient**. Since a deformation is orientation-preserving by definition, the determinant of the deformation gradient satisfies the **orientation-preserving condition:**

$$\boxed{\det \nabla\boldsymbol{\varphi}(x) > 0 \text{ for all } x \in \bar{\Omega} .}$$

In particular, the matrix $\nabla\boldsymbol{\varphi}(x)$ is invertible at all points x of the reference configuration $\bar{\Omega}$.

Remarks. (1) The notations

$$\boldsymbol{F} = \nabla\boldsymbol{\varphi} \text{ and } J = \det \nabla\boldsymbol{\varphi}$$

are commonly used in the literature.

(2) As already mentioned in Sect. 1.2, the notation $\nabla\boldsymbol{\varphi}$ is confusing, since the *gradient of a real-valued function f* is the *column* vector formed by the first partial derivative $\partial_i f$, while $(\nabla\boldsymbol{\varphi})_{ij} = \partial_j \varphi_i$ (this explains why we used the notation **grad** f, and not ∇f, in Sect. 1.2). Indeed, *the deformation gradient is simply the matrix representing the Fréchet derivative of the mapping* $\boldsymbol{\varphi}$, which for real-valued functions, is to be identified with the *transpose* of the gradient. ∎

Together with a deformation $\boldsymbol{\varphi}$, it is often convenient to introduce the **displacement** $\boldsymbol{u}$, which is the vector field

$$\boldsymbol{u} : \bar{\Omega} \to \mathbb{R}^3$$

defined by the relation

$$\boxed{\boldsymbol{\varphi} = \boldsymbol{id} + \boldsymbol{u} ,}$$

where ***id*** denotes the (restriction to $\bar{\Omega}$ of the) identity map from $\mathbb{R}^3$ onto $\mathbb{R}^3$. Notice that the **displacement gradient**

$$\boxed{\nabla u := \begin{pmatrix} \partial_1 u_1 & \partial_2 u_1 & \partial_3 u_1 \\ \partial_1 u_2 & \partial_2 u_2 & \partial_3 u_2 \\ \partial_1 u_3 & \partial_2 u_3 & \partial_3 u_3 \end{pmatrix}}$$

and the deformation gradient are related by the equation

$$\boxed{\nabla \varphi = I + \nabla u\,.}$$

Given a reference configuration $\bar{\Omega}$ and a deformation $\varphi : \bar{\Omega} \to \mathbb{R}^3$, the set $\varphi(\bar{\Omega})$ is called a **deformed configuration.** At each point

$$\boxed{x^{\varphi} := \varphi(x)}$$

of a deformed configuration, we define the three vectors (Fig. 1.4-1)

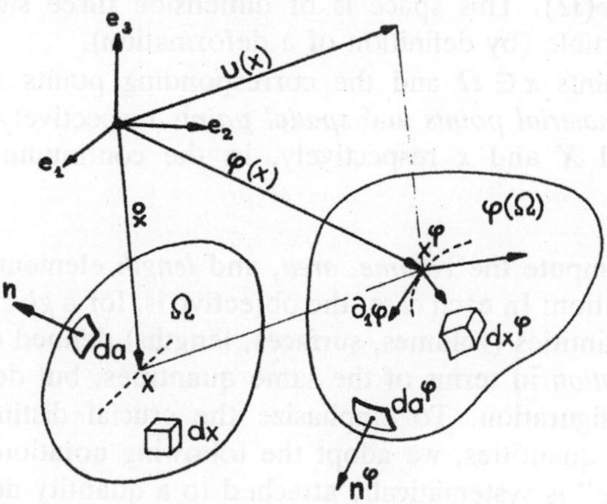

Fig. 1.4-1. Geometry of a deformation: The volume element, the area element, and the unit outer normal, are denoted dx, da, n in the reference configuration $\bar{\Omega}$, and dx^{φ}, da^{φ}, n^{φ} in the deformed configuration $\varphi(\bar{\Omega})$. The vectors $\partial_j \varphi(x)$ define the deformation at a point $x \in \bar{\Omega}$ to within the first order.

$$\partial_j \boldsymbol{\varphi}(x) = \partial_j \varphi_i(x) \boldsymbol{e}_i \,.$$

Each vector $\partial_j \boldsymbol{\varphi}(x)$ measures the "local deformation in the direction of the vector $\boldsymbol{e}_j$" in the sense that, *to within the first order with respect to* $\mathrm{d}t$, *the vector* $\mathrm{d}t\, \boldsymbol{e}_j$ *is transformed into the vector* $\mathrm{d}t\, \partial_j \boldsymbol{\varphi}(x)$. Equivalently, the vector $\partial_j \boldsymbol{\varphi}(x)$ is the tangent vector to the jth *coordinate line* passing through the point x^{φ} (i.e., the image by the deformation $\boldsymbol{\varphi}$ of a segment parallel to the vector $\boldsymbol{e}_j$ containing the point x in its interior, and parametrized by t). Since the vector $\partial_j \boldsymbol{\varphi}(x)$ is precisely the jth column of the matrix $\boldsymbol{\nabla\varphi}(x)$, *the knowledge of the deformation gradient completely defines the local deformation to within the first order.*

Remarks. (1) While the deformation gradient $\boldsymbol{\nabla\varphi}(x)$ clearly depends upon the basis $(\boldsymbol{e}_i)$, it is possible to exhibit the intrinsic geometrical character of the deformation at the point x, by means of the polar factorization (Theorem 3.2-2) of the matrix $\boldsymbol{\nabla\varphi}(x)$, which then appears as the product of a "rotation tensor" by a "stretch tensor". For details about this classical result, see for instance Germain [1972, p. 97], Gurtin [1981b, p. 46], Truesdell & Noll [1965, p. 52].

(2) If the point $x^{\varphi} = \boldsymbol{\varphi}(x)$ belongs to the interior of the deformed configuration $\boldsymbol{\varphi}(\bar{\Omega})$, the three vectors $\partial_j \boldsymbol{\varphi}$ define in the terminology of differential geometry the *tangent vector space* at the point x of the *manifold* int $\boldsymbol{\varphi}(\bar{\Omega})$. This space is of dimension three since the matrix $\boldsymbol{\nabla\varphi}(x)$ is invertible (by definition of a deformation).

(3) The points $x \in \bar{\Omega}$ and the corresponding points $x^{\varphi} \in \boldsymbol{\varphi}(\bar{\Omega})$ are often called *material points* and *spatial points* respectively, and they are often denoted X and x respectively, in the continuum mechanics literature. ■

We next compute the *volume*, *area*, and *length* elements in the deformed configuration: In each case, the objective is, for a *given* deformation, to express quantities (volumes, surfaces, lengths) defined over the *deformed configuration* in terms of the same quantities, but defined over the *reference* configuration. To emphasize the crucial distinction between both types of quantities, we adopt the following notational device: The superscript "$\boldsymbol{\varphi}$" is systematically attached to a quantity defined over the deformed configuration, while the related quantity over the reference configuration is designated by the same letter, but without the superscript "$\boldsymbol{\varphi}$"; this rule has already been applied, for denoting a generic point $x \in \bar{\Omega}$ and the corresponding point $x^{\varphi} = \boldsymbol{\varphi}(x) \in \boldsymbol{\varphi}(\bar{\Omega})$.

This correspondence between a quantity defined as a function of the

Lagrange variable x, and a similar quantity defined as a function of the **Euler variable** $x^{\varphi} = \varphi(x)$, can be extended to other quantities than volumes, surfaces, and lengths: As we shall see, it applies equally well to divergences of tensor fields (Theorem 1.7-1) and applied forces (Sects. 2.6 and 2.7).

Remark. This idea can be systematized through the notions of "pull-back" and "push-forward", familiar in differential geometry. In this respect, see for instance Choquet-Bruhat, Dewitt-Morette & Dillard-Bleick [1977], or Marsden & Hughes [1983]. ■

1.5. VOLUME ELEMENT IN THE DEFORMED CONFIGURATION

Let φ be a deformation. If $\mathrm{d}x$ denotes the **volume element** at the point x of the reference configuration, *the volume element* $\mathrm{d}x^{\varphi}$ *at the point* $x^{\varphi} = \varphi(x)$ *of the deformed configuration* (Fig. 1.4-1) *is given by*

$$\boxed{\mathrm{d}x^{\varphi} = \det \nabla\varphi(x)\,\mathrm{d}x\,,}$$

since $|\det \nabla\varphi(x)| = \det \nabla\varphi(x) > 0$ by assumption.

The volume element $\mathrm{d}x^{\varphi}$ *is used for computing volumes in the deformed configuration*: If A denotes a measurable subset of the reference configuration $\bar{\Omega}$, the **volume** of the set A and the volume of the deformed set $A^{\varphi} := \varphi(A)$ are respectively given by

$$\operatorname{vol} A := \int_A \mathrm{d}x, \quad \operatorname{vol} A^{\varphi} := \int_{A^{\varphi}} \mathrm{d}x^{\varphi} = \int_A \det \nabla\varphi(x)\,\mathrm{d}x\,.$$

Notice that the last equality is nothing but a special case of the *formula for changes of variables in multiple integrals*: Let $\varphi : A \rightarrow \varphi(A) = A^{\varphi}$ be an *injective*, continuously differentiable mapping with a continuous inverse $\varphi^{-1} : A^{\varphi} \rightarrow A$. Then a function $u : x^{\varphi} \in A^{\varphi} \rightarrow \mathbb{R}$ is $\mathrm{d}x^{\varphi}$-integrable over the set A^{φ} if and only if the function

$$x \in A \rightarrow (u \circ \varphi)(x) |\det \nabla\varphi(x)|$$

is $\mathrm{d}x$-integrable over the set A and if this is the case,

$$\boxed{\int_{A^{\varphi} = \varphi(A)} u(x^{\varphi})\,\mathrm{d}x^{\varphi} = \int_A (u \circ \varphi)(x) |\det \nabla\varphi(x)|\,\mathrm{d}x\,.}$$

It should be remembered that *the validity of this formula hinges critically on the assumption that the mapping φ is injective*. Otherwise, it must be replaced by the more general relation:

$$\int_{\varphi(A)} u(x') \operatorname{card} \varphi^{-1}(x') \, dx' = \int_A (u \circ \varphi)(x) |\det \nabla\varphi(x)| \, dx ,$$

where card B denotes in general the number of elements in a set B. For details, see Schwartz [1967, Corollaire 2, p. 675], Rado & Reichelderfer [1955, p. 438], Federer [1969, p. 241 ff.], Smith [1983, Ch. 16], and also Bojarski & Iwaniec [1983, Sect. 8], Marcus & Mizel [1973], Vodopyanov, Goldshtein & Reshetnyak [1979] for its extension to Sobolev space-valued mappings (we shall use this extension in Chapter 7).

These properties hold in $\mathbb{R}^n$, for arbitrary n. The *volume* $\int_A dx$ of a dx-measurable subset *of* $\mathbb{R}^n$ is denoted dx-meas A.

*1.6. SURFACE INTEGRALS; GREEN'S FORMULAS

We essentially follow here the presentation of Nečas [1967, p. 119 ff.]. Let $|\cdot|$ stand for the Euclidean norm, $X - A$ denote the complement of a subset A of a set X, and supp ψ denote the support of a real-valued function ψ. The boundary $\partial\Omega$ of an open subset Ω of $\mathbb{R}^n$ is said to be **Lipschitz-continuous** if the following conditions are simultaneously satisfied: There exist constants $\alpha > 0$, $\beta > 0$, λ, and a finite number of *local coordinate systems*, with origins o_r, coordinates $\zeta'_r = (\xi^r_1, \xi^r_2, \ldots, \xi^r_{n-1})$ and $\zeta_r = \xi^r_n$, and corresponding maps a_r, $1 \leq r \leq R$, such that (cf. Fig. 1.6-1 and the counter-examples of Fig. 1.6-2 in the case $n = 2$):

$$\partial\Omega = \bigcup_{r=1}^{R} \{(\zeta'_r, \zeta_r);\ \zeta_r = a_r(\zeta'_r);\ |\zeta'_r| < \alpha\} ,$$

$$\{(\zeta'_r, \zeta_r);\ a_r(\zeta'_r) < \zeta_r < a_r(\zeta'_r) + \beta;\ |\zeta'_r| \leq \alpha\} \subset \Omega,\ 1 \leq r \leq R ,$$

$$\{(\zeta'_r, \zeta_r);\ a_r(\zeta'_r) - \beta < \zeta_r < a_r(\zeta'_r);\ |\zeta'_r| \leq \alpha\} \subset \mathbb{R}^3 - \bar{\Omega},\ 1 \leq r \leq R ,$$

$$|a_r(\zeta'_r) - a_r(\eta'_r)| \leq |\zeta'_r - \eta'_r| \quad \text{for all } |\zeta'_r| \leq \alpha,\ |\eta'_r| \leq \alpha,\ 1 \leq r \leq R ,$$

where the last inequalities express the *Lipschitz-continuity of the maps* a_r. Notice that, while a Lipschitz-continuous boundary $\partial\Omega$ is necessarily bounded, this is not necessarily true of the set Ω, which can be interchanged with the set $\mathbb{R}^3 - \bar{\Omega}$ in the definition.

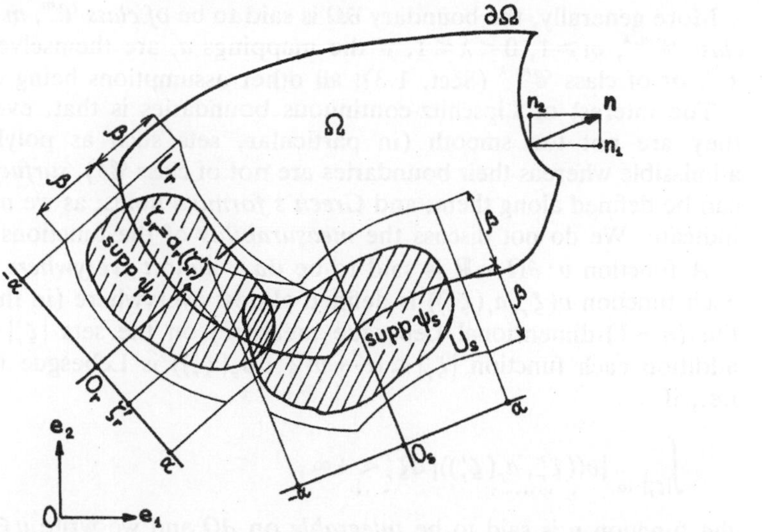

Fig. 1.6-1. An open subset $\Omega \subset \mathbb{R}^2$ with a Lipschitz-continuous boundary $\partial\Omega$.

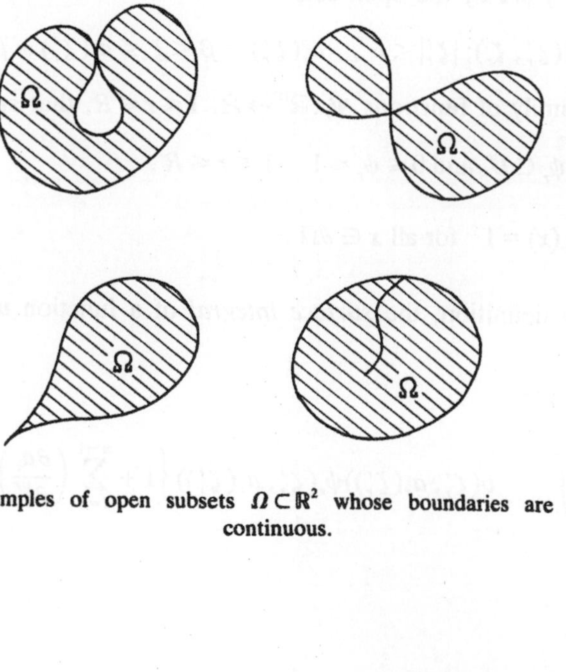

Fig. 1.6-2. Examples of open subsets $\Omega \subset \mathbb{R}^2$ whose boundaries are not Lipschitz-continuous.

More generally, the boundary $\partial\Omega$ is said to be *of class* $\mathscr{C}^m$, $m \geq 1$, or *of class* $\mathscr{C}^{m,\lambda}$, $m \geq 1$, $0 < \lambda \leq 1$, if the mappings a_r are themselves of class $\mathscr{C}^m$, or of class $\mathscr{C}^{m,\lambda}$ (Sect. 1.3), all other assumptions being unaltered.

The interest of Lipschitz-continuous boundaries is that, even though they are not too smooth (in particular, sets such as polyhedra are admissible whereas their boundaries are not of class $\mathscr{C}^1$), *surface integrals* can be defined along them, and *Green's formula* holds, as we now briefly indicate. We do not discuss the *measurability* of the functions involved.

A function $v : \partial\Omega \to \mathbb{R}$ is said to be $\mathrm{d}a$ – *almost everywhere* defined if each function $v(\zeta_r', a_r(\zeta_r'))$ is defined almost everywhere (in the sense of the $(n-1)$-dimensional Lebesgue measure) on the sets $|\zeta_r'| < \alpha$. If in addition each function $|\zeta_r'| < \alpha \to v(\zeta_r', a_r(\zeta_r'))$ is Lebesgue integrable, i.e., if

$$\int_{|\zeta_r'|<\alpha} |v((\zeta_r', a_r(\zeta_r'))| \,\mathrm{d}\zeta_r' < +\infty ,$$

the function v is said to be *integrable* on $\partial\Omega$ and we write $v \in \mathscr{L}^1(\partial\Omega)$. More generally, if, for some $p \geq 1$,

$$\int_{|\zeta_r'|<\alpha} |v((\zeta_r', a_r(\zeta_r'))|^p \,\mathrm{d}\zeta_r' < +\infty , \quad 1 \leq r \leq R ,$$

we write $v \in \mathscr{L}^p(\partial\Omega)$. In order to define the integral of a function $v \in \mathscr{L}^1(\partial\Omega)$, we need a *partition of unity* associated with the covering of the boundary $\partial\Omega$ by the open sets

$$U_r := \{(\zeta_r', \zeta_r); |\zeta_r'| < \alpha , \quad a_r(\zeta_r') - \beta < \zeta_r < a_r(\zeta_r') + \beta\} ,$$

that is, a family of functions $\Psi_r : \mathbb{R}^n \to \mathbb{R}$, $1 \leq r \leq R$, that satisfy:

$$\begin{cases} \operatorname{supp} \psi_r \subset U_r \text{ and } 0 \leq \psi_r \leq 1, \quad 1 \leq r \leq R , \\ \displaystyle\sum_{r=1}^{R} \psi_r(x) = 1 \quad \text{for all } x \in \partial\Omega . \end{cases}$$

Then, by definition, the *surface integral* of a function $v \in \mathscr{L}^1(\partial\Omega)$ is

$$\int_{\partial\Omega} v \,\mathrm{d}a := \sum_{r=1}^{R} \int_{|\zeta_r'|<\alpha} v(\zeta_r', a_r(\zeta_r'))\psi_r(\zeta_r', a_r(\zeta_r')) \left\{1 + \sum_{i=1}^{n-1} \left(\frac{\partial a_r}{\partial \xi_i^r}\right)^2\right\}^{1/2} \mathrm{d}\zeta_r' ,$$

and da denotes the **area element** along $\partial\Omega$. This definition makes sense: First, being Lipschitz-continuous, the functions a_r are almost everywhere (in the sense of the $(n-1)$-dimensional Lebesgue measure) Fréchet differentiable since they are Lipschitz-continuous (cf. e.g. Nečas [1967, p. 88]), and each partial derivative satisfies

$$\left|\frac{\partial a_r}{\partial \zeta_i^r}(\zeta_r')\right| \leq \lambda \quad \text{for almost all } |\zeta_r'| < \alpha \,;$$

secondly, it can be shown that the number $\int_{\partial\Omega} v \, \mathrm{d}a$ defined in this fashion is independent of the local coordinate systems considered and independent of the associated partition of unity. The functions $v \in \mathcal{L}^p(\partial\Omega)$ that are da-almost everywhere equal are identified, and the space formed by the equivalence classes, which are denoted by the same letter v, is denoted $L^p(\partial\Omega)$. Equipped with the norm

$$\|v\|_{L^p(\partial\Omega)} := \left\{\int_{\partial\Omega} |v|^p \, \mathrm{d}a\right\}^{1/p},$$

the space $L^p(\partial\Omega)$ is a Banach space.

The **area** of a da-measurable subset Δ of $\partial\Omega$ is denoted and defined by

$$\mathrm{d}a\text{-meas } \Delta = \int_{\partial\Omega} \chi_\Delta \, \mathrm{d}a = \int_\Delta \mathrm{d}a, \text{ or simply area } \Delta \text{ if } n = 3\,,$$

where χ_Δ denotes the characteristic functions of the set Δ.

Another consequence of the almost everywhere differentiability of the functions a_r is that a **unit outer normal vector** $\boldsymbol{n} = n_i\boldsymbol{e}_i$ *exists* da*-almost everywhere along* $\partial\Omega$; "unit" means that its Euclidean norm $|\boldsymbol{n}|$ is 1, and "outer" means that it is directed outward.

A **domain** in $\mathbb{R}^n$ is an *open*, *bounded*, *connected* subset with a *Lipschitz-continuous boundary*, and a **subdomain** is a domain contained in a domain. An important property of domains is the validity of the **fundamental Green's formula**: Given a domain Ω in $\mathbb{R}^n$ with normal vector $\boldsymbol{n} = (n_i)$ along $\partial\Omega$, and a smooth enough function $u : \bar{\Omega} \to \mathbb{R}$, then

$$\boxed{\int_\Omega \partial_i u \, \mathrm{d}x = \int_{\partial\Omega} u n_i \, \mathrm{d}a\,, \quad 1 \leq i \leq n\,.}$$

This formula holds for example if the function u is continuously differentiable over the set $\bar{\Omega}$, but we shall see (Theorem 6.1-9) that such a smoothness assumption can be significantly relaxed. Notice that the fundamental Green's formula is nothing but the multi-dimensional extension of the one-dimensional formula for integration by parts $\int_a^b v'(t)\,\mathrm{d}t = v(b) - v(a)$. Using it, one can prove other **Green's formulas** where, in essence, *a particular combination of integrals over Ω is written as a combination of surface integrals over $\partial\Omega$.* For example, replacing the function u by a product of two functions u and v in the fundamental Green's formula yields another well-known formula:

$$\int_\Omega u\partial_i v\,\mathrm{d}x = -\int_\Omega \partial_i u\, v\,\mathrm{d}x + \int_{\partial\Omega} uvn_i\,\mathrm{d}a\,.$$

As a second example, let there be given a vector field $\boldsymbol{u}:\bar{\Omega}\to\mathbb{R}^n$ with components $u_i:\bar{\Omega}\to\mathbb{R}$; then the fundamental Green's formula yields

$$\int_\Omega \partial_i u_i\,\mathrm{d}x = \int_{\partial\Omega} u_i n_i\,\mathrm{d}a$$

or, in vector form,

$$\int_\Omega \operatorname{div}\boldsymbol{u}\,\mathrm{d}x = \int_{\partial\Omega} \boldsymbol{u}\cdot\boldsymbol{n}\,\mathrm{d}a,\ \operatorname{div}\boldsymbol{u} := \partial_i u_i\,.$$

This Green formula is the **divergence theorem for vector fields.**

Recall that the reference configuration $\bar{\Omega}$ has been defined (Sect. 1.4) as the closure of an open subset Ω of $\mathbb{R}^3$. *We shall assume in what follows that the set Ω is a domain.* This implies that the interior of the set $\bar{\Omega}$ coincides with the set Ω (Exercise 1.7), and thus we can apply Theorem 1.2-8: *If a mapping $\boldsymbol{\varphi}\in\mathscr{C}^0(\bar{\Omega})$ is injective on $\bar{\Omega}$, we have*

$$\boldsymbol{\varphi}(\bar{\Omega}) = \{\boldsymbol{\varphi}(\Omega)\}^-,\ \operatorname{int}\boldsymbol{\varphi}(\bar{\Omega}) = \boldsymbol{\varphi}(\Omega),\ \boldsymbol{\varphi}(\partial\Omega) = \partial\boldsymbol{\varphi}(\Omega)\,.$$

These relations justify in particular the notations

$$\boxed{\bar{\Omega}^\varphi = \boldsymbol{\varphi}(\bar{\Omega}),\ \Omega^\varphi = \boldsymbol{\varphi}(\Omega),\ \partial\Omega^\varphi = \boldsymbol{\varphi}(\partial\Omega)\,,}$$

which we shall henceforth adopt for the deformed configuration, its interior, and its boundary, whenever the deformation $\boldsymbol{\varphi}$ is injective on $\bar{\Omega}$ (recall that in general, a deformation may lose its injectivity on $\partial\Omega$).

We shall also assume that the deformations $\varphi : \bar{\Omega} \to \mathbb{R}^3$ that we shall consider are such that the set Ω^φ is also a domain (this is true if φ is sufficiently smooth; cf. Exercise 1.10). This being the case, an area element da^φ can be defined along the boundary $\partial\Omega^\varphi$ of the deformed configuration, and a unit outer normal vector $\boldsymbol{n}^\varphi = n_i^\varphi \boldsymbol{e}_i$ can be defined da^φ-almost everywhere along $\partial\Omega^\varphi$.

1.7. THE PIOLA TRANSFORM; AREA ELEMENT IN THE DEFORMED CONFIGURATION

As a preparation for computing the area element in the deformed configuration in terms of the area element in the reference configuration, it is convenient to introduce a particular *transformation between tensors defined over the reference configuration* $\bar{\Omega}$ *and tensors defined over the deformed configuration* $\bar{\Omega}^\varphi$. Besides, this transform plays a crucial rôle in the definition of the first Piola–Kirchhoff stress tensor, to be introduced in Sect. 2.5.

Let us first review some definitions and results pertaining to tensor fields defined over either sets $\bar{\Omega}$ or $\bar{\Omega}^\varphi$. By a **tensor**, we mean here a *second-order tensor*

$$\boldsymbol{T} = (T_{ij}),\ i\text{: row index},\ j\text{: column index}\,.$$

Since we ignore the distinction between covariant and contravariant components, the set of all such tensors will be identified with the set $\mathbb{M}^3$ of all square matrices of order three.

Given a smooth enough tensor field $\boldsymbol{T} : \bar{\Omega} \to \mathbb{M}^3$ defined over the reference configuration $\bar{\Omega}$, we define at each point of $\bar{\Omega}$ its **divergence** $\mathbf{div}\,\boldsymbol{T}$ as the vector whose components are the divergences of the *transposes* of the row vectors of the matrix $\boldsymbol{T}$. More explicitly,

$$\boldsymbol{T} = (T_{ij}) = \begin{pmatrix} T_{11} & T_{12} & T_{13} \\ T_{21} & T_{22} & T_{23} \\ T_{31} & T_{32} & T_{33} \end{pmatrix} \Rightarrow \mathbf{div}\,\boldsymbol{T} := \begin{pmatrix} \partial_1 T_{11} + \partial_2 T_{12} + \partial_3 T_{13} \\ \partial_1 T_{21} + \partial_2 T_{22} + \partial_3 T_{23} \\ \partial_1 T_{31} + \partial_2 T_{32} + \partial_3 T_{33} \end{pmatrix}$$

$$= \partial_j T_{ij} \boldsymbol{e}_i\,.$$

Of course, a similar definition holds for the divergence $\mathbf{div}^\varphi \boldsymbol{T}^\varphi$ of tensor fields $\boldsymbol{T}^{\varphi}: \boldsymbol{B}^{\varphi} \to \mathbb{M}^3$ defined over the deformed configuration:

$$\boldsymbol{T}^{\varphi} = (T_{ij}^{\varphi}) \Rightarrow \mathbf{div}^{\varphi}\, \boldsymbol{T}^{\varphi} := \partial_j^{\varphi} T_{ij}^{\varphi} \boldsymbol{e}_i\,,$$

where

$$\partial_j^{\varphi} := \partial / \partial x_j^{\varphi}$$

denote the partial derivatives with respect to the variables x_j^{φ}.

A simple application of the fundamental Green's formula over the set $\bar{\Omega}$ shows that the divergence of a tensor field satisfies:

$$\int_{\Omega} \mathbf{div}\, \boldsymbol{T}\, \mathrm{d}X = \left(\int_{\Omega} \partial_j T_{ij}\, \mathrm{d}x\right) \boldsymbol{e}_i = \left(\int_{\partial\Omega} T_{ij} n_j\, \mathrm{d}a\right) \boldsymbol{e}_i\,,$$

or equivalently in matrix form:

$$\boxed{\int_{\Omega} \mathbf{div}\, \boldsymbol{T}\, \mathrm{d}x = \int_{\partial\Omega} \boldsymbol{T}\boldsymbol{n}\, \mathrm{d}a\,.}$$

Recall that a vector is always understood as a *column vector* when viewed as a matrix; thus the notation $\boldsymbol{T}\boldsymbol{n}$ in the previous formula represents the column vector obtained by applying the matrix $\boldsymbol{T}$ to the column vector $\boldsymbol{n}$. This *Green formula* is called the **divergence theorem for tensor fields** (compare with the divergence theorem for vector fields established in Sect. 1.6). A tensor field $\boldsymbol{T}^{\varphi}: \bar{\Omega}^{\varphi} \to \mathbb{M}^3$ likewise satisfies:

$$\int_{\Omega^{\varphi}} \mathbf{div}^{\varphi}\, \boldsymbol{T}^{\varphi}\, \mathrm{d}x^{\varphi} = \int_{\partial\Omega^{\varphi}} \boldsymbol{T}^{\varphi} \boldsymbol{n}^{\varphi}\, \mathrm{d}a^{\varphi}\,,$$

where $\boldsymbol{n}^{\varphi}$ denotes the unit outer normal vector along the boundary of the deformed configuration.

We now come to an important definition. Let $\boldsymbol{\varphi}$ be a deformation that is injective on $\bar{\Omega}$, so that the matrix $\boldsymbol{\nabla\varphi}$ is invertible at all points of the reference configuration. Then if $\boldsymbol{T}^{\varphi}(x^{\varphi})$ is a tensor defined at the point $x^{\varphi} = \boldsymbol{\varphi}(x)$ of the deformed configuration, we associate with $\boldsymbol{T}^{\varphi}(x^{\varphi})$ a tensor $\boldsymbol{T}(x)$ defined at the point x of the reference configuration by:

$$\boxed{\begin{aligned} \boldsymbol{T}(x) := (\det \boldsymbol{\nabla\varphi}(x)) \boldsymbol{T}^{\varphi}(x^{\varphi}) \boldsymbol{\nabla\varphi}(x)^{-\mathrm{T}} = \boldsymbol{T}^{\varphi}(x^{\varphi})\, \mathbf{Cof}\, \boldsymbol{\varphi}(x)\,, \\ x^{\varphi} = \boldsymbol{\varphi}(x)\,. \end{aligned}}$$

In this fashion, a correspondence, called the **Piola transform**, is established between tensor fields defined over the deformed and reference configurations, respectively.

Remark. It would be equally conceivable, and somehow more natural, to start with a tensor field $\boldsymbol{T} : \bar{\Omega} \to \mathbb{M}^3$ and to associate with it its "inverse Piola transform" $\boldsymbol{T}^\varphi : \bar{\Omega}^\varphi \to \mathbb{M}^3$ defined by

$$\boldsymbol{T}^\varphi(x^\varphi) := (\det \boldsymbol{\nabla}\boldsymbol{\varphi}(x))^{-1}\boldsymbol{T}(x)\boldsymbol{\nabla}\boldsymbol{\varphi}(x)^{\mathrm{T}},\ x \in \bar{\Omega}\,.$$

As we shall see in Chapter 2, the reason we proceed the other way is that the "starting point" in elasticity is a tensor field defined over the *deformed configuration* (the Cauchy stress tensor field), and it is its Piola transform over the reference configuration (the first Piola–Kirchhoff stress tensor field) that subsequently plays a key role. ■

As shown in the next theorem, the main interest of the Piola transform is that it yields a *simple relation between the divergences of the tensors* $\boldsymbol{T}^\varphi$ *and* $\boldsymbol{T}$ *and* (as a corollary) *the desired relation between corresponding area elements* $\mathrm{d}a^\varphi$ *and* $\mathrm{d}a$.

Theorem 1.7-1 (properties of the Piola transform). *Let* $\boldsymbol{T} : \bar{\Omega} \to \mathbb{M}^3$ *denote the Piola transform of* $\boldsymbol{T}^\varphi : \bar{\Omega}^\varphi \to \mathbb{M}^3$. *Then*

$$\boxed{\mathbf{div}\,\boldsymbol{T}(x) = (\det \boldsymbol{\nabla}\boldsymbol{\varphi}(x))\,\mathbf{div}^\varphi\,\boldsymbol{T}^\varphi(x^\varphi) \quad \text{for all } x^\varphi = \boldsymbol{\varphi}(x),\ x \in \bar{\Omega}\,,}$$

$$\boxed{\boldsymbol{T}(x)\boldsymbol{n}\,\mathrm{d}a = \boldsymbol{T}^\varphi(x^\varphi)\boldsymbol{n}^\varphi\,\mathrm{d}a^\varphi \quad \text{for all } x^\varphi = \boldsymbol{\varphi}(x),\ x \in \partial\Omega\,.}$$

The area elements $\mathrm{d}a$ *and* $\mathrm{d}a^\varphi$ *at the points* $x \in \partial\Omega$ *and* $x^\varphi = \boldsymbol{\varphi}(x) \in \partial\Omega^\varphi$, *with unit outer normal vectors* $\boldsymbol{n}$ *and* $\boldsymbol{n}^\varphi$ *respectively, are related by*

$$\boxed{\det \boldsymbol{\nabla}\boldsymbol{\varphi}(x)|\boldsymbol{\nabla}\boldsymbol{\varphi}(x)^{-T}\boldsymbol{n}|\,\mathrm{d}a = |\mathbf{Cof}\,\boldsymbol{\nabla}\boldsymbol{\varphi}(x)\boldsymbol{n}|\,\mathrm{d}a = \mathrm{d}a^\varphi\,.}$$

Proof. The key to the proof is the *Piola identity*

$$\mathbf{div}\{(\det \boldsymbol{\nabla}\boldsymbol{\varphi})\boldsymbol{\nabla}\boldsymbol{\varphi}^{-\mathrm{T}}\} = \mathbf{div}\,\mathbf{Cof}\,\boldsymbol{\nabla}\boldsymbol{\varphi} = \boldsymbol{o}\,,$$

which we first prove: Counting the indices modulo 3, the elements of the matrix $\mathbf{Cof}\,\nabla\varphi$ are given by (Sect. 1.1):

$$(\mathbf{Cof}\,\nabla\varphi)_{ij} = \partial_{j+1}\varphi_{i+1}\partial_{j+2}\varphi_{i+2} - \partial_{j+2}\varphi_{i+1}\partial_{j+1}\varphi_{i+2} \quad \text{(no summation)}\,,$$

and a direct computation shows that

$$\partial_j((\det \nabla\varphi)\nabla\varphi^{-\mathrm{T}})_{ij} = \partial_j(\mathbf{Cof}\,\nabla\varphi)_{ij} = 0\,.$$

Then the relations

$$T_{ij}(x) = (\det \nabla\varphi(x))T^{\varphi}_{ik}(x^{\varphi})(\nabla\varphi(x)^{-\mathrm{T}})_{kj}$$

imply that

$$\partial_j T_{ij}(x) = (\det \nabla\varphi(x))\partial_j T^{\varphi}_{ik}(\varphi(x))(\nabla\varphi(x)^{-\mathrm{T}})_{kj}\,,$$

since the other term vanishes as a consequence of the Piola identity. Next, by the chain rule,

$$\partial_j T^{\varphi}_{ik}(x^{\varphi}) = \partial^{\varphi}_l T^{\varphi}_{ik}(\varphi(x))\partial_j\varphi_l(x) = \partial^{\varphi}_l T^{\varphi}_{ik}(x^{\varphi})(\nabla\varphi(x))_{lj}\,,$$

and the relation between $\mathbf{div}\, T(x)$ and $\mathbf{div}^{\varphi}\, T^{\varphi}(x^{\varphi})$ follows by noting that

$$(\nabla\varphi(x))_{li}(\nabla\varphi(x)^{-\mathrm{T}})_{ki} = \delta_{lk}\,.$$

Combining with the relation $\mathrm{d}x^{\varphi} = \det \nabla\varphi(x)\,\mathrm{d}x$, the divergence theorem for tensor fields expressed over arbitrary subdomains A of $\bar{\Omega}$, and the formula for changes of variables in multiple integrals, we obtain

$$\begin{aligned}\int_{\partial A} T(x)n\,\mathrm{d}a &= \int_A \mathbf{div}\, T(x)\,\mathrm{d}x = \int_A \mathbf{div}^{\varphi}\, T^{\varphi}(\varphi(x))\det\nabla\varphi(x)\,\mathrm{d}x\\ &= \int_{\varphi(A)} \mathbf{div}^{\varphi}\, T^{\varphi}(x^{\varphi})\,\mathrm{d}x^{\varphi} = \int_{\partial\varphi(A)} T^{\varphi}(x^{\varphi})n^{\varphi}\,\mathrm{d}a^{\varphi}\,,\end{aligned}$$

which proves the relation $Tn\,\mathrm{d}a = T^{\varphi}n^{\varphi}\,\mathrm{d}a^{\varphi}$ since the domains A are arbitrary. As a special case, we obtain the relation $(\det\nabla\varphi)\nabla\varphi^{-\mathrm{T}}n\,\mathrm{d}a = n^{\varphi}\,\mathrm{d}a^{\varphi}$ between the area elements $\mathrm{d}a$ and $\mathrm{d}a^{\varphi}$ by taking the Piola transform $(\det\nabla\varphi)\nabla\varphi^{-\mathrm{T}}$ of the unit tensor I. The relation $(\det\nabla\varphi)|\nabla\varphi^{-\mathrm{T}}n|\,\mathrm{d}a = \mathrm{d}a^{\varphi}$ then follows by expressing that, since $|n^{\varphi}| = 1$, $\mathrm{d}a^{\varphi}$ is also the Euclidean norm of the vector that appears in the left-hand side of the relation $(\det\nabla\varphi)\,\nabla\varphi^{-\mathrm{T}}n\,\mathrm{d}a = n^{\varphi}\,\mathrm{d}a^{\varphi}$. ■

Remarks. (1) Of course, the conclusions of Theorem 1.7-1 still hold if we replace the set Ω by any *subdomain* A of Ω, in which case the corresponding area elements and outer normal vectors are to be understood as being defined along the corresponding boundaries ∂A and $\partial A^{\varphi} = \varphi(\partial A)$.

(2) A weaker version of the Piola identity, *in the sense of distributions*, will be needed in the proof of Theorem 7.6-1.

(3) While the relation between the vectors $\mathbf{div}\, \boldsymbol{T}$ and $\mathbf{div}^{\varphi}\, \boldsymbol{T}^{\varphi}$ has been established here for deformations $\boldsymbol{\varphi}$ that are twice differentiable, the relations between the area elements established in Theorem 1.7-1 still hold under weaker regularity assumptions on the deformation. In this respect, see Exercise 1.13.

(4) The last equation in Theorem 1.7-1 shows that the unit outer normal vectors at the points $x^{\varphi} = \varphi(x)$ and x are related by

$$\boldsymbol{n}^{\varphi} = \frac{\mathbf{Cof}\, \boldsymbol{\nabla}\boldsymbol{\varphi}(x)\boldsymbol{n}}{|\mathbf{Cof}\, \boldsymbol{\nabla}\boldsymbol{\varphi}(x)\boldsymbol{n}|} . \qquad \blacksquare$$

We now have everything at our disposal to specify how *areas are transformed*: If Δ is a measurable subset of the boundary ∂A of a subdomain A, the area of the deformed set $\Delta^{\varphi} = \varphi(\Delta)$ is given by

$$\text{area}\, \Delta^{\varphi} := \int_{\Delta^{\varphi}} \mathrm{d}a^{\varphi} = \int_{\Delta} (\det \boldsymbol{\nabla}\boldsymbol{\varphi})|\boldsymbol{\nabla}\boldsymbol{\varphi}^{-\mathrm{T}}\boldsymbol{n}|\, \mathrm{d}a .$$

1.8. LENGTH ELEMENT IN THE DEFORMED CONFIGURATION; STRAIN TENSORS

If a deformation $\boldsymbol{\varphi}$ is differentiable at a point $x \in \bar{\Omega}$, then (by definition of differentiability) we can write, for all points $x + \boldsymbol{\delta x} \in \bar{\Omega}$:

$$\boldsymbol{\varphi}(x + \boldsymbol{\delta x}) - \boldsymbol{\varphi}(x) = \boldsymbol{\nabla}\boldsymbol{\varphi}(x)\boldsymbol{\delta x} + o(|\boldsymbol{\delta x}|) ,$$

and whence

$$|\boldsymbol{\varphi}(x + \boldsymbol{\delta x}) - \boldsymbol{\varphi}(x)|^2 = \boldsymbol{\delta x}^{\mathrm{T}}\boldsymbol{\nabla}\boldsymbol{\varphi}^{\mathrm{T}}(x)\boldsymbol{\nabla}\boldsymbol{\varphi}(x)\boldsymbol{\delta x} + o(|\boldsymbol{\delta x}|^2) .$$

The *symmetric* tensor

$$\boxed{\boldsymbol{C} := \boldsymbol{\nabla}\boldsymbol{\varphi}^{\mathrm{T}}\boldsymbol{\nabla}\boldsymbol{\varphi}}$$

found in the above expression is called in elasticity the **right Cauchy–Green strain tensor**. Notice that the associated quadratic form:

$$(\xi, \xi) \in \mathbb{R}^3 \times \mathbb{R}^3 \to \xi^T C(x) \xi = |\nabla\varphi(x)\xi^2|$$

is positive definite at all points $x \in \bar{\Omega}$, since the deformation gradient $\nabla\varphi$ is everywhere invertible by assumption. As expected, this quadratic form is used for computing lengths: Let

$$\gamma = f(I),\ f: I \to \bar{\Omega}\ ,\quad I\text{: compact interval of } \mathbb{R}\ ,$$

be a curve in the reference configuration (Fig. 1.8-1). Denoting by f_i the components of the mapping f, the *length* of the curve γ is given by ($f' = \mathrm{d}f/\mathrm{d}t$):

$$\text{length } \gamma := \int_I |f'(t)|\,\mathrm{d}t = \int_I \{f_i'(t) f_i'(t)\}^{1/2}\,\mathrm{d}t\ ,$$

while the length of the *deformed* curve $\gamma^\varphi := \varphi(\gamma)$ is given by

$$\text{length } \gamma^\varphi := \int_I |(\varphi \circ f)'(t)|\,\mathrm{d}t = \int_I \{C_{ij}(f(t)) f_i'(t) f_j'(t)\}^{1/2}\,\mathrm{d}t\ .$$

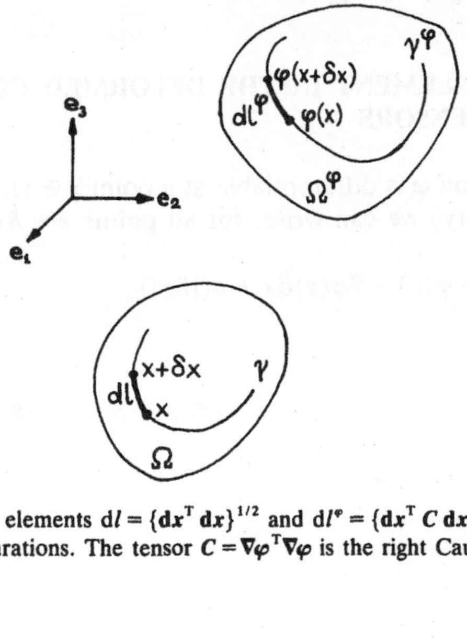

Fig. 1.8-1. The length elements $\mathrm{d}l = \{\mathbf{d}x^T\,\mathbf{d}x\}^{1/2}$ and $\mathrm{d}l^\varphi = \{\mathbf{d}x^T\,C\,\mathbf{d}x\}^{1/2}$ in the reference and deformed configurations. The tensor $C = \nabla\varphi^T\nabla\varphi$ is the right Cauchy–Green tensor.

Consequently, the **length elements** $\mathrm{d}l$ and $\mathrm{d}l^{\varphi}$ in the reference and deformed configurations may be symbolically written as:

$$\mathrm{d}l = \{\mathbf{dx}^{\mathrm{T}}\,\mathbf{dx}\}^{1/2},\ \mathrm{d}l^{\varphi} = \{\mathbf{dx}^{\mathrm{T}}\,\boldsymbol{C}\,\mathbf{dx}\}^{1/2}\,.$$

If in particular $\mathbf{d}\boldsymbol{x} = \mathrm{d}t\,\boldsymbol{e}_j$, the corresponding length element in the deformed configuration is $\{C_{jj}\}^{1/2}\,\mathrm{d}t = |\partial_j\varphi|\,\mathrm{d}t$. This observation is helpful in interpreting Fig. 1.4-1.

Remark. In the language of differential geometry, the manifold $\bar{\Omega}$ is equipped with a *Riemannian structure* through the data of the *metric tensor* $\boldsymbol{C} = (C_{ij})$, often denoted $\boldsymbol{g} = (g_{ij})$ in differential geometry, whose associated quadratic form, often denoted $\mathrm{d}s^2$, is called the *first fundamental form* of the manifold. For details, see e.g. Lelong-Ferrand [1963], Malliavin [1972]. ■

Although it has no immediate geometric interpretation, the **left Cauchy–Green strain tensor**

$$\boldsymbol{B} := \boldsymbol{\nabla}\varphi\,\boldsymbol{\nabla}\varphi^{\mathrm{T}}\,,$$

which is also *symmetric*, is equally important; in particular, it plays an essential role in the representation theorem for the response function of the Cauchy stress tensor (Theorem 3.6-2). For the time being, we simply notice that *the two matrices $\boldsymbol{C} = \boldsymbol{F}^{\mathrm{T}}\boldsymbol{F}$ and $\boldsymbol{B} = \boldsymbol{F}\boldsymbol{F}^{\mathrm{T}}$ have the same characteristic polynomial*, since this is true in general of the products $\boldsymbol{FG}$ and $\boldsymbol{GF}$ of two arbitrary matrices $\boldsymbol{F}$ and $\boldsymbol{G}$ of the same order. When $\boldsymbol{G} = \boldsymbol{F}^{\mathrm{T}}$, this result is a direct consequence of the polar factorization theorem (Theorem 3.2-2).

In view of showing that the tensor $\boldsymbol{C}$ is indeed a good measure of "strain", understood here in its intuitive sense of "change in *form* or *size*", let us first consider a class of deformations that induce no "strain": A deformation is called a **rigid deformation** if it is of the form

$$\varphi(x) = \boldsymbol{a} + \boldsymbol{Q}ox,\ \boldsymbol{a} \in \mathbb{R},\ \boldsymbol{Q} \in \mathbb{O}^3_+\,, \quad \text{for all } x \in \bar{\Omega}\,,$$

where $\mathbb{O}^3_+$ denotes the set of *rotations* in $\mathbb{R}^3$, i.e., the set of orthogonal matrices of order 3 whose determinant is +1. In other words, *the* *corresponding deformed configuration is obtained by rotating the reference configuration around the origin by the rotation Q and by translating it by the vector a*: this indeed corresponds to the idea of a "rigid" deformation, where the reference configuration is "moved", but without any "strain" (Fig. 1.8-2). Observe that the rotation Q may be performed around any point $\tilde{x} \in \mathbb{R}^3$ (Fig. 1.8-2), since we can also write

$$\varphi(x) = \varphi(\tilde{x}) + Q\tilde{x}x .$$

If φ is a rigid deformation, then $\nabla\varphi(x) = Q \in \mathbb{O}^3_+$ at all points $x \in \bar{\Omega}$, and therefore

$$C = I \text{ in } \bar{\Omega}, \quad \text{i.e., } \nabla\varphi(x)^{\mathrm{T}}\nabla\varphi(x) = I \quad \text{for all } x \in \bar{\Omega} .$$

It is remarkable that *conversely*, if $C = I$ in $\bar{\Omega}$ and $\det \nabla\varphi > 0$, the corresponding deformation is necessarily rigid, as we now prove under mild assumptions (for various complements, see Exercises 1.14, 1.15, 1.16). We let $\mathbb{O}^n$ denote the set of all orthogonal matrices of order n.

Theorem 1.8-1 (characterization of rigid deformations). *Let Ω be an open connected subset of $\mathbb{R}^n$, and let there be given a mapping*

$$\varphi \in \mathscr{C}^1(\Omega; \mathbb{R}^n)$$

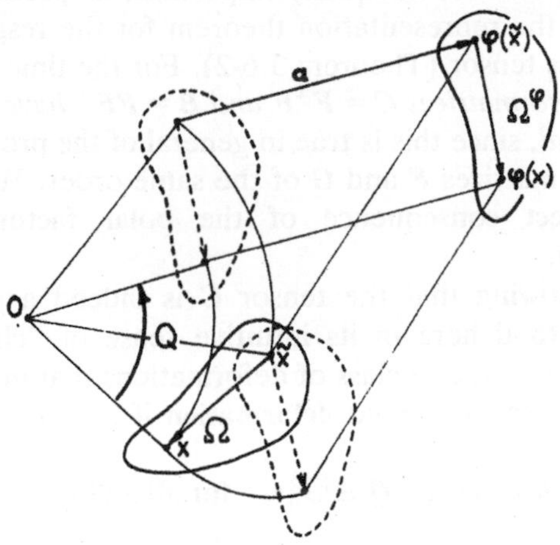

Fig. 1.8-2. A rigid deformation is a translation, followed by a rotation (or *vice versa*), of the reference configuration.

that satisfies

$$\nabla\varphi(x)^{\mathrm{T}}\nabla\varphi(x) = \boldsymbol{I} \quad \text{for all } x \in \Omega .$$

Then there exists a vector $\boldsymbol{a} \in \mathbb{R}^n$ *and an orthogonal matrix* $\boldsymbol{Q} \in \mathbb{O}^n$ *such that*

$$\varphi(x) = \boldsymbol{a} + \boldsymbol{Q}ox \quad \text{for all } x \in \Omega .$$

Proof. (i) Let us first establish that *locally, the mapping* φ *is an isometry, i.e., given any point* $x_0 \in \Omega$, *there exists an open subset* V *such that*

$$x_0 \in V \subset \Omega, \text{ and } |\varphi(y) - \varphi(x)| = |y - x| \quad \text{for all } x, y \in V .$$

Since Ω is open, there exists $\rho > 0$ such that the open ball

$$B_\rho(x_0) := \{x \in \mathbb{R}^n;\, |x - x_0| < \rho\}$$

is contained in Ω. The spectral norm of an orthogonal matrix is equal to 1, since it is defined for an arbitrary square matrix A by

$$|A| := \sup_{v \neq o} \frac{|Av|}{|v|} = \max_i \{\lambda_i(A^{\mathrm{T}}A)\}^{1/2} .$$

Thus we deduce from the mean value theorem (cf. Theorem 1.2-2; the ball $B_\rho(x_0)$ is a convex subset of Ω) that

$$|\varphi(y) - \varphi(x)| \leq \sup_{z \in]x,y[} |\nabla\varphi(z)|\,|y - x| = |y - x|$$

for all $x, y \in B_\rho(x_0)$.

In view of proving the opposite inequality, we observe that, by the local inversion theorem (Theorem 1.2-4), the mapping φ is locally invertible in Ω since the matrix $\nabla\varphi(x)$ is invertible for all $x \in \Omega$. In particular then, there exist open sets V and V^φ containing the points x_0 and $x_0^\varphi = \varphi(x_0)$ respectively, such that the restriction of the mapping φ to the set V is a $\mathscr{C}^1$-diffeomorphism from V onto V^φ, i.e., the mapping $\varphi : V \to V^\varphi$ is bijective, and its inverse mapping $\psi : V^\varphi \to V$ is also continuously differentiable.

Without loss of generality, we may assume that the set V is contained in the ball $B_\rho(x_0)$ and that the set V^φ is convex (otherwise we replace the set V by the inverse image by φ of an open ball contained in $V^\varphi \cap \varphi(B_\rho(x_0))$ and centered at the point x_0^φ). Since

$$\psi(\varphi(x)) = x \quad \text{for all } x \in V ,$$

we deduce that the mapping ψ satisfies

$$\nabla\psi(x^\varphi) = \nabla\varphi(x)^{-1} \quad \text{for all } x^\varphi = \varphi(x),\ x \in V .$$

As a consequence, the matrix $\nabla\psi(x^\varphi)$ is also orthogonal for all points $x^\varphi \in V^\varphi$. Since the set V^φ is convex, another application of the mean value theorem shows that

$$|\psi(y^\varphi) - \psi(x^\varphi)| \leq |y^\varphi - x^\varphi| \quad \text{for all } x^\varphi,\ y^\varphi \in V .$$

This inequality can be equivalently written as

$$|y - x| \leq |\varphi(y) - \varphi(x)| \quad \text{for all } x,\ y \in V ,$$

and since the set V^φ is contained in the set $B_\rho(x_0)$, we have therefore proved that

$$|\varphi(y) - \varphi(x)| = |y - x| \text{ for all } x,\ y \in V .$$

(ii) Let us next show that *locally, the matrix* $\nabla\varphi$ *is a constant*, in the sense that *the matrix* $\nabla\varphi(x)$ *is independent of* x *for all* $x \in V$. To do this, we write the property established in step (i) in the equivalent form

$$F(x, y) := (\varphi_k(y) - \varphi_k(x))(\varphi_k(y) - \varphi_k(x)) - (y_k - x_k)(y_k - x_k) = 0$$

for all $x, y \in V$. For each $x \in V$, the function $y \in V \to F(x, y)$ is differentiable, with

$$G_i(x, y) := \frac{1}{2}\frac{\partial F}{\partial y_i}(x, y) = \frac{\partial \varphi_k}{\partial y_i}(y)(\varphi_k(y) - \varphi_k(x)) - \delta_{ik}(y_k - x_k) = 0$$

for all $x, y \in V$. For each $y \in V$, each function $x \in V \to G_i(x, y)$ is differentiable, with

$$\frac{\partial G_i}{\partial x_j}(x, y) = -\frac{\partial \varphi_k}{\partial y_i}(y)\frac{\partial \varphi_k}{\partial x_j}(x) + \delta_{ij} = 0\,,$$

i.e.,

$$\nabla\boldsymbol{\varphi}(y)^{\mathrm{T}}\nabla\boldsymbol{\varphi}(x) = \boldsymbol{I} \quad \text{for all } x, y \in V\,.$$

Letting $y = x_0$, we obtain

$$\nabla\boldsymbol{\varphi}(x) = \nabla\boldsymbol{\varphi}(x_0) \quad \text{for all } x \in V\,.$$

(iii) It follows from step (ii) that the mapping $\nabla\boldsymbol{\varphi} : \Omega \to \mathbb{M}^3$ is differentiable and that its derivative vanishes in Ω; equivalently, *the mapping* $\boldsymbol{\varphi}$ *is twice differentiable and its second Fréchet derivative vanishes*. Because the set Ω is *connected* (this assumption has not been used so far), a classical result from differential calculus (see e.g. Schwartz [1967, p. 266]) shows that the mapping $\boldsymbol{\varphi}$ is necessarily of the form

$$\boldsymbol{\varphi}(x) = \boldsymbol{a} + \boldsymbol{Q}\boldsymbol{o}\boldsymbol{x},\ \boldsymbol{a} \in \mathbb{R}^n,\ \boldsymbol{Q} \in \mathbb{M}^n\,, \quad \text{for all } x \in \Omega\,,$$

so that its gradient $\boldsymbol{Q}$ is the same constant $\nabla\boldsymbol{\varphi}(x_0)$ in all the set Ω. Thus the matrix $\boldsymbol{Q} = \nabla\boldsymbol{\varphi}(x_0)$ is orthogonal since $\nabla\boldsymbol{\varphi}(x_0)^{\mathrm{T}}\nabla\boldsymbol{\varphi}(x_0) = \boldsymbol{I}$ by assumption. ■

Remark. If we are provided with the additional assumption that $\det \nabla\boldsymbol{\varphi}(x) > 0$ for at least one point $x \in \Omega$ (and consequently for all points in the connected set Ω), we can conclude that the orthogonal matrix $\boldsymbol{Q}$ obtained in the theorem is a rotation (i.e., $\det \boldsymbol{Q} = 1$). Notice also that if the mapping $\boldsymbol{\varphi}$ is continuous up to the boundary, the relation $\boldsymbol{\varphi}(x) = \boldsymbol{a} + \boldsymbol{Q}\boldsymbol{o}\boldsymbol{x}$ holds for all points $x \in \bar{\Omega}$. ■

The result of Theorem 1.8-1 can be viewed as a special case (let $\boldsymbol{\psi}$ be any rigid deformation in Theorem 1.8-2) of the following result, which shows that *two deformations corresponding to the same tensor* $\boldsymbol{C}$ *can be obtained from one another by composition with a rigid deformation*.

Theorem 1.8-2. *Let* Ω *be an open connected subset of* $\mathbb{R}^n$, *and let there be given two mappings*

$$\boldsymbol{\varphi}, \boldsymbol{\psi} \in \mathscr{C}^1(\Omega; \mathbb{R}^n)$$

such that

$$\nabla\varphi(x)^{\mathrm{T}}\,\nabla\varphi(x) = \nabla\psi(x)^{\mathrm{T}}\,\nabla\psi(x) \quad \text{for all } x \in \Omega\,,$$

$$\psi : \Omega \to \mathbb{R}^n \text{ is injective, and } \det \nabla\psi(x) \neq 0 \quad \text{for all } x \in \Omega\,.$$

Then there exist a vector $a \in \mathbb{R}^n$ *and an orthogonal matrix* $Q \in \mathbb{O}^n$ *such that*

$$\varphi(x) = a + Q\psi(x) \quad \text{for all } x \in \Omega\,.$$

Proof. Since the mapping ψ is injective and $\det \nabla\psi(x) \neq 0$ for all $x \in \Omega$, the inverse mapping

$$\psi^{-1} : \psi(\Omega) \to \Omega$$

is also continuously differentiable and it satisfies

$$\nabla\psi^{-1}(\xi)\,\nabla\psi(x) = I \quad \text{for all } \xi = \psi(x),\ x \in \Omega\,.$$

Besides, the set $\psi(\Omega)$ is open by the invariance of domain theorem (Theorem 1.2-6), and connected since it is the image of a connected set by a continuous mapping. The composite mapping

$$\theta := \varphi \circ \psi^{-1} : \psi(\Omega) \to \mathbb{R}^n$$

is such that

$$\nabla\theta(\xi) = \nabla\varphi(x)\,\nabla\psi(\xi)^{-1} = \nabla\varphi(x)\,\nabla\psi(x)^{-1} \quad \text{for all } \xi = \psi(x),\ x \in \Omega\,,$$

and consequently, by assumption,

$$\nabla\theta(\xi)^{\mathrm{T}}\,\nabla\theta(\xi) = \nabla\psi(x)^{-\mathrm{T}}\,\nabla\varphi(x)^{\mathrm{T}}\,\nabla\varphi(x)\,\nabla\psi(x)^{-1} = I \quad \text{for all } \xi = \psi(x),\ x \in \Omega\,.$$

We may therefore apply Theorem 1.8-1 to the mapping θ: There exists a vector $a \in \mathbb{R}^n$ and an orthogonal matrix $Q \in \mathbb{O}^n$ such that

$$\theta(\xi) = a + Qo\xi \quad \text{for all } \xi = \psi(x),\ x \in \Omega\,,$$

but this is just another equivalent statement of the conclusion of the theorem. ■

The previous two theorems are useful for understanding the rôle played by the tensor $\boldsymbol{C}$. First, Theorem 1.8-1 shows that the difference

$$\boxed{2\boldsymbol{E} := \boldsymbol{C} - \boldsymbol{I}}$$

is a measure of the "deviation" between a given deformation and a rigid deformation, since $\boldsymbol{C} = \boldsymbol{I}$ if and only if the deformation is rigid. Secondly, Theorem 1.8-2 shows that *the knowledge of the tensor field* $\boldsymbol{C} : \Omega \to \mathbb{S}^3_>$ *completely determines the deformation, up to composition with rigid deformations* (the question of proving the *existence* of deformations for which the associated tensor field $\boldsymbol{C} : \Omega \to \mathbb{S}^3_>$ is equal to a *given* tensor field is quite another matter; see Exercise 1.18). These considerations are illustrated in Figure 1.8-3.

The tensor $\boldsymbol{E}$ is called the **Green–St Venant strain tensor**. Expressed in terms of the *displacement gradient* $\nabla \boldsymbol{u}$, in lieu of the deformation gradient $\nabla \boldsymbol{\varphi} = \boldsymbol{I} + \nabla \boldsymbol{u}$ (recall that $\boldsymbol{\varphi} = \boldsymbol{id} + \boldsymbol{u}$), the strain tensor $\boldsymbol{C}$ becomes

$$\boxed{\boldsymbol{C} = \nabla \boldsymbol{\varphi}^{\mathrm{T}} \nabla \boldsymbol{\varphi} = \boldsymbol{I} + \nabla \boldsymbol{u}^{\mathrm{T}} + \nabla \boldsymbol{u} + \nabla \boldsymbol{u}^{\mathrm{T}} \nabla \boldsymbol{u} = \boldsymbol{I} + 2\boldsymbol{E}\,,}$$

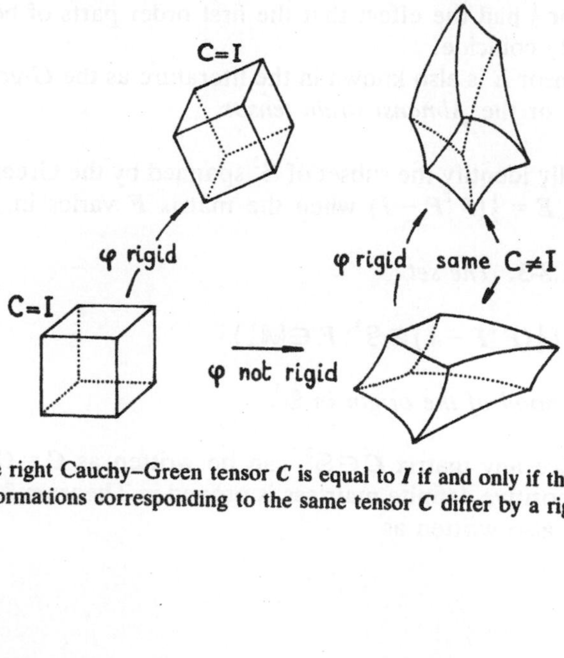

Fig. 1.8-3. The right Cauchy–Green tensor $\boldsymbol{C}$ is equal to $\boldsymbol{I}$ if and only if the deformation is rigid. Two deformations corresponding to the same tensor $\boldsymbol{C}$ differ by a rigid deformation.

with

$$\boxed{E(u) := E = \tfrac{1}{2}(\nabla u^{\mathrm{T}} + \nabla u + \nabla u^{\mathrm{T}} \nabla u)\,.}$$

For future use, we record the formulas:

$$C_{ij} = \partial_i \varphi_k \partial_j \varphi_k\,, \quad E_{ij} = \tfrac{1}{2}(\partial_i u_j + \partial_j u_i + \partial_i u_k \partial_j u_k)\,,$$

where $\varphi = \varphi_i e_i$, $u = u_i e_i$.

Remarks. (1) Both Theorems 1.8-1 and 1.8-2 can be rephrased in terms of the Green–St Venant strain tensor, using the equivalences

$$\nabla \varphi^{\mathrm{T}} \nabla \varphi = I \text{ in } \Omega \Leftrightarrow E(u) = 0 \text{ in } \Omega,\ \varphi = id + u\,,$$
$$\nabla \varphi^{\mathrm{T}} \nabla \varphi = \nabla \psi^{\mathrm{T}} \nabla \psi \text{ in } \Omega \Leftrightarrow E(u) = E(v) \text{ in } \Omega,\ \varphi = id + u,\ \psi = id + v\,.$$

(2) The introduction of the factor $\frac{1}{2}$ in the definition of the tensor E is motivated by the requirement that its "first order" part $\frac{1}{2}(\nabla u^{\mathrm{T}} + \nabla u)$ coincide with the *linearized strain tensor* (cf. Chapter 6), which played a key role in the earlier linearized theories that prevailed in elasticity. Besides, the tensor $(C^{1/2} - I)$ was sometimes advocated as an alternative measure of strain (cf. Sect. 3.2 for the definition of the square root $C^{1/2}$), and the factor $\frac{1}{2}$ had the effect that the first order parts of both tensor E and $(C^{1/2} - I)$ coincide.

(3) The tensor E is also known in the literature as the *Green–Lagrange strain tensor*, or the *Almansi strain tensor*. ■

Let us finally identify the subset of $\mathbb{S}^3$ spanned by the Green–St Venant strain tensor $E = \frac{1}{2}(F^{\mathrm{T}}F - I)$ when the matrix F varies in the set $\mathbb{M}^3_+$:

Theorem 1.8-3. *The set*

$$\mathbb{V}(0) := \{\tfrac{1}{2}(F^{\mathrm{T}}F - I) \in \mathbb{S}^3;\ F \in \mathbb{M}^3_+\}$$

is a neighborhood of the origin in $\mathbb{S}^3$.

Proof. Since any matrix $C \in \mathbb{S}^3_>$ can be written as $C = C^{1/2}C^{1/2}$ (this property of positive definite matrices is proved in Theorem 3.2-1), the set $\mathbb{V}(0)$ can be also written as

$$\mathbb{V}(\boldsymbol{0}) = \{\tfrac{1}{2}(\boldsymbol{C} - \boldsymbol{I}) \in \mathbb{S}^3; \boldsymbol{C} \in \mathbb{S}^3_>\} = f^{-1}(\mathbb{S}^3_>) ,$$

where f is the continuous mapping $f: \boldsymbol{E} \in \mathbb{S}^3 \to (\boldsymbol{I} + 2\boldsymbol{E}) \in \mathbb{S}^3$. Since the set $\mathbb{S}^3_>$ is open in $\mathbb{S}^3$ (Exercise 1.1), and since $\boldsymbol{0} \in \mathbb{V}(\boldsymbol{0})$, the conclusion follows. ■

EXERCISES

1.1. (1) Show that the set of all invertible matrices of order n is open in $\mathbb{M}^n$.

(2) Show that the set $\mathbb{S}^n_>$ is open in $\mathbb{S}^n$.

(3) Show that the set $\mathbb{S}^n$ is closed in $\mathbb{M}^n$.

(4) What can be said to the set $\mathbb{S}^n_>$ as a subset of $\mathbb{M}^n$?

1.2. (1) Let $\boldsymbol{A}$ be an invertible matrix, with eigenvalues λ_i. Using the relation $\mathbf{Cof}\,\boldsymbol{A} = (\det \boldsymbol{A})\boldsymbol{A}^{-\mathrm{T}}$, compute the eigenvalues of the matrix $\mathbf{Cof}\,\boldsymbol{A}$.

(2) Show that any square matrix is the limit of a sequence of invertible matrices of the same order.

(3) Let $\boldsymbol{A}$ be an arbitrary matrix of order n, with eigenvalues λ_i, $1 \leq i \leq n$, in an arbitrary order. Let $\varepsilon > 0$ be given. Show that there exists $\delta = \delta(\boldsymbol{A}, \varepsilon) > 0$ with the following property: Let $\boldsymbol{B}$ be any matrix of order n that satisfies $\|\boldsymbol{A} - \boldsymbol{B}\| \leq \delta$, and let μ_i, $1 \leq i \leq n$, be its eigenvalues in an arbitrary order. Then there exists a permutation $\sigma \in \mathfrak{S}_n$ such that

$$|\lambda_i - \mu_{\sigma(i)}| \leq \varepsilon , \quad 1 \leq i \leq n .$$

(4) Let $\boldsymbol{A}$ be an arbitrary matrix of order n with eigenvalues λ_i, $1 \leq i \leq n$. Deduce from (1), (2), (3) that the eigenvalues of the matrix $\mathbf{Cof}\,\boldsymbol{A}$ are $\Pi_{j \neq i} \lambda_j$, $1 \leq i \leq n$.

Remarks. The proof of (3) is delicate in the case of multiple eigenvalues (see e.g. Ostrowski [1966, p. 282]). The result of (4) provides another proof to Theorem 1.1-1.

1.3. (1) Let $\boldsymbol{A}$ and $\boldsymbol{B}$ be arbitrary square matrices of the same order. Show that:

$$\det \boldsymbol{A} = \tfrac{1}{3}\boldsymbol{A} : \mathbf{Cof}\,\boldsymbol{A} ,$$

$$\det(\boldsymbol{A} + \boldsymbol{B}) = \det \boldsymbol{A} + \mathbf{Cof}\,\boldsymbol{A} : \boldsymbol{B} + \boldsymbol{A} : \mathbf{Cof}\,\boldsymbol{B} + \det \boldsymbol{B} .$$

(2) Show that the second relation can be used for computing the derivative of the mapping $A \in \mathbb{M}^n \to \det A$ (Sect. 1.2).

1.4. It has been proved in Section 1.2 that the derivative of the function $\iota_{n-1} : A \in \mathbb{M}^n \to \operatorname{tr} \operatorname{Cof} A$ is given by

$$\frac{\partial \iota_{n-1}}{\partial A}(A) = (\operatorname{tr}(A^{-1})I - A^{-T}) \operatorname{Cof} A$$

when the matrix A is invertible; it is also clear that the function ι_{n-1} is differentiable for all $A \in \mathbb{M}^n$. What does the combination of these two results imply?

1.5. The *principal invariants* of a matrix A of order n are the coefficients $\iota_1, \ldots, \iota_n$ appearing in its characteristic polynomial:

$$\det(A - \lambda I) = (-1)^n \lambda^n + (-1)^{n-1} \iota_1(A) \lambda^{n-1} + \cdots - \iota_{n-1}(A)\lambda + \iota_n(A),$$

so that $\iota_1(A) = \operatorname{tr} A$, $\iota_{n-1}(A) = \operatorname{tr} \operatorname{Cof} A$, $\iota_n(A) = \det A$, as in Sect. 1.2. Show that the gradients of the principal invariants satisfy the following recursion formulas (cf. e.g. Truesdell & Noll [1965, p. 26], Carlson & Hoger [1986]):

$$\frac{\partial \iota_k}{\partial A}(A) = \left\{ \sum_{j=0}^{k-1} (-1)^j \iota_{k-j-1}(A) A^j \right\}^{T}, \quad 1 \leq k \leq n, \; \iota_0(A) = 1.$$

1.6. Show that the mapping $A \to A^{-1}$ is infinitely differentiable on the open subset of $\mathbb{M}^n$ formed by all invertible matrices. Compute its first and second derivatives.

1.7. Show that, if an open subset Ω of $\mathbb{R}^n$ is a domain, then $\operatorname{int} \bar{\Omega} = \Omega$.

1.8. Show that if an open subset Ω of $\mathbb{R}^n$ satisfies all the assumptions required for being a domain, except that of connectedness, it has a finite number of connected components.

1.9. Let Ω be a domain in $\mathbb{R}^n$. Show that there exists a number $c(\Omega)$ such that, given any points x_1 and x_2 in the set $\bar{\Omega}$, there exists a finite number of points y_k, $1 \leq k \leq l+1$, such that

$$y_1 = x_1, \; y_k \in \Omega \text{ for } 2 \leq k \leq l, \; y_{l+1} = x_2,$$

$$]y_k, y_{k+1}[\subset \Omega \text{ for } 1 \leq k \leq l, \sum_{k=1}^{l} |y_k - y_{k+1}| \leq c(\Omega)|x_1 - x_2| .$$

Remark. The result of this exercise will be used in the proof of Theorem 5.5-1.

1.10. Let Ω be a domain in $\mathbb{R}^n$, and let $\varphi : \bar{\Omega} \to \varphi(\bar{\Omega})$ be a $\mathscr{C}^1$-diffeomorphism. Show that the set $\varphi(\Omega)$ is also a domain.

1.11. Let Ω be a bounded convex open subset of $\mathbb{R}^2$. Is it a domain?

1.12. Given a vector field $\boldsymbol{v}^\varphi : \bar{\Omega}^\varphi \to \mathbb{R}^3$ over the deformed configuration $\bar{\Omega}^\varphi$, its *Piola transform* is the vector field $\boldsymbol{v} : \bar{\Omega} \to \mathbb{R}^3$ defined over the reference configuration $\bar{\Omega}$ by the relation

$$\boldsymbol{v}(x) = (\det \nabla\varphi(x))\nabla\varphi^{-1}(x)\boldsymbol{v}^\varphi(x^\varphi),\ x^\varphi = \varphi(x) .$$

Show that the divergence of these two vector fields are related by:

$$\operatorname{div} \boldsymbol{v}(x) = (\det \nabla\varphi(x)) \operatorname{div}^\varphi \boldsymbol{v}^\varphi(x^\varphi) ,$$

i.e., by a relation similar in spirit to the first relation of Theorem 1.7-1.

1.13. (1) Show that the formula giving the area element in the definition of a surface integral can be understood *a posteriori* as a special case of the relation $\det \nabla\varphi |\nabla\varphi^{-\mathrm{T}}\boldsymbol{n}|\, \mathrm{d}a = \mathrm{d}a^\varphi$ established in Theorem 1.7-1.

(2) The last two relations relating the area elements $\mathrm{d}a$ and $\mathrm{d}a^\varphi$ in Theorem 1.7-1 have been established under the assumption that the deformation φ is twice differentiable. Do they still hold under weaker regularity assumptions on the deformation?

1.14. The object of this exercise is to provide an easier, more computational (but perhaps less illuminating . . .) proof of Theorem 1.8-1, under the *additional* assumption that the mapping φ is *twice* differentiable. To this end, show first that the assumption $\nabla\varphi(x)^{\mathrm{T}}\nabla\varphi(x) = \boldsymbol{I}$ implies that relations of the form

$$\partial_{ij}\varphi(x) = \alpha_{ijm}(x)\partial_m\varphi(x)$$

hold, and then that the coefficients $\alpha_{ijm}(x)$ vanish.

1.15. Given a number $l > 0$, a mapping $\varphi : \mathbb{R}^n \to \mathbb{R}^n$ is said to *preserve the distance* l if

$$|\varphi(y) - \varphi(x)| = |y - x| \quad \text{for all } x, y \in \mathbb{R}^n \quad \text{such that } |y - x| = l .$$

(1) Show that if a mapping $\varphi : \mathbb{R}^n \to \mathbb{R}^n$, $n \geq 2$, preserves one distance $l_0 > 0$, then it preserves all distances l (for a proof, see Cabane [1981], or Beckman & Quarles [1953]); notice that it is not necessary to assume that φ is continuous.

(2) Is the result true for $n = 1$?

(3) Show that, if a mapping $\varphi : \mathbb{R}^n \to \mathbb{R}^n$ preserves all distances $l > 0$, it is of the form $\varphi(x) = \boldsymbol{a} + \boldsymbol{Q}\boldsymbol{o}\boldsymbol{x}$, $\boldsymbol{a} \in \mathbb{R}^n$, $\boldsymbol{Q} \in \mathbb{O}^n$, or equivalently, that it is the product of no more than $(n + 1)$ reflexions across hyperplanes (see e.g. Yale [1968, p. 60]; the last equivalence is a classical property of orthogonal matrices, proved for instance in Ciarlet [1983, Theorem 4.5-2]).

Remark. Property (3) is established in part (ii) of the proof of Theorem 1.8-1 for mappings of class $\mathscr{C}^1$.

1.16. Show that (the Sobolev spaces $\boldsymbol{H}^1(\Omega)$ and $\boldsymbol{W}^{1,p}(\Omega)$ are defined in Sect. 6.1):

$$\boldsymbol{u} \in \boldsymbol{H}^1(\Omega) \text{ and } \boldsymbol{E}(\boldsymbol{u}) \in \boldsymbol{L}^r(\Omega)\,, \quad r \geq 1 \Rightarrow \boldsymbol{u} \in \boldsymbol{W}^{1,2r}(\Omega)$$

(the opposite implication clearly holds). This observation is due to Luc Tartar.

1.17. The object of this exercise is to study to what extent the characterization of rigid deformations established in Theorem 1.8-1 for mappings $\varphi \in \mathscr{C}^1(\Omega; \mathbb{R}^n)$ can be generalized under *weaker regularity assumptions* on the mappings φ. The Sobolev spaces $\boldsymbol{H}^1(\Omega)$ and $\boldsymbol{W}^{1,p}(\Omega)$ are defined in Sect. 6.1.

(1) Let $\varphi \in \boldsymbol{H}^1(\Omega)$ satisfy $\nabla\varphi(x)^{\mathrm{T}}\nabla\varphi(x) = \boldsymbol{I}$ for almost all $x \in \Omega$. Show that $\varphi \in \boldsymbol{W}^{1,\infty}(\Omega)$.

(2) With the same assumptions as in (1), show that φ is not necessarily of the form $\varphi(x) = \boldsymbol{a} + \boldsymbol{Q}\boldsymbol{o}\boldsymbol{x}$, $\boldsymbol{a} \in \mathbb{R}^n$, $\boldsymbol{Q} \in \mathbb{O}^n$, even if $\varphi = \boldsymbol{id}$ on a subset Γ_0 of $\Gamma = \partial\Omega$ with area $\Gamma_0 > 0$.

(3) Assume that Ω is bounded, and let $\varphi \in \boldsymbol{H}^1(\Omega)$ satisfy

$$\nabla\varphi(x)^{\mathrm{T}}\nabla\varphi(x) = \boldsymbol{I} \quad \text{for almost } x \in \Omega\,,$$

$$\varphi = \varphi_0 \text{ on } \Gamma\,, \quad \text{where } \varphi_0(x) = a_0 + Qox,\ a_0 \in \mathbb{R}^n,\ Q \in \mathbb{O}^n\,.$$

Show that $\varphi = \varphi_0$ in Ω.

Remark: These results are due to Florian Laurent. See also Reshetnyak [1967], Kohn [1982] for related results.

1.18. The object of this exercise is to state necessary and sufficient compatibility conditions that a tensor field $C: \Omega \to \mathbb{S}^n_>$ (Ω open in $\mathbb{R}^n$) should satisfy in order that there exist a deformation field $\varphi: \Omega \to \mathbb{R}^3$ such that

$$\nabla\varphi(x)^{\mathrm{T}}\nabla\varphi(x) = C(x) \quad \text{for all } x \in \Omega\,.$$

(1) Let there be given a twice differentiable vector field $\varphi: \Omega \to \mathbb{R}^n$, and assume that the symmetric matrix $C = (C_{ij}) := \nabla\varphi^{\mathrm{T}}\nabla\varphi$ is everywhere positive definite. Let

$$\Gamma_{ikj} = \tfrac{1}{2}(\partial_j C_{ik} + \partial_i C_{jk} - \partial_k C_{ij})\,, \quad \Gamma^k_{ij} = (C^{-1})_{kl}\Gamma_{ilj}\,.$$

Show that the following *compatibility equations* hold:

$$\partial_l \Gamma^j_{ik} - \partial_k \Gamma^j_{il} + \Gamma^m_{ik}\Gamma^j_{ml} - \Gamma^m_{il}\Gamma^j_{mk} = 0\,, \quad 1 \le i, j, k, l \le n\,.$$

Remarks: In the language of differential geometry, the functions Γ_{ikj} and Γ^k_{ij} are the *Christoffel symbols of the first and second kind* of the manifold Ω equipped with the *metric tensor* C, and the above relations express that the *Riemann–Christoffel curvature tensor* of the manifold vanishes (see e.g. Choquet-Bruhat, Dewitt-Morette & Dillard-Bleick [1977, p. 303]).

A useful survey about representations of orthogonal matrices is given in Guo [1981] (see also the references therein). These may be used to derive the compatibility equations satisfied by the tensor C (see Signorini [1943], Shamina [1974], Guo [1963]).

(2) Conversely, let Ω be a bounded, simply connected, open subset of $\mathbb{R}^n$, and let there be given a twice continuously differentiable tensor field $C: \Omega \to \mathbb{S}^n_> \subset \mathbb{M}^n$ that satisfies the above compatibility equations. Show that there exists a vector field $\varphi \in \mathscr{C}^3(\Omega; \mathbb{R}^n)$ such that $C = \nabla\varphi^{\mathrm{T}}\nabla\varphi$ in Ω (such a vector field is only determined up to composition with mappings of the form $\theta(x) = a + Qox$, $a \in \mathbb{R}^n$, $Q \in \mathbb{O}^n$; cf. Theorem 1.8-2).

Remarks. The existence of such a "global" solution φ is due to Ciarlet & Laurent [1987]. In texts on differential geometry, this problem is usually only given a "local" solution as an application of the theory of totally integrable Pfaff systems (see e.g. Malliavin [1972, p. 133]). The main difficulty lies in obtaining a "global" solution, i.e., in all of Ω. See also Pietraszkiewicz [1982], Pietraszkiewicz & Badur [1983a, 1983b], and Deturck & Yang [1983] for the problem of finding deformations whose strain tensor has prescribed eigenvalues.

For the related problem of finding various *a priori* bounds for deformations in terms of strains, see notably John [1961, 1972, 1975], Kohn [1982].

CHAPTER 2

THE EQUATIONS OF EQUILIBRIUM AND THE PRINCIPLE OF VIRTUAL WORK

INTRODUCTION

A body occupying a deformed configuration $\bar{\Omega}^{\varphi}$, and subjected to applied body forces in its interior Ω^{φ} and to applied surface forces on a portion $\Gamma_1^{\varphi} = \varphi(\Gamma_1)$ of its boundary (Sect. 2.1), is in static equilibrium if the fundamental *stress principle of Euler and Cauchy* (Sect. 2.2) is satisfied. This axiom, which is the basis of continuum mechanics, implies the celebrated *Cauchy theorem* (Theorem 2.3-1), according to which there exists a symmetric tensor field $\boldsymbol{T}^{\varphi} : \bar{\Omega}^{\varphi} \to \mathbb{S}^3$ such that

$$\begin{cases} -\mathbf{div}^{\varphi}\, \boldsymbol{T}^{\varphi} = \boldsymbol{f}^{\varphi} \text{ in } \Omega^{\varphi}, \\ \boldsymbol{T}^{\varphi}\boldsymbol{n}^{\varphi} = \boldsymbol{g}^{\varphi} \text{ on } \Gamma_1^{\varphi}, \end{cases}$$

where $\boldsymbol{f}^{\varphi}$ and $\boldsymbol{g}^{\varphi}$ denote the densities of the applied body and surface forces respectively, and $\boldsymbol{n}^{\varphi}$ is the unit outer normal vector along Γ_1^{φ}. These equations are called the *equilibrium equations over the deformed configuration* (Sect. 2.4), and the tensor $\boldsymbol{T}^{\varphi}$ is called the *Cauchy stress tensor*.

A remarkable feature of these equations is their "*divergence structure*", which makes them amenable to a variational formulation (Theorem 2.4-1); a disadvantage is that they are expressed in terms of the *unknown* $x^{\varphi} = \varphi(x)$. In order to obviate this difficulty while retaining the divergence structure of the equations, we use the *Piola transform* $\boldsymbol{T} : \bar{\Omega} \to \mathbb{M}^3$ of the Cauchy stress tensor field, which is defined by $\boldsymbol{T}(x) = \boldsymbol{T}^{\varphi}(x^{\varphi})\, \mathbf{Cof}\, \boldsymbol{\nabla}\varphi(x)$ (Sect. 2.5). In this fashion, it is found (Theorem 2.6-1) that the equilibrium equations over $\bar{\Omega}^{\varphi}$ are equivalent to the *equilibrium equations over the reference configuration* $\bar{\Omega}$, *viz*.,

$$\begin{cases} -\mathbf{div}\, \boldsymbol{T} = \boldsymbol{f} \text{ in } \Omega, \\ \boldsymbol{T}\boldsymbol{n} = \boldsymbol{g} \text{ on } \Gamma_1, \end{cases}$$

where $\boldsymbol{n}$ denotes the unit outer normal vector along Γ_1, and the fields $\boldsymbol{f} : \Omega \to \mathbb{R}^3$ and $\boldsymbol{g} : \Gamma_1 \to \mathbb{R}^3$ are related to the fields $\boldsymbol{f}^{\varphi} : \Omega^{\varphi} \to \mathbb{R}^3$ and

$g^\varphi : \Gamma_1^\varphi \to \mathbb{R}^3$ by the simple formulas $f\,dx = f^\varphi\,dx^\varphi$ and $g\,da = g^\varphi\,da^\varphi$. Because they are still in divergence form, these equations can be given a variational formulation, known as the *principle of virtual work* (Theorem 2.6-1). This principle plays a key role as the starting point of the theory of hyperelastic materials (Chapter 4), as well as in the asymptotic theory of two-dimensional plate models (Vol. II).

The tensor $\boldsymbol{T}$ is called the *first Piola–Kirchhoff stress tensor*. We also introduce the symmetric *second Piola–Kirchhoff stress tensor* $\boldsymbol{\Sigma} = \nabla\varphi^{-1}\boldsymbol{T}$, which naturally arises in the expression of the constitutive equations of elastic materials (Chapter 3).

We conclude this chapter by describing various realistic examples of applied forces (Sect. 2.7), corresponding to densities f and g of the form

$$f(x) = \hat{f}(x, \varphi(x)),\ x \in \Omega,\ \text{and}\ g(x) = \hat{g}(x, \nabla\varphi(x)),\ x \in \Gamma_1\,,$$

for given mappings $\hat{f}$ and $\hat{g}$.

2.1. APPLIED FORCES

We assume that in the deformed configuration $\bar{\Omega}^\varphi$ associated with an arbitrary deformation φ, the body is subjected to **applied forces** of two types:

(i) **applied body forces**, defined by a vector field

$$f^\varphi := \Omega^\varphi \to \mathbb{R}^3\,,$$

called the **density of the applied body force** *per unit volume in the deformed configuration*;

(ii) **applied surface forces**, defined by a vector field

$$g^\varphi : \Gamma_1^\varphi \to \mathbb{R}^3$$

on a da^φ-measurable subset Γ_1^φ of the boundary

$$\Gamma^\varphi := \partial\Omega^\varphi\,,$$

called the **density of the applied surface force** *per unit area in the deformed configuration.*

Let $\rho^\varphi : \Omega^\varphi \to \mathbb{R}$ denote the *mass density in the deformed configuration*, so that the mass of every dx^φ-measurable subset A^φ of $\bar{\Omega}^\varphi$ is given by the integral $\int_{A^\varphi} \rho^\varphi(x^\varphi)\,dx^\varphi$. We assume that

$$\rho^\varphi(x^\varphi) > 0 \text{ for all } x^\varphi \in \Omega^\varphi .$$

The applied body forces can be equivalently defined by their *density* $b^\varphi : \Omega^\varphi \to \mathbb{R}^3$ *per unit mass in the deformed configuration*, which is related to the density f^φ by the equation

$$f^\varphi = \rho^\varphi b^\varphi .$$

The applied forces describe the action of the outside world on the body: An elementary force $f(x^\varphi)\,dx^\varphi$ is exerted on the elementary volume dx^φ at each point x^φ of the deformed configuration (Fig. 2.1-1). For example, this is the case of the *gravity field*, for which $f^\varphi(x^\varphi) = -g\rho^\varphi(x^\varphi)e_3$ for all $x^\varphi \in \Omega^\varphi$ (assuming that the basis vector e_3 is "vertical" and oriented "upward"), where g is the gravitational constant. Another example is given by the action of electrostatic forces.

Likewise, an elementary force $g^\varphi(x^\varphi)\,da^\varphi$ is exerted on the elementary area da^φ at each point x^φ of the subset Γ_1^φ of the boundary of the deformed configuration (Fig. 2.1-1). Such forces generally represent the action of another body (whatever its nature may be) along the portion Γ_1^φ of the boundary. Specific examples of applied forces are given in Sect. 2.7.

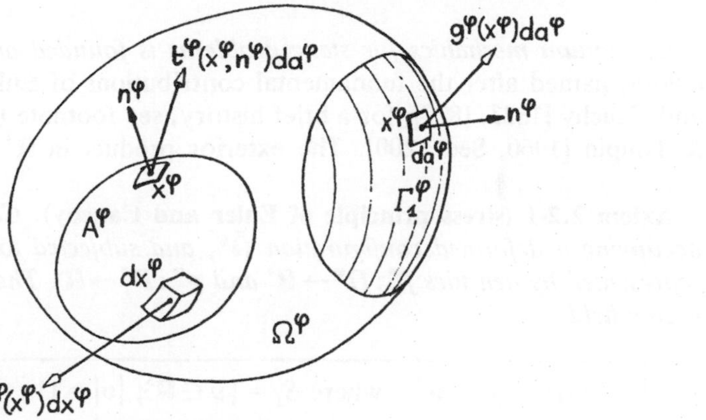

Fig. 2.1-1. Applied forces comprise applied body forces $f^\varphi(x^\varphi)\,dx^\varphi$, $x^\varphi \in \Omega^\varphi$, and applied surface forces $g^\varphi(x^\varphi)\,da^\varphi$, $x^\varphi \in \Gamma_1^\varphi$. The stress principle of Euler and Cauchy asserts in addition the existence of elementary surface forces $t^\varphi(x^\varphi, n^\varphi)\,da^\varphi$, $x^\varphi \in \partial A^\varphi$, along the boundary ∂A^φ, with unit outer normal vector n^φ, of any subdomain A^φ of the deformed configuration $\bar{\Omega}^\varphi$.

Remark. In order to avoid introducing too many notations, we use the same symbol to denote distinct quantities in the same figure. For instance in Fig. 2.1-1, the symbol x^φ stands for three different points, and the symbols $\mathrm{d}a^\varphi$ and n^φ stand for two different area elements and normal vectors. ■

Applied surface forces that are only "partially" specified (for instance, only the normal component $g^\varphi(x^\varphi) \cdot n^\varphi$ could be prescribed along Γ_1^φ) are not excluded from our analysis, and indeed, examples of such "intermediate" cases are considered in Sect. 5.2; but in order to simplify the exposition, we solely consider at this stage the "extreme" cases where either the density g^φ is fully known on Γ_1^φ, or is left completely unspecified, as on the remaining portion

$$\Gamma_0^\varphi := \Gamma^\varphi - \Gamma_1^\varphi$$

of the boundary of the deformed configuration. This being the case, we shall see that it is the deformation itself that should be specified on the corresponding portion $\Gamma_0 := \varphi^{-1}(\Gamma_0^\varphi)$ of the boundary of the reference configuration, in order that the problem be well posed.

2.2. THE STRESS PRINCIPLE OF EULER AND CAUCHY

Continuum mechanics for static problems is founded on the following axiom, named after the fundamental contributions of Euler [1757, 1771] and Cauchy [1823, 1827]; for a brief history, see footnote ([1]) in Truesdell & Toupin [1960, Sect. 200]. The exterior product in $\mathbb{R}^3$ is denoted $\wedge$.

Axiom 2.2-1 (stress principle of Euler and Cauchy). *Consider a body occupying a deformed configuration $\bar{\Omega}^\varphi$, and subjected to applied forces represented by densities $f^\varphi : \Omega^\varphi \to \mathbb{R}^3$ and $g^\varphi : \Gamma_1^\varphi \to \mathbb{R}^3$. Then there exists a vector field*

$$\boxed{t^\varphi : \bar{\Omega}^\varphi \times S_1 \to \mathbb{R}^3, \quad \text{where } S_1 = \{v \in \mathbb{R}^3;\ |v| = 1\}\,,}$$

such that:

(a) *For any subdomain A^φ of $\bar{\Omega}^\varphi$, and at any point $x^\varphi \in \Gamma_1^\varphi \cap \partial A^\varphi$*

where the unit outer normal vector n^{φ} *to* $\Gamma_1^{\varphi} \cap \partial A^{\varphi}$ *exists,*

$$t^{\varphi}(x^{\varphi}, n^{\varphi}) = g^{\varphi}(x^{\varphi}) .$$

(b) **Axiom of force balance:** *For any subdomain* A^{φ} *of* $\bar{\Omega}^{\varphi}$,

$$\int_{A^{\varphi}} f^{\varphi}(x^{\varphi})\, dx^{\varphi} + \int_{\partial A^{\varphi}} t^{\varphi}(x^{\varphi}, n^{\varphi})\, da^{\varphi} = o ,$$

where n^{φ} *denotes the unit outer normal vector along* ∂A^{φ}.

(c) **Axiom of moment balance:** *For any subdomain* A^{φ} *of* Ω^{φ},

$$\int_{A^{\varphi}} ox^{\varphi} \wedge f^{\varphi}(x^{\varphi})\, dx^{\varphi} + \int_{\partial A^{\varphi}} ox^{\varphi} \wedge t^{\varphi}(x^{\varphi}, n^{\varphi})\, da^{\varphi} = o .$$ ■

The stress principle thus first asserts the *existence* of elementary surface forces $t^{\varphi}(x^{\varphi}, n^{\varphi})\, da^{\varphi}$ along the boundaries of all domains of the reference configuration (Fig. 2.1-1).

Secondly, the stress principle asserts that at a point x^{φ} of the boundary ∂A^{φ} of a subdomain A^{φ}, the elementary surface force depends on the subdomain A^{φ} *only* via the normal vector n^{φ} to ∂A^{φ} at x^{φ}. While it would be equally conceivable *a priori* that the elementary surface force at x^{φ} be also dependent on other geometrical properties of the subdomain A^{φ}, for instance the curvature of ∂A^{φ} at x^{φ}, etc., it is possible to rigorously rule out such further geometrical dependences by constructing a general theory of surfaces forces, as shown by Noll [1959] (see also Gurtin & Williams [1967], Ziemer [1983]).

Thirdly, the stress principle asserts that any subdomain A^{φ} of the deformed configuration $\bar{\Omega}^{\varphi}$, including $\bar{\Omega}^{\varphi}$ itself, is in *static equilibrium*, in the sense that *the torsor formed by the elementary forces* $t^{\varphi}(x^{\varphi}, n^{\varphi})\, da^{\varphi}$, $x^{\varphi} \in \partial A^{\varphi}$, n^{φ} *normal to* ∂A^{φ} *at* x^{φ}, *and the body forces* $f^{\varphi}(x^{\varphi})\, dx^{\varphi}$, $x^{\varphi} \in A^{\varphi}$, *is equivalent to zero*. This means that its resultant vector vanishes (axiom of force balance) and that its resulting moment with respect to the origin (and thus with respect to any other point, by a classical property of torsors) vanishes (axiom of moment balance).

Hence the stress principle mathematically expresses, in the form of an axiom, the intuitive idea that the static equilibrium of any subdomain A^{φ}

of $\bar{\Omega}^\varphi$, already subjected to given applied body forces $f^\varphi(x^\varphi)\,dx^\varphi$, $x^\varphi \in A^\varphi$, and (possibly) to given applied surface forces $g^\varphi(x^\varphi)\,da^\varphi$ at those points $x^\varphi \in \Gamma_1^\varphi \cap \partial A^\varphi$ where the outer normal vector to $\Gamma_1^\varphi \cap \partial A^\varphi$ exists, is made possible by the added effect of elementary surfaces forces of the specific form indicated, acting on the remaining part of the boundary ∂A^φ.

Remark. Gurtin [1981a, 1981b] calls *system of forces* the set formed by the applied body forces, corresponding to the vector field $f^\varphi : \Omega^\varphi \to \mathbb{R}^3$, and by the surface forces, corresponding to the vector field t^φ: $\bar{\Omega}^\varphi \times S_1 \to \mathbb{R}^3$. ■

Let x^φ be a point of the deformed configuration. The vector $t^\varphi(x^\varphi, n^\varphi)$ is called the **Cauchy stress vector** *across an oriented surface element with normal* n^φ, or the *density of the surface force per unit area in the deformed configuration*.

2.3. CAUCHY'S THEOREM; THE CAUCHY STRESS TENSOR

We now derive consequences of paramount importance from the stress principle. The first one, due to Cauchy [1823, 1827a], is *one of the most important results in continuum mechanics*. It asserts that *the dependence of the Cauchy stress vector $t^\varphi(x^\varphi, n)$ with respect to its second argument $n \in S_1$ is linear*, i.e., at each point $x^\varphi \in \bar{\Omega}^\varphi$, there exists a tensor $T^\varphi(x^\varphi) \in \mathbb{M}^3$ such that $t^\varphi(x^\varphi, n) = T^\varphi(x^\varphi)n$ for all $n \in S_1$; the second one asserts that at each point $x^\varphi \in \bar{\Omega}^\varphi$, the tensor $T^\varphi(x^\varphi)$ is *symmetric*; the third one, again due to Cauchy [1827b, 1828], is that the tensor field $T^\varphi : \Omega^\varphi \to \mathbb{M}^3$ and the vector fields $f^\varphi : \Omega^\varphi \to \mathbb{R}^3$ and $g^\varphi : \Gamma_1^\varphi \to \mathbb{R}^3$ are related by a *partial differential equation in* Ω^φ, and by a *boundary condition on* Γ_1^φ, respectively.

Theorem 2.3-1 (Cauchy's theorem). *Assume that the applied body force density $f^\varphi : \bar{\Omega}^\varphi \to \mathbb{R}^3$ is continuous, and that the Cauchy stress vector field*

$$t^\varphi : (x^\varphi, n) \in \bar{\Omega}^\varphi \times S_1 \to t^\varphi(x^\varphi, n) \in \mathbb{R}^3$$

is continuously differentiable with respect to the variable $x^\varphi \in \bar{\Omega}^\varphi$ for each $n \in S_1$ and continuous with respect to the variable $n \in S_1$ for each $x^\varphi \in \bar{\Omega}^\varphi$. Then the axioms of force and moment balance imply that there exists a continuously differentiable tensor field

$$T^\varphi : x^\varphi \in \bar{\Omega}^\varphi \to T^\varphi(x^\varphi) \in \mathbb{M}^3 ,$$

such that the Cauchy stress vector satisfies

$$\boxed{t^\varphi(x^\varphi, n) = T^\varphi(x^\varphi)n \quad \text{for all } x^\varphi \in \bar{\Omega}^\varphi \text{ and all } n \in S_1 ,}$$

and such that

$$\boxed{-\mathbf{div}^\varphi\, T^\varphi(x^\varphi) = f^\varphi(x^\varphi) \quad \text{for all } x^\varphi \in \Omega^\varphi ,}$$

$$\boxed{T^\varphi(x^\varphi) = T^\varphi(x^\varphi)^{\mathrm{T}} \quad \text{for all } x^\varphi \in \bar{\Omega}^\varphi ,}$$

$$\boxed{T^\varphi(x^\varphi)n^\varphi = g^\varphi(x^\varphi) \quad \text{for all } x^\varphi \in \Gamma_1^\varphi ,}$$

where n^φ is the unit outer normal vector along Γ_1^φ.

Proof. Let x^φ be a fixed point in Ω^φ. Because the set Ω^φ is open, we can find, as a particular subdomain of $\bar{\Omega}^\varphi$, a tetrahedron T with vertex x^φ, with three faces parallel to the coordinate planes, and with a face F whose normal vector $n = n_i e_i$ has all its components $n_i > 0$ (Fig. 2.3-1). Let v_i denote the vertices other than x^φ, as indicated in the figure, and let F_i denote the face opposite to the vertex v_i, so that area $F_i = n_i$ area F. The axiom of force balance over the tetrahedron T reads

$$\int_T f^\varphi(y^\varphi)\,\mathrm{d}y^\varphi + \int_{\partial T} t^\varphi(y^\varphi, n^\varphi)\,\mathrm{d}a^\varphi = o .$$

Writing this relation componentwise, with

$$f^\varphi(y^\varphi) = f_i^\varphi(y^\varphi)e_i, \quad t^\varphi(y^\varphi, n^\varphi) = t_i^\varphi(y^\varphi, n^\varphi)e_i ,$$

and using the mean value theorem for integrals (which can be applied to the integral on ∂T since the four functions $y^\varphi \in \partial T \to t^\varphi(y^\varphi, n^\varphi)$ and $y^\varphi \in \partial T \to t^\varphi(y^\varphi, e_j)$ are continuous by assumption), we obtain for each index i,

$$|t_i^\varphi(y_i, n) + t_i^\varphi(y_{ij}, -e_j)n_j| \text{ area } F \leq \sup_{y \in T} |f_i^\varphi(y)| \text{ vol } T ,$$

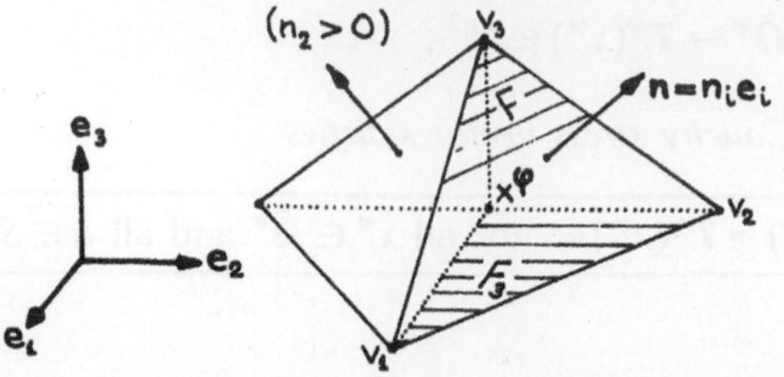

Fig. 2.3-1. Cauchy's celebrated proof of Cauchy's theorem.

for appropriate points $y_i \in F$, $y_{ij} \in F_j$. Keeping the vector $\boldsymbol{n}$ fixed, let the vertices v_i coalesce into the vertex x^φ. Using again the continuity of the vector field $\boldsymbol{t}^\varphi(x^\varphi, \boldsymbol{n})$ with respect to the first variable x^φ, and using the relation $\operatorname{vol} T = c(\boldsymbol{n})\ (\operatorname{area} F)^{3/2}$ coupled with the boundedness of the applied body force density, we obtain:

$$\boldsymbol{t}^\varphi(x^\varphi, \boldsymbol{n}) = -n_j \boldsymbol{t}^\varphi(x^\varphi, -\boldsymbol{e}_j)\,.$$

Using next the continuity with respect to the second variable, let $\boldsymbol{n}$ approach a particular basis vector $\boldsymbol{e}_j$ in the above relation; we obtain in this fashion:

$$\boldsymbol{t}^\varphi(x^\varphi, \boldsymbol{e}_j) = -\boldsymbol{t}^\varphi(x^\varphi, -\boldsymbol{e}_j)\,.$$

It follows that the relation

$$\boldsymbol{t}^\varphi(x^\varphi, \boldsymbol{n}) = n_j \boldsymbol{t}^\varphi(x^\varphi, \boldsymbol{e}_j)$$

also holds if some, or all, components n_i are <0 (the case $n_2 < 0$ is shown in Fig. 2.3-1); hence it holds for all points $x^\varphi \in \bar{\Omega}^\varphi$ and all unit vectors $\boldsymbol{n} \in S_1$. We now define functions $T^\varphi_{ij} : \bar{\Omega}^\varphi \to \mathbb{R}$ by letting

$$\boldsymbol{t}^\varphi(x^\varphi, \boldsymbol{e}_j) = T^\varphi_{ij}(x^\varphi)\boldsymbol{e}_i \quad \text{for all } x^\varphi \in \bar{\Omega}^\varphi\,,$$

so that $\boldsymbol{t}^\varphi(x^\varphi, \boldsymbol{n}) = T^\varphi_{ij}(x^\varphi) n_j \boldsymbol{e}_i$, and thus

$$t^\varphi_i(x^\varphi, \boldsymbol{n}) = T^\varphi_{ij}(x^\varphi) n_j \quad \text{for all } x^\varphi \in \bar{\Omega}^\varphi,\ \boldsymbol{n} \in S_1\,.$$

Therefore if we define a tensor $\boldsymbol{T}^\varphi(x^\varphi)$ by

$$\boldsymbol{T}^\varphi(x^\varphi) := (T^\varphi_{ij}(x^\varphi))\,,$$

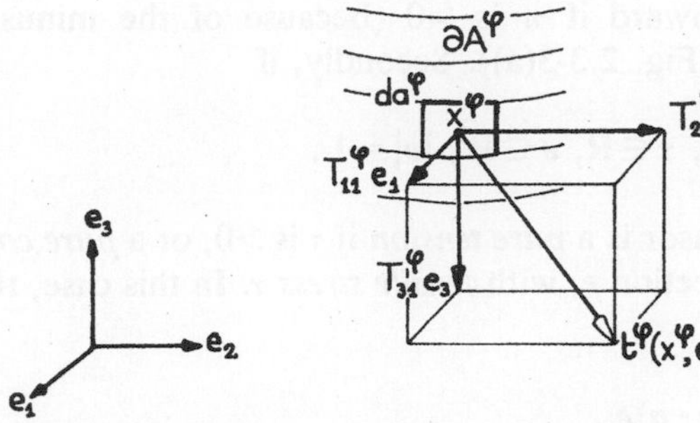

Fig. 2.3-2. Interpretation of the elements T^{φ}_{i1} of the Cauchy stress tensor $T^{\varphi} = (T^{\varphi}_{ij})$.

the Cauchy stress vector is often represented on three mutually perpendicular faces of a rectangular parallelepiped.

The following three special cases of Cauchy stress tensors are particularly worthy of interest (cf. Fig. 2.3-3, where in each case it is assumed that the Cauchy stress tensor is constant in the particular region considered). First, if

$$T^{\varphi}(x^{\varphi}) = -\pi I, \quad \pi \in \mathbb{R},$$

the Cauchy stress tensor is a *pressure*, and the real number π is also called a *pressure*. In this case, the Cauchy stress vector

$$t^{\varphi}(x^{\varphi}, n) = -\pi n$$

is always normal to the elementary surface element, its length is constant,

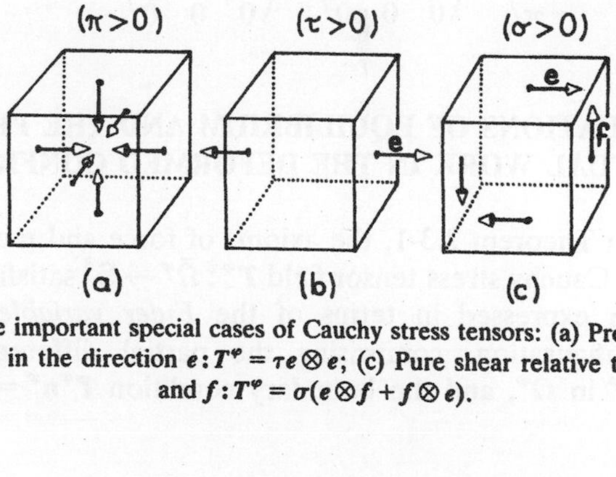

Fig. 2.3-3. Three important special cases of Cauchy stress tensors: (a) Pressure: $T^{\varphi} = -\pi I$; (b) Pure tension in the direction $e: T^{\varphi} = \tau e \otimes e$; (c) Pure shear relative to the directions e and $f: T^{\varphi} = \sigma(e \otimes f + f \otimes e)$.

and it is directed inward if π is >0 (because of the minus sign), or outward if π is <0 (Fig. 2.3-3(a)). Secondly, if

$$T^{\varphi}(x^{\varphi}) = \tau e \otimes e,\ \tau \in \mathbb{R},\ e \in \mathbb{R}^3,\ |e| = 1\,,$$

the Cauchy stress tensor is a *pure tension* if τ is >0, or a *pure compression* if τ is <0, in the direction e, with *tensile stress* τ. In this case, the Cauchy stress vector

$$t^{\varphi}(x^{\varphi}, n) = -\tau(e \cdot n)e\,,$$

which is always parallel to the vector e, is directed outward if $\tau > 0$, or inward if $\tau < 0$, on the faces with normals $n = e$ or $n = -e$, and it vanishes on the faces whose normal is orthogonal to the vector e (Fig. 2.3-3(b)). Thirdly (Fig. 2.3-3(c)), if

$$T^{\varphi}(x^{\varphi}) = \sigma(e \otimes f + f \otimes e),\ \sigma \in \mathbb{R},\ e, f \in \mathbb{R}^3,\ |e| = |f| = 1,\ e \cdot f = 0\,,$$

the Cauchy stress tensor is a *pure shear*, with *shear stress* τ, *relative to the directions e and f*. In this case, the Cauchy stress vector is given by

$$t^{\varphi}(x^{\varphi}, n) = \sigma\{(f \cdot n)e + (e \cdot n)f\}\,.$$

The Cauchy stress tensors corresponding to these three special cases are respectively given by (for definiteness, we assume that $e = e_1$ and $f = e_2$):

$$\begin{pmatrix} -\pi & 0 & 0 \\ 0 & -\pi & 0 \\ 0 & 0 & -\pi \end{pmatrix},\ \begin{pmatrix} \tau & 0 & 0 \\ 0 & 0 & 0 \\ 0 & 0 & 0 \end{pmatrix},\ \begin{pmatrix} 0 & \sigma & 0 \\ \sigma & 0 & 0 \\ 0 & 0 & 0 \end{pmatrix}.$$

2.4. THE EQUATIONS OF EQUILIBRIUM AND THE PRINCIPLE OF VIRTUAL WORK IN THE DEFORMED CONFIGURATION

As shown in Theorem 2.3-1, the axioms of force and moment balance imply that the Cauchy stress tensor field $T^{\varphi} : \bar{\Omega}^{\varphi} \to \mathbb{S}^3$ satisfies a *boundary value problem* expressed in terms of the *Euler variable* x^{φ} over the deformed configuration, comprising the partial differential equation $-\mathbf{div}^{\varphi}\, T^{\varphi} = f^{\varphi}$ in Ω^{φ}, and the boundary condition $T^{\varphi}n^{\varphi} = g^{\varphi}$ on Γ_1^{φ}. A remarkable property of this boundary value problem, due to its "divergence form", is that it can be given a *variational formulation*, as we now show (this terminology is justified in Sect. 2.6). In what follows, $\boldsymbol{u} \cdot \boldsymbol{v} = u_i v_i$ denotes the Euclidean vector inner product, $\boldsymbol{A} : \boldsymbol{B} = A_{ij} B_{ij} = \operatorname{tr} \boldsymbol{A}^{\mathrm{T}} \boldsymbol{B}$ denotes the matrix inner product, and $\boldsymbol{\nabla}^{\varphi} \boldsymbol{\theta}^{\varphi}$ denotes the matrix $(\partial_j^{\varphi} \theta_i^{\varphi})$.

Theorem 2.4-1. *The boundary value problem*

$$\boxed{\begin{aligned} -\mathbf{div}^{\varphi}\, \boldsymbol{T}^{\varphi} &= \boldsymbol{f}^{\varphi} \text{ in } \Omega^{\varphi}, \\ \boldsymbol{T}^{\varphi} \boldsymbol{n}^{\varphi} &= \boldsymbol{g}^{\varphi} \text{ on } \Gamma_1^{\varphi}, \end{aligned}}$$

is formally equivalent to the variational equations:

$$\boxed{\int_{\Omega^{\varphi}} \boldsymbol{T}^{\varphi} : \boldsymbol{\nabla}^{\varphi} \boldsymbol{\theta}^{\varphi} \, \mathrm{d}x^{\varphi} = \int_{\Omega^{\varphi}} \boldsymbol{f}^{\varphi} \cdot \boldsymbol{\theta}^{\varphi} \, \mathrm{d}x^{\varphi} + \int_{\Gamma_1^{\varphi}} \boldsymbol{g}^{\varphi} \cdot \boldsymbol{\theta}^{\varphi} \, \mathrm{d}x^{\varphi} \,,}$$

valid for all smooth enough vector fields: $\boldsymbol{\theta}^{\varphi} : \Omega^{\varphi} \to \mathbb{R}^3$ *that satisfy*

$$\boxed{\boldsymbol{\theta}^{\varphi} = \boldsymbol{o} \text{ on } \Gamma_0^{\varphi} := \Gamma^{\varphi} - \Gamma_1^{\varphi} \,.}$$

Proof. The equivalence with the variational equations rests on another *Green's formula* (whose proof is again a direct application of the fundamental Green formula; cf. Sect. 1.6): *For any smooth enough tensor field* $\boldsymbol{T}^{\varphi} : \bar{\Omega}^{\varphi} \to \mathbb{M}^3$ *and vector field* $\boldsymbol{\theta}^{\varphi} : \bar{\Omega}^{\varphi} \to \mathbb{R}^3$,

$$\int_{\Omega^{\varphi}} \mathbf{div}^{\varphi} \boldsymbol{T}^{\varphi} \cdot \boldsymbol{\theta}^{\varphi} \, \mathrm{d}x^{\varphi} = -\int_{\Omega^{\varphi}} \boldsymbol{T}^{\varphi} : \boldsymbol{\nabla}^{\varphi} \boldsymbol{\theta}^{\varphi} \, \mathrm{d}x^{\varphi} + \int_{\Gamma^{\varphi}} \boldsymbol{T}^{\varphi} \boldsymbol{n}^{\varphi} \cdot \boldsymbol{\theta}^{\varphi} \, \mathrm{d}a^{\varphi} \,.$$

Thus, if we integrate over the set Ω^{φ} the inner product of the equation $\mathbf{div}^{\varphi}\, \boldsymbol{T}^{\varphi} + \boldsymbol{f}^{\varphi} = \boldsymbol{o}$ with a vector field $\boldsymbol{\theta}^{\varphi}$ that vanishes on Γ_0^{φ}, we obtain:

$$\begin{aligned} \boldsymbol{o} &= \int_{\Omega^{\varphi}} (\mathbf{div}^{\varphi}\, \boldsymbol{T}^{\varphi} + \boldsymbol{f}^{\varphi}) \cdot \boldsymbol{\theta}^{\varphi} \, \mathrm{d}x^{\varphi} \\ &= \int_{\Omega^{\varphi}} \{-\boldsymbol{T}^{\varphi} : \boldsymbol{\nabla}^{\varphi} \boldsymbol{\theta}^{\varphi} + \boldsymbol{f}^{\varphi} \cdot \dot{\boldsymbol{\theta}}^{\varphi}\} \, \mathrm{d}x^{\varphi} + \int_{\Gamma_1^{\varphi}} \boldsymbol{T}^{\varphi} \boldsymbol{n}^{\varphi} \cdot \boldsymbol{\theta}^{\varphi} \, \mathrm{d}a^{\varphi} \,, \end{aligned}$$

and the variational equations follow, since $\boldsymbol{T}^{\varphi} \boldsymbol{n}^{\varphi} = \boldsymbol{g}^{\varphi}$ on Γ_1^{φ}. Conversely, assume that the variational equations are satisfied. They reduce to

$$\int_{\Omega^\varphi} \boldsymbol{T}^\varphi : \boldsymbol{\nabla}^\varphi \boldsymbol{\theta}^\varphi \, \mathrm{d}x^\varphi = \int_{\Omega^\varphi} \boldsymbol{f}^\varphi \cdot \boldsymbol{\theta}^\varphi \, \mathrm{d}x^\varphi \text{ if } \boldsymbol{\theta}^\varphi = \boldsymbol{o} \text{ on } \Gamma^\varphi,$$

and since, by the above Green formula,

$$\int_{\Omega^\varphi} \boldsymbol{T}^\varphi : \boldsymbol{\nabla}^\varphi \boldsymbol{\theta} \, \mathrm{d}x^\varphi = -\int_{\Omega^\varphi} \mathbf{div}^\varphi \, \boldsymbol{T}^\varphi \, \mathrm{d}x^\varphi \text{ if } \boldsymbol{\theta}^\varphi = \boldsymbol{o} \text{ on } \Gamma^\varphi,$$

we deduce that $\mathbf{div}^\varphi \, \boldsymbol{T}^\varphi + \boldsymbol{f}^\varphi = \boldsymbol{o}$ in Ω^φ. Taking this equation into account and using the same Green formula, we find that the variational equations reduce to the equations

$$\int_{\Gamma_1^\varphi} \boldsymbol{T}^\varphi \boldsymbol{n}^\varphi \cdot \boldsymbol{\theta}^\varphi \, \mathrm{d}a^\varphi = \int_{\Gamma_1^\varphi} \boldsymbol{g}^\varphi \cdot \boldsymbol{\theta}^\varphi \, \mathrm{d}a^\varphi,$$

which imply that the boundary condition $\boldsymbol{T}^\varphi \boldsymbol{n}^\varphi = \boldsymbol{g}^\varphi$ holds on Γ_1^φ. ∎

The equations

$$\boxed{\begin{aligned} -\mathbf{div}^\varphi \, \boldsymbol{T}^\varphi &= \boldsymbol{f}^\varphi \text{ in } \Omega^\varphi, \\ \boldsymbol{T}^\varphi &= (\boldsymbol{T}^\varphi)^{\mathrm{T}} \text{ in } \Omega^\varphi, \\ \boldsymbol{T}^\varphi \boldsymbol{n}^\varphi &= \boldsymbol{g}^\varphi \text{ on } \Gamma_1^\varphi, \end{aligned}}$$

are called the **equations of equilibrium in the deformed configuration**, while the associated variational equations of Theorem 2.4-1 constitute the **principle of virtual work in the deformed configuration.**

Remark. In both the axiom of force balance and the principle of virtual work, the required smoothness on the field $\boldsymbol{T}^\varphi : \bar{\Omega}^\varphi \to \mathbb{S}^3$ is very mild (it suffices that all integrals make sense). By contrast, *a significant additional smoothness is required for writing the equations of equilibrium* (in order that $\mathbf{div}^\varphi \, \boldsymbol{T}^\varphi$ makes sense), which are only used as an intermediary between the axiom and the principle. Hence the question naturally arises as to whether the equations of equilibrium can be by-passed in this process, and the regularity requirements be reduced accordingly. In this direction, Antman & Osborn [1979] have shown that the principle of virtual work can be indeed directly deduced from the axiom of force balance. Their basic idea is to put on an equivalent basis the fact that the axiom is valid "for all subdomains A^φ" while the principle holds "for all mappings $\boldsymbol{\theta}^\varphi$", by associating special classes of subdomains (cubes and their bi-Lipschitz continuous images) with special families of variations (basically piecewise linear functions). The methods of proof are reminiscent of those used for proving Green's formulas in the theory of integration. ∎

2.5. THE PIOLA–KIRCHHOFF STRESS TENSORS

Our final objective is to determine the deformation field and the Cauchy stress tensor field that arise in a body subjected to a given system of applied forces. In this respect, the equations of equilibrium in the deformed configuration are of not much avail, since they are expressed in terms of the *Euler variable* $x^{\varphi} = \varphi(x)$, *which is precisely one of the unknowns*. To obviate this difficulty, we shall rewrite these equations in terms of the *Lagrange variable* x that is attached to the reference configuration, which is considered as being given once and for all. More specifically, we shall transform the left-hand sides $\mathbf{div}^{\varphi}\, \boldsymbol{T}^{\varphi}$ and $\boldsymbol{T}^{\varphi}\boldsymbol{n}^{\varphi}$ and the right-hand sides $\boldsymbol{f}^{\varphi}$ and $\boldsymbol{g}^{\varphi}$ appearing in the equations of equilibrium over $\bar{\Omega}^{\varphi}$ into similar expressions over $\bar{\Omega}$.

We have laid the ground for transforming the left-hand sides in Sect. 1.7, where we defined the Piola transform $\boldsymbol{T} : \bar{\Omega} \rightarrow \mathbb{M}^3$ of a tensor field $\boldsymbol{T}^{\varphi} : \bar{\Omega}^{\varphi} = \varphi(\bar{\Omega}) \rightarrow \mathbb{M}^3$ by letting

$$\boxed{\boldsymbol{T}(x) = (\det \boldsymbol{\nabla}\varphi(x))\boldsymbol{T}^{\varphi}(x^{\varphi})\boldsymbol{\nabla}\varphi(x)^{-\mathrm{T}}, \quad x^{\varphi} = \varphi(x)\,.}$$

We shall therefore apply this transform to the Cauchy stress tensor $\boldsymbol{T}^{\varphi}$, in which case its Piola transform $\boldsymbol{T}$ is called the **first Piola–Kirchhoff stress tensor**. As shown in Theorem 1.7-1, the main advantage of this transform is to induce a particularly simple relation between the divergences of both tensors:

$$\boxed{\mathbf{div}\, \boldsymbol{T}(x) = (\det \boldsymbol{\nabla}\varphi(x))\, \mathbf{div}^{\varphi}\, \boldsymbol{T}^{\varphi}(x^{\varphi}), \quad x^{\varphi} = \varphi(x)\,.}$$

As a consequence, the equations of equilibrium over the deformed configuration will be transformed (Theorem 2.6-1) into equations over the reference configuration that have a similar "*divergence structure*". This property in turn makes it possible to write these partial differential equations in *variational form*, as shown in Theorem 2.4-1 for the equations of equilibrium in the deformed configuration, and in Theorem 2.6-1 below for the equations of equilibrium over the reference configuration.

One can likewise transform the Cauchy stress vector $\boldsymbol{t}^\varphi(x^\varphi, \boldsymbol{n}^\varphi) = \boldsymbol{T}^\varphi(x^\varphi)\boldsymbol{n}^\varphi$ into a vector $\boldsymbol{t}(x, \boldsymbol{n})$ in such a way that the relation

$$\boxed{\boldsymbol{t}(x, \boldsymbol{n}) = \boldsymbol{T}(x)\boldsymbol{n}\,,}$$

holds, where $\boldsymbol{T}(x)$ is the first Piola–Kirchhoff stress tensor and where $\boldsymbol{n}$ and $\boldsymbol{n}^\varphi$ are the corresponding normal vectors at the points x and $x^\varphi = \boldsymbol{\varphi}(x)$ of the boundaries of corresponding subdomains A and $A^\varphi = \boldsymbol{\varphi}(A)$. Notice that there is no ambiguity in this process since the normal vector $\boldsymbol{n}^\varphi$ at the point $x^\varphi = \boldsymbol{\varphi}(x)$ is the same for *all* subdomains whose boundary passes through the point x with $\boldsymbol{n}$ as the normal vector there. In view of the relation $\boldsymbol{T}(x)\boldsymbol{n}\,\mathrm{d}a = \boldsymbol{T}^\varphi(x^\varphi)\boldsymbol{n}^\varphi\,\mathrm{d}a^\varphi$ established in Theorem 1.7-1, it suffices to define the vector $\boldsymbol{t}(x, \boldsymbol{n})$ by the relation:

$$\boxed{\boldsymbol{t}(x, \boldsymbol{n})\,\mathrm{d}a = \boldsymbol{t}^\varphi(x^\varphi, \boldsymbol{n}^\varphi)\,\mathrm{d}a^\varphi\,.}$$

Since $\boldsymbol{t}^\varphi(x^\varphi, \boldsymbol{n}^\varphi) = \boldsymbol{T}^\varphi(x^\varphi)\boldsymbol{n}^\varphi$ by Cauchy's theorem, the desired relation $\boldsymbol{t}(x, \boldsymbol{n}) = \boldsymbol{T}(x)\boldsymbol{n}$ holds.

The vector $\boldsymbol{t}(x, \boldsymbol{n})$ is called the **first Piola–Kirchhoff stress vector** *at the point x of the reference configuration, across the oriented surface element with normal* $\boldsymbol{n}$. The vector field $\boldsymbol{t}: \bar{\Omega} \times S_1 \to \mathbb{R}^3$ defined in this fashion thus measures the *density of the surface force per unit area in the reference configuration*.

While the Cauchy stress tensor $\boldsymbol{T}^\varphi(x^\varphi)$ is symmetric (Theorem 2.3-1), the first Piola–Kirchhoff stress tensor $\boldsymbol{T}(x)$ is not symmetric in general; instead one has:

$$\boldsymbol{T}(x)^{\mathrm{T}} = \boldsymbol{\nabla}\boldsymbol{\varphi}(x)^{-1}\boldsymbol{T}(x)\boldsymbol{\nabla}\boldsymbol{\varphi}(x)^{-\mathrm{T}}\,.$$

It is nevertheless desirable to define a symmetric stress tensor in the reference configuration, essentially because *the constitutive equation in the reference configuration then takes a simpler form*, as we shall see in the next chapter (see notably Theorem 3.6-2). More specifically, we define the **second Piola–Kirchhoff stress tensor** $\boldsymbol{\Sigma}(x)$ by letting

$$\boxed{\begin{aligned}\boldsymbol{\Sigma}(x) = \boldsymbol{\nabla}\boldsymbol{\varphi}(x)^{-1}\boldsymbol{T}(x) = (\det \boldsymbol{\nabla}\boldsymbol{\varphi}(x))\boldsymbol{\nabla}\boldsymbol{\varphi}(x)^{-1}\boldsymbol{T}^\varphi(x^\varphi)\boldsymbol{\nabla}\boldsymbol{\varphi}(x)^{-\mathrm{T}},\\ x^\varphi = \boldsymbol{\varphi}(x)\,.\end{aligned}}$$

Remarks. (1) In fact, the question of whether or not the matrix $\boldsymbol{T}(x)$ is symmetric does not make sense for, as a tensor, it has one index attached to the reference configuration and one index attached to the deformed configuration. A complete discussion of these aspects can be found in Marsden & Hughes [1983].

(2) Historical reference on the Piola–Kirchhoff stress tensors are given in Truesdell & Toupin [1960, Sect. 210]. ■

The Piola–Kirchhoff stress tensors $\boldsymbol{T}(x)$ *and* $\boldsymbol{\Sigma}(x)$ *both depend on the deformation* $\boldsymbol{\varphi}$, first through the Piola transform itself, secondly because the Cauchy stress tensor is also dependent on $\boldsymbol{\varphi}$. The study of these dependences is the object of Chapter 3.

2.6. THE EQUATIONS OF EQUILIBRIUM AND THE PRINCIPLE OF VIRTUAL WORK IN THE REFERENCE CONFIGURATION

It remains to transform the applied force densities that appear in the equilibrium equations over the deformed configuration: First, with the density $\boldsymbol{f}^{\varphi}: \Omega^{\varphi} \rightarrow \mathbb{R}^3$ of the applied body force per unit volume in the deformed configuration, we associate a vector field $\boldsymbol{f}: \Omega \rightarrow \mathbb{R}^3$ in such a way that

$$\boxed{\boldsymbol{f}(x)\,\mathrm{d}x = \boldsymbol{f}^{\varphi}(x^{\varphi})\,\mathrm{d}x^{\varphi} \quad \text{for all } x^{\varphi} = \boldsymbol{\varphi}(x) \in \Omega^{\varphi}\,,}$$

where $\mathrm{d}x$ and $\mathrm{d}x^{\varphi}$ denote the corresponding volume elements. Since $\mathrm{d}x^{\varphi} = \det \boldsymbol{\nabla\varphi}(x)\,\mathrm{d}x$, we thus have

$$\boxed{\boldsymbol{f}(x) = (\det \boldsymbol{\nabla\varphi}(x))\boldsymbol{f}^{\varphi}(x^{\varphi}), \quad x^{\varphi} = \boldsymbol{\varphi}(x)\,,}$$

so that *the vector* $\boldsymbol{f}(x)$ *depends on the deformation* $\boldsymbol{\varphi}$, via the factor $\det \boldsymbol{\nabla\varphi}(x)$ on the one hand, and via the possible dependence of the density $\boldsymbol{f}^{\varphi}$ on the deformation $\boldsymbol{\varphi}$ on the other hand. Notice that this relation displays the *same* factor $\det \boldsymbol{\nabla\varphi}(x)$ as the relation between the vectors $\mathbf{div}\,\boldsymbol{T}(x)$ and $\mathbf{div}^{\varphi}\,\boldsymbol{T}^{\varphi}(x^{\varphi})$ (this observation will be used in the proof of Theorem 2.6-1).

The vector field $\boldsymbol{f}: \Omega \rightarrow \mathbb{R}^3$ measures the **density of the applied body force** *per unit volume in the reference configuration*; the vector $\boldsymbol{f}(x)$ is defined in such a way that the elementary vector $\boldsymbol{f}(x)\,\mathrm{d}x$ is equal to the elementary body force $f(x^\varphi)\,dx^\varphi$ acting on the corresponding volume element dx^φ at the point $x^\varphi = \varphi(x)$ (Fig. 2.6-1).

Let $\rho : \Omega \to \mathbb{R}$ denote the *mass density in the reference configuration*. Expressing that the mass of the elementary volumes dx and $dx^\varphi = \det \nabla\varphi(x)\,dx$ is the same, we find that the mass densities $\rho : \Omega \to \mathbb{R}$ and $\rho^\varphi : \Omega^\varphi \to \mathbb{R}$ are related by the equation

$$\rho(x) = \det \nabla\varphi(x)\rho^\varphi(x^\varphi), \quad x^\varphi = \varphi(x).$$

Incidentally, this relation also shows that, regardless of any consideration concerning the preservation of orientation, the Jacobian $\det \nabla\varphi(x)$ should not vanish in an actual deformation, since a mass density is always >0, at least macroscopically.

Then if we define the *density* $\boldsymbol{b} : \Omega \to \mathbb{R}^3$ *of the applied body force per unit mass in the reference configuration* by letting

$$\boldsymbol{f}(x) = \rho(x)\boldsymbol{b}(x) \quad \text{for all } x \in \Omega,$$

it follows that the densities of the applied body force per unit mass are related by

$$\boldsymbol{b}(x) = \boldsymbol{b}^\varphi(x^\varphi), \quad x^\varphi = \varphi(x).$$

Secondly, in order to transform the boundary condition $\boldsymbol{T}^\varphi \boldsymbol{n}^\varphi = \boldsymbol{g}^\varphi$ over $\Gamma_1^\varphi = \varphi(\Gamma_1)$ into a similar condition over Γ_1, it suffices to use the *first Piola–Kirchhoff stress vector*, which was precisely defined for this purpose in Sect. 2.5: With the density $\boldsymbol{g}^\varphi : \Gamma_1^\varphi \to \mathbb{R}^3$ of the applied surface force per unit area in the deformed configuration, we associate the vector field $\boldsymbol{g} : \Gamma_1 \to \mathbb{R}^3$ defined by

$$\boxed{\boldsymbol{g}(x)\,da = \boldsymbol{g}^\varphi(x^\varphi)\,da^\varphi \quad \text{for all } x^\varphi = \varphi(x) \in \Gamma_1^\varphi,}$$

where da and da^φ are the corresponding area elements. Hence by Theorem 1.7-1, the vector $\boldsymbol{g}(x)$ is given by

$$\boxed{\boldsymbol{g}(x) = \det \nabla\varphi(x) |\nabla\varphi(x)^{-T}\boldsymbol{n}| \boldsymbol{g}^\varphi(x^\varphi).}$$

Notice that the vector $\boldsymbol{g}(x)$ *depends on the deformation* φ, via the formula relating the corresponding area elements on the one hand, and via the possible dependence of the density g^{φ} on the deformation φ on the other hand. The vector field $g: \Gamma_1 \to \mathbb{R}^3$ measures the **density of the applied surface force** *per unit area in the reference configuration*; it is defined in such a way that the elementary vector $g(x)\,\mathrm{d}a$ is equal to the elementary surface force $g^{\varphi}(x^{\varphi})\,\mathrm{d}a^{\varphi}$ acting on the corresponding area element $\mathrm{d}a^{\varphi}$ at the point $x^{\varphi} = \varphi(x)$ (Fig. 2.6-1).

We can now establish the analog of Theorem 2.4-1 over the reference configuration:

Theorem 2.6-1. *The first Piola–Kirchhoff stress tensor* $T(x) = (\det \nabla\varphi(x)) T^{\varphi}(x^{\varphi}) \nabla\varphi(x)^{-\mathrm{T}}$ *satisfies the following equations in the reference configuration* Ω:

$$\boxed{\begin{aligned} &-\mathbf{div}\, T(x) = f(x),\ x \in \Omega\,, \\ &\nabla\varphi(x) T(x)^{\mathrm{T}} = T(x) \nabla\varphi(x)^{\mathrm{T}},\ x \in \Omega\,, \\ &T(x) n = g(x),\ x \in \Gamma_1\,, \end{aligned}}$$

where $f\,\mathrm{d}x = f^{\varphi}\,\mathrm{d}x^{\varphi}$, $g\,\mathrm{d}a = g^{\varphi}\,\mathrm{d}a^{\varphi}$. *The first and third equations are together equivalent to the variational equations*:

$$\boxed{\int_{\Omega} T : \nabla\theta\,\mathrm{d}x = \int_{\Omega} f \cdot \theta\,\mathrm{d}x + \int_{\Gamma_1} g \cdot \theta\,\mathrm{d}a\,,}$$

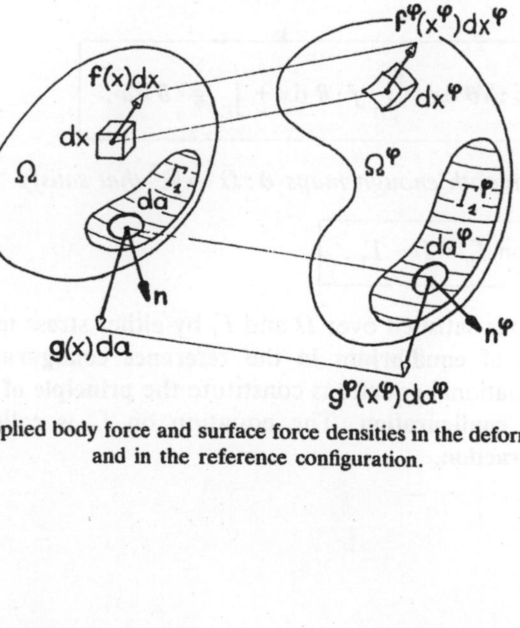

Fig. 2.6-1. The applied body force and surface force densities in the deformed configuration and in the reference configuration.

valid for all smooth enough vector fields $\boldsymbol{\theta} : \bar{\Omega} \to \mathbb{R}^3$ *that satisfy*

$$\boldsymbol{\theta} = \boldsymbol{o} \text{ on } \Gamma_0 := \Gamma - \Gamma_1 .$$

Proof. The first equation follows from the equations $-\mathbf{div}^{\varphi} \boldsymbol{T}^{\varphi} = \boldsymbol{f}^{\varphi}$ in Ω^{φ}, $\mathbf{div}^{\varphi} \boldsymbol{T}^{\varphi} = (\det \boldsymbol{\nabla}\boldsymbol{\varphi})\, \mathbf{div}\, \boldsymbol{T}$, and $\boldsymbol{f} = (\det \boldsymbol{\nabla}\boldsymbol{\varphi})\boldsymbol{f}^{\varphi}$; the second follows from the definition of the tensor $\boldsymbol{T}$ and the symmetry of the tensor $\boldsymbol{T}^{\varphi}$; the third one follows from the equations $\boldsymbol{T}^{\varphi}\boldsymbol{n}^{\varphi} = \boldsymbol{g}^{\varphi}$, $\boldsymbol{T}^{\varphi}\boldsymbol{n}^{\varphi}\, \mathrm{d}a^{\varphi} = \boldsymbol{T}\boldsymbol{n}\, \mathrm{d}a$, and $\boldsymbol{g}^{\varphi}\, \mathrm{d}a^{\varphi} = \boldsymbol{g}\, \mathrm{d}a$. The equivalence with the variational equations is then established as in Theorem 2.4-1. ■

In terms of the second Piola–Kirchhoff stress tensor, the above result becomes:

Theorem 2.6-2. *The second Piola–Kirchhoff stress tensor* $\boldsymbol{\Sigma}(x) = (\det \boldsymbol{\nabla}\boldsymbol{\varphi}(x))\boldsymbol{\nabla}\boldsymbol{\varphi}(x)^{-1}\boldsymbol{T}^{\varphi}(x^{\varphi})\boldsymbol{\nabla}\boldsymbol{\varphi}(x)^{-\mathrm{T}}$ *satisfies the following equations in the reference configuration* $\bar{\Omega}$:

$$\begin{aligned} -\mathbf{div}(\boldsymbol{\nabla}\boldsymbol{\varphi}(x)\boldsymbol{\Sigma}(x)) &= \boldsymbol{f}(x),\ x \in \bar{\Omega} , \\ \boldsymbol{\Sigma}(x) &= \boldsymbol{\Sigma}(x)^{\mathrm{T}},\ x \in \Omega , \\ \boldsymbol{\nabla}\boldsymbol{\varphi}(x)\boldsymbol{\Sigma}(x)\boldsymbol{n} &= \boldsymbol{g}(x),\ x \in \Gamma_1 . \end{aligned}$$

The first and third equations are together equivalent to the variational equations

$$\int_{\Omega} \boldsymbol{\nabla}\boldsymbol{\varphi}\boldsymbol{\Sigma} : \boldsymbol{\nabla}\boldsymbol{\theta}\, \mathrm{d}x = \int_{\Omega} \boldsymbol{f} \cdot \boldsymbol{\theta}\, \mathrm{d}x + \int_{\Gamma_1} \boldsymbol{g} \cdot \boldsymbol{\theta}\, \mathrm{d}a ,$$

valid for all smooth enough maps $\boldsymbol{\theta} : \bar{\Omega} \to \mathbb{R}^3$ *that satisfy*

$$\boldsymbol{\theta} = \boldsymbol{o} \text{ on } \Gamma_0 = \Gamma - \Gamma_1 .$$

■

The equations satisfied over Ω and Γ_1 by either stress tensor are called the **equations of equilibrium in the reference configuration**, and their associated variational equations constitute the **principle of virtual work in the reference configuration**. The equation on Γ_1 is called a **boundary condition of traction**.

As we already mentioned in Sect. 2.1, a **boundary condition of place** of the form

$$\varphi = \varphi_0 \text{ on } \Gamma_0 ,$$

where $\varphi_0 : \Gamma_0 \to \mathbb{R}^3$ is a given mapping, will be later adjoined to the equations of equilibrium in the reference configuration. This being the case, we may think of each vector field $\boldsymbol{\theta} : \bar{\Omega} \to \mathbb{R}^3$ occurring in the principle of virtual work as a *"virtual" variation of a deformation consistent with the boundary condition of place*. More specifically, if we define the set

$$\boldsymbol{\Phi} := \{\boldsymbol{\psi} : \bar{\Omega} \to \mathbb{R}^3;\quad \det \nabla\boldsymbol{\psi} > 0 \text{ in } \bar{\Omega};\ \boldsymbol{\psi} = \varphi_0 \text{ on } \Gamma_0\}$$

(at this stage, we do not require that the vector fields $\boldsymbol{\psi} : \Omega \to \mathbb{R}^3$ be injective on Ω, a condition that is part of the definition of a deformation; this assumption will be taken into account in Chapter 5), we remark that *the tangent space at a point* φ *of the manifold* $\boldsymbol{\Phi}$ is precisely

$$T_{\varphi}\boldsymbol{\Phi} := \{\boldsymbol{\theta} : \bar{\Omega} \to \mathbb{R}^3;\ \boldsymbol{\theta} = \boldsymbol{o} \text{ on } \Gamma_0\} .$$

It is thus as elements of this tangent space that the vector fields occurring in the principle of virtual work are to be correctly understood as *variations*; this observation is also the basis for attaching the label "variational" to the equations themselves. The adjective "virtual", borrowed from classical continuum mechanics, reflects the fact that the vector fields $\boldsymbol{\theta} \in T_{\varphi}\boldsymbol{\Phi}$ appearing in the principle are essentially mathematical quantities, which need not be given a physical interpretation.

Remarks. (1) A more transparent interpretation of these "variations" in the sense of the calculus of variations will be provided in Chapters 4 and 5, where the principle of virtual work will be understood as a requirement for a certain functional to be stationary.

(2) The introduction of a tangent space can prove quite useful in more complex situations where the set of admissible deformations include other geometrical constraints, such as incompressibility (Marsden & Hughes [1983, p. 279]).

(3) Another form of the principle of virtual work, involving "infinitesimal rigid displacements" is proposed in Exercise 2.2.

(4) The regularity assumptions on the applied force densities, on the

boundary of the body, etc., can be relaxed in various ways that still guarantee that the axioms of force and moment balance and the principle of virtual work make sense. In this direction, see notably Noll [1959, 1966, 1978], Gurtin & Williams [1967], Truesdell [1977], Antman & Osborn [1979]. ■

2.7. EXAMPLES OF APPLIED FORCES; CONSERVATIVE FORCES

The applied forces appear at two places in the equations of equilibrium in the reference configuration: The density field $\boldsymbol{f}:\Omega \to \mathbb{R}^3$ enters the equation

$$-\mathbf{div}\, \boldsymbol{T}(x) = \boldsymbol{f}(x), \quad x \in \Omega,$$

while the density field $\boldsymbol{g}:\Gamma_1 \to \mathbb{R}^3$ enters the boundary condition of traction

$$\boldsymbol{T}(x)\boldsymbol{n}(x) = \boldsymbol{g}(x), \quad x \in \Gamma_1.$$

We recall that these fields are respectively related to the densities $\boldsymbol{f}^{\varphi}:\Omega^{\varphi} \to \mathbb{R}^3$ and $\boldsymbol{g}:\Gamma_1^{\varphi} \to \mathbb{R}^3$ by $\boldsymbol{f}\, \mathrm{d}x = \boldsymbol{f}^{\varphi}\, \mathrm{d}x^{\varphi}$ and $\boldsymbol{g}\, \mathrm{d}a = \boldsymbol{g}^{\varphi}\, \mathrm{d}a^{\varphi}$.

An applied body force is a **dead load** if its associated density $\boldsymbol{f}:\Omega \to \mathbb{R}^3$ per unit volume in the reference configuration is independent of the particular deformation $\boldsymbol{\varphi}$ considered. This is the case of the *gravity field* (Sect. 2.1), for which

$$\boldsymbol{f}(x) = -g\rho(x)\boldsymbol{e}_3, \quad x \in \Omega.$$

Likewise, an applied surface force is a **dead load** if its associated density $\boldsymbol{g}$ per unit area in the reference configuration is independent of the particular deformation $\boldsymbol{\varphi}$ considered. A simple example corresponds to the case where $\boldsymbol{g}^{\varphi} = \boldsymbol{o}$ on Γ_1^{φ}, in which case $\boldsymbol{g} = \boldsymbol{o}$ on Γ_1: A portion of the boundary of the body is held fixed, while the remaining portion Γ_1^{φ} is considered as "free" from all external actions (the arc de triomphe, a day with a gentle breeze).

Remark. Some authors call *live loads* applied forces that are not dead loads. ■

The assumption that the applied forces are dead loads is a simplification from the *mathematical* standpoint, since the right-hand sides of the resulting boundary value problem in the reference configuration become *known* functions of $x \in \Omega$ in this case. On the other hand, it should be kept in mind that, with the exceptions of the two special cases mentioned above, *actual applied forces can seldom be modeled as dead loads*; *the densities* f, *or* g, *appear instead not only as functions of* $x \in \Omega$, *or* $x \in \Gamma_1$, *but also of the deformation* φ *itself.*

Let us consider an example: An applied surface force is a **pressure load** if the density g^φ in the *deformed* configuration is of the form

$$\boxed{g^\varphi(x^\varphi) = -\pi n^\varphi(x^\varphi), \quad x^\varphi \in \Gamma_1^\varphi,}$$

where π is a constant on Γ_1^φ, called a *pressure*; the minus sign insures that the vector g is directed inward for $\pi > 0$ (observe that this definition is consistent with that of the Cauchy stress tensor being also a "pressure"; cf. Sect. 2.3).

Except if $\pi = 0$, *in which case* $g = o$, *a pressure load is not a dead load.* This is intuitively clear: Think for example of a deflated *balloon* as a reference configuration and of the same balloon, but inflated, as the deformed configuration (Fig. 2.7-1); clearly the vectors $g^{id}(x^{id})$ and $g^\varphi(x^\varphi)$ have different directions in general. A more convincing argument consists in combining the relation between the area elements $\mathrm{d}a$ and $\mathrm{d}a^\varphi$ given in Theorem 1.7-1, the relation $g\,\mathrm{d}a = g^\varphi\,\mathrm{d}a^\varphi$, and the definition of a pressure load. In this fashion, we find that

$$\boxed{\begin{aligned} g(x) = -\pi(\mathbf{Cof}\,\nabla\varphi(x))n(x) = -\pi(\det \nabla\varphi(x))\nabla\varphi(x)^{-\mathrm{T}}n(x),& \\ x \in \Gamma_1 .& \end{aligned}}$$

The corresponding **boundary condition of pressure** thus takes the form

$$T(x)n(x) = g(x) := \hat{g}(x, \nabla\varphi(x)), \quad x \in \Gamma_1,$$

where the mapping $\hat{g} : \Gamma_1 \times \mathbb{M}_+^3 \to \mathbb{R}^3$ is given by

$$\hat{g}(x, F) = -\pi(\mathbf{Cof}\,F)n(x) = -\pi(\det F)F^{-\mathrm{T}}n(x), \quad x \in \Gamma_1, \quad F \in \mathbb{M}_3^+ .$$

As another example, consider the *centrifugal force* acting on a body

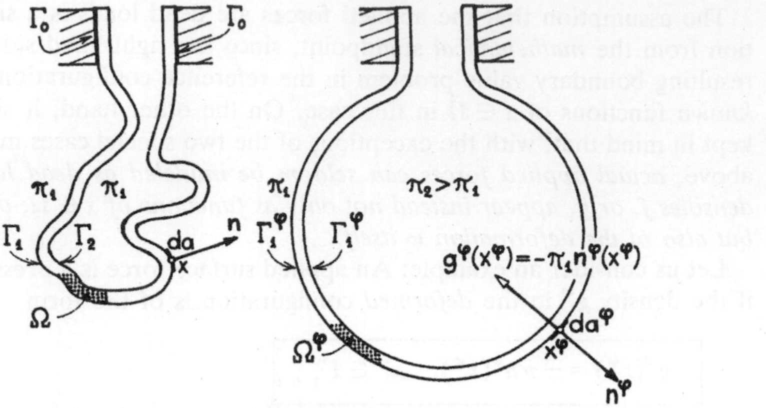

Fig. 2.7-1. A pressure load is an example of applied surface force that is not a dead load.

rotating with a constant angular velocity ω around the axis e_1 (Fig. 2.7-2), and *assume that the axes* e_2 *and* e_3 *also rotate around* e_1 *with the same angular velocity*. Then the density f^φ is given by

$$f^\varphi(x^\varphi) = \omega^2 \rho^\varphi(x^\varphi)(x_2^\varphi e_2 + x_3^\varphi e_3), \quad x^\varphi \in \Omega^\varphi ,$$

and thus

$$f(x) = \omega^2 \rho(x)(\varphi_2(x) e_2 + \varphi_3(x) e_3) .$$

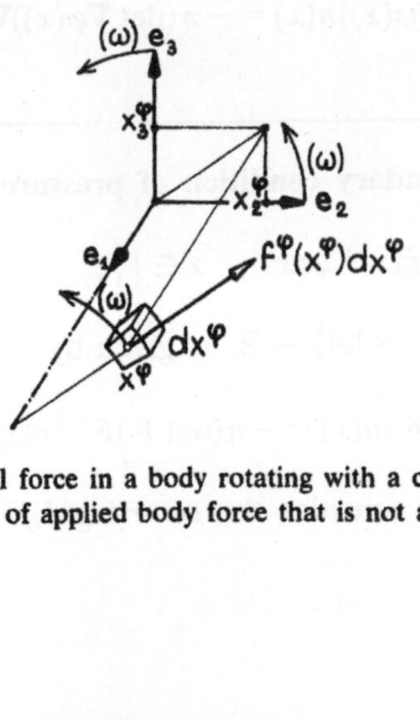

Fig. 2.7-2. The centrifugal force in a body rotating with a constant angular velocity is an instance of applied body force that is not a dead load.

Hence if there are no other body forces, the equations of equilibrium take the form

$$-\mathbf{div}\, \boldsymbol{T}(x) = \boldsymbol{f}(x) := \hat{\boldsymbol{f}}(x, \boldsymbol{\varphi}(x)), \quad x \in \Omega ,$$

where the mapping $\hat{\boldsymbol{f}} : \Omega \times \mathbb{R}^3 \to \mathbb{R}^3$ is given by

$$\hat{\boldsymbol{f}}(x, \boldsymbol{\eta}) = \omega^2 \rho(x)(\eta_2 \boldsymbol{e}_2 + \eta_3 \boldsymbol{e}_3), \quad x \in \Omega,\ \boldsymbol{\eta} \in \mathbb{R}^3 .$$

Motivated by the above examples, *we shall consistently assume in the sequel that the applied forces are either dead loads, or that their densities in the reference configuration are of the form*

$$\boxed{\boldsymbol{f}(x) = \hat{\boldsymbol{f}}(x, \boldsymbol{\varphi}(x)),\ x \in \Omega, \text{ and } \boldsymbol{g}(x) = \hat{\boldsymbol{g}}(x, \boldsymbol{\nabla}\boldsymbol{\varphi}(x)),\ x \in \Gamma_1 ,}$$

for given mappings $\hat{\boldsymbol{f}} : \Omega \times \mathbb{R}^3 \to \mathbb{R}^3$ *and* $\hat{\boldsymbol{g}} : \Gamma_1 \times \mathbb{M}^3_+ \to \mathbb{R}^3$.

These specific forms are essentially convenient ways of fixing ideas while being reasonably general, but they are not meant to cover all possible cases. For instance, the *balloon problem* (Exercise 2.5) gives rise to an applied surface force that is *nonlocal*, in that its value at a point depends on the values of the deformations at other points; taking into account mutual gravitational forces within the deformed configuration likewise results in a nonlocal applied body force density (Exercise 2.6), etc.

Remark. Applied forces described by more general densities are discussed in Noll [1978], Podio-Guidugli & Vergara-Caffarelli [1984], Spector [1980, 1982], Podio-Guidugli [1986b]. ■

Let us conclude by two important definitions: An applied body force with density $\boldsymbol{f} : \Omega \to \mathbb{R}^3$ in the reference configuration is **conservative** if the integral

$$\int_\Omega \boldsymbol{f}(x) \cdot \boldsymbol{\theta}(x)\, \mathrm{d}x = \int_\Omega \hat{\boldsymbol{f}}(x, \boldsymbol{\varphi}(x)) \cdot \boldsymbol{\theta}(x)\, \mathrm{d}x$$

that appears in the principle of virtual work in the reference configuration (Theorems 2.6-1 or 2.6-2) can also be written as the *Gâteaux derivative*

$$\boxed{F'(\varphi)\theta = \int_\Omega \hat{f}(x, \varphi(x)) \cdot \theta(x)\,dx}$$

of a functional F of the form

$$\boxed{F : \{\psi : \bar{\Omega} \to \mathbb{R}^3\} \to F(\psi) = \int_\Omega \hat{F}(x, \psi(x))\,dx\,.}$$

If this is the case the function $\hat{F} : \Omega \times \mathbb{R}^3 \to \mathbb{R}$ is called the *potential of the applied body force*. Clearly, *an applied body force that is a dead load is conservative, with*

$$\hat{F}(x, \eta) = f(x) \cdot \eta \quad \text{for all } x \in \Omega \text{ and } \eta \in \mathbb{R}^3\,.$$

More generally, a density of the applied body force of the form

$$f(x) = \hat{f}(x, \varphi(x)) \quad \text{for all } x \in \Omega\,,$$

where $\hat{f} : \Omega \times \mathbb{R}^3 \to \mathbb{R}^3$ is a given mapping, is conservative if

$$\hat{f}(x, \eta) = \mathbf{grad}_\eta\, \hat{F}(x, \eta) \quad \text{for all } x \in \Omega,\ \eta \in \mathbb{R}^3\,.$$

For example, *the centrifugal force described in Fig. 2.7-2 is conservative* (Exercise 2.7).

Similarly, an applied surface force with density $g : \Gamma_1 \to \mathbb{R}^3$ in the reference configuration is **conservative** if the integral

$$\int_{\Gamma_1} g(x) \cdot \theta(x)\,dx = \int_{\Gamma_1} \hat{g}(x, \nabla\varphi(x)) \cdot \theta(x)\,dx$$

that appears in the principle of virtual work in the reference configuration can also be written as the Gâteaux derivative

$$\boxed{G'(\varphi)\theta = \int_{\Gamma_1} \hat{g}(x, \nabla\varphi(x)) \cdot \theta(x)\,dx}$$

of a functional of the form

$$G:\{\psi:\bar{\Omega}\to\mathbb{R}^3\}\to G(\psi)=\int_{\Gamma_1}\hat{G}(x,\psi(x),\nabla\psi(x))\,\mathrm{d}a\,.$$

If this is the case, the function $\hat{G}:\Gamma_1\times\mathbb{R}^3\times\mathbb{M}^3_+$ is called the *potential of the applied surface force*. Clearly, *an applied surface force that is a dead load is conservative, with*

$$G(x,\eta,F)=\hat{g}(x)\cdot\eta\quad\text{for all } x\in\Gamma_1,\ \eta\in\mathbb{R}^3,\ F\in\mathbb{M}^3_+\,.$$

Let us consider another example.

Theorem 2.7-1. *Let $\pi\in\mathbb{R}$ be given, and let the functional G be defined by*

$$G(\psi)=-\frac{\pi}{3}\int_{\Gamma}\{(\mathbf{Cof}\,\nabla\psi)n\}\cdot\psi\,\mathrm{d}a\,,$$

for all smooth enough mappings $\psi:\bar{\Omega}\to\mathbb{R}^3$. Then one also has

$$G(\psi)=-\pi\int_{\Omega}\det\nabla\psi\,\mathrm{d}x\,,$$

and the Gâteaux derivative of the functional G is given by

$$G'(\varphi)\theta=-\pi\int_{\Gamma}\{(\mathbf{Cof}\,\nabla\varphi)n\}\cdot\theta\,\mathrm{d}a\,.$$

Hence a pressure load, corresponding to the boundary condition

$$T(x)n(x)=-\pi(\mathbf{Cof}\,\nabla\varphi(x))n(x),\quad x\in\Gamma_1\,,$$

is a conservative applied surface force.

Proof. Let $\pi=-1$. We first show that one also has $G(\psi)=\int_{\Omega}\det\nabla\psi\,\mathrm{d}x$. Using the expressions of $\det\nabla\psi$ and $\mathbf{Cof}\,\nabla\psi$ in terms of the orientation tensor (ε_{ijk}) (Sect. 1.1), the Piola identity (cf. the proof of Theorem 1.7-1) and the fundamental Green formula, we obtain:

$$\int_\Omega \det \nabla\psi \, dx = \tfrac{1}{6}\int_\Omega \varepsilon_{ijk}\varepsilon_{pqr}\partial_p\psi_i\partial_q\psi_j\partial_r\psi_k \, dx$$

$$= \tfrac{1}{6}\int_\Omega \partial_p\{\varepsilon_{ijk}\varepsilon_{pqr}\psi_i\partial_q\psi_j\partial_r\psi_k\} \, dx$$

$$= \tfrac{1}{6}\int_\Gamma (\varepsilon_{ijk}\varepsilon_{pqr}\partial_q\psi_j\partial_r\psi_k)n_p\psi_i \, da$$

$$= \tfrac{1}{3}\int_\Gamma \{(\mathbf{Cof}\,\nabla\psi)n\}\cdot\psi \, da\,.$$

We next compute the derivative of the functional

$$G(\psi) = \int_\Omega \det \nabla\psi \, dx = \int_\Omega \iota_3(\nabla\psi) \, dx\,,$$

with $\iota_3(F) = \det F$. In Sect. 1.2, we have shown that

$$\iota_3'(F)G = \mathbf{Cof}\,F : G\,,$$

for arbitrary matrices F and G. Hence

$$G'(\varphi)\theta = \int_\Omega \iota_3'(\nabla\varphi) : \nabla\theta \, dx = \int_\Omega \mathbf{Cof}\,\nabla\varphi : \nabla\theta \, dx\,.$$

Combining the Green formula

$$\int_\Omega H : \nabla\theta \, dx = -\int_\Omega \mathbf{div}\,H\cdot\theta \, dx + \int_\Gamma Hn\cdot\theta \, da$$

with another application of the Piola identity, we find that

$$G'(\varphi)\theta = \int_\Gamma \{(\mathbf{Cof}\,\nabla\varphi)n\}\cdot\theta \, da\,,$$

and the proof is complete. ■

Remarks. (1) There may be several subsets of the boundary Γ, on each of which the applied surface force is a pressure load, as in Fig. 2.7-1; for details, see Exercise 2.8.

(2) A general presentation of conservative applied surface forces is given in Podio-Guidugli [1987b]. ■

The interest of considering applied forces that are conservative will be explained in Chapter 4, where the remaining integral $\int_\Omega \boldsymbol{T}(x):\boldsymbol{\nabla\theta}(x)\,\mathrm{d}x$ in the principle of virtual work will be also written as the Gâteaux derivative of a functional, under the assumption of hyperelasticity.

EXERCISES

2.1. Let there be given a system of applied forces $\boldsymbol{f}^\varphi:\Omega^\varphi\to\mathbb{R}^3$ and $\boldsymbol{t}^\varphi:\Gamma^\varphi\to\mathbb{R}^3$ that does not necessarily satisfy the axiom of moment balance. Show that there exists an orthogonal matrix $\boldsymbol{Q}$ such that

$$\int_{\Omega^\varphi} \boldsymbol{ox}^\varphi \wedge \boldsymbol{Q f}^\varphi(x^\varphi)\,\mathrm{d}x^\varphi + \int_{\Gamma^\varphi} \boldsymbol{ox}^\varphi \wedge \boldsymbol{Q g}^\varphi(x^\varphi)\,\mathrm{d}a^\varphi = \boldsymbol{o}\,,$$

$$\int_{\Omega^\varphi} \boldsymbol{Q}^{\mathrm{T}}\boldsymbol{ox}^\varphi \wedge \boldsymbol{f}^\varphi(x^\varphi)\,\mathrm{d}x^\varphi + \int_{\Gamma^\varphi} \boldsymbol{Q}^{\mathrm{T}}\boldsymbol{ox}^\varphi \wedge \boldsymbol{g}^\varphi(x^\varphi)\,\mathrm{d}a^\varphi = \boldsymbol{o}\,.$$

This result, known as *Da Silva's theorem*, is used in the analysis of the *pure traction problem* (Sect. 5.1); cf. the discussion in Truesdell & Noll [1965, p. 128] and Marsden & Hughes [1983, p. 466].

2.2. (1) Let $\boldsymbol{f}:\Omega\to\mathbb{R}^3$ and $\boldsymbol{t}:\bar{\Omega}\times S_1\to\mathbb{R}^3$ denote respectively the applied body force density in the reference configuration and the first Piola–Kirchhoff stress vector. Show that

$$\int_A \boldsymbol{f}(x)\,\mathrm{d}x + \int_{\partial A} \boldsymbol{t}(x,\boldsymbol{n})\,\mathrm{d}a = \boldsymbol{o}\,,$$

$$\int_A \boldsymbol{\varphi}(x)\wedge\boldsymbol{f}(x)\,\mathrm{d}x + \int_{\partial A} \boldsymbol{\varphi}(x)\wedge\boldsymbol{t}(x,\boldsymbol{n})\,\mathrm{d}a = \boldsymbol{o}\,,$$

for all subdomains $A\subset\bar{\Omega}$. These relations constitute the *axioms of force and moment balance in the reference configuration*.

(2) Show that these relations imply that the analog of Cauchy's theorem, expressed now in terms of the first Piola–Kirchhoff stress tensor, holds on the reference configuration.

Remark. This approach, which consists in developing the theory entirely on the reference configuration, is notably advocated by Antman [1984].

(3) Show that the axioms of force and moment balance are satisfied if

and only if

$$\int_A f(x) \cdot v(x) \, dx + \int_{\partial A} t(x, n) \cdot v(x) \, dx = 0$$

for all subdomains $A \subset \bar{\Omega}$ and all vector fields $v : \bar{\Omega} \to \mathbb{R}^3$ of the form

$$v(x) = a + b \wedge ox \quad \text{for all } x \in \bar{\Omega} ,$$

with $a \in \mathbb{R}^3$, $b \in \mathbb{R}^3$. Such vector fields are called *infinitesimal rigid displacements*, following a terminology that will be explained in Sect. 6.3. This necessary and sufficient condition is called the *principle of virtual work* by Gurtin [1981b, p. 100].

2.3. Is the first Piola–Kirchhoff stress vector field the Piola transform, according to the definition given in Exercise 1.12, of the Cauchy stress vector field?

2.4. Some authors (e.g. Washizu [1975, p. 64]) write the principle of virtual work in the reference configuration as:

$$\int_\Omega \boldsymbol{\Sigma} : \delta \boldsymbol{E} \, dx = \int_\Omega \boldsymbol{f} \cdot \delta \boldsymbol{u} \, dx + \int_{\Gamma_1} \boldsymbol{g} \cdot \delta \boldsymbol{u} \, da$$

for all "variations" δu. Justify these expressions.

2.5. Following Noll [1978], consider the *balloon problem*, where the exterior boundary of a balloon is subjected to a constant pressure load while the interior boundary is subjected to a pressure that is a given function of the enclosed volume. Write down the corresponding boundary conditions, in both the deformed and reference configurations.

2.6. Give the expression of the densities, in both the deformed and the reference configurations, of the following applied body forces:

(1) The gravitational field, when the curvature of the earth is taken into account;

(2) Mutual gravitational forces within the deformed configuration;

(3) Mutual and exterior electrostatic forces.

2.7. Show that the centrifugal force acting on a body rotating with a constant angular velocity around a fixed axis (Fig. 2.7-2) is conservative.

2.8. This exercise is a complement to Theorem 2.7-1. Let $r \geq 2$, let $\Gamma = \bigcup_{\rho=0}^{r} \Gamma_\rho$ with $\Gamma_\rho \cap \Gamma_{\rho'} = \emptyset$ if $\rho \neq \rho'$, let π_ρ, $1 \leq \rho \leq r$, be given constants, and assume that there exists a smooth enough function $\pi : \bar{\Omega} \to \mathbb{R}$ such that $\pi | \bar{\Gamma}_\rho = \pi_\rho$, $1 \leq \rho \leq r$, so that in particular $\pi_\rho \neq \pi_{\rho'}$ implies $\bar{\Gamma}_\rho \cap \bar{\Gamma}_{\rho'} = \emptyset$ if $1 \leq \rho < \rho' \leq r$. Compute the Gâteaux derivative of the functional

$$Q(\boldsymbol{\psi}) = \pi \int_\Omega \det \boldsymbol{\nabla}\boldsymbol{\psi} \, dx + \tfrac{1}{3} \int_\Gamma \det \boldsymbol{\nabla}\boldsymbol{\psi} (\boldsymbol{\nabla}\boldsymbol{\psi}^{-T} \, \mathbf{grad} \, \pi) \cdot \boldsymbol{\psi} \, da ,$$

and conclude that the applied surface force consisting of the simultaneous pressure loads

$$\boldsymbol{T}(x)\boldsymbol{n}(x) = -\pi_\rho \, \mathbf{Cof} \, \boldsymbol{\nabla}\boldsymbol{\varphi}(x)\boldsymbol{n}(x), \quad x \in \Gamma_\rho, \ 1 \leq \rho \leq r ,$$

is conservative. For further considerations, see Ball [1977], Beatty [1970], Bufler [1984], Romano [1972], Pearson [1956], Sewell [1967].

CHAPTER 3

ELASTIC MATERIALS AND THEIR CONSTITUTIVE EQUATIONS

INTRODUCTION

The *three* equations of equilibrium over the reference configuration, which are valid regardless of the macroscopic continuum (gaz, liquid, solid) that they are supposed to model, form an undetermined system since there are *nine* unknown functions, namely the three components of the deformation and the six components of the first Piola–Kirchhoff stress tensor (taking into account the symmetry of the Cauchy stress tensor). The six missing equations are provided by assumptions regarding the nature of the constituting material that is considered.

In particular, we shall consider in this book materials that behave according to the following definition (Sect. 3.1): A material is *elastic* if at each point $x^{\varphi} = \varphi(x)$ of the deformed configuration, the Cauchy stress tensor $T^{\varphi}(x^{\varphi})$ is solely a function of x and of the deformation gradient $\nabla\varphi(x)$. Equivalently, since $T(x) = T^{\varphi}(x^{\varphi}) \operatorname{Cof} \nabla\varphi(x)$ and $\Sigma(x) = \nabla\varphi(x)^{-1}T(x)$, a material is elastic if each Piola–Kirchhoff stress tensor is expressed in terms of x and $\nabla\varphi(x)$ through a *constitutive equation* of the form

$$T(x) = \hat{T}(x, \nabla\varphi(x)), \text{ or } \Sigma(x) = \hat{\Sigma}(x, \nabla\varphi(x)), \quad \text{for all } x \in \bar{\Omega},$$

where the *response functions* $\hat{T}: \bar{\Omega} \times \mathbb{M}^3_+ \to \mathbb{M}^3$ and $\hat{\Sigma}: \bar{\Omega} \times \mathbb{M}^3_+ \to \mathbb{S}^3$ characterize the elastic material.

We first show (Sect. 3.3) that the consideration of the general *axiom of material frame-indifference* implies that at each point $x \in \bar{\Omega}$, the response function $\hat{\Sigma}(x, \cdot)$ is only a function of the strain tensor $C = F^{\mathrm{T}}F$, in the sense that there exists a mapping $\tilde{\Sigma}(x, \cdot)$ such that

$$\hat{\Sigma}(x, F) = \tilde{\Sigma}(x, F^{\mathrm{T}}F) \quad \text{for all } F \in \mathbb{M}^3_+ .$$

Combining next the *property of isotropy* with the *Rivlin–Ericksen representation theorem* (Theorem 3.6-1), we show (Sect. 3.6) that the form of the response function $\hat{\Sigma}$ can be further simplified: If a material is isotropic at a point $x \in \bar{\Omega}$, the mapping $\tilde{\boldsymbol{\Sigma}}(x, \cdot)$ reduces to:

$$\tilde{\boldsymbol{\Sigma}}(x, \boldsymbol{C}) = \gamma_0(x, \iota_C)\boldsymbol{I} + \gamma_1(x, \iota_C)\boldsymbol{C} + \gamma_2(x, \iota_C)\boldsymbol{C}^2 ,$$

where $\gamma_0(x, \cdot)$, $\gamma_1(x, \cdot)$, $\gamma_2(x, \cdot)$ are real-valued functions of the principal invariants of the strain tensor $\boldsymbol{C}$.

If we assume that the material is *homogeneous*, i.e., that its response function is independent of $x \in \bar{\Omega}$, and that *the reference configuration is a natural state*, i.e., that $\tilde{\boldsymbol{\Sigma}}(\boldsymbol{0}) = \boldsymbol{0}$ (both assumptions are realistic for common elastic materials, such as steel, iron, aluminum), we arrive at a striking result (cf. Theorems 3.7-1 and 3.8-1): Near the reference configuration (for which $\boldsymbol{C} = \boldsymbol{I}$), the response function $\tilde{\boldsymbol{\Sigma}}$ satisfies

$$\tilde{\boldsymbol{\Sigma}}(\boldsymbol{C}) = \lambda(\operatorname{tr} \boldsymbol{E})\boldsymbol{I} + 2\mu\boldsymbol{E} + o(\|\boldsymbol{E}\|), \quad \boldsymbol{C} = \boldsymbol{I} + 2\boldsymbol{E} ,$$

i.e., in this case there are only *two* arbitrary constants λ, μ in the first order term of the expansion of the response function with respect to the Green–St Venant strain tensor $\boldsymbol{E}$. We then describe (Sect. 3.8) how the two constants λ and μ, which are called the *Lamé constants* of the material, are determined experimentally, together with two other elastic constants, the *Young modulus* and the *Poisson ratio*.

Finally, we discuss (Sect. 3.9) *St Venant–Kirchhoff materials*, which obey the simplest constitutive equation compatible with the various requirements set so far, namely,

$$\boldsymbol{\Sigma} = \tilde{\boldsymbol{\Sigma}}(\boldsymbol{C}) = \lambda(\operatorname{tr} \boldsymbol{E})\boldsymbol{I} + 2\mu\boldsymbol{E}, \quad \boldsymbol{C} = \boldsymbol{I} + 2\boldsymbol{E} .$$

Although such simple models suffer otherwise from various drawbacks, they are commonly used in the numerical computation of equilibriums of nonlinearly elastic structures near their reference configuration.

3.1. ELASTIC MATERIALS

If we consider the equations of equilibrium in the reference configuration (written in terms of either Piola–Kirchhoff stress tensor) as part of a boundary value problem whose unknowns are the six components of the stress tensor (taking either equation $\boldsymbol{\nabla}\varphi \boldsymbol{T}^{\mathrm{T}} = \boldsymbol{T}\boldsymbol{\nabla}\varphi^{\mathrm{T}}$ or $\boldsymbol{\Sigma} = \boldsymbol{\Sigma}^{\mathrm{T}}$ into account) and the three components of the deformation, it is clear that there is a discrepancy between the total number of unknown functions (nine) and the number of available equations (three). Hence *six equations must be supplied.*

That the mathematical model developed so far is incomplete is also evident on *physical* grounds. *While the equations of equilibrium are valid regardless of the particular material the body in consideration is made of* (it may be a solid, a liquid, a gas) *it is clear that the nature of the underlying material should be taken into account*: For given applied forces (e.g., dead loads), the resulting deformations from a given reference configuration surely differ if the body is made of lead or of steel. In the same spirit, in order to produce the same deformation in a body made of ply-wood in one case and in a body made of iron in another case, which both occupy the same reference configuration, it is clear that different systems of forces must be applied, and that different stress tensor fields must arise.

In this book, we shall consider exclusively one category of materials, for which the required additional equations can be supplied in a very simple form, according to the *assumption* that *the Cauchy stress tensor* $T^{\varphi}(x^{\varphi})$ *at any point* $x^{\varphi} = \varphi(x) \in \bar{\Omega}^{\varphi}$ *is completely determined by the deformation gradient* $\nabla\varphi(x)$ *at the corresponding point* $x \in \bar{\Omega}$. Let us express this assumption as a mathematical definition.

A material is **elastic** if there exists a mapping

$$\hat{T}^{D} : (x, F) \in \bar{\Omega} \times \mathbb{M}^3_+ \rightarrow \hat{T}^{D}(x, F) \in \mathbb{S}^3 ,$$

called the **response function for the Cauchy stress**, such that in *any* deformed configuration that a body made of this material occupies, the Cauchy stress tensor $T^{\varphi}(x^{\varphi})$ at any point $x^{\varphi} = \varphi(x)$ of the deformed configuration is related to the deformation gradient $\nabla\varphi(x)$ at the corresponding point x of the reference configuration by the equation

$$\boxed{T^{\varphi}(x^{\varphi}) = \hat{T}^{D}(x, \nabla\varphi(x)), \quad x^{\varphi} = \varphi(x) .}$$

This relation is called the **constitutive equation** of the material. We recall that $\mathbb{M}^3_+$ denotes the set of all matrices of order 3 whose determinant is >0 (the determinant of a deformation gradient is >0 by definition) and that $\mathbb{S}^3$ denotes the set of all symmetric matrices of order 3 (the Cauchy stress tensor is always symmetric; cf. Theorem 2.3-1).

By definition, the response function at each point of an elastic material must be defined *for all matrices* $F \in \mathbb{M}^3_+$. Implicit in the definition is thus the property that, *given any point* $x \in \bar{\Omega}$ *and any matrix* $\boldsymbol{F} \in \mathbb{M}^3_+$, *there exists a deformation* $\boldsymbol{\varphi}$ *of the body that satisfies* $\boldsymbol{\nabla\varphi}(x) = \boldsymbol{F}$ (as the result of the application of appropriate applied forces and boundary conditions, which are left unspecified). Therefore *this definition rules out materials subjected to internal constraints*, i.e., materials that can only undergo a restricted class of deformations (Sect. 5.7). This is notably the case of the *incompressible materials*, defined in Sect. 5.7.

Observe that, by virtue of the relations $\boldsymbol{T} = (\det \boldsymbol{\nabla\varphi})\boldsymbol{T}^{\varphi}\boldsymbol{\nabla\varphi}^{-\mathrm{T}}$ and $\boldsymbol{\Sigma} = \boldsymbol{\nabla\varphi}^{-1}\boldsymbol{T}$ relating the first and second Piola–Kirchhoff stress tensors to the Cauchy stress tensor, there exist mappings

$$\hat{\boldsymbol{T}} : \bar{\Omega} \times \mathbb{M}^3_+ \to \mathbb{M}^3 \text{ and } \hat{\boldsymbol{\Sigma}} : \bar{\Omega} \times \mathbb{M}^3_+ \to \mathbb{S}^3 ,$$

given respectively by

$$\hat{\boldsymbol{T}}(x, \boldsymbol{F}) = (\det \boldsymbol{F})\hat{\boldsymbol{T}}^{\mathrm{D}}(x, \boldsymbol{F})\boldsymbol{F}^{-\mathrm{T}}, \text{ and}$$
$$\hat{\boldsymbol{\Sigma}}(x, \boldsymbol{F}) = (\det \boldsymbol{F})\boldsymbol{F}^{-1}\hat{\boldsymbol{T}}^{\mathrm{D}}(x, \boldsymbol{F})\boldsymbol{F}^{-\mathrm{T}} \quad \text{for all } x \in \bar{\Omega},\ \boldsymbol{F} \in \mathbb{M}^3_+,$$

such that

$$\boxed{\boldsymbol{T}(x) = \hat{\boldsymbol{T}}(x, \boldsymbol{\nabla\varphi}(x)), \text{ and } \boldsymbol{\Sigma}(x) = \hat{\boldsymbol{\Sigma}}(x, \boldsymbol{\nabla\varphi}(x)), \quad \text{for all } x \in \bar{\Omega} .}$$

Since such relations can be taken as equivalent definitions of elastic materials, they are also called **constitutive equations**, and the mappings $\hat{\boldsymbol{T}}$ and $\hat{\boldsymbol{\Sigma}}$ are called **response functions for the first and second Piola–Kirchhoff stress.**

A material in a reference configuration Ω is called **homogeneous** if its response function is independent of the particular point $x \in \bar{\Omega}$ considered; otherwise the material is said to be **nonhomogeneous**. Thus the constitutive equation of a homogeneous elastic material takes the simpler form

$$\boldsymbol{T}^{\varphi}(x^{\varphi}) = \hat{\boldsymbol{T}}^{\mathrm{D}}(\boldsymbol{\nabla\varphi}(x)) \quad \text{for all } x^{\varphi} = \boldsymbol{\varphi}(x) \in \bar{\Omega}^{\varphi} .$$

For a given material, homogeneity is a property satisfied *in a given reference configuration*, which may no longer hold if a "deformed" state is chosen as the reference configuration.

Notice that *the response function* $\hat{\boldsymbol{T}}^{\mathrm{D}}$ *of an elastic material is by definition independent of the particular deformation considered.* This is why the symbol "$\boldsymbol{\varphi}$" does not appear in the notation used for the response function. The superscript "D" reminds us that this function is used for computing a quantity in a *deformed* configuration.

Notice also that, *by definition, the Cauchy stress tensor* $T^\varphi(x^\varphi)$ *at a point* $x^\varphi = \varphi(x)$ *of an elastic material depends on the deformation solely through its deformation gradient, i.e., through its first order partial derivatives* $\partial_j\varphi_i(x)$. It is clear on the one hand that the tensor $T^\varphi(x^\varphi)$ should not be a function of the values $\varphi_i(x)$ themselves, for otherwise the Cauchy stress tensor field would vary if the deformed configuration were rigidly translated. On the other hand, experimental evidence suggests that the tensor $T^\varphi(x^\varphi)$ at a point $x^\varphi = \varphi(x)$ may also depend on the deformation gradients $\nabla\varphi(y)$ evaluated at *all* other points $y \in \bar{\Omega}$, this dependence being however a very rapidly decreasing function of $|y - x|$. This observation leads to the theory of *nonlocal elasticity*, as advocated by Eringen [1966], who also takes into account the *history* of the material; see also Edelen [1969a, 1969b, 1970], Eringen & Edelen [1972], Eringen [1978]. In the same spirit, it would be equally conceivable that the tensor $T^\varphi(x^\varphi)$ be also dependent on *higher-order derivatives* of the deformation φ at the point x. For instance, a theory of *elastic materials of second grade* can be developed, where $T^\varphi(x^\varphi)$ is a function of x, $\nabla\varphi(x)$, and of all second partial derivatives $\partial_{ij}\varphi_k(x)$; see Murdoch [1979], Triantafyllidis & Aifantis [1986], and the earlier contributions of Toupin [1962, 1964], Mindlin [1964, 1965], Green & Rivlin [1964].

It can also be noted that the response function, which is essentially determined *by experiment* in a necessarily narrow range of possible "values" of the variable F (this adjustment is discussed in Sect. 3.8), is then used mathematically as if it were valid "for all F", while for "large" deformations, plasticity, fracture, and various other phenomena occur that are not taken into account by the definition of elastic materials.

These brief comments already indicate why, by contrast with the axioms of force and moment balance, which are universally agreed upon in macroscopic physics, the definition of an elastic material is questionable on many grounds. Its use has nevertheless led to so many achievements in the analysis of structures, and its mathematical analysis has led to so many challenging problems (some of them yet unsolved as we shall see in this book), that the theory of elasticity, as imperfect a model as it may be, stands as one of the major achievements of continuum mechanics.

Notice that the response function of an elastic material is *a priori* dependent of the particular *orthonormal basis* chosen, *and* on the particular *reference configuration* considered since any deformed configuration can be chosen as a new reference configuration. The study of these dependences, which are respectively governed by the axiom of material frame-indifference and by the properties of isotropy of the material, is the object of Sects. 3.3 and 3.4. As a preparation, we need a brief, but important, incursion in matrix theory.

*3.2. THE POLAR FACTORIZATION AND THE SINGULAR VALUES OF A MATRIX

All definitions and results of this section are stated and proved for real matrices (the only ones found in this book), but they can be extended to complex matrices. We begin by a preparatory result, which is also important *per se*.

Theorem 3.2-1. *Let* $\boldsymbol{K}$ *be a symmetric positive definite matrix. Then there exists one, and only one, symmetric positive definite matrix* $\boldsymbol{H}$ *such that* $\boldsymbol{H}^2 = \boldsymbol{K}$.

Proof. (i) We first show that, *if* $\boldsymbol{H}$ *is a symmetric positive definite matrix, then any eigenvector of the matrix* $\boldsymbol{H}^2$, *associated with an eigenvalue* μ, *is also an eigenvector of the matrix* $\boldsymbol{H}$, *associated with the eigenvalue* $\sqrt{\mu}$. *In other words,*

$$\boldsymbol{H}^2\boldsymbol{v} = \mu\boldsymbol{v}, \quad \boldsymbol{v} \neq \boldsymbol{o} \Rightarrow \boldsymbol{H}\boldsymbol{v} = \sqrt{\mu}\boldsymbol{v} .$$

Observe first that the matrix $\boldsymbol{H}^2$ is also symmetric and positive definite. Then $\boldsymbol{H}^2\boldsymbol{v} = \mu\boldsymbol{v}$ implies

$$(\boldsymbol{H} + \sqrt{\mu}\boldsymbol{I})(\boldsymbol{H} - \sqrt{\mu}\boldsymbol{I})\boldsymbol{v} = \boldsymbol{o} ,$$

and we must have

$$\boldsymbol{w} := (\boldsymbol{H} - \sqrt{\mu}\boldsymbol{I})\boldsymbol{v} = \boldsymbol{o} ,$$

for otherwise $\boldsymbol{w}$ would be an eigenvector of the matrix $\boldsymbol{H}$ that would correspond to the eigenvalue $-\sqrt{\mu} < 0$ (this particularly short proof is due to Stephenson [1980]).

(ii) Let now $\boldsymbol{K}$ be a symmetric positive definite matrix. Then the *existence* of a symmetric positive definite matrix $\boldsymbol{H}$ satisfying $\boldsymbol{H}^2 = \boldsymbol{K}$ is clear: Let $\boldsymbol{P}$ be an orthogonal matrix that diagonalizes the matrix $\boldsymbol{K}$, i.e.,

$$\boldsymbol{K} = \boldsymbol{P}^{\mathrm{T}}\boldsymbol{D}\boldsymbol{P}, \quad \text{with } \boldsymbol{D} = \mathbf{Diag}\ \mu_i,\ \mu_i > 0\,.$$

Then the matrix

$$\boldsymbol{H} = \boldsymbol{P}^{\mathrm{T}}\mathbf{Diag}\ \sqrt{\mu_i}\,\boldsymbol{P}$$

is symmetric, positive definite, and it satisfies $\boldsymbol{H}^2 = \boldsymbol{K}$. *Uniqueness* is less immediate, since it relies on property (i): Let $\boldsymbol{H}_1$ and $\boldsymbol{H}_2$ be two symmetric, positive definite matrices that satisfy

$$\boldsymbol{K} = \boldsymbol{H}_1^2 = \boldsymbol{H}_2^2\,.$$

By (i),

$$\boldsymbol{K}\boldsymbol{v} = \mu\boldsymbol{v}, \quad \boldsymbol{v} \neq \boldsymbol{o} \Rightarrow \boldsymbol{H}_\alpha^2\boldsymbol{v} = \mu\boldsymbol{v} \Rightarrow \boldsymbol{H}_\alpha\boldsymbol{v} = \sqrt{\mu}\,\boldsymbol{v}, \ \alpha = 1, 2\,.$$

The matrices $\boldsymbol{H}_1$ and $\boldsymbol{H}_2$, which have the same eigenvectors and the same eigenvalues, are therefore equal. ∎

The matrix $\boldsymbol{H} \in \mathbb{S}^n_>$ is called the **square root** of the matrix $\boldsymbol{K} = \boldsymbol{H}^2 \in \mathbb{S}^n_>$. It is denoted

$$\boxed{\boldsymbol{H} = \boldsymbol{K}^{1/2}\,.}$$

The following result plays a crucial role in the characterization of material frame-indifference established in the next section (Theorem 3.3-1). It extends to matrices the factorization $z = |z|\,\mathrm{e}^{i\theta}$ of complex numbers (see Exercise 3.2 for another proof).

Theorem 3.2-2 (polar factorization of an invertible matrix). *A real invertible matrix* $\boldsymbol{F}$ *can be factored in a unique fashion as*

$$\boxed{\boldsymbol{F} = \boldsymbol{R}\boldsymbol{U}, \quad \text{or } \boldsymbol{F} = \boldsymbol{V}\boldsymbol{S}\,,}$$

where $\boldsymbol{R}, \boldsymbol{S}$ *are orthogonal matrices, and* $\boldsymbol{U}, \boldsymbol{V}$ *are symmetric positive definite matrices. One has*

$$\boxed{U=(F^{\mathrm{T}}F)^{1/2},\ V=(FF^{\mathrm{T}})^{1/2},\ R=S=FU^{-1}=V^{-1}F\,.}$$

Proof. (i) Assume that we have found an orthogonal matrix R and a symmetric positive definite matrix U such that

$$F = RU\,.$$

Then necessarily,

$$F^{\mathrm{T}}F = U^{\mathrm{T}}R^{\mathrm{T}}RU = U^2, \quad \text{and } R = FU^{-1}\,.$$

Consequently, it suffices to let U be the unique (by Theorem 3.2-1) symmetric, positive definite matrix that satisfies $U^2 = F^{\mathrm{T}}F$ (the symmetric matrix $F^{\mathrm{T}}F$ is positive definite if F is invertible) and then to let $R = FU^{-1}$, since the matrix FU^{-1} is automatically orthogonal:

$$(FU^{-1})^{\mathrm{T}}FU^{-1} = U^{-1}F^{\mathrm{T}}FU^{-1} = U^{-1}U^2U^{-1} = I\,.$$

We can similarly prove that the matrix F can be uniquely factorized as

$$F = VS\,,$$

where V is a symmetric positive definite matrix and S is orthogonal. It suffices to let

$$V = (FF^{\mathrm{T}})^{1/2} \quad \text{and } S = V^{-1}F\,.$$

(ii) It remains to show that $R = S$, or equivalently that

$$V = FUF^{-1}\,.$$

On the one hand, we have

$$(FUF^{-1})^2 = FU^2F^{-1} = FF^{\mathrm{T}} = V^2\,,$$

and on the other both matrices V and FUF^{-1} are symmetric, positive definite: To prove these assertions for the matrix FUF^{-1}, we first note that

$$U^2 = F^{\mathrm{T}}F \Rightarrow F^{\mathrm{T}}FU = UF^{\mathrm{T}}F \Rightarrow FUF^{-1} = F^{-\mathrm{T}}UF^{\mathrm{T}} = (FUF^{-1})^{\mathrm{T}}\,,$$

and secondly that, given any vector $\boldsymbol{w} \neq \boldsymbol{o}$,

$$\boldsymbol{w}^{\mathrm{T}}\boldsymbol{F}\boldsymbol{U}\boldsymbol{F}^{-1}\boldsymbol{w} = \boldsymbol{w}^{\mathrm{T}}\boldsymbol{F}\boldsymbol{U}^{-1}\boldsymbol{F}^{\mathrm{T}}\boldsymbol{w} = (\boldsymbol{F}^{\mathrm{T}}\boldsymbol{w})^{\mathrm{T}}\boldsymbol{U}^{-1}(\boldsymbol{F}^{\mathrm{T}}\boldsymbol{w}) > 0\,,$$

since $\boldsymbol{U}^2 = \boldsymbol{F}^{\mathrm{T}}\boldsymbol{F}$ implies $\boldsymbol{U}\boldsymbol{F}^{-1} = \boldsymbol{U}^{-1}\boldsymbol{F}^{\mathrm{T}}$, and since the matrix $\boldsymbol{U}^{-1}$ is positive definite. Consequently we conclude that $\boldsymbol{V} = \boldsymbol{F}\boldsymbol{U}\boldsymbol{F}^{-1}$ by using once more Theorem 3.2-1. ■

Notice that *when the polar factorization theorem is applied to a matrix* $\boldsymbol{F}$ *satisfying* $\det \boldsymbol{F} > 0$ (the gradient of a deformation in particular), *the orthogonal matrix* $\boldsymbol{R}$ *satisfies* $\det \boldsymbol{R} = 1$, *so that* $\boldsymbol{R}$ *is a rotation.*

Remark: The matrix $\boldsymbol{F}$ can be also uniquely written as the *Cartesian sum* $\boldsymbol{F} = \boldsymbol{I} + \boldsymbol{E} + \boldsymbol{W}$, where the matrix $\boldsymbol{E}$ is symmetric and the matrix $\boldsymbol{W}$ is skew-symmetric ($\boldsymbol{W} = -\boldsymbol{W}^{\mathrm{T}}$). The relations between the Cartesian sum and the polar factorization and their applications to deformations are studied in Martins, Oliveira & Podio-Guidugli [1987]. ■

The two matrices

$$\boxed{\boldsymbol{C} := \boldsymbol{F}^{\mathrm{T}}\boldsymbol{F} = \boldsymbol{U}^2 \quad \text{and } \boldsymbol{B} := \boldsymbol{F}\boldsymbol{F}^{\mathrm{T}} = \boldsymbol{V}^2}$$

that appeared in the above proof play a key role in the representation theorem for general response functions (Theorem 3.6-2). For the time being, we notice that the matrices $\boldsymbol{U}$ and $\boldsymbol{V}$ are orthogonally equivalent since

$$\boldsymbol{F} = \boldsymbol{R}\boldsymbol{U} = \boldsymbol{V}\boldsymbol{R} \Rightarrow \boldsymbol{V} = \boldsymbol{R}\boldsymbol{U}\boldsymbol{R}^{\mathrm{T}}\,;$$

therefore *the matrices* $\boldsymbol{B} = \boldsymbol{V}^2$ *and* $\boldsymbol{C} = \boldsymbol{U}^2$ *are likewise orthogonally equivalent*:

$$\boxed{\boldsymbol{B} = \boldsymbol{R}\boldsymbol{C}\boldsymbol{R}^{\mathrm{T}}\,.}$$

Hence they have the same characteristic polynomial and consequently, the same *principal invariants* (Sect. 3.5).

Let $\boldsymbol{F}$ be an arbitrary (i.e., not necessarily invertible) matrix of order n, and let $\lambda_i(\boldsymbol{F}^{\mathrm{T}}\boldsymbol{F})$, $1 \leq i \leq n$, denote the n eigenvalues (which are all ≥ 0) of the symmetric, positive semidefinite, matrix $\boldsymbol{F}^{\mathrm{T}}\boldsymbol{F}$. The n numbers

$$\boxed{v_i(F) := \{\lambda_i(F^{\mathrm{T}}F)\}^{1/2}, \quad 1 \leqslant i \leqslant n,}$$

are called the **singular values** of the matrix F. Using the next two theorems, we shall show later (Sect. 4.9) that they play an important role in the construction of a wide class of *polyconvex stored energy functions*.

Theorem 3.2-3 (singular value decomposition of a matrix). *Let F be an arbitrary real square matrix, with singular values $v_i(F)$, counted in an arbitrary order. Then there exist orthogonal matrices P and Q such that*

$$F = P\{\mathbf{Diag}\, v_i(F)\}Q^{\mathrm{T}}.$$

Proof. We first show that *a singular matrix $F \in \mathbb{M}^n$ can also be factored as $F = RU$, with $R \in \mathbb{O}^n$, $U \in \mathbb{S}^n_{\geqslant}$*, thus extending to arbitrary matrices the existence (but not the uniqueness) of the polar factorization established in Theorem 3.2-2 for invertible matrices.

Let (F_k) be a sequence of invertible matrices converging to F. By Theorem 3.2-2, each matrix F_k can be factored as $F_k = R_k U_k$, with $R_k \in \mathbb{O}^n$, $U_k \in \mathbb{S}^n_{>}$. Since the sequence (R_k) is bounded (the spectral norm of an orthogonal matrix is one), there exists a subsequence (R_l) that converges to an orthogonal matrix R. Hence the sequence $(U_l) = (R_l^{\mathrm{T}} F_l)$ converges and $U = \lim_{l\to\infty} U_l = R^{\mathrm{T}}F$ is a symmetric positive semidefinite matrix.

Given $F \in \mathbb{M}^n$, let $F = RU$ with $R \in \mathbb{O}^n$, $U \in \mathbb{S}^n_{\geqslant}$; then there exists $Q \in \mathbb{O}^n$ such that $U = Q\, \mathbf{Diag}\, \lambda_i(U) Q^{\mathrm{T}}$, and thus

$$F = P\, \mathbf{Diag}\, \lambda_i(U) Q^{\mathrm{T}}, \quad \text{with } P = RQ,\ Q \in \mathbb{O}^n.$$

Since $F^{\mathrm{T}}F = P\, \mathbf{Diag}\, \lambda_i^2(U) Q^{\mathrm{T}}$, and since $\lambda_i(U) \geqslant 0$, we deduce that $\lambda_i(U) = v_{\sigma(i)}(F)$ for some permutation $\sigma \in \mathfrak{S}_n$. In order to rearrange the numbers $\lambda_i(U)$ in the proper order, define the permutation matrix $P_\sigma := (\delta_{i\sigma(j)})$, which is also orthogonal. Then

$$P_\sigma\, \mathbf{Diag}\, \lambda_i(U) P_\sigma^{\mathrm{T}} = \mathbf{Diag}\, \lambda_{\sigma^{-1}(i)} = \mathbf{Diag}\, v_i(F),$$

and the proof is complete. ■

It follows from the singular value decomposition theorem that the matrices $F^{\mathrm{T}}F$ and FF^{T} are always orthogonally equivalent (this has been so far proved only for an invertible matrix $\boldsymbol{F}$), since $\boldsymbol{F} = \boldsymbol{P}\boldsymbol{D}\boldsymbol{Q}^{\mathrm{T}}$ with $\boldsymbol{D} := \mathbf{Diag}\, v_i(\boldsymbol{F})$ implies

$$\boldsymbol{F}^{\mathrm{T}}\boldsymbol{F} = \boldsymbol{Q}\boldsymbol{D}^2\boldsymbol{Q}^{\mathrm{T}}, \quad \text{and } \boldsymbol{F}\boldsymbol{F}^{\mathrm{T}} = \boldsymbol{P}\boldsymbol{D}^2\boldsymbol{P}^{\mathrm{T}} = (\boldsymbol{P}\boldsymbol{Q}^{\mathrm{T}})(\boldsymbol{F}^{\mathrm{T}}\boldsymbol{F})(\boldsymbol{P}\boldsymbol{Q}^{\mathrm{T}})^{\mathrm{T}} .$$

Further properties of the singular value decomposition may be found in Exercise 3.3.

The following result, which relates the trace of the product of two matrices to their singular values, will be the key to proving that certain functions of matrices are convex (Theorem 4.9-1). It was first proved by von Neumann [1937], then by Mirsky [1959] (Exercise 3.4) and again, but differently, by Mirsky [1975], which we follow here (yet another proof, using the Lagrange multipliers, is suggested in Exercise 3.5). Unexpectedly, finding a decent proof of this seemingly simple result turns out to be anything but trivial, and we cannot help recommending the reader to verify this assertion by devising his, or her, own proof.

Theorem 3.2-4. *Let $\boldsymbol{A}$ and $\boldsymbol{B}$ be two matrices of order n, with singular values $\alpha_i := v_i(\boldsymbol{A})$ and $\beta_i := v_i(\boldsymbol{B})$, ordered as*

$$\alpha_1 \geqslant \alpha_2 \geqslant \cdots \geqslant \alpha_n \geqslant 0 \text{ and } \beta_1 \geqslant \beta_2 \geqslant \cdots \geqslant \beta_n \geqslant 0 .$$

Then

$$|\mathrm{tr}\, \boldsymbol{A}\boldsymbol{B}| \leqslant \sum_{i=1}^{n} \alpha_i \beta_i .$$

Proof. (i) By Theorem 3.2-3, there exist matrices $\boldsymbol{P}, \boldsymbol{Q}, \boldsymbol{R}, \boldsymbol{S} \in \mathbb{O}^n$ such that

$$\boldsymbol{A} = \boldsymbol{P}\boldsymbol{D}_\alpha\boldsymbol{Q}^{\mathrm{T}} \text{ and } \boldsymbol{B} = \boldsymbol{R}\boldsymbol{D}_\beta\boldsymbol{S}^{\mathrm{T}}, \quad \text{with } \boldsymbol{D}_\alpha = \mathbf{Diag}\, \alpha_i,\ \boldsymbol{D}_\beta = \mathbf{Diag}\, \beta_i .$$

Let

$$\boldsymbol{M} = (m_{ij}) := \boldsymbol{P}^{\mathrm{T}}\boldsymbol{S} \in \mathbb{O}^n \text{ and } \boldsymbol{N} = (n_{ij}) := \boldsymbol{Q}^{\mathrm{T}}\boldsymbol{R} \in \mathbb{O}^n .$$

Then

$$\mathrm{tr}\, \boldsymbol{A}\boldsymbol{B} = \mathrm{tr}\, \boldsymbol{P}\boldsymbol{D}_\alpha\boldsymbol{Q}^{\mathrm{T}}\boldsymbol{R}\boldsymbol{D}_\beta\boldsymbol{S}^{\mathrm{T}} = \mathrm{tr}\, \boldsymbol{M}^{\mathrm{T}}\boldsymbol{D}_\alpha\boldsymbol{N}\boldsymbol{D}_\beta = \sum_{i,j=1}^{n} m_{ij} n_{ij} \alpha_i \beta_j ,$$

and thus

$$|\mathrm{tr}\, \boldsymbol{AB}| \leq \sum_{i,j=1}^{n} |m_{ij}n_{ij}|\alpha_i\beta_j \leq \tfrac{1}{2} \sum_{i,j=1}^{n} |m_{ij}|^2\alpha_i\beta_j + \tfrac{1}{2} \sum_{i,j=1}^{n} |n_{ij}|^2\alpha_i\beta_j \,.$$

(ii) Let

$$|m_{ij}|^2 = \mu_{ij},\ \zeta_n = \alpha_n \text{ and } \zeta_i = \alpha_i - \alpha_{i+1},\ 1 \leq i \leq n-1,$$

$$\eta_n = \beta_n \text{ and } \eta_j = \beta_j - \beta_{j+1},\ 1 \leq j \leq n-1\,.$$

Then

$$\begin{aligned}\sum_{i=1}^{n} \alpha_i\beta_i - \sum_{i,j=1}^{n} \mu_{ij}\alpha_i\beta_j &= \sum_{i,j=1}^{n} (\delta_{ij} - \mu_{ij}) \sum_{i\leq k\leq n} \zeta_k \sum_{j\leq l\leq n} \eta_l \\ &= \sum_{i,j=1}^{n} \zeta_i\eta_j \sum_{k=1}^{i}\sum_{l=1}^{j} (\delta_{kl} - \mu_{kl}) \geq \sum_{1\leq i\leq j\leq n} \zeta_i\eta_j \sum_{k=1}^{i}\sum_{l=1}^{n} (\delta_{kl} - \mu_{kl}) \\ &\quad + \sum_{1\leq j<i\leq n} \zeta_i\eta_j \sum_{j=1}^{l}\sum_{k=1}^{n} (\delta_{kl} - \mu_{kl}) = 0\,,\end{aligned}$$

since all numbers $\zeta_i, \eta_j, \mu_{kl}$ are ≥ 0, and since

$$\sum_{l=1}^{n} \mu_{kl} = 1 \quad \text{for all } k, \text{ and } \sum_{k=1}^{n} \mu_{kl} = 1 \text{ for all } l\,,$$

by the orthogonality of the matrix $\boldsymbol{M}$. We thus have

$$\tfrac{1}{2} \sum_{i,j=1}^{n} |m_{ij}|^2\alpha_i\beta_j \leq \tfrac{1}{2} \sum_{i=1}^{n} \alpha_i\beta_i \,. \qquad \blacksquare$$

Remarks. (1) The result established in part (ii) of the proof is a property of *stochastic matrices* (a stochastic matrix (μ_{ij}) of order n is such that $\mu_{ij} \geq 0$, $\Sigma_{i=1}^{n} \mu_{ij} = 1$, $\Sigma_{j=1}^{n} \mu_{ij} = 1$ for all i and j).

(2) Similar inequalities involving singular values of completely continuous linear operators in infinite-dimensional vector spaces are found in Gohberg & Krejn [1971]. ■

3.3. MATERIAL FRAME-INDIFFERENCE

A general axiom in physics asserts that *any "observable quantity" i.e., any quantity with an intrinsic character*, such as a mass density, an acceleration vector, etc., *must be independent of the particular orthogonal basis in which it is computed*. Rather than stating this principle in its most general form, we simply apply it to elastic materials, where *the "observable" quantity computed through a constitutive equation is the Cauchy stress vector*. We first note that, instead of considering another orthogonal basis (for this viewpoint, see Exercise 3.6), we can equivalently keep the basis fixed and rotate the deformed configuration around the origin (translations of the origin may be ignored since they have no effect on the deformation gradient). It thus suffices to express that the Cauchy stress vectors rotate accordingly; in this fashion, we are led to the following axiom (for more general presentations, see Noll [1955, 1958], Truesdell & Noll [1965, Sects. 19 & 19A]):

Axiom 3.3-1 (axiom of material frame-indifference). *Let the deformed configuration $\bar{\Omega}^{\varphi}$ be rotated into another deformed configuration $\bar{\Omega}^{\psi}$, i.e., $\psi = Q\varphi$ for some $Q \in \mathbb{O}^3_+$* (Fig. 3.3-1). *Then*

$$\boxed{t^{\psi}(x^{\psi}, Qn) = Qt^{\varphi}(x^{\varphi}, n) \quad \text{for all } x \in \bar{\Omega},\ n \in S_1\,,}$$

where $x^{\psi} = \psi(x)$, $x^{\varphi} = \varphi(x)$, and $t^{\psi} := \bar{\Omega}^{\psi} \times S_1 \to \mathbb{R}^3$ and $t^{\varphi} : \bar{\Omega}^{\varphi} \times S_1 \to \mathbb{R}^3$ denote the Cauchy stress vector fields in the deformed configurations $\bar{\Omega}^{\psi}$ and $\bar{\Omega}^{\varphi}$, respectively. ■

It can therefore be predicted, and we now show, that the effect of this axiom, also known as the *axiom of invariance under a change of observer*, or the *axiom of objectivity*, is to *reduce* the class of mappings $\hat{T}^{D} = (\hat{T}^{D}_{ij})$ that may be used for expressing the constitutive equation

$$\begin{cases} T^{\varphi}_{11}(x^{\varphi}) = \hat{T}^{D}_{11}(x, \partial_1\varphi_1(x), \ldots, \partial_3\varphi_3(x))\,, \\ \quad\vdots \\ T^{\varphi}_{33}(x^{\varphi}) = \hat{T}^{D}_{33}(x, \partial_1\varphi_1(x), \ldots, \partial_3\varphi_3(x))\,, \end{cases}$$

of an arbitrary elastic material. We recall that $\mathbb{O}^3_+$ denotes the set of all *rotations* in $\mathbb{R}^3$, i.e., orthogonal matrices Q of order three with $\det Q = +1$, and that $\mathbb{S}^3_>$ denotes the set of all symmetric, positive definite, matrices of order three.

Theorem 3.3-1. *The response function $\hat{T}^{D} : \bar{\Omega} \times \mathbb{M}^3_+ \to \mathbb{S}^3$ for the Cauchy stress satisfies the axiom of material frame-indifference if and only if, for all $x \in \bar{\Omega}$,*

$$\boxed{\hat{T}^{D}(x, QF) = Q\hat{T}^{D}(x, F)Q^{T} \quad \text{for all } F \in \mathbb{M}^3_+,\ Q \in \mathbb{O}^3_+ ;}$$

or equivalently, if and only if, for all $x \in \bar{\Omega}$,

$$\hat{T}^{D}(x, F) = R\hat{T}^{D}(x, U)R^{T} \quad \text{for all } F = RU \in \mathbb{M}^3_+ ,$$

where $F = RU$ *is the polar factorization of* F; *or equivalently, if and only if there exists a mapping* $\tilde{\Sigma} : \bar{\Omega} \times \mathbb{S}^3_> \to \mathbb{S}^3$ *such that, for all* $x \in \bar{\Omega}$,

$$\boxed{\hat{\Sigma}(x, F) = \tilde{\Sigma}(x, F^{T}F) \quad \text{for all } F \in \mathbb{M}^3_+ ,}$$

where $\hat{\Sigma} : \bar{\Omega} \times \mathbb{M}^3_+ \to \mathbb{S}^3$ *is the response function for the second Piola–Kirchhoff stress.*

Proof. Let $T^{\varphi}(x^{\varphi})$ and $T^{\psi}(x^{\psi})$ denote the Cauchy stress tensors at a point $x^{\varphi} \in \bar{\Omega}^{\varphi}$ and at the rotated point $x^{\psi} \in \bar{\Omega}^{\psi}$. The axiom of material frame-indifference then implies that

$$t^{\psi}(x^{\psi}, Qn) = T^{\psi}(x^{\psi})Qn = Qt^{\varphi}(x^{\varphi}, n) = QT^{\varphi}(x^{\varphi})n .$$

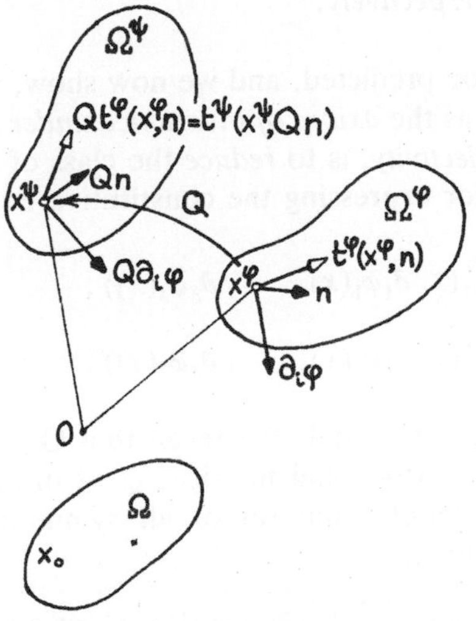

Fig. 3.3-1. The axiom of material frame-indifference: If the deformed configuration is rotated by a matrix $Q \in \mathbb{O}^3_+$, the Cauchy stress vector is rotated by the same matrix Q.

Since this relation must hold for all unit vectors $\boldsymbol{n}$, we deduce that the Cauchy stress tensors $\boldsymbol{T}^{\varphi}(x^{\varphi})$ and $\boldsymbol{T}^{\psi}(x^{\psi})$ are related by

$$\boldsymbol{T}^{\psi}(x^{\psi}) = \boldsymbol{Q}\boldsymbol{T}^{\varphi}(x^{\varphi})\boldsymbol{Q}^{\mathrm{T}} .$$

On the other hand, the geometric interpretation of the deformation gradient (Fig. 1.4-1) shows that the matrix $\nabla\varphi(x)$ becomes the matrix $\nabla\psi(x) = \boldsymbol{Q}\nabla\varphi(x)$ (this also follows from the relation $ox^{\psi} = \boldsymbol{Q}ox^{\varphi}$).

The axiom of material frame-indifference is therefore satisfied if and only if

$$\hat{\boldsymbol{T}}^{\mathrm{D}}(x, \nabla\psi(x)) = \hat{\boldsymbol{T}}^{\mathrm{D}}(x, \boldsymbol{Q}\nabla\varphi(x)) = \boldsymbol{Q}\hat{\boldsymbol{T}}^{\mathrm{D}}(x, \nabla\varphi(x))\boldsymbol{Q}^{\mathrm{T}} .$$

Given an arbitrary matrix $\boldsymbol{F} \in \mathbb{M}^3_+$, there exist deformations that satisfy $\nabla\varphi(x) = \boldsymbol{F}$; therefore, the first equivalence stated in the theorem is established.

To prove the second equivalence, let $\boldsymbol{F} = \boldsymbol{R}\boldsymbol{U}$ be the polar factorization of a matrix $\boldsymbol{F} \in \mathbb{M}_3^+$. Since in this case, $\boldsymbol{R} \in \mathbb{O}^3_+$, and $\boldsymbol{U} \in \mathbb{S}^3_> \subset \mathbb{M}^3_+$, the first equivalence immediately implies (for notational simplicity, we drop the dependence on the variable $x \in \Omega$ in the remainder of the proof):

$$\hat{\boldsymbol{T}}^{\mathrm{D}}(\boldsymbol{F}) = \hat{\boldsymbol{T}}^{\mathrm{D}}(\boldsymbol{R}\boldsymbol{U}) = \boldsymbol{R}\hat{\boldsymbol{T}}^{\mathrm{D}}(\boldsymbol{U})\boldsymbol{R}^{\mathrm{T}} .$$

Conversely, the polar factorization $\boldsymbol{F} = \boldsymbol{R}\boldsymbol{U}$ implies that the polar factorization of any matrix $\boldsymbol{Q}\boldsymbol{F}$ with $\mathrm{Q} \in \mathbb{O}^3_+$ is necessarily $\boldsymbol{Q}\boldsymbol{F} = (\boldsymbol{Q}\boldsymbol{R})\boldsymbol{U}$, by virtue of its uniqueness (Theorem 3.2-2), and therefore

$$\hat{\boldsymbol{T}}^{\mathrm{D}}(\boldsymbol{Q}\boldsymbol{F}) = \hat{\boldsymbol{T}}^{\mathrm{D}}((\boldsymbol{Q}\boldsymbol{R})\boldsymbol{U}) = \boldsymbol{Q}\boldsymbol{R}\hat{\boldsymbol{T}}^{\mathrm{D}}(\boldsymbol{U})\boldsymbol{R}^{\mathrm{T}}\boldsymbol{Q}^{\mathrm{T}} = \boldsymbol{Q}\hat{\boldsymbol{T}}^{\mathrm{D}}(\boldsymbol{F})\boldsymbol{Q}^{\mathrm{T}} .$$

The second equivalence can also be written as

$$\hat{\boldsymbol{T}}^{\mathrm{D}}(\boldsymbol{F}) = \boldsymbol{R}\hat{\boldsymbol{T}}^{\mathrm{D}}(\boldsymbol{U})\boldsymbol{R}^{\mathrm{T}} = \boldsymbol{F}\boldsymbol{U}^{-1}\hat{\boldsymbol{T}}^{\mathrm{D}}(\boldsymbol{U})\boldsymbol{U}^{-1}\boldsymbol{F}^{\mathrm{T}} ,$$

and thus,

$$\hat{\boldsymbol{\Sigma}}(\boldsymbol{F}) = (\det \boldsymbol{F})\boldsymbol{F}^{-1}\hat{\boldsymbol{T}}^{\mathrm{D}}(\boldsymbol{F})\boldsymbol{F}^{-\mathrm{T}} = \tilde{\boldsymbol{\Sigma}}(\boldsymbol{F}^{\mathrm{T}}\boldsymbol{F}) ,$$

with

$$\tilde{\boldsymbol{\Sigma}}(\boldsymbol{C}) := (\det \boldsymbol{U})\boldsymbol{U}^{-1}\hat{\boldsymbol{T}}^{\mathrm{D}}(\boldsymbol{U})\boldsymbol{U}^{-1}, \quad \boldsymbol{U} = \boldsymbol{C}^{1/2}, \quad \text{for all } \boldsymbol{C} \in \mathbb{S}^3_> .$$

Conversely, the third equivalence implies

$$\hat{T}^{\mathrm{D}}(F) = (\det F)^{-1} F \tilde{\Sigma}(F^{\mathrm{T}}F) F^{\mathrm{T}} = (\det U)^{-1} R U \tilde{\Sigma}(U^2) U R^{\mathrm{T}}$$
$$= R \hat{T}^{\mathrm{D}}(U) R^{\mathrm{T}},$$

and the proof is complete. ■

Expressed in terms of the response function $\hat{T}$ and $\hat{\Sigma}$ for the first and second Piola–Kirchhoff stress, the first equivalence $\hat{T}^{\mathrm{D}}(x, QF) = Q\hat{T}^{\mathrm{D}}(x, F)Q^{\mathrm{T}}$ respectively becomes

$$\boxed{\begin{aligned} \hat{T}(x, QF) &= Q\hat{T}(x, F) \quad \text{for all } F \in \mathbb{M}^3_+,\ Q \in \mathbb{O}^3_+\,, \\ \hat{\Sigma}(x, QF) &= \hat{\Sigma}(x, F) \quad \text{for all } F \in \mathbb{M}^3_+,\ Q \in \mathbb{O}^3_+\,. \end{aligned}}$$

As a natural extension of the definition, we shall say that *the response functions $\hat{T}$ and $\hat{\Sigma}$ satisfy the axiom of material frame-indifference*, or simply are **frame-indifferent,** if and only if the response function $\hat{T}^{\mathrm{D}}$ satisfies this axiom, i.e., if and only if the above relations are satisfied.

The second equivalence $\hat{T}^{\mathrm{D}}(x, F) = RT^{\mathrm{D}}(x, U)R^{\mathrm{T}}$, which is known as *Richter's theorem*, means that *the response function $\hat{T}^{\mathrm{D}}$ is determined at a point $x \in \bar{\Omega}$ as long as its restriction to the set of symmetric, positive definite, matrices, is known*, or, to put it differently, "the contribution of the rotation R is independent of the particular response function". The third equivalence expresses an analogous idea, but in terms of the second Piola–Kirchhoff stress tensor: *A constitutive equation then appears as a functional dependence between a "measure of the deformation", the strain tensor $C = \nabla\varphi^{\mathrm{T}}\nabla\varphi$, and a "measure of the stress", the stress tensor Σ.* This explains why constitutive equations are often referred to as *stress–strain laws* in the literature.

3.4. ISOTROPIC ELASTIC MATERIALS

We just saw how an *axiom* (material frame-indifference) restricts the form of the response function. We now examine how its form can be further restricted by a *property* that a given material may possess. The property in question, which is called *isotropy*, corresponds to the intuitive idea that at a given point, the response of the material "is the same in all directions". To give a precise mathematical statement of this property (and this is not so obvious as it may seem *a priori*), consider an arbitrary point $x^{\varphi} = \varphi(x)$ of a body occupying a deformed configuration $\bar{\Omega}^{\varphi} = \varphi(\bar{\Omega})$. If the material is elastic, then by definition, the Cauchy stress tensor at the point x^{φ} is given by

$$\boldsymbol{T}^{\varphi}(x^{\varphi}) = \hat{\boldsymbol{T}}^{\mathrm{D}}(x, \boldsymbol{\nabla}\varphi(x)) .$$

Let us rotate the reference configuration around the point x by a rotation $\boldsymbol{Q}^{\mathrm{T}}$ (Fig. 3.4-1). Then the same deformed configuration can be obtained as the image of the new reference configuration $\boldsymbol{\theta}(\bar{\Omega})$, where

$$\boldsymbol{\theta}(y) = x + \boldsymbol{Q}^{\mathrm{T}}\boldsymbol{xy} \quad \text{for all } y \in \bar{\Omega} ,$$

through the mapping

$$\tilde{\varphi} = \varphi \cdot \boldsymbol{\theta}^{-1} : \tilde{y} \in \boldsymbol{\theta}(\bar{\Omega}) \rightarrow \varphi(x + \boldsymbol{Q}\boldsymbol{x}\tilde{\boldsymbol{y}}) ,$$

which is also a deformation. The Cauchy stress tensor at the same point $x^{\varphi} = x^{\tilde{\varphi}}$ is now given by

$$\boldsymbol{T}^{\tilde{\varphi}}(x^{\tilde{\varphi}}) = \hat{\boldsymbol{T}}^{\mathrm{D}}(x, \boldsymbol{\nabla}\tilde{\varphi}(x)) = \hat{\boldsymbol{T}}^{\mathrm{D}}(x, \boldsymbol{\nabla}\varphi(x)\boldsymbol{Q}) .$$

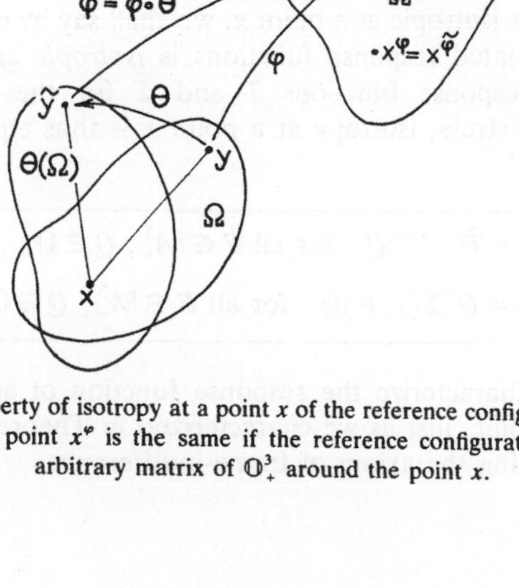

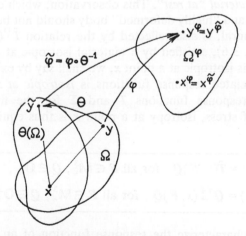

Fig. 3.4-1. The property of isotropy at a point x of the reference configuration: The Cauchy stress tensor at the point x^{φ} is the same if the reference configuration is rotated by an arbitrary matrix of $\mathbb{O}^3_+$ around the point x.

We are thus led to the following definition: An elastic material is **isotropic at a point** x if its response function for the Cauchy stress satisfies

$$\boxed{\hat{\boldsymbol{T}}^{\mathrm{D}}(x, \boldsymbol{F}\boldsymbol{Q}) = \boldsymbol{T}^{\mathrm{D}}(x, \boldsymbol{F}) \quad \text{for all } \boldsymbol{F} \in \mathbb{M}^3_+, \boldsymbol{Q} \in \mathbb{O}^3_+ ,}$$

i.e., if the Cauchy stress tensor (and consequently the Cauchy stress vector) is left unaltered when the reference configuration is subjected to an arbitrary rotation around the point x. If this is not the case, i.e., if the above relation remains valid only for matrices $\boldsymbol{Q}$ in a strict subset $\mathbb{G}_x$ of the group $\mathbb{O}^3_+$, the material is said to be **anisotropic at** x.

Remarks. (1) It can be shown (Exercise 3.8) that the subset $\mathbb{G}_x$ is always a *subgroup* of the group $\mathbb{O}^3_+$, called the *symmetry group* at the point x.

(2) Details about anisotropic materials are found in Truesdell & Noll [1965, Sect. 33], Ogden [1984, Sect. 4.2.5]. ■

An elastic material occupying a reference configuration $\bar{\Omega}$ is **isotropic** if it is isotropic at all points of $\bar{\Omega}$.

Since a reference configuration is often thought of as being occupied by a body "at rest", i.e., before any deformation other than a rigid one has taken place under the action of applied forces, *isotropy thus appears as a property of a material "at rest"*. This observation, which corroborates the intuitive idea that a "highly deformed" body should not be expected to be isotropic in general, is also reflected by the relation $\hat{\boldsymbol{T}}^{\mathrm{D}}(x, \boldsymbol{I}) = -\pi(x)\boldsymbol{I}$, $\pi(x) \in \mathbb{R}$ (Sect. 3.6), satisfied by a material isotropic at a point x.

If a material is isotropic at a point x, we shall say by extension that any one of its associated response functions is *isotropic at* x. Expressed in terms of the response functions $\hat{\boldsymbol{T}}$ and $\hat{\boldsymbol{\Sigma}}$ for the first and second Piola–Kirchhoff stress, isotropy at a point x is thus equivalent to either relation:

$$\boxed{\begin{aligned} &\hat{\boldsymbol{T}}(x, \boldsymbol{F}\boldsymbol{Q}) = \hat{\boldsymbol{T}}(x, \boldsymbol{F})\boldsymbol{Q} \quad \text{for all } \boldsymbol{F} \in \mathbb{M}^3_+, \boldsymbol{Q} \in \mathbb{O}^3_+ , \\ &\hat{\boldsymbol{\Sigma}}(x, \boldsymbol{F}\boldsymbol{Q}) = \boldsymbol{Q}^{\mathrm{T}}\hat{\boldsymbol{\Sigma}}(x, \boldsymbol{F})\boldsymbol{Q} \quad \text{for all } \boldsymbol{F} \in \mathbb{M}^3_+, \boldsymbol{Q} \in \mathbb{O}^3_+ . \end{aligned}}$$

Let us now characterize the response function of an elastic material isotropic at a point, just as we characterized in Theorem 3.3-1 response functions satisfying the axiom of frame-indifference.

Theorem 3.4-1. *A response function* $\hat{\boldsymbol{T}}^{\mathrm{D}}: \bar{\Omega} \times \mathbb{M}^3_+ \to \mathbb{S}^3$ *is isotropic at a point* $x \in \bar{\Omega}$, *i.e., it satisfies*

$$\boxed{\hat{\boldsymbol{T}}^{\mathrm{D}}(x, \boldsymbol{F}\boldsymbol{Q}) = \hat{\boldsymbol{T}}^{\mathrm{D}}(x, \boldsymbol{F}) \quad \text{for all } \boldsymbol{F} \in \mathbb{M}^3_+,\ \boldsymbol{Q} \in \mathbb{O}^3_+,}$$

if and only if there exists a mapping $\bar{\boldsymbol{T}}^{\mathrm{D}}(x, \cdot): \mathbb{S}^3_> \to \mathbb{S}^3$ *such that*

$$\boxed{\hat{\boldsymbol{T}}^{\mathrm{D}}(x, \boldsymbol{F}) = \bar{\boldsymbol{T}}^{\mathrm{D}}(x, \boldsymbol{F}\boldsymbol{F}^{\mathrm{T}}) \text{ for all } \boldsymbol{F} \in \mathbb{M}^3_+ .}$$

Proof. We first recall a simple result (cf. e.g. Bourbaki [1970, p. E II 20]): Let X, Y, Z be three sets and let $f: X \to Y$ and $g: X \to Z$ be two mappings such that

$$\{x, x' \in X \text{ and } g(x) = g(x')\} \Rightarrow f(x) = f(x') .$$

Then "f is a function of g only", i.e., there exists a mapping

$$h: g(X) \subset Z \to Y$$

such that

$$f(x) = h(g(x)) \quad \text{for all } x \in X .$$

For notational brevity, the explicit dependence on x is dropped in the proof. The existence of the mapping $\bar{\boldsymbol{T}}^{\mathrm{D}}$ amounts to proving the implication

$$\boldsymbol{F}\boldsymbol{F}^{\mathrm{T}} = \boldsymbol{G}\boldsymbol{G}^{\mathrm{T}} \quad \text{with } \boldsymbol{F}, \boldsymbol{G} \in \mathbb{M}^3_+ \Rightarrow \hat{\boldsymbol{T}}^{\mathrm{D}}(\boldsymbol{F}) = \hat{\boldsymbol{T}}^{\mathrm{D}}(\boldsymbol{G}) .$$

The equivalence

$$\boldsymbol{F}\boldsymbol{F}^{\mathrm{T}} = \boldsymbol{G}\boldsymbol{G}^{\mathrm{T}} \Leftrightarrow (\boldsymbol{G}^{-1}\boldsymbol{F})(\boldsymbol{G}^{-1}\boldsymbol{F})^{\mathrm{T}} = \boldsymbol{I}$$

shows that the matrix $\boldsymbol{G}^{-1}\boldsymbol{F}$ is orthogonal. Since its determinant is >0, the definition of isotropy implies that

$$\hat{\boldsymbol{T}}^{\mathrm{D}}(\boldsymbol{G}) = \hat{\boldsymbol{T}}^{\mathrm{D}}(\boldsymbol{G}(\boldsymbol{G}^{-1}\boldsymbol{F})) = \hat{\boldsymbol{T}}^{\mathrm{D}}(\boldsymbol{F}) .$$

Conversely,

$$\hat{T}^{D}(F) = \bar{T}^{D}(FF^{T}) \Rightarrow \hat{T}^{D}(FQ) = \bar{T}^{D}(FQQ^{T}F^{T})$$
$$= \bar{T}^{D}(FF^{T}) = \hat{T}^{D}(F) . \quad \blacksquare$$

Notice a certain parallel between Theorems 3.3-1 (where the response functions of frame-indifferent materials were characterized) and 3.4-1. In the first case, it is the *deformed* configuration that has been rotated, with the effect of multiplying F on the *left* by Q and of reducing the response function to a function of the product $F^{T}F$. In the second case, it is the *reference* configuration that has been rotated, with the effect of multiplying F on the *right* by Q and of reducing the response function to a function of the product FF^{T}.

*3.5. PRINCIPAL INVARIANTS OF A MATRIX OF ORDER THREE

The **principal invariants** of a matrix A of order 3 are the coefficients $\iota_1, \iota_2, \iota_3$, also denoted $\iota_1(A)$, $\iota_2(A)$, $\iota_3(A)$ if we wish to make the dependence on A explicit, appearing in the *characteristic polynomial* of A:

$$\det(A - \lambda I) = -\lambda^{3} + \iota_1 \lambda^{2} - \iota_2 \lambda + \iota_3 .$$

From the definition, we readily deduce the following relations, where $\lambda_1, \lambda_2, \lambda_3$ denote the eigenvalues of the matrix $A = (a_{ij})$:

$$\boxed{\begin{aligned} \iota_1 &= a_{ii} = \operatorname{tr} A = \lambda_1 + \lambda_2 + \lambda_3 , \\ \iota_2 &= \tfrac{1}{2}(a_{ii}a_{jj} - a_{ij}a_{ji}) = \tfrac{1}{2}\{(\operatorname{tr} A)^2 - \operatorname{tr} A^2\} \\ &= \operatorname{tr} \mathbf{Cof}\, A = \lambda_1\lambda_2 + \lambda_2\lambda_3 + \lambda_3\lambda_1 \\ &= \det A \operatorname{tr} A^{-1} \text{ if the matrix } A \text{ is invertible,} \\ \iota_3 &= \det A = \tfrac{1}{6}\{(\operatorname{tr} A)^3 - 3 \operatorname{tr} A \operatorname{tr} A^2 + 2 \operatorname{tr} A^3\} = \lambda_1\lambda_2\lambda_3 . \end{aligned}}$$

We shall denote by

$$\boldsymbol{\iota}_A = (\iota_1(A), \iota_2(A), \iota_3(A))$$

the triple formed by the three principal invariants of the matrix A.

More generally, an *invariant* of a matrix A is any real-valued function $\omega(A)$ with the property that

$$\omega(A) = \omega(B^{-1}AB) \quad \text{for all invertible matrices } B .$$

The functions $\operatorname{tr} A^2$ or $\operatorname{tr} A^3$ that appeared above are examples of invariants that are not principal (for another example, see Exercice 5.10).

We also recall the **Cayley–Hamilton theorem**, which asserts that "a matrix is a root of its characteristic polynomial", in the sense that

$$-A^3 + \iota_1 A^2 - \iota_2 A + \iota_3 I = 0 .$$

Consequently, for all integers $p \geqslant 0$, and also for all integers $p \leqslant -1$ if the matrix A is invertible, the matrix A^p can be written as

$$A^p = \alpha_{0p}(\iota_A)I + \alpha_{1p}(\iota_A)A + \alpha_{2p}(\iota_A)A^2 ,$$

where the coefficients α_{0p}, α_{1p}, α_{2p} are polynomial functions of the principal invariants ι_1, ι_2, ι_3 for $p \geqslant 0$, and polynomial functions multiplied by ι_3^p for $p < 0$.

Remarks. (1) That $\iota_2 = \operatorname{tr} \mathbf{Cof}\, A$ follows from Theorem 1.1-1.

(2) The mappings ι_1, ι_{n-1}, $\iota_n : \mathbb{M}^n \to \mathbb{R}$ introduced in Sect. 1.2 generalize to matrices of arbitrary order the principal invariants of a matrix of order 3. ∎

3.6. THE RESPONSE FUNCTION OF AN ISOTROPIC ELASTIC MATERIAL

The simultaneous consideration of the axiom of material frame-indifference and of the property of isotropy at a point will yield a remarkably simple form for the response functions of a general isotropic elastic material, as a consequence of the following *representation theorem for matrix functions of matrices* (the superscript D and the dependence on x are momentarily dropped), due to Rivlin & Ericksen [1955, §39].

Theorem 3.6-1 (Rivlin–Ericksen representation theorem). *A mapping* $\hat{T} : \mathbb{M}^3_+ \to \mathbb{S}^3$ *satisfies*

$$\hat{T}(QF) = Q\hat{T}(F)Q^{\mathrm{T}} \text{ and } \hat{T}(FQ) = \hat{T}(F) \quad \text{for all } F \in \mathbb{M}^3_+,\ Q \in \mathbb{O}^3_+ ,$$

if and only if

$$\hat{T}(F) = \bar{T}(FF^{\mathrm{T}}) \quad \text{for all } F \in \mathbb{M}^3_+ ,$$

where the mapping $\bar{T} : \mathbb{S}^3_> \to \mathbb{S}^3$ *is of the form*

$$\bar{T}(B) = \beta_0(\iota_B)I + \beta_1(\iota_B)B + \beta_2(\iota_B)B^2 \quad \text{for all } B \in \mathbb{S}^3_> ,$$

β_0, β_1, β_2 *being real-valued functions of the three principal invariants of the matrix* B.

Proof. (i) By Theorem 3.4-1, a mapping $\hat{T} : \mathbb{M}^3_+ \to \mathbb{S}^3$ satisfies $\hat{T}(FQ) = \hat{T}(F)$ for all $F \in \mathbb{M}^3_+$, $Q \in \mathbb{O}^3_+$, if and only if there exists a mapping $\bar{T} : \mathbb{S}^3_> \to \mathbb{S}^3$ such that $\hat{T}(F) = \bar{T}(FF^{\mathrm{T}})$ for all $F \in \mathbb{M}^3_+$. We show that the *mapping* $\hat{T}$ *satisfies the additional condition* $\hat{T}(QF) = Q\hat{T}(F)Q^{\mathrm{T}}$ *for all* $F \in \mathbb{M}^3_+$, $Q \in \mathbb{O}^3_+$, *if and only if the mapping* $\bar{T}$ *satisfies*

$$\bar{T}(QBQ^{\mathrm{T}}) = Q\bar{T}(B)Q^{\mathrm{T}} \quad \text{for all } B \in \mathbb{S}^3_>, \ Q \in \mathbb{O}^3_+ .$$

To see this, let there be given two matrices $B \in \mathbb{S}^3_>$, $Q \in \mathbb{O}^3_+$. Then

$$\bar{T}(QBQ^{\mathrm{T}}) = \bar{T}(QB^{1/2}(QB^{1/2})^{\mathrm{T}}) = \hat{T}(QB^{1/2})$$

on the one hand, and

$$\hat{T}(QB^{1/2}) = Q\hat{T}(B^{1/2})Q^{\mathrm{T}} = Q\bar{T}(B)Q^{\mathrm{T}}$$

on the other hand, by Theorem 3.3-1 (for the definition of the matrix $B^{1/2}$, see Sect. 3.2). Conversely, let $F \in \mathbb{M}^3_+$ and $Q \in \mathbb{O}^3_+$. Since $FF^{\mathrm{T}} \in \mathbb{S}^3_>$,

$$\hat{T}(QF) = \bar{T}(QFF^{\mathrm{T}}Q^{\mathrm{T}}) = Q\bar{T}(FF^{\mathrm{T}})Q^{\mathrm{T}} = Q\hat{T}(F)Q^{\mathrm{T}} .$$

(ii) It therefore suffices to *characterize those mappings* $\bar{T} : \mathbb{S}^3_> \to \mathbb{S}^3$ *that satisfy*

$$\bar{T}(QBQ^{\mathrm{T}}) = Q\bar{T}(B)Q^{\mathrm{T}} \quad \text{for all } B \in \mathbb{S}^3_>, \ Q \in \mathbb{O}^3_+ .$$

In this direction, we first notice that if the mapping $\bar{T}$ is of the form given in the theorem, we have

$$\bar{T}(QBQ^{\mathrm{T}}) = \beta_0(\iota_B)I + \beta_1(\iota_B)QBQ^{\mathrm{T}} + \beta_2(\iota_B)QB^2Q^{\mathrm{T}} = Q\bar{T}(B)Q^{\mathrm{T}},$$

for all $B \in \mathbb{S}^3_>$ and $Q \in \mathbb{O}^3_+$, since $\iota_B = \iota_{QBQ^{\mathrm{T}}}$.

In order to provide a simple motivation for our final result, we first consider a special class of functions that satisfy $\bar{T}(QBQ^{\mathrm{T}}) = Q\bar{T}(B)Q^{\mathrm{T}}$ for all $B \in \mathbb{S}^3_>$ and all $Q \in \mathbb{O}^3_+$, viz., functions $\bar{T}$ that are *polynomials in* B:

$$\bar{T}(B) = a_0 I + a_1 B + \cdots + a_p B^p \quad \text{for all } B \in \mathbb{S}^3_> .$$

First, we notice that, *for each matrix* $B \in \mathbb{S}^3_>$, *any orthogonal matrix that diagonalizes the matrix* B *also diagonalizes the matrix* $\bar{T}(B)$; this property is established in general in the next step of the proof. Secondly, by the Cayley–Hamilton theorem, each power B^p, $p \geqslant 3$, can be expanded as a polynomial function of the three matrices I, B, B^2, whose coefficients are functions of the three principal invariants of the matrix B. Hence the theorem is proved in this case. This proof could be extended to all functions $\bar{T}$ that can be expanded as infinite series of powers of B, provided all series involved are convergent. But this would put severe restrictions on the *regularity* of the admissible functions $\bar{T}$, a property that is irrelevant here. Let us therefore proceed differently.

(iii) Let us show that, *if a mapping* $\bar{T}: \mathbb{S}^3_> \to \mathbb{S}^3$ *satisfies*

$$\bar{T}(QBQ^{\mathrm{T}}) = Q\bar{T}(B)Q^{\mathrm{T}} \quad \text{for all } B \in \mathbb{S}^3_>,\ Q \in \mathbb{O}^3_+ ,$$

then, for any $B \in \mathbb{S}^3_>$, *any matrix that diagonalizes the matrix* B *also diagonalizes the matrix* $\bar{T}(B)$.

Given a matrix $B \in \mathbb{S}^3_>$, let Q be any orthogonal matrix that diagonalizes the matrix B (such matrices Q depend on the matrix B):

$$Q^{\mathrm{T}}BQ = \mathbf{Diag}\ \lambda_i .$$

Without loss of generality, we may assume that det $Q = +1$, so that $Q \in \mathbb{O}^3_+$ (otherwise replace one column of the matrix Q by its opposite). Consider the two matrices

$$Q_1 = \begin{pmatrix} 1 & 0 & 0 \\ 0 & -1 & 0 \\ 0 & 0 & -1 \end{pmatrix}, \qquad Q_2 = \begin{pmatrix} -1 & 0 & 0 \\ 0 & 1 & 0 \\ 0 & 0 & -1 \end{pmatrix},$$

which clearly belong to the set $\mathbb{O}^3_+$. The relation $Q^{\mathrm{T}}BQ = \mathbf{Diag}\ \lambda_i$ means that for $j = 1, 2, 3$, the jth column of the matrix $\boldsymbol{Q} \in \mathbb{O}^3_+$ is an eigenvector of the matrix $\boldsymbol{B}$ corresponding to the eigenvalue λ_j. Hence we likewise have

$$(\boldsymbol{Q}\boldsymbol{Q}_\beta)^{\mathrm{T}}\boldsymbol{B}(\boldsymbol{Q}\boldsymbol{Q}_\beta) = \mathbf{Diag}\,\lambda_i = \boldsymbol{Q}^{\mathrm{T}}\boldsymbol{B}\boldsymbol{Q}, \quad \beta = 1, 2,$$

since the effect of multiplying $\boldsymbol{Q}$ by $\boldsymbol{Q}_\beta$ on the right is to replace two of its column vectors by their opposites. By the assumed relation, we must therefore have

$$\begin{aligned}\boldsymbol{Q}^{\mathrm{T}}_\beta(\boldsymbol{Q}^{\mathrm{T}}\bar{\boldsymbol{T}}(\boldsymbol{B})\boldsymbol{Q})\boldsymbol{Q}_\beta &= (\boldsymbol{Q}\boldsymbol{Q}_\beta)^{\mathrm{T}}\bar{\boldsymbol{T}}(\boldsymbol{B})(\boldsymbol{Q}\boldsymbol{Q}_\beta) = \bar{\boldsymbol{T}}((\boldsymbol{Q}\boldsymbol{Q}_\beta)^{\mathrm{T}}\boldsymbol{B}(\boldsymbol{Q}\boldsymbol{Q}_\beta)) \\ &= \bar{\boldsymbol{T}}(\boldsymbol{Q}^{\mathrm{T}}\boldsymbol{B}\boldsymbol{Q}) = \boldsymbol{Q}^{\mathrm{T}}\bar{\boldsymbol{T}}(\boldsymbol{B})\boldsymbol{Q}, \quad \beta = 1, 2,\end{aligned}$$

and these two relations imply that *the matrix* $\boldsymbol{Q}^{\mathrm{T}}\bar{\boldsymbol{T}}(\boldsymbol{B})\boldsymbol{Q}$ *must be also diagonal*, as a straightforward computation shows.

(iv) Let us now establish that *the function* $\bar{\boldsymbol{T}}$ *is necessarily of the form*

$$\bar{\boldsymbol{T}}(\boldsymbol{B}) = b_0(\boldsymbol{B})\boldsymbol{I} + b_1(\boldsymbol{B})\boldsymbol{B} + b_2(\boldsymbol{B})\boldsymbol{B}^2 \quad \text{for all } \boldsymbol{B} \in \mathbb{S}^3_>,$$

where b_0, b_1, b_2 *are real-valued functions of* $\boldsymbol{B}$. Three cases have to be distinguished.

Assume first that *the matrix* $\boldsymbol{B}$ *has three distinct eigenvalues* λ_i, with associated orthormalized eigenvectors $\boldsymbol{p}_i$. Then *the two sets* $\{\boldsymbol{I}, \boldsymbol{B}, \boldsymbol{B}^2\}$ *and* $\{\boldsymbol{p}_1\boldsymbol{p}_1^{\mathrm{T}}, \boldsymbol{p}_2\boldsymbol{p}_2^{\mathrm{T}}, \boldsymbol{p}_3\boldsymbol{p}_3^{\mathrm{T}}\}$ *span the same subspace of the vector space* $\mathbb{S}^3$. To see this, we observe that

$$\begin{aligned}\boldsymbol{I} &= \boldsymbol{p}_1\boldsymbol{p}_1^{\mathrm{T}} + \boldsymbol{p}_2\boldsymbol{p}_2^{\mathrm{T}} + \boldsymbol{p}_3\boldsymbol{p}_3^{\mathrm{T}}, \\ \boldsymbol{B} &= \lambda_1\boldsymbol{p}_1\boldsymbol{p}_1^{\mathrm{T}} + \lambda_2\boldsymbol{p}_2\boldsymbol{p}_2^{\mathrm{T}} + \lambda_3\boldsymbol{p}_3\boldsymbol{p}_3^{\mathrm{T}}, \\ \boldsymbol{B}^2 &= \lambda_1^2\boldsymbol{p}_1\boldsymbol{p}_1^{\mathrm{T}} + \lambda_2^2\boldsymbol{p}_2\boldsymbol{p}_2^{\mathrm{T}} + \lambda_3^2\boldsymbol{p}_3\boldsymbol{p}_3^{\mathrm{T}},\end{aligned}$$

and that the van der Monde determinant

$$\det\begin{pmatrix} 1 & 1 & 1 \\ \lambda_1 & \lambda_2 & \lambda_3 \\ \lambda_1^2 & \lambda_2^2 & \lambda_3^2 \end{pmatrix}$$

does not vanish, since the three eigenvalues are assumed to be distinct.

If we denote by μ_i the eigenvalues of the symmetric matrix $\bar{T}(B)$, the result of step (iii) shows that we can expand $\bar{T}(B)$ as

$$\bar{T}(B) = \mu_1 p_1 p_1^T + \mu_2 p_2 p_2^T + \mu_3 p_3 p_3^T ,$$

and consequently also as

$$\bar{T}(B) = b_0(B)I + b_1(B)B + b_2(B)B^2 .$$

The components $b_0(B)$, $b_1(B)$, $b_2(B)$ are uniquely determined since the matrices I, B, B^2 are linearly independent in this case.

Assume next that *the matrix B has a double eigenvalue, say* $\lambda_2 = \lambda_3 \neq \lambda_1$. Then *the two sets* $\{I, B\}$ *and* $\{p_1 p_1^T, p_2 p_2^T + p_3 p_3^T\}$ *span the same subspace of the space* $\mathbb{S}^3$, since in this case, we can write

$$I = p_1 p_1^T + (p_2 p_2^T + p_3 p_3^T)$$
$$B = \lambda_1 p_1 p_1^T + \lambda_2 (p_2 p_2^T + p_3 p_3^T)$$

and

$$\det\begin{pmatrix} 1 & 1 \\ \lambda_1 & \lambda_2 \end{pmatrix} \neq 0 .$$

Besides, *the matrix $\bar{T}(B)$ has also a double eigenvalue $\mu_2 = \mu_3$ associated with the eigenvectors p_2 and p_3*, since all nonzero vectors in the subspace spanned by p_2 and p_3 are eigenvectors of the matrix B, hence also of the matrix $\bar{T}(B)$ by the result of step (iii); to prove this last assertion, observe that

$$\bar{T}(B)p_2 = \mu_2 p_2, \ \bar{T}(B)p_3 = \mu_3 p_3,$$
$$\bar{T}(B)(p_2 + p_3) = \mu(p_2 + p_3) \Rightarrow \mu_2 = \mu_3 = \mu .$$

Therefore in this case the matrix

$$\bar{T}(B) = \mu_1 p_1 p_1^T + \mu(p_2 p_2^T + p_3 p_3^T)$$

can be expanded as

$$\bar{T}(B) = b_0(B)I + b_1(B)B .$$

Assume finally that *the matrix B has a triple eigenvalue*. Since in this case all nonzero vectors are eigenvectors of $\boldsymbol{B}$, whence of $\bar{T}(\boldsymbol{B})$, we deduce that $\bar{T}(\boldsymbol{B})$ is a multiple of the identity, i.e., there exists a number $b_0(\boldsymbol{B})$ such that

$$\bar{T}(\boldsymbol{B}) = b_0(\boldsymbol{B})\boldsymbol{I} ,$$

and the assertion is proved in all cases.

(v) It remains to show that *the functions* $b_0, b_1, b_2 : \boldsymbol{B} \in \mathbb{S}^3_> \to \mathbb{R}$ *are in fact only functions of* $\iota_{\boldsymbol{B}}$, *i.e., of the three principal invariants of the matrix* $\boldsymbol{B}$. We first notice that, since these functions must be such that

$$\begin{aligned}\bar{T}(\boldsymbol{QBQ}^{\mathrm{T}}) &= \boldsymbol{Q}(b_0(\boldsymbol{QBQ}^{\mathrm{T}})\boldsymbol{I} + b_1(\boldsymbol{QBQ}^{\mathrm{T}})\boldsymbol{B} + b_2(\boldsymbol{QBQ}^{\mathrm{T}})\boldsymbol{B}^2)\boldsymbol{Q}^{\mathrm{T}} \\ &= \boldsymbol{Q}\bar{T}(\boldsymbol{B})\boldsymbol{Q}^{\mathrm{T}} = \boldsymbol{Q}(b_0(\boldsymbol{B})\boldsymbol{I} + b_1(\boldsymbol{B})\boldsymbol{B} + b_2(\boldsymbol{B})\boldsymbol{B}^2)\boldsymbol{Q}^{\mathrm{T}}\end{aligned}$$

for all $\boldsymbol{B} \in \mathbb{S}^3_>$, $\boldsymbol{Q} \in \mathbb{O}^3_+$, *they necessarily satisfy*

$$b_\alpha(\boldsymbol{QBQ}^{\mathrm{T}}) = b_\alpha(\boldsymbol{B}) \quad \text{for all } \boldsymbol{B} \in \mathbb{S}^3_>,\ \boldsymbol{Q} \in \mathbb{O}^3_+, \quad \text{for } \alpha = 0, 1, 2 ,$$

again by the uniqueness of the expansion of the matrix $\bar{T}(\boldsymbol{B})$ in the spaces spanned by either sets $\{\boldsymbol{I}, \boldsymbol{B}, \boldsymbol{B}^2\}$, $\{\boldsymbol{I}, \boldsymbol{B}\}$, or $\{\boldsymbol{I}\}$, according to which case is considered.

Let b denote any one of the functions b_α, $\alpha = 0, 1, 2$. We show that

$$\boldsymbol{A}, \boldsymbol{B} \in \mathbb{S}^3_>, \text{ and } \iota_{\boldsymbol{A}} = \iota_{\boldsymbol{B}} \Rightarrow b(\boldsymbol{A}) = b(\boldsymbol{B}) .$$

If the matrices $\boldsymbol{A}$ and $\boldsymbol{B}$ have the same principal invariants, their eigenvalues are the same. Hence after a possible re-ordering of their eigenvalues, we can write

$$\boldsymbol{A} = \sum_{i=1}^{3} \lambda_i \boldsymbol{r}_i \boldsymbol{r}_i^{\mathrm{T}} \quad \text{with } \boldsymbol{r}_i^{\mathrm{T}} \boldsymbol{r}_j = \delta_{ij} ,$$

$$\boldsymbol{B} = \sum_{i=1}^{3} \lambda_i \boldsymbol{s}_i \boldsymbol{s}_i^{\mathrm{T}} \quad \text{with } \boldsymbol{s}_i^{\mathrm{T}} \boldsymbol{s}_i = \delta_{ij} ,$$

and there is no loss of generality in assuming that there exists a matrix $\boldsymbol{Q} \in \mathbb{O}^3_+$ such that $\boldsymbol{Q}\boldsymbol{s}_i = \boldsymbol{r}_i$, $1 \leqslant i \leqslant 3$. This being the case,

$$\boldsymbol{A} = \sum_{i=1}^{3} \lambda_i \boldsymbol{Q}\boldsymbol{s}_i \boldsymbol{s}_i^{\mathrm{T}} \boldsymbol{Q}^{\mathrm{T}} = \boldsymbol{QBQ}^{\mathrm{T}} ,$$

and thus $b(\boldsymbol{A}) = b(\boldsymbol{QBQ}^{\mathrm{T}}) = b(\boldsymbol{B})$ by the above invariance.

By virtue of an argument already used in the proof of Theorem 3.4-1, we conclude that there exists a function $\beta : \iota(\mathbb{S}^3_>) \to \mathbb{R}$ such that

$$b(\boldsymbol{B}) = \beta(\iota_{\boldsymbol{B}}) \quad \text{for all } \boldsymbol{B} \in \mathbb{S}^3_> . \qquad \blacksquare$$

Among the features of the above theorem, the absence of any *regularity* assumption on the mapping $\bar{\boldsymbol{T}}$ is noteworthy. By contrast, such an assumption is needed in some subsequent developments, as in the expansion of the constitutive equation near the reference configuration (Sect. 3.7). Notice also the essential rôle played in the proof by the *symmetry* of both the matrix variable $\boldsymbol{B}$ and the matrix function $\bar{\boldsymbol{T}}(\boldsymbol{B})$.

As a corollary of the Rivlin–Ericksen representation theorem, we obtain remarkably simple expressions of the response functions $\hat{\boldsymbol{T}}^{\mathrm{D}}$ and $\hat{\boldsymbol{\Sigma}}$ for the Cauchy stress and the second Piola–Kirchhoff stress of an isotropic elastic material that satisfies the axiom of material frame-indifference. See Boehler [1978] for an extension to anisotropic materials.

Theorem 3.6-2. *Let there be given an elastic material whose response function is frame-indifferent and isotropic at a point* $x \in \bar{\Omega}$. *Given an arbitrary deformation* $\boldsymbol{\varphi} : \bar{\Omega} \to \mathbb{R}^3$, *the Cauchy stress tensor at the point* $x^{\varphi} = \boldsymbol{\varphi}(x)$ *is given by*

$$\boxed{\boldsymbol{T}^{\varphi}(x^{\varphi}) = \hat{\boldsymbol{T}}^{\mathrm{D}}(x, \boldsymbol{\nabla}\boldsymbol{\varphi}(x)) = \bar{\boldsymbol{T}}^{\mathrm{D}}(x, \boldsymbol{\nabla}\boldsymbol{\varphi}(x)\boldsymbol{\nabla}\boldsymbol{\varphi}(x)^{\mathrm{T}}) ,}$$

where the response function $\bar{\boldsymbol{T}}^{\mathrm{D}}(x, \cdot) : \mathbb{S}^3_> \to \mathbb{S}^3$ *is of the form*

$$\boxed{\begin{aligned} &\bar{\boldsymbol{T}}^{\mathrm{D}}(x, \boldsymbol{B}) = \beta_0(x, \iota_{\boldsymbol{B}})\boldsymbol{I} + \beta_1(x, \iota_{\boldsymbol{B}})\boldsymbol{B} + \beta_2(x, \iota_{\boldsymbol{B}})\boldsymbol{B}^2, \\ &\quad \text{for all } \boldsymbol{B} \in \mathbb{S}^3_> , \end{aligned}}$$

$\beta_0(x, \cdot)$, $\beta_1(x, \cdot)$, $\beta_2(x, \cdot)$ *being real-valued functions of the three principal invariants of the matrix* $\boldsymbol{B}$; *the second Piola–Kirchhoff stress tensor at the point* x *is given by*

$$\boxed{\boldsymbol{\Sigma}(x) = \hat{\boldsymbol{\Sigma}}(x, \boldsymbol{\nabla}\boldsymbol{\varphi}(x)) = \tilde{\boldsymbol{\Sigma}}(x, \boldsymbol{\nabla}\boldsymbol{\varphi}(x)^{\mathrm{T}}\boldsymbol{\nabla}\boldsymbol{\varphi}(x)) ,}$$

where the response function $\tilde{\boldsymbol{\Sigma}}(x, \cdot) : \mathbb{S}^3_> \to \mathbb{S}^3$ *is of the form*

$$\boxed{\begin{aligned} &\tilde{\boldsymbol{\Sigma}}(x, \boldsymbol{C}) = \gamma_0(x, \iota_{\boldsymbol{C}})\boldsymbol{I} + \gamma_1(x, \iota_{\boldsymbol{C}})\boldsymbol{C} + \gamma_2(x, \iota_{\boldsymbol{C}})\boldsymbol{C}^2 \\ &\quad \text{for all } \boldsymbol{C} \in \mathbb{S}^3_> , \end{aligned}}$$

$\gamma_0(x, \cdot)$, $\gamma_1(x, \cdot)$, $\gamma_2(x, \cdot)$ being real-valued functions of the three principal invariants of the matrix C. Conversely, if either one of the response functions $\bar{T}^{D}$ and $\tilde{\Sigma}$ is of the above form, the axiom of material frame-indifference is satisfied and the material is isotropic at the point x.

Proof. That the response function $\bar{T}^{D}$ is of the form indicated follows from Theorems 3.3-1, 3.4-1 and 3.6-1. Next, for an arbitrary matrix $F \in \mathbb{M}^3_+$, let

$$B = FF^{T} \text{ and } C = F^{T}F .$$

Observing that $\iota(B) = \iota(C)$ (since $B = RCR^{T}$ by the polar factorization $F = RU$; cf. Theorem 3.2-2) and that

$$\det F = \iota_3(F) = \{\iota_3(B)\}^{1/2} = \{\iota_3(C)\}^{1/2} ,$$

we obtain (the explicit dependence on x is dropped for notational convenience)

$$\begin{aligned}\hat{\Sigma}(F) &= (\det F)F^{-1}\hat{T}^{D}(F)F^{-T} \\ &= (\det F)(\beta_0(\iota_B)F^{-1}F^{-T} + \beta_1(\iota_B)I + \beta_2(\iota_B)F^{T}F) \\ &= \{\iota_3(C)\}^{1/2}(\beta_0(\iota_C)C^{-1} + \beta_1(\iota_C)I + \beta_2(\iota_C)C) .\end{aligned}$$

Since

$$C^{-1} = \iota_3^{-1}(C)(\iota_2(C)I - \iota_1(C)C + C^2) ,$$

the assertion follows. ■

Arguing as above, we could also derive various equivalent forms of the response function $\bar{T}^{D}$ or $\tilde{\Sigma}$, for instance:

$$\begin{aligned}&\bar{T}^{D}(x, B) = \beta'_{-1}(x, \iota_B)B^{-1} + \beta'_0(x, \iota_B)I + \beta'_1(x, \iota_B)B , \\ &\bar{T}^{D}(x, B) = \kappa_0(x, \iota_V)I + \kappa_1(x, \iota_V)V + \kappa_2(x, \iota_V)V^2, \quad V = B^{1/2} , \\ &\tilde{\Sigma}(x, C) = \gamma'_{-1}(x, \iota_C)C^{-1} + \gamma'_0(x, \iota_C)I + \gamma'_1(x, \iota_C)C, \text{ etc.}\end{aligned}$$

By contrast, the first Piola–Kirchhoff stress tensor $T(x)$ cannot be expressed in terms of either symmetric tensors $B = FF^{T}$ or $C = F^{T}F$. Using for instance the last expression of the response function $\tilde{\Sigma}$, we obtain

$$T(x) = \hat{T}(x, \nabla\varphi(x)),$$

with

$$\hat{T}(x, F) = \gamma'_{-1}(x, \iota_C)F^{-T} + \gamma'_0(x, \iota_C)F + \gamma'_1(x, \iota_C)FF^{T}F.$$

Nevertheless, if a constitutive equation in the reference configuration is more conveniently expressed in terms of the *second*, rather than the first, Piola–Kirchhoff stress tensor, it is the *first* Piola–Kirchhoff stress tensor that appears naturally in the *equations of equilibrium* in the reference configuration (Chapter 2), as well as in the *constitutive equation of a hyperelastic material* (Sect. 4.1).

Let

$$T_R(x) := \hat{T}^D(x, I) = \hat{T}(x, I) = \hat{\Sigma}(x, I) = \tilde{\Sigma}(x, I)$$

denote the **residual stress tensor** at a point x of the reference configuration, thus viewed as the particular deformed configuration that corresponds to $\varphi = id$. Then it follows from Theorem 3.6-2 that *if an elastic material is isotropic at a point $x \in \bar{\Omega}$, the residual stress tensor at x is a pressure* (i.e., a multiple of the unit matrix, following the terminology introduced in Sect. 2.3), since

$$\nabla\varphi(x) = I \Rightarrow T^{\varphi}(x^{\varphi}) = \hat{T}^D(x, I) = -\pi(x)I,$$

with $-\pi(x) = \beta_0(x, \iota_1) + \beta_1(x, \iota_1) + \beta_2(x, \iota_1)$.

This result is a consequence of the assumed property of isotropy: In order that the properties of the material be "identical in all directions", it is natural that the Cauchy stress tensor be a pressure. However, if the same elastic material undergoes an arbitrary deformation, it usually loses its isotropy, since by Theorem 3.6-2 there is no reason to expect the matrix $\hat{T}^D(x, F)$ to be equal to a multiple of the unit matrix for an arbitrary matrix $F \in \mathbb{M}^3_+$. Thus, if it is *a priori* possible to choose as a new reference configuration an arbitrary deformed configuration, it is no longer licit in general to assume that the material is isotropic in an arbitrary reference configuration: *Isotropy is a property that only holds in particular reference configurations.*

Remark. The residual stress tensor of an *anisotropic* elastic material is not arbitrary either; see Coleman & Noll [1964], Hoger [1985, 1986]. ■

A reference configuration $\bar{\Omega}$ is called a **natural state** if the residual stress tensor $\boldsymbol{T}_{\mathrm{R}}(x)$ vanishes at all points $x \in \bar{\Omega}$. This definition corresponds to the *assumption* that there exist "unstressed states" of a given body (in which case all applied forces vanish).

If a reference configuration is a natural state, any deformed configuration corresponding to a rigid deformation is again a natural state if it is chosen as a new reference configuration. To see this, let $\boldsymbol{\varphi}: \bar{\Omega} \to \mathbb{R}^3$ be a rigid deformation, so that $\nabla\boldsymbol{\varphi}(x) = \boldsymbol{Q} \in \mathbb{O}^3_+$ for all $x \in \bar{\Omega}$. By Theorem 3.6-2, we infer that

$$\hat{\boldsymbol{T}}^{\mathrm{D}}(x, \boldsymbol{Q}) = \bar{\boldsymbol{T}}^{\mathrm{D}}(x, \boldsymbol{I}) = \boldsymbol{0}\,.$$

Remark. The existence of unstressed states is a reasonable assumption for *elastic solids*. A gas is an *elastic fluid* that has no natural state. ■

3.7. THE CONSTITUTIVE EQUATION NEAR THE REFERENCE CONFIGURATION

It was shown in Sect. 1.8 that the Green–St Venant strain tensor

$$\boldsymbol{E} = \tfrac{1}{2}(\boldsymbol{C} - \boldsymbol{I}), \quad \boldsymbol{C} = \nabla\boldsymbol{\varphi}^{\mathrm{T}}\nabla\boldsymbol{\varphi}\,,$$

is in a sense a measure of the discrepancy between a given deformation $\boldsymbol{\varphi}$ and a rigid deformation, for which $\boldsymbol{C} = \boldsymbol{I}$. It is therefore natural to compute, up to a specified order with respect to $\|\boldsymbol{E}\|$, the difference $(\tilde{\boldsymbol{\Sigma}}(x, \boldsymbol{I} + 2\boldsymbol{E}) - \tilde{\boldsymbol{\Sigma}}(x, \boldsymbol{I}))$ (where $\tilde{\boldsymbol{\Sigma}}(x, \cdot)$ is the response function for the second Piola–Kirchhoff stress at a point $x \in \bar{\Omega}$) in terms of the right Cauchy–Green strain tensor $\boldsymbol{C}$. Equivalently we wish to compute, to within a given order in terms of the tensor $\boldsymbol{E}$, an expansion of the stress tensor $\boldsymbol{\Sigma}$ corresponding to a deformed configuration that is "near" the reference configuration, the latter corresponding to the particular rigid deformation ***id***.

As shown in the next theorem, the result is startling: *The first-order term involves only two constants*, while there are *a priori* thirty-six constants $(\partial\tilde{A}_{ij}/\partial C_{kl})(\boldsymbol{I})$ in the first-order term of the expansion

$$\tilde{\boldsymbol{A}}(\boldsymbol{I} + 2\boldsymbol{E}) = \tilde{\boldsymbol{A}}(\boldsymbol{I}) + 2\,\frac{\partial\tilde{\boldsymbol{A}}}{\partial C_{ij}}(\boldsymbol{I})E_{ij} + o(\boldsymbol{E}) \in \mathbb{S}^3, \quad \boldsymbol{I} + 2\boldsymbol{E} \in \mathbb{S}^3_>\,,$$

of an arbitrary matrix function $\tilde{\boldsymbol{A}}$. One can likewise show that the second-order term involve only four constants (Exercise 3.11), instead of the expected two hundred and sixteen constants in the general case. The origin of these striking reductions lies of course in the Rivlin–Ericksen representation theorem, which implies (Theorems 3.6-1 and 3.6-2) that the function $\tilde{\Sigma}(x, \cdot)$ is of the form:

$$\tilde{\Sigma}(x, C) = \gamma_0(x, \iota_C)I + \gamma_1(x, \iota_C)C + \gamma_2(x, \iota_C)C^2 \quad \text{for all } C \in \mathbb{S}^3_> .$$

We recall that the set $\{E = \frac{1}{2}\{F^{\mathrm{T}}F - I\} \in \mathbb{S}^3; F \in \mathbb{M}^3_+\}$ is a neighborhood of the origin in $\mathbb{S}^3$ (Theorem 1.8-3), and that the notation $f^\varepsilon(x) = o(\varepsilon; x)$ means $\lim_{\varepsilon\to 0} \{\|f^\varepsilon(x)\| / \varepsilon\} = 0$.

Theorem 3.7-1. *Let there be given an elastic material whose response function is frame-indifferent and isotropic at a point* $x \in \bar{\Omega}$. *Assume further that the functions* $\gamma_\alpha(x, \cdot)$, $\alpha = 0, 1, 2$, *are differentiable at the point* $\iota_I = (3, 3, 1)$. *Then there exist constants* $\pi(x)$, $\lambda(x)$, $\mu(x) \in \mathbb{R}$ *such that*

$$\tilde{\Sigma}(x, C) = -\pi(x)I + \lambda(x)(\operatorname{tr} E)I + 2\mu(x)E + o(E; x)$$
$$\text{for all } C = I + 2E \in \mathbb{S}^3_> .$$

Proof. The relation $C = I + 2E$ implies

$$\operatorname{tr} C = 3 + 2 \operatorname{tr} E ,$$
$$\operatorname{tr} C^2 = 3 + 4 \operatorname{tr} E + o(E) ,$$
$$\operatorname{tr} C^3 = 3 + 6 \operatorname{tr} E + o(E) ,$$

so that the first-order part of each principal invariant of the matrix C is simply a multiple of $\operatorname{tr} E$:

$$\iota_1(C) = \operatorname{tr} C = 3 + 2 \operatorname{tr} E ,$$
$$\iota_2(C) = \tfrac{1}{2}\{(\operatorname{tr} C)^2 - \operatorname{tr} C^2\} = 3 + 4 \operatorname{tr} E + o(E) ,$$
$$\iota_3(C) = \tfrac{1}{6}\{(\operatorname{tr} C)^3 - 3 \operatorname{tr} C \operatorname{tr} C^2 + 2 \operatorname{tr} C^3\} = 1 + 2 \operatorname{tr} E + o(E) .$$

By the assumed differentiability of the functions $\gamma_0, \gamma_1, \gamma_2$ at the point ι_I, the above relations in turn imply that each function $\gamma_0, \gamma_1, \gamma_2$, abbreviated as γ, can be expanded as (the dependence on the variable x is omitted)

$$\gamma(\iota_C) = \gamma(\iota_I) + \dot{\gamma}(\iota_I) \operatorname{tr} E + o(E) ,$$

where

$$\dot{\gamma}(\iota_I) := 2\frac{\partial\gamma}{\partial\iota_1}(\iota_I) + 4\frac{\partial\gamma}{\partial\iota_2}(\iota_I) + 2\frac{\partial\gamma}{\partial\iota_3}(\iota_I),$$

and $\iota_I = (3, 3, 1)$ denotes the set formed by the three principal invariants of the unit matrix. Combining these relations and noting that $C^2 = I + 4E + o(E)$, we obtain

$$\tilde{\Sigma}(C) = \tilde{\Sigma}(I + 2E) = \tilde{\Sigma}(I) + \{\dot{\gamma}_0(\iota_I) + \dot{\gamma}_1(\iota_I) + \dot{\gamma}_2(\iota_I)\}(\operatorname{tr} E)I$$
$$+ \{2\gamma_1(\iota_I) + 4\gamma_2(\iota_I)\}E + o(E),$$

and the proof is completed by observing that

$$\tilde{\Sigma}(I) = \{\gamma_0(\iota_I) + \gamma_1(\iota_I) + \gamma_2(\iota_I)\}I. \qquad \blacksquare$$

Remarks. (1) By letting $\varphi = id$ in the relation

$$T^{\varphi}(x^{\varphi}) = (\det \nabla\varphi(x))^{-1}\nabla\varphi(x)\tilde{\Sigma}(x, \nabla\varphi(x)^{\mathrm{T}}\nabla\varphi(x))\nabla\varphi(x)^{-\mathrm{T}},$$

we infer from Theorem 3.7-1 that

$$T_{\mathrm{R}}(x) = \tilde{\Sigma}(x, I) = -\pi(x)I,$$

i.e., the residual stress tensor $T_{\mathrm{R}}(x)$ is a pressure. This result was already established at the end of Sect. 3.6, without any differentiability assumption on the functions $\gamma_0, \gamma_1, \gamma_2$.

(2) A shorter proof can be given in the special case where the response function $\tilde{\Sigma}$ is a linear function of E; cf. Exercise 3.12. ∎

3.8. THE LAMÉ CONSTANTS OF A HOMOGENEOUS ISOTROPIC ELASTIC MATERIAL WHOSE REFERENCE CONFIGURATION IS A NATURAL STATE

If we make the simultaneous assumptions that the elastic material is isotropic, homogeneous, and that the reference configuration is a natural state, we obtain the following corollary of Theorem 3.7-1:

Theorem 3.8-1. *Let there be given a homogeneous, isotropic, elastic* *material, whose reference configuration is a natural state. If the functions* γ_α, $\alpha = 0, 1, 2$, *of Theorem* 3.6-2 *are differentiable at the point* $\boldsymbol{\iota}_I = (3, 3, 1)$, *there exist two constants* λ *and* μ *such that the response function* $\hat{\boldsymbol{\Sigma}} : \mathbb{M}^3_+ \to \mathbb{S}^3$ *is of the form*

$$\hat{\boldsymbol{\Sigma}}(\boldsymbol{F}) = \tilde{\boldsymbol{\Sigma}}(\boldsymbol{C}) = \check{\boldsymbol{\Sigma}}(\boldsymbol{E}) = \lambda(\operatorname{tr} \boldsymbol{E})\boldsymbol{I} + 2\mu\boldsymbol{E} + o(\boldsymbol{E}),$$
$$\boldsymbol{C} = \boldsymbol{F}^{\mathrm{T}}\boldsymbol{F} = \boldsymbol{I} + 2\boldsymbol{E},\ \boldsymbol{F} \in \mathbb{M}^3_+ .$$ ■

In this case, and *only in this case*, the constants λ and μ are called the **Lamé constants** of the material under consideration.

By appealing to our physical intuition concerning three "ideal" experiments, we can simply impose restrictions on the admissible *numerical values* of the Lamé constants of any "real" elastic homogeneous isotropic material. In each experiment, we proceed as follows:

(i) We consider a body that occupies a reference configuration $\bar{\Omega}$ of a simple geometric form (a rectangular block, a sphere, a circular cylinder) and we assume that $\bar{\Omega}$ is a natural state.

(ii) We assume that the body can undergo a *family of deformations* of a particularly simple form: The family is indexed by a "small" parameter, and for a given parameter, *the deformation gradient is a constant in* $\bar{\Omega}$ *to within the first order with respect to the parameter*, the principal part of the deformation gradient being suggested by experience and physical intuition. More specifically, the deformations are of the form

$$\boldsymbol{\varphi}^\varepsilon : x \in \bar{\Omega} \to \boldsymbol{\varphi}^\varepsilon(x) = x + \boldsymbol{u}^\varepsilon(x) = x + \varepsilon\boldsymbol{\zeta}(x) + o(\varepsilon; x) ,$$

where ε is the "small" parameter, *the vector field* $\boldsymbol{\zeta} : \bar{\Omega} \to \mathbb{R}^3$ *is independent of* ε, *and its gradient*

$$\boldsymbol{G} := \boldsymbol{\nabla}\boldsymbol{\zeta}$$

is constant in $\bar{\Omega}$. The notation $f^\varepsilon(x) = o(\varepsilon; x)$ means that

$$\lim_{\substack{\varepsilon \to 0 \\ \varepsilon \neq 0}} \frac{\|f^\varepsilon(x)\|}{\varepsilon} = 0$$

for each $x \in \bar{\Omega}$.

Remark. To within the first order, such deformations are special cases of *homogeneous deformations*, characterized by a constant deformation gradient throughout the reference configuration; some properties of homogeneous deformations are studied in Exercise 4.7. ∎

Assume that the functions $\{\varphi^\varepsilon - (id + \varepsilon\zeta)\}$ are twice differentiable in $\bar{\Omega}$ for all ε and that their first and second partial derivatives are also of the order $o(\varepsilon; x)$ at each $x \in \bar{\Omega}$. If the response function $\hat{T}^{\mathrm{D}}$ is twice differentiable in a neighborhood of the unit matrix, the Cauchy stress tensor

$$T^\varepsilon(x^\varepsilon) := \hat{T}^{\mathrm{D}}(\nabla\varphi^\varepsilon(x)), \quad x^\varepsilon = \varphi^\varepsilon(x),$$

satisfies

$$T^\varepsilon(x^\varepsilon) = \hat{T}^{\mathrm{D}}(I + \varepsilon G + o(\varepsilon; x)) = \hat{T}^{\mathrm{D}}(I + \varepsilon G) + o(\varepsilon; x) \quad \text{for all } x \in \bar{\Omega}$$

(to avoid cumbersome notations, the exponents φ^ε are abbreviated as ε). A simple computation likewise shows that the associated first Piola–Kirchhoff stress tensor satisfies

$$T(x) = \hat{T}(\nabla\varphi^\varepsilon(x)) = \det(I + \varepsilon G)\hat{T}^{\mathrm{D}}(I + \varepsilon G)(I + \varepsilon G)^{-1} + o(\varepsilon; x).$$

Thus *both stress tensors are constant throughout the reference configuration to within the first order*. In particular then,

$$\operatorname{div} \hat{T}(\nabla\varphi^\varepsilon(x)) = o(\varepsilon; x),$$

so that $\operatorname{div} \hat{T}(\nabla\varphi^\varepsilon) = o$ in Ω to within the first order with respect to the parameter ε. This means that we may consider that *the above deformations φ^ε are caused by applied surface forces only*, thereby neglecting the influence of body forces, and hence that the *assumed form of the Cauchy stress tensor is solely dependent on the form of the applied surface forces*.

(iii) The assumed form of the deformation φ^ε implies that

$$\nabla\varphi^\varepsilon(x)^{\mathrm{T}}\nabla\varphi^\varepsilon(x) = I + \varepsilon(G + G^{\mathrm{T}}) + o(\varepsilon; x) \quad \text{for all } x \in \bar{\Omega},$$

so that the second Piola–Kirchhoff stress tensor is given by

$$\Sigma(x) = \varepsilon(\lambda(\operatorname{tr} G)I + \mu(G^{\mathrm{T}} + G)) + o(\varepsilon; x),$$

and thus the constitutive equation takes the similar form:

$$\begin{aligned} \boldsymbol{T}^{\varepsilon}(x^{\varepsilon}) &= (\det \boldsymbol{\nabla}\boldsymbol{\varphi}^{\varepsilon}(x))^{-1}\boldsymbol{\nabla}\boldsymbol{\varphi}^{\varepsilon}(x)\boldsymbol{\Sigma}(x)\boldsymbol{\nabla}\boldsymbol{\varphi}^{\varepsilon}(x)^{-\mathrm{T}} \\ &= \varepsilon(\lambda(\operatorname{tr}\boldsymbol{G})\boldsymbol{I} + \mu(\boldsymbol{G}^{\mathrm{T}} + \boldsymbol{G})) + o(\varepsilon; x) \\ &\text{for all } x^{\varepsilon} = \boldsymbol{\varphi}^{\varepsilon}(x),\ x \in \bar{\Omega}\,. \end{aligned}$$

On the other hand, the relations $\boldsymbol{T}^{\varepsilon}(x^{\varepsilon}) = \hat{\boldsymbol{T}}^{\mathrm{D}}(\boldsymbol{I} + \varepsilon\boldsymbol{G}) + o(\varepsilon; x)$ and $\hat{\boldsymbol{T}}^{\mathrm{D}}(\boldsymbol{I}) = \boldsymbol{0}$ (the reference configuration is assumed to be a natural state) imply that the Cauchy stress tensor is of the form

$$\boldsymbol{T}^{\varepsilon}(x^{\varepsilon}) = \varepsilon\boldsymbol{T} + o(\varepsilon; x), \quad \text{with } (\boldsymbol{T})_{ij} = \frac{\partial \hat{T}_{ij}^{\mathrm{D}}}{\partial \boldsymbol{F}}(\boldsymbol{I})\,,$$

so that $\boldsymbol{T}$ is a symmetric tensor independent of x. Thus by equating the first-order parts of each side of the constitutive equation, we obtain the relation

$$\boldsymbol{T} = \lambda(\operatorname{tr}\boldsymbol{G})\boldsymbol{I} + \mu(\boldsymbol{G}^{\mathrm{T}} + \boldsymbol{G})\,.$$

(iv) To within the first order with respect to the parameter ε, we assume that the corresponding Cauchy stress tensor presents some simple features, suggested by experience and physical intuition. This means that one of its components has a given sign, as in the first experiment, or that it coincides with one of the special cases considered in Fig. 2.3-3 as in the second and third experiments. We then express that *the direction of the applied forces is related in some natural manner to the direction in which the resulting displacement occurs*. Taking such a relation into account in the equation $\boldsymbol{T} = \lambda(\operatorname{tr}\boldsymbol{G})\boldsymbol{I} + \mu(\boldsymbol{G}^{\mathrm{T}} + \boldsymbol{G})$ results in an *inequality involving the Lamé constants* λ *and* μ.

Let us now describe the three experiments. In the *first experiment*, the set $\bar{\Omega}$ is a rectangular block, and the assumed displacement is of the form

$$\boldsymbol{u}^{\varepsilon}(x) = \varepsilon\begin{pmatrix} 0 \\ x_3 \\ 0 \end{pmatrix} + o(\varepsilon; x)\,,$$

as indicated in Fig. 3.8-1.

Remarks. (1) To within the first order, such a deformation is called a *simple shear* (for complements, see Exercise 3.13).

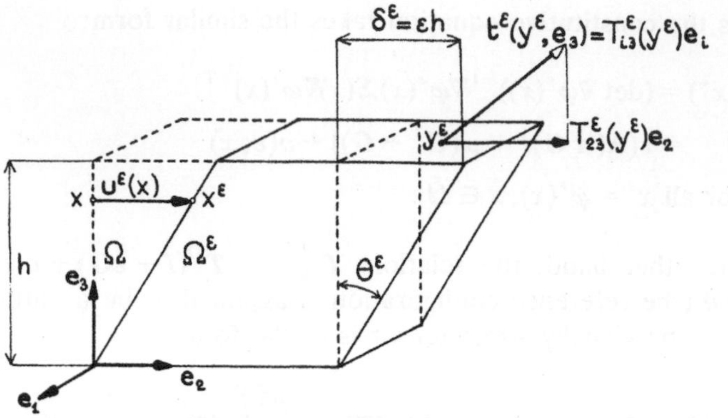

Fig. 3.8-1. "Simple shear" of a rectangular block. This experiment shows that the Lamé constant $\mu = \frac{T^{\varepsilon}_{23}}{\varepsilon}$ is >0, where $\varepsilon = \frac{\delta^{\varepsilon}}{h}$.

(2) In Fig. 3.8-1, as in the next two figures, only the principal part of the deformation, i.e., the linear part with respect to ε, is represented. Note that each figure is drawn for a strictly positive value of the parameter, which purposely far exceeds values that could be observed in actual experiments. ■

We then express the natural assumption that the component $T^{\varepsilon}_{23}(x^{\varepsilon})$ of the corresponding Cauchy stress tensor is of the form

$$T^{\varepsilon}_{23}(x^{\varepsilon}) = \varepsilon T_{23} + o(\varepsilon; x) .$$

Since in this case the matrix $\boldsymbol{G}$ is given by

$$\boldsymbol{G} = \begin{pmatrix} 0 & 0 & 0 \\ 0 & 0 & 1 \\ 0 & 0 & 0 \end{pmatrix},$$

the constitutive equation yields

$$\varepsilon T_{23} = \varepsilon\mu + o(\varepsilon; x), \quad x \in \bar{\Omega} ,$$

and thus we obtain a *first inequality*:

$$\boxed{\mu > 0 .}$$

In the *second experiment* (Fig. 3.8-2), the set $\bar{\Omega}$ is a sphere, and the assumed displacement is of the form

$$u^{\varepsilon}(x) = -\varepsilon\begin{pmatrix} x_1 \\ x_2 \\ x_3 \end{pmatrix} + o(\varepsilon; x) .$$

We then express the natural assumption that the principal part of the corresponding Cauchy stress tensor is a *pressure* for $\varepsilon > 0$ (according to the terminology of Fig. 2.3-3), i.e., that it is of the form

$$T^{\varepsilon}(x^{\varepsilon}) = -\pi\varepsilon I + o(\varepsilon; x), \quad \pi > 0 .$$

Since $G = -I$, the constitutive equation yields

$$-\pi\varepsilon I = -\varepsilon(3\lambda + 2\mu)I + o(\varepsilon; x) ,$$

and from this equation we deduce a *second inequality*:

$$\boxed{3\lambda + 2\mu > 0 .}$$

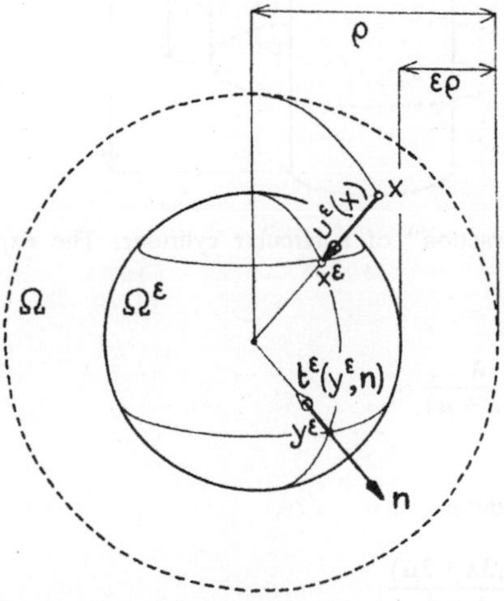

Fig. 3.8-2. "Uniform compression" of a sphere. This experiment shows that the bulk modulus $\kappa = \frac{1}{3}(3\lambda + 2\mu) = \dfrac{|T^{\varepsilon}|}{3\varepsilon}$ is >0, where $\varepsilon = \dfrac{\rho - \rho^{\varepsilon}}{\rho}$.

In the *third experiment* (Fig. 3.8-3), the set $\bar{\Omega}$ is a circular cylinder, and the assumed displacement is of the form

$$u^\varepsilon(x) = \varepsilon \begin{pmatrix} -\nu x_1 \\ -\nu x_2 \\ x_3 \end{pmatrix} + o(\varepsilon; x),$$

where the constant $\nu > 0$ is *to be determined* (it will turn out to be a

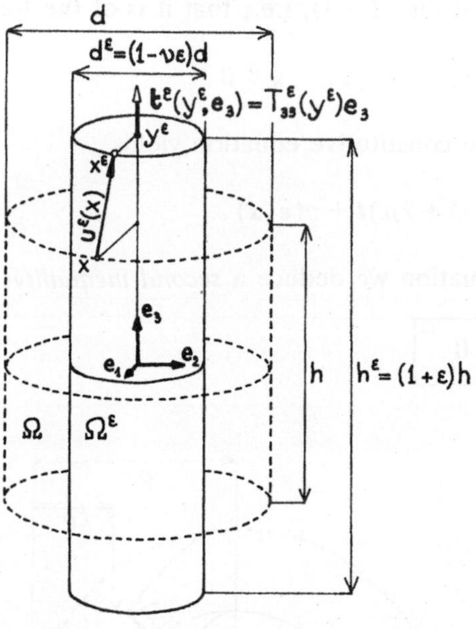

Fig. 3.8-3. "Uniform traction" of a circular cylinder. The experiment shows that the Poisson ratio

$$\nu = \frac{\left(\dfrac{d - d^\varepsilon}{d}\right)}{\left(\dfrac{h^\varepsilon - h}{h}\right)} = \frac{\lambda}{2(\lambda + \mu)}$$

and that the Young modulus

$$E = \frac{T_{33}^\varepsilon}{\left(\dfrac{h^\varepsilon - h}{h}\right)} = \frac{\mu(3\lambda + 2\mu)}{\lambda + \mu}$$

are both >0.

function of λ and μ). We then express the natural assumption that the corresponding Cauchy stress tensor is a *pure tension in the direction* $\boldsymbol{e}_3$ (according to the terminology of Fig. 2.3-3) to within the first order, i.e., that

$$T^{\varepsilon}(x^{\varepsilon}) = \varepsilon\begin{pmatrix} 0 & 0 & 0 \\ 0 & 0 & 0 \\ 0 & 0 & E \end{pmatrix} + o(\varepsilon; x) \quad \text{for some constant } E > 0$$

(the constant E will again be shown to be a function of λ and μ). Since in this case the matrix G is given by

$$G = \begin{pmatrix} -\nu & 0 & 0 \\ 0 & -\nu & 0 \\ 0 & 0 & 1 \end{pmatrix},$$

the constitutive equation reads, componentwise,

$$T^{\varepsilon}_{\alpha\alpha}(x^{\varepsilon}) = \varepsilon(\lambda(1-2\nu) - 2\nu\mu) + o(\varepsilon; x), \quad \alpha = 1, 2,$$
$$T^{\varepsilon}_{ij}(x^{\varepsilon}) = o(\varepsilon; x), \quad i \neq j,$$
$$T^{\varepsilon}_{33}(x^{\varepsilon}) = \varepsilon(\lambda(1-2\nu) + 2\mu) + o(\varepsilon; x).$$

Expressing that $T^{\varepsilon}_{\alpha\alpha}(x^{\varepsilon}) = o(\varepsilon; x)$ yields the relation $-2\nu(\lambda + \mu) + \lambda = 0$. Since the inequalities $\mu > 0$ and $(3\lambda + 2\mu) > 0$ imply $(\lambda + \mu) > 0$, we can solve the preceding relation with respect to the constant ν:

$$\nu = \frac{\lambda}{2(\lambda + \mu)},$$

which thus appears as a well-determined function of the Lamé constants. Therefore the natural assumption $\nu > 0$ yields a *third inequality*:

$$\boxed{\lambda > 0.}$$

Of course this last inequality, coupled with the first one ($\mu > 0$), renders the second one ($3\lambda + 2\mu > 0$) redundant *a posteriori*. Yet the inequality $(3\lambda + 2\mu) > 0$ was a needed intermediary in that it guaranteed that the number $(\lambda + \mu)$ was >0.

If we now compute the component $T^{\varepsilon}_{33}(x^{\varepsilon})$ of the Cauchy stress tensor, we obtain, upon replacing the constant ν by its expression in terms of λ and μ:

$$T^{\varepsilon}_{33}(x^{\varepsilon}) = \varepsilon E + o(\varepsilon; x),$$

where the number

$$E = \frac{\mu(3\lambda + 2\mu)}{\lambda + \mu}$$

is thus automatically >0, since we already know that the Lamé constants are >0.

Each constant ν and E has a remarkable physical interpretation, which is clear from the last experiment: The constant ν, which is called the **Poisson ratio** of the material, measures to within the first order the ratio between the relative decrease of the diameter of the cylinder and the relative increase of its length (Fig. 3.8-3). The constant E, which is called the **Young modulus** of the material, measures to within the first order the ratio between the component T^{ε}_{33} of the Cauchy stress tensor and the relative increase in length $\varepsilon = (h^{\varepsilon} - h)/h$ (Fig. 3.8-3).

In the same vein, the Lamé constant μ measures, to within the first order, half of the ratio between the component T^{ε}_{23} of the Cauchy stress tensor and $\varepsilon = \operatorname{tg}\theta^{\varepsilon}$ (with the notation of Fig. 3.8-1). For this reason, the constant μ is also called the *shear modulus* of the material. Finally, the number $(3\lambda + 2\mu)$ measures to within the first order the ratio between the pressure $\pi^{\varepsilon} = \pi\varepsilon$ and the relative decrease ε of the diameter of the sphere (Fig. 3.8-2). The number

$$\kappa = \tfrac{1}{3}(3\lambda + 2\mu)$$

is called the *bulk modulus* of the given material.

Such considerations are indeed the basis for the experimental determination of the Lamé constants λ, μ, the Poisson ratio ν, and the Young modulus E, whose average values for some common elastic materials are given in Fig. 3.8-4. Observe that the physical interpretation of each constant specifies the unit in which it should be expressed.

Since the Lamé constants, the Poisson ratio, and the Young modulus, are related by the following equations:

$$\nu = \frac{\lambda}{2(\lambda + \mu)}, \qquad E = \frac{\mu(3\lambda + 2\mu)}{\lambda + \mu},$$

$$\lambda = \frac{E\nu}{(1 + \nu)(1 - 2\nu)}, \qquad \mu = \frac{E}{2(1 + \nu)},$$

	E (10^5 kg/cm^2)	ν	λ (10^5 kg/cm^2)	μ (10^5 kg/cm^2)	$\kappa = \frac{1}{3}(3\lambda + 2\mu)$ (10^5 kg/cm^2)
Steel	21	0.28	10	8.2	16
Iron	20	0.28	9.9	7.8	15
Copper	11	0.34	8.7	4.1	11
Bronze	10	0.31	6.2	3.8	8.8
Aluminium	7.0	0.34	5.6	2.6	7.3
Glass	5.5	0.25	2.2	2.2	3.7
Nickel	2.2	0.30	1.3	0.85	1.8
Lead	1.8	0.44	4.6	0.63	5.0
Rubber	0.037	0.485	0.40	0.012	0.41

Fig. 3.8-4. Average values of the constants E, ν, λ, μ for common elastic materials.

it is easily seen that the following equivalence holds:

$$\lambda > 0 \text{ and } \mu > 0 \Leftrightarrow 0 < \nu < \tfrac{1}{2} \text{ and } E > 0 .$$

Remarks. (1) Whilst the first and third experiment can be realized with incompressible materials as well, since the corresponding deformations are volume-preserving to within the first order (provided $\nu = \frac{1}{2}$ in the third case), *the assumed form of the displacement precludes incompressible materials in the second experiment*, since vol Ω^{ε} = vol $\Omega\,(1 - 3\varepsilon + o(\varepsilon))$ in this case. In this direction, see the discussion given in Scott [1986].

(2) So far, we have been concerned with the behavior of the constitutive equation of an elastic material for small values of $\|E\|$, i.e., for "moderate" strains only. The behavior for "large" strains will be studied in Sect. 4.6.

(3) An extensive discussion of "special" deformations, and of the applied forces that induce them, such as those considered in this section for the determination of the Lamé constants, is given in Green & Zerna [1968, Chapter 3]. See also Ogden [1984, Sect. 5.2].

(4) There are many specialized treatments of constitutive equations for elastic materials in the literature; see in particular Murnaghan [1951], Varga [1966], Bell [1973], Chen & Saleeb [1982], Ogden [1984]. ■

To conclude, we wish to emphasize that the existence and the experimental determination of the Lamé constants have been derived here from an expansion of the constitutive equation with respect to the Green-St Venant strain tensor $\boldsymbol{E} = \frac{1}{2}(\nabla \boldsymbol{u}^{\mathrm{T}} + \nabla \boldsymbol{u} + \nabla \boldsymbol{u}^{\mathrm{T}} \nabla \boldsymbol{u})$, and not, as is often done, with respect to the *linearized strain tensor* $\frac{1}{2}(\nabla \boldsymbol{u}^{\mathrm{T}} + \nabla \boldsymbol{u})$. This second approach is unduly restrictive, since it tends to erroneously indicate that the Lamé constants are restricted to linearized elasticity.

3.9. ST VENANT–KIRCHHOFF MATERIALS

If we neglect the higher-order terms in the expansion of the second Piola–Kirchhoff stress tensor, we obtain a first candidate for a response function, as proposed by St Venant [1844] and Kirchhoff [1852]. An elastic material is a **St Venant–Kirchhoff material** if its response function for the second Piola–Kirchhoff stress is of the form

$$\boxed{\check{\boldsymbol{\Sigma}}(\boldsymbol{E}) = \tilde{\boldsymbol{\Sigma}}(\boldsymbol{I} + 2\boldsymbol{E}) = \lambda(\operatorname{tr} \boldsymbol{E})\boldsymbol{I} + 2\mu \boldsymbol{E}, \quad \boldsymbol{I} + 2\boldsymbol{E} \in \mathbb{S}^3_>\,,}$$

where λ and μ are constants. Clearly such a material is homogeneous and the reference configuration is a natural state. We also have

$$\tilde{\boldsymbol{\Sigma}}(\boldsymbol{C}) = \left\{\frac{\lambda}{2}(\iota_1(\boldsymbol{C}) - 3) - \mu\right\}\boldsymbol{I} + \mu\boldsymbol{C}, \quad \boldsymbol{C} = \boldsymbol{I} + 2\boldsymbol{E}\,,$$

where $\iota_1(\boldsymbol{C}) = \operatorname{tr} \boldsymbol{C}$, so that the material is frame-indifferent and isotropic, by Theorem 3.6-2. Thus the constants λ and μ are precisely the Lamé constants of a St Venant–Kirchhoff material. Notice the equivalent expression of the constitutive equation, in terms of the Young modulus E and of the Poisson ratio ν:

$$\check{\boldsymbol{\Sigma}}(\boldsymbol{E}) = \frac{E}{(1+\nu)}\left\{\frac{\nu}{(1-2\nu)}(\operatorname{tr} \boldsymbol{E})\boldsymbol{I} + \boldsymbol{E}\right\}.$$

Hence by definition, the mapping $\boldsymbol{E} \rightarrow \check{\boldsymbol{\Sigma}}(\boldsymbol{E})$ corresponding to a St Venant–Kirchhoff material is linear. Notice however that the associated mapping

$$\boldsymbol{u} \rightarrow \check{\boldsymbol{\Sigma}}(\boldsymbol{E}(\boldsymbol{u})) = \lambda(\operatorname{tr} \nabla\boldsymbol{u})\boldsymbol{I} + \mu(\nabla\boldsymbol{u}^{\mathrm{T}} + \nabla\boldsymbol{u}) + \frac{\lambda}{2}(\operatorname{tr} \nabla\boldsymbol{u}^{\mathrm{T}}\nabla\boldsymbol{u})\boldsymbol{I} + \mu\nabla\boldsymbol{u}^{\mathrm{T}}\nabla\boldsymbol{u}\,,$$

where the displacement vector $\boldsymbol{u}$ is related to the strain tensor $\boldsymbol{E}(\boldsymbol{u})$ by

$$2\boldsymbol{E}(\boldsymbol{u}) = \nabla\boldsymbol{u}^{\mathrm{T}} + \nabla\boldsymbol{u} + \nabla\boldsymbol{u}^{\mathrm{T}}\nabla\boldsymbol{u}\,,$$

is no longer linear, since it also contains quadratic terms. This is equally evident from the componentwise expression of the constitutive equation, which we record here for future uses; if we let $\check{\Sigma}(E) = (\check{\sigma}_{ij}(E))$ and $E = (E_{ij})$, we have:

$$\check{\sigma}_{ij}(E) = \lambda E_{kk}\delta_{ij} + 2\mu E_{ij}, \quad E_{ij} = \tfrac{1}{2}(\partial_i u_j + \partial_j u_i + \partial_i u_k \partial_j u_k)\,,$$

or equivalently,

$$\check{\sigma}_{ij}(E) = a_{ijkl}E_{kl} \quad \text{with } a_{ijkl} := \lambda\delta_{ik}\delta_{kl} + \mu(\delta_{ik}\delta_{jl} + \delta_{il}\delta_{jk})\,.$$

Since St Venant–Kirchhoff materials are the simplest among the nonlinear models (in the sense that they are the simplest that are compatible with Theorem 3.8-1), *they are quite popular in actual computations*, where they are often used to model engineering structures in conjunction with finite element methods (see in particular Oden [1972] and Washizu [1975]).

On the other hand the relative simplicity of their practical implementation is more than compensated by various shortcomings. One inadequacy is, perhaps unexpectedly, the invertibility of the associated linear mapping

$$\check{\Sigma} : E \in \mathbb{S}^3 \to \Sigma = \check{\Sigma}(E) = \lambda(\operatorname{tr} E)I + 2\mu E \in \mathbb{S}^3\,.$$

It is easily seen that such a mapping is invertible if and only if $\mu(3\lambda + 2\mu) \neq 0$ (and we have seen in Sect. 3.8 that the Lamé constants of actual materials satisfy $\lambda > 0$, $\mu > 0$), in the following form:

$$E = -\frac{\lambda}{2\mu(3\lambda + 2\mu)}(\operatorname{tr}\Sigma)I + \frac{1}{2\mu}\Sigma = \frac{1}{E}\{-\nu(\operatorname{tr}\Sigma)I + (1+\nu)\Sigma\}$$

or equivalently, letting $\Sigma = (\sigma_{ij})$:

$$E_{ij} = A_{ijkl}\sigma_{kl}, \quad \text{with } A_{ijkl} := -\frac{\nu}{E}\delta_{ij}\delta_{kl} + \frac{(1+\nu)}{2E}(\delta_{ik}\delta_{jl} + \delta_{il}\delta_{jk})\,.$$

But, on the other hand, some "eversion problems" (Sect. 5.8) indicate that large strains are also possible when the stress is small (Antman [1979], Truesdell [1978]), while a linear relation implies that the stress is small if and only if the strain is small. In the same direction, Ogden [1977] has also established the surprising result that, for general isotropic elastic solids, any given first Piola–Kirchhoff stress tensor $T = \hat{T}(F)$ should correspond to at least four distinct deformation gradients F.

Another drawback is that, although such materials are hyperelastic, their associated stored energy function is not polyconvex (the notions of hyperelasticity and polyconvexity are introduced in the next Chapter); as a consequence, only partial existence results are as yet available for such materials as we shall see. The lack of any term preventing det $\nabla\boldsymbol{\varphi}$ to approach zero, let alone to become negative, in the associated stored energy function is another serious disadvantage, both mathematically and empirically, since any reasonable model should hinder this possibility (Sect. 4.6).

At their best, St Venant–Kirchhoff materials can be only expected to be useful in a narrow range of "small" strains $\boldsymbol{E}$, as indeed they should be from their very definition; this is why such materials are often referred to as "large displacement–small strain" models. In spite of these various inadequacies, St Venant–Kirchhoff materials can be nevertheless expected to perform better than the linearized models (Chapter 6) that are so often used.

EXERCISES

3.1. (1) Show that the mappings $\boldsymbol{F} \to \boldsymbol{R}$ and $\boldsymbol{F} \to \boldsymbol{U}$ defined by the polar factorization $\boldsymbol{F} = \boldsymbol{RU}$ are continuous when $\boldsymbol{F}$ varies in the set of all invertible matrices. Are they differentiable?

(2) Show that the mapping $\boldsymbol{C} \in \mathbb{S}^n_> \to \boldsymbol{C}^{1/2} \in \mathbb{S}^n_>$ is infinitely differentiable; compute its first and second derivatives.

(3) Show that the mapping $\boldsymbol{C} \in \mathbb{S}^n_> \to \{\boldsymbol{C}^{1/2}\}^{-1} \in \mathbb{S}^n$ is infinitely differentiable; compute its first and second derivatives.

Remark. For these and further results, see Guo [1984], Hoger & Carlson [1984a, 1984b], Ting [1985].

3.2. Let a matrix $\boldsymbol{F} \in \mathbb{M}^n_+$ be given.

(1) Show that the orthogonal matrix found in the polar factorization of the matrix $\boldsymbol{F}$ (Theorem 3.2-2) is the unique solution of the minimization problem: Find $\boldsymbol{R} \in \mathbb{O}^n_+$ such that

$$\|\boldsymbol{F} - \boldsymbol{R}\| = \inf_{\boldsymbol{S} \in \mathbb{O}^n_+} \|\boldsymbol{F} - \boldsymbol{S}\| .$$

(2) Deduce from (1) an alternate proof of the polar factorization theorem.

Remark. These results are due to Martins & Podio-Guidugli [1979]. See also Martins & Podio-Guidugli [1980].

3.3. (1) Show that the polar factorization of a singular matrix $\boldsymbol{F}$ is no longer unique; specify as precisely as possible how two polar factorizations $\boldsymbol{F} = \boldsymbol{R}_1\boldsymbol{U}_1$ and $\boldsymbol{F} = \boldsymbol{R}_2\boldsymbol{U}_2$ may differ in this case.

(2) Show that the rank of a matrix is equal to the number of its singular values that are >0.

3.4. The purpose of this exercise is to provide another proof to Theorem 3.2-4.

(1) Let $\boldsymbol{A}$ and $\boldsymbol{B}$ be two symmetric matrices, with eigenvalues λ_i and μ_i respectively, ordered as $\lambda_1 \geqslant \lambda_2 \geqslant \cdots \geqslant \lambda_n$ and $\mu_1 \geqslant \mu_2 \geqslant \cdots \geqslant \mu_n$. Show that $\operatorname{tr} \boldsymbol{AB} \leqslant \Sigma_{i=1}^{n} \lambda_i \mu_i$ (Mirsky [1959]).

(2) Show that the result of Theorem 3.2-4 is a consequence of (1), by relating the singular values of an arbitrary square matrix $\boldsymbol{A}$ to the eigenvalues of the symmetric matrix

$$\begin{pmatrix} \boldsymbol{0} & \boldsymbol{A} \\ \boldsymbol{A}^{\mathrm{T}} & \boldsymbol{0} \end{pmatrix}.$$

3.5. The purpose of this exercise is to provide yet another proof to Theorem 3.2-4.

(1) Let $\boldsymbol{D}_\alpha = \mathbf{Diag}\, \alpha_i$, $\boldsymbol{D}_\beta = \mathbf{Diag}\, \beta_i$, where the numbers α_i and β_i are ordered as

$$\alpha_1 \geqslant \alpha_2 \geqslant \cdots \geqslant \alpha_n \geqslant 0, \quad \beta_1 \geqslant \beta_2 \geqslant \cdots \geqslant \beta_n \geqslant 0.$$

By using the Lagrange multipler technique (see e.g. Ciarlet [1983, p. 149]), show that

$$\max_{\boldsymbol{P},\boldsymbol{Q} \in \mathbb{O}^n} |\operatorname{tr}(\boldsymbol{P}\boldsymbol{D}_\alpha \boldsymbol{Q}\boldsymbol{D}_\beta)| = \sum_{i=1}^{n} \alpha_i \beta_i .$$

(2) Show that the result of Theorem 3.2-4 follows from (1).

3.6. Give another proof of the first equivalence of Theorem 3.3-1, by keeping the deformed configuration fixed and rotating the basis: Given a rotation $\boldsymbol{Q} \in \mathbb{O}^3_+$, find the expressions of the Cauchy stress tensor and of the deformation gradient with respect to the new basis $(\boldsymbol{e}_i')$, where $\boldsymbol{e}_i' = \boldsymbol{Q}^{\mathrm{T}}\boldsymbol{e}_i$.

3.7. (1) Show that a linear mapping from $\mathbb{M}^3$ into $\mathbb{M}^3$ whose restriction to the set $\mathbb{O}^3_+$ is constant is the null mapping.

(2) Assume that the reference configuration is a natural state. Show that neither response functions $\hat{\boldsymbol{T}}$ and $\hat{\boldsymbol{T}}^{\mathrm{D}}$ can be simultaneously frame-indifferent and linear in $\boldsymbol{F}$ (Fosdick & Serrin [1979]).

(3) Show that the response function $\hat{\boldsymbol{T}}$ can be simultaneously frame-indifferent and linear in $\boldsymbol{F}$ when the reference configuration is not a natural state (Podio-Guidugli [1987a]).

(4) Show that the response function $\hat{\boldsymbol{T}}^{\mathrm{D}}$ cannot be simultaneously frame-indifferent and linear in $\boldsymbol{F}$, even if the reference configuration is not a natural state (Podio-Guidugli [1987a]).

Remark. For related results, see Bampi & Morro [1982], Dunn [1981].

3.8. Consider an elastic material that is not necessarily isotropic at a point $x \in \bar{\Omega}$. Show that the set of all rotations $\boldsymbol{Q} \in \mathbb{O}^3_+$ that satify

$$\hat{\boldsymbol{T}}^{\mathrm{D}}(x, \boldsymbol{F}) = \hat{\boldsymbol{T}}^{\mathrm{D}}(x, \boldsymbol{F}\boldsymbol{Q}) \quad \text{for all } \boldsymbol{F} \in \mathbb{M}^3_+ ,$$

is a subgroup of the group $\mathbb{O}^3_+$.

3.9. Consider a homogeneous isotropic elastic material whose response function $\tilde{\boldsymbol{\Sigma}} : \mathbb{S}^3_> \to \mathbb{S}^3$ is of the form (Theorem 3.6-2):

$$\tilde{\boldsymbol{\Sigma}}(\boldsymbol{C}) = \gamma_0(\boldsymbol{\iota}_C)\boldsymbol{I} + \gamma_1(\boldsymbol{\iota}_C)\boldsymbol{C} + \gamma_2(\boldsymbol{\iota}_C)\boldsymbol{C}^2 \quad \text{for all } \boldsymbol{C} \in \mathbb{S}^3_> .$$

Clearly, the function $\tilde{\boldsymbol{\Sigma}}$ is differentiable at $\boldsymbol{C} \in \mathbb{S}^3_>$ if the functions $\gamma_0, \gamma_1, \gamma_2$ are differentiable at $\boldsymbol{\iota}_C$. Is the converse true?

Remark. The differentiability of the functions γ_α at the particular point $(3, 3, 1)$ was an assumption in both Theorems 3.7-1 and 3.8-1.

3.10. Let

$$\boldsymbol{\iota}(\mathbb{S}^3_>) = \{\boldsymbol{\iota}_A \in \mathbb{R}^3; \boldsymbol{A} \in \mathbb{S}^3_>\} .$$

Describe as precisely as possible this subset of the set $]0, +\infty[^3$, notably by specifying its boundary. Conclude in particular that the invariance of domain theorem (Theorem 1.2-5) does not apply to the set $\boldsymbol{\iota}(\mathbb{S}^3_>)$; yet $\boldsymbol{\iota}(\mathbb{S}^3_>)$ is the image of an open subset of $\mathbb{S}^3$ by a continuous mapping.

Remark. The set $\iota(\mathbb{S}^3_>)$ is the domain of the functions $\beta_\alpha(x,\cdot)$ and $\gamma_\alpha(x,\cdot)$, $\alpha = 0, 1, 2$, appearing in Theorem 3.6-2.

3.11. The notations and assumptions are as in Theorem 3.7-1 (we drop the dependence on x for notational brevity). Show that, if the functions $\gamma_0, \gamma_1, \gamma_2$ are twice differentiable at the point ι_I, the second Piola–Kirchhoff stress tensor is of the form

$$\begin{aligned}\boldsymbol{\Sigma} = \check{\boldsymbol{\Sigma}}(\boldsymbol{E}) = &-\pi \boldsymbol{I} + \lambda(\operatorname{tr} \boldsymbol{E})\boldsymbol{I} + 2\mu \boldsymbol{E} \\ &+ \nu_1(\operatorname{tr} \boldsymbol{E}^2)\boldsymbol{I} + \nu_2(\operatorname{tr} \boldsymbol{E})^2\boldsymbol{I} + \nu_3(\operatorname{tr} \boldsymbol{E})\boldsymbol{E} + \nu_4\boldsymbol{E}^2 \\ &+ o(\|\boldsymbol{E}\|^2)\,,\end{aligned}$$

where $\nu_1, \nu_2, \nu_3, \nu_4$ are constants. If the $o(\|\boldsymbol{E}\|^2)$ term is neglected in the above expansion, the resulting constitutive equation corresponds to a *Murnaghan's material* (named after Murnaghan [1937], although it seems to have been first proposed by Voigt [1893–1894]). Such a model, which in essence represents one step further than a St Venant–Kirchhoff material, seems to be of limited interest, both practically or theoretically. In particular the argument given in Novozhilov [1953, p. 126] indicates that an actual response function $\check{\boldsymbol{\Sigma}}$ should only contain odd functions of the tensor $\boldsymbol{E}$.

3.12. Let $\boldsymbol{G}: \mathbb{S}^3 \to \mathbb{S}^3$ be a mapping that satisfies

$$(*) \quad \boldsymbol{G}(\boldsymbol{Q}\boldsymbol{A}\boldsymbol{Q}^{\mathrm{T}}) = \boldsymbol{Q}\boldsymbol{G}(\boldsymbol{A})\boldsymbol{Q}^{\mathrm{T}} \quad \text{for all } \boldsymbol{A} \in \mathbb{S}^3_>,\ \boldsymbol{Q} \in \mathbb{O}^3_+\,.$$

(1) Show that $(*)$ holds for all $\boldsymbol{A} \in \mathbb{S}^3_>$ and all $\boldsymbol{Q} \in \mathbb{O}^3$.
(2) If $\boldsymbol{G}$ is linear, show that $(*)$ holds for all $\boldsymbol{A} \in \mathbb{S}^3$ and all $\boldsymbol{Q} \in \mathbb{O}^3$.
(3) If $\boldsymbol{G}$ is linear, show directly that there exist two constants λ and μ such that

$$\boldsymbol{G}(\boldsymbol{A}) = \lambda(\operatorname{tr} \boldsymbol{A})\boldsymbol{I} + 2\mu \boldsymbol{A} \quad \text{for all } \boldsymbol{A} \in \mathbb{S}^3\,.$$

Remark. These results are proved in Gurtin [1972, Sect. 22; 1974; 1981b, p. 235], Martins & Podio-Guidugli [1978]. See also Guo [1983a, 1983b], Telega [1984], de Boor [1985], for the generalization to mappings $\boldsymbol{G}: \mathbb{M}^3 \to \mathbb{M}^3$ that satisfy $(*)$.

(4) Show that the response function $\boldsymbol{\Sigma}$ satisfies $(*)$ and that the result

of (3) thus provides a short proof of Theorem 3.8-1 when the response $\check{\Sigma}$ is a linear function of E.

3.13. Consider a homogeneous, isotropic, rectangular block occupying the reference configuration $\bar{\Omega}$ indicated in Fig. 3.8-1, and subjected to a *simple shear*, i.e., a deformation of the form

$$\varphi : x \in \bar{\Omega} \rightarrow \varphi(x) = x + \alpha x_3 e_2 , \quad \alpha \in \mathbb{R} .$$

(1) Show that the Cauchy stress tensor is independent of x^φ, that $T^\varphi_{12} = T^\varphi_{13} = 0$, and that T^φ_{23} is an odd function of α.

(2) Show that $T^\varphi_{22} - T^\varphi_{33} = \alpha T^\varphi_{23}$.

Remarks. The relation $T^\varphi_{22} \neq T^\varphi_{33}$ for $\alpha \neq 0$ (which follows from (2)) is called the *Poynting effect*. Also note that the relation established in (2) is valid regardless of the material constituting the body; it is called *Rivlin's universal relation*. Discussions on these matters are found in Wang & Truesdell [1973, p. 280 ff], Gurtin [1981a, Chapter 6]. See also Beatty [1987], Rajagopal & Wineman [1987].

CHAPTER 4

HYPERELASTICITY

INTRODUCTION

If a material is elastic, we can replace the first Piola–Kirchhoff stress tensor $T(x)$ by $\hat{T}(x, \nabla\varphi(x))$ in the equations of equilibrium, which then form a system of three nonlinear partial differential equations, and boundary conditions, with respect to the three unknown components of the deformation φ, *viz.*,

$$\begin{cases} -\operatorname{\mathbf{div}} \hat{T}(x, \nabla\varphi(x)) = \hat{f}(x, \varphi(x)) , \quad x \in \Omega , \\ \hat{T}(x, \nabla\varphi(x))n = \hat{g}(x, \nabla\varphi(x)) , \quad x \in \Gamma_1 , \\ \varphi(x) = \varphi_0(x) , \quad x \in \Gamma_0 . \end{cases}$$

An elastic material is *hyperelastic* if there exists a *stored energy function* $\hat{W} : \bar{\Omega} \times \mathbb{M}^3_+ \to \mathbb{R}$ such that (Sect. 4.1)

$$\hat{T}(x, F) = \frac{\partial \hat{W}}{\partial F}(x, F) \quad \text{for all } x \in \bar{\Omega},\ F \in \mathbb{M}^3_+ .$$

If this is the case, and if the applied forces are conservative, solving the above boundary value problem is formally equivalent to finding the stationary point of a functional, called the *total energy* (Theorems 4.1-1 and 4.1-2) when the *admissible deformations* $\psi : \bar{\Omega} \to \mathbb{R}^3$ satisfy the constraints $\det \nabla\psi > 0$ in Ω and $\psi = \varphi_0$ on Γ_0 (for ease of exposition, the injectiveness constraint will not be taken into account until Chapter 5). If the applied forces are dead loads, the total energy is given by

$$I(\psi) = \int_\Omega \hat{W}(x, \nabla\psi(x))\, dx - \left\{ \int_\Omega f \cdot \psi \, dx + \int_{\Gamma_1} g \cdot \psi \, da \right\} .$$

We show (Theorem 4.2-1) that the *axiom of material frame-indifference* implies that, at each point $x \in \bar{\Omega}$, the stored energy function $\hat{W}(x, \cdot)$ is only a function of the strain tensor $C = F^T F$, in the sense that there exists a mapping $\tilde{W}(x, \cdot)$ such that

$$\hat{W}(x, F) = \tilde{W}(x, F^{\mathrm{T}}F) \quad \text{for all } F \in \mathbb{M}^3_+ \,,$$

and that the second Piola–Kirchhoff stress tensor is then given by (Theorem 4.2-2)

$$\tilde{\Sigma}(x, C) = 2 \frac{\partial \tilde{W}}{\partial C}(x, C) \quad \text{for all } C \in \mathbb{S}^3_> \,.$$

We next show (Theorem 4.4-1) that the form of the stored energy function at a point $x \in \bar{\Omega}$ can be further reduced if the material is *isotropic at* x, since in this case there is a function $\dot{W}(x, \cdot)$ such that

$$\tilde{W}(x, C) = \dot{W}(x, \iota_C) \quad \text{for all } C \in \mathbb{S}^3_> \,.$$

We also show (Theorem 4.5-1) that, if the hyperelastic material is homogeneous and isotropic, and if the reference configuration is a natural state, the terms of lowest order in the expansion of the stored energy function for small strain tensors E are of the form

$$\tilde{W}(C) = \frac{\lambda}{2}(\operatorname{tr} E)^2 + \mu \operatorname{tr} E^2 + o(\|E\|^2)\,, \quad C = I + 2E\,,$$

where λ and μ are the Lamé constants of the material.

We then examine further properties of the stored energy function that will all play a crucial rôle in the existence theory of Chapter 7. To begin with, we describe the *behavior of the stored energy function for large strains* (Sect. 4.6), which mathematically reflects the idea that "infinite stress should accompany extreme strains", both in the form of the *behavior as* $\det F \to 0^+$:

$$\hat{W}(x, F) \to +\infty \text{ as } \det F \to 0^+ \,,$$

and in the form of a *coerciveness inequality*

$$\hat{W}(x, F) \geq \alpha\{\|F\|^p + \|\mathrm{Cof}\, F\|^q + (\det F)^r\} + \beta\,,$$
$$\text{for all } F \in \mathbb{M}^3_+ \,,$$

where α is >0, $\beta \in \mathbb{R}$, and the numbers p, q, r are sufficiently large. The *lack of convexity of the stored energy function with respect to the variable* $F \in \mathbb{M}^3_+$ (Theorem 4.8-1) is the root of a major difficulty in the mathematical analysis of the associated minimization problem. To obviate this difficulty, J. Ball has introduced the fundamental notion of *polyconvex stored energy functions*, which we discuss in Sect. 4.9.

We next present (Theorem 4.9-2) a useful class of polyconvex stored energy functions, proposed by R.W. Ogden in the form

$$\hat{W}(F) = \sum_{i=1}^{M} a_i \operatorname{tr}(F^{\mathrm{T}}F)^{\gamma_i/2} + \sum_{j=1}^{N} b_j \operatorname{tr} \mathbf{Cof}(F^{\mathrm{T}}F)^{\delta_j/2} + \Gamma(\det F)\,,$$

where $a_i > 0$, $b_j > 0$, $\gamma_i \geq 1$, $\delta_j \geq 1$, and $\Gamma :]0, +\infty[\to \mathbb{R}$ is a convex function that satisfies $\lim_{\delta \to 0^+} \Gamma(\delta) = +\infty$ and $\Gamma(\delta) \geq c\delta^r + d$ with $c > 0$ and r large enough. Our last result (Theorem 4.10-1) shows that it is possible to find a class of very simple Ogden's stored energy functions whose lower order terms of the expansion in terms of the Green–St Venant strain tensor E is equal to $\{\frac{1}{2}\lambda(\operatorname{tr} E)^2 + \mu \operatorname{tr} E^2\}$, for arbitrary Lamé constants $\lambda > 0$ and $\mu > 0$. We conclude this chapter by listing stored energy functions that are commonly used in practice.

4.1. HYPERELASTIC MATERIALS

Combining the equations of equilibrium in the reference configuration, expressed in terms of the first Piola–Kirchhoff stress tensor (Theorem 2.6-1) with the definition of an elastic material, and assuming that a boundary condition of place is specified on the portion $\Gamma_0 = \Gamma - \Gamma_1$ of the boundary, we find that the deformation φ satisfies the following *boundary value problem*:

$$-\mathbf{div}\, \hat{T}(x, \nabla\varphi(x)) = \hat{f}(x, \varphi(x))\,, \quad x \in \Omega\,,$$
$$\hat{T}(x, \nabla\varphi(x))n = \hat{g}(x, \nabla\varphi(x))\,, \quad x \in \Gamma_1\,,$$
$$\varphi(x) = \varphi_0(x)\,, \quad x \in \Gamma_0\,,$$

where $\hat{T} : \bar{\Omega} \times \mathbb{M}^3_+ \to \mathbb{M}^3$ is the response function for the first Piola–Kirchhoff stress, and where the functions $\hat{f}$ and $\hat{g}$ are chosen according to the examples given in Sect. 2.7. The remaining equations of equilibrium, which express the symmetry of the Cauchy stress tensor, place a restriction on the response function $\hat{T}$. We provisionally ignore this restriction, as it will be automatically obtained as a by-product of the assumption of hyperelasticity (Theorem 4.2-2).

We have seen in Theorem 2.6-1 that the first and second equations are together equivalent, at least formally, to the principle of virtual work in the reference configuration, expressed by the equations:

$$\int_\Omega \hat{T}(x, \hat{\nabla}\varphi(x)) : \nabla\theta(x)\, dx = \int_\Omega \hat{f}(x, \varphi(x)) \cdot \theta(x)\, dx + \int_{\Gamma_1} \hat{g}(x, \nabla\varphi(x)) \cdot \theta(x)\, da\,,$$

valid for all sufficiently regular vector fields $\theta : \bar{\Omega} \to \mathbb{R}^3$ that vanish on Γ_0. In Sect. 2.7, we singled out the *conservative applied body forces* and *conservative applied surface forces*, for which the integrals appearing in the right-hand side can be written as Gâteaux derivatives:

$$\int_\Omega \hat{f}(x, \varphi(x)) \cdot \theta(x)\, dx = F'(\varphi)\theta\,,$$

$$\int_{\Gamma_1} \hat{g}(x, \nabla\varphi(x)) \cdot \theta(x)\, da = G'(\varphi)\theta\,,$$

of functionals F and G of the form

$$F(\psi) = \int_\Omega \hat{F}(x, \psi(x))\, dx\,, \quad G(\psi) = \int_{\Gamma_1} \hat{G}(x, \psi(x), \nabla\psi(x))\, da\,.$$

It is therefore natural to ask whether the left-hand side in the principle of virtual work can be similarly written as the Gâteaux derivative of an appropriate functional W, i.e., as

$$\int_\Omega \hat{T}(x, \nabla\varphi(x)) : \nabla\theta(x)\, dx = W'(\varphi)\theta\,.$$

When this is the case, the principle of virtual work is equivalent to expressing that the Gâteaux derivative of the functional $\{W - (F + G)\}$ is zero for all "variations" that vanish on Γ_0. Motivated by these considerations, we set the following definition, whose justification is the object of Theorem 4.1-1.

An elastic material with response function $\hat{T}: \bar{\Omega} \times \mathbb{M}^3_+ \to \mathbb{M}^3$ is **hyperelastic** if there exists a function

$$\hat{W}: \bar{\Omega} \times \mathbb{M}^3_+ \to \mathbb{R},$$

differentiable with respect to the variable $F \in \mathbb{M}^3_+$ for each $x \in \bar{\Omega}$, such that

$$\boxed{\hat{T}(x, F) = \frac{\partial \hat{W}}{\partial F}(x, F) \quad \text{for all } x \in \bar{\Omega},\ F \in \mathbb{M}^3_+,}$$

i.e., componentwise,

$$T_{ij}(x, F) = \frac{\partial \hat{W}}{\partial F_{ij}}(x, F).$$

The notation $\partial/\partial F$ for the gradient of a real-valued function of matrices has been introduced at the end of Sect. 1.2.

The function $\hat{W}$ is called the **stored energy function**. Naturally, if the material is homogeneous, it is a function of $F \in \mathbb{M}^3_+$ only.

Remarks. (1) For an in-depth presentation of the principle of virtual work and its relation to the concept of energy, see Germain [1972, 1973].

(2) The stored energy function is sometimes called the *strain energy function*; also, some authors call stored energy function the above function multiplied by the mass density in the reference configuration.

(3) The stored energy function of a given hyperelastic material is only defined up to the addition of an arbitrary function of $x \in \bar{\Omega}$.

(4) While the definition of a hyperelastic material given above appears as being only motivated by mathematical considerations, it can also be given a more "mechanical" interpretation: More precisely it can be shown that an elastic material is hyperelastic if and only if "the work is $\geqslant 0$ in closed processes", as in Gurtin [1973; 1981b, p. 186]; see also Marques [1984] for a related result. ∎

Theorem 4.1-1. *Let there be given a hyperelastic material subjected to conservative applied body forces and conservative applied surface forces.*

Then the equations

$$-\operatorname{div} \frac{\partial \hat{W}}{\partial F}(x, \nabla\varphi(x)) = \hat{f}(x, \varphi(x)), \quad x \in \Omega,$$

$$\frac{\partial \hat{W}}{\partial F}(x, \nabla\varphi(x))n = \hat{g}(x, \nabla\varphi(x)), \quad x \in \Gamma_1,$$

are formally equivalent to the equations

$$I'(\varphi)\theta = 0,$$

for all smooth maps $\theta : \bar{\Omega} \to \mathbb{R}^3$ *that vanish on* Γ_0, *where the functional I is defined for smooth enough mappings* $\psi : \bar{\Omega} \to \mathbb{R}^3$ by

$$I(\psi) = \int_\Omega \hat{W}(x, \nabla\psi(x))\,dx - \{F(\psi) + G(\psi)\}.$$

Proof. Letting

$$W(\psi) := \int_\Omega \hat{W}(x, \nabla\psi(x))\,dx$$

for any smooth enough mapping $\psi : \bar{\Omega} \to \mathbb{R}^3$, we formally compute the Gâteaux derivative $W'(\psi)\theta$, where $\theta : \bar{\Omega} \to \mathbb{R}^3$ is an arbitrary vector field. Using the assumption of hyperelasticity, we obtain

$$\begin{aligned}
W(\psi + \theta) - W(\psi) &= \int_\Omega \{\hat{W}(x, \nabla\psi(x) + \nabla\theta(x)) - W(x, \nabla\psi(x))\}\,dx \\
&= \int_\Omega \left\{\frac{\partial \hat{W}}{\partial F}(x, \nabla\psi(x)) : \nabla\theta(x) + o(|\nabla\theta(x)|; x)\right\} dx \\
&= \int_\Omega \hat{T}(x, \nabla\psi(x)) : \nabla\theta(x)\,dx + \int_\Omega o(|\nabla\theta(x)|; x)\,dx\,.
\end{aligned}$$

Hence we conclude that

$$W'(\psi)\theta = \int_\Omega \hat{T}(x, \nabla\psi(x)) : \nabla\theta(x)\,dx$$

whenever the space of mappings ψ is equipped with a norm $\|\cdot\|$ that renders the linear form

$$\theta \to \int_\Omega \frac{\partial \hat{W}}{\partial F}(x, \nabla\psi(x)) : \nabla\theta(x)\, dx$$

continuous, and for which

$$\int_\Omega o(|\nabla\theta(x)|; x)\, dx = o(\|\theta\|) .$$

For example, this is the case if the partial derivatives $\partial \hat{W}/\partial F_{ij}$ are Lipschitz-continuous with respect to the argument F and if the space of mappings ψ is endowed with the norm of the space $\mathscr{C}^1(\bar{\Omega}; \mathbb{R}^3)$ (for details, see Exercise 4.1). We have thus established that

$$I'(\varphi)\theta = \int_\Omega \hat{T}(x, \nabla\varphi(x)) : \nabla\theta(x)\, dx - \left\{ \int_\Omega \hat{f}(x, \varphi(x)) \cdot \theta(x)\, dx + \int_{\Gamma_1} \hat{g}(x, \nabla\varphi(x)) \cdot \theta(x)\, da \right\},$$

for an arbitrary vector field $\theta : \bar{\Omega} \to \mathbb{R}^3$, and the conclusion follows from the principle of virtual work in the reference configuration (Theorem 2.6-1). ∎

The functional W defined for any smooth enough mapping ψ by

$$\boxed{W(\psi) = \int_\Omega \hat{W}(x, \nabla\psi(x))\, dx ,}$$

is called the **strain energy**, and the functional I is called the **total energy**.

In order to be in the usual framework where both the functions and their variations (here, φ and θ respectively) lie in the *same* vector space, assume that there exists a mapping from $\bar{\Omega}$ into $\mathbb{R}^3$ that coincides with the given function φ_0 on Γ_0. Denoting such a mapping by the same letter φ_0, consider the functional I_0 defined for an arbitrary mapping χ by

$$I_0(\chi) = I(\chi + \varphi_0) .$$

Since the functions $(\varphi - \varphi_0)$ and θ now belong to the *same* vector space, we can conclude that *the functional I_0 is stationary* (Sect. 1.2) at the point $(\boldsymbol{\varphi} - \boldsymbol{\varphi}_0)$, since

$$I_0'(\boldsymbol{\varphi} - \boldsymbol{\varphi}_0) = 0 \Leftrightarrow I_0'(\boldsymbol{\varphi} - \boldsymbol{\varphi}_0)\boldsymbol{\theta} = I'(\boldsymbol{\varphi})\boldsymbol{\theta} = 0$$

for all $\boldsymbol{\theta}$ that vanish on Γ_0. By extension, we shall say that *the total energy I is stationary at the deformation $\boldsymbol{\varphi}$ with respect to variations vanishing on Γ_0*.

The observation that a minimum of the total energy I is a particular stationary point of the functional I_0 leads to the following crucial corollary of Theorem 4.1-1:

Theorem 4.1-2. *Let the assumptions and notation be as in Theorem 4.1-1. Then any smooth enough mapping $\boldsymbol{\varphi}$ that satisfies*

$$\boldsymbol{\varphi} \in \boldsymbol{\Phi} := \{\boldsymbol{\psi} : \bar{\Omega} \to \mathbb{R}^3 ; \quad \boldsymbol{\psi} = \boldsymbol{\varphi}_0 \text{ on } \Gamma_0\}$$
$$\text{and} \quad I(\boldsymbol{\varphi}) = \inf_{\boldsymbol{\psi} \in \boldsymbol{\Phi}} I(\boldsymbol{\psi}) ,$$
$$\text{with } I(\boldsymbol{\psi}) = \int_\Omega \hat{W}(x, \boldsymbol{\nabla}\boldsymbol{\psi}(x))\,\mathrm{d}x - \{F(\boldsymbol{\psi}) + G(\boldsymbol{\psi})\} ,$$

solves the following boundary value problem:

$$-\mathbf{div}\, \frac{\partial \hat{W}}{\partial \boldsymbol{F}}(x, \boldsymbol{\nabla}\boldsymbol{\varphi}(x)) = \boldsymbol{f}(x, \boldsymbol{\varphi}(x)) , \quad x \in \Omega ,$$
$$\boldsymbol{\varphi}(x) = \boldsymbol{\varphi}_0(x) , \quad x \in \Gamma_0 ,$$
$$\frac{\partial \hat{W}}{\partial \boldsymbol{F}}(x, \boldsymbol{\nabla}\boldsymbol{\varphi}(x))\boldsymbol{n} = \boldsymbol{g}(x, \boldsymbol{\nabla}\boldsymbol{\varphi}(x)) , \quad x \in \Gamma_1 . \qquad \blacksquare$$

In the language of the calculus of variations, this boundary value problem forms the **Euler–Lagrange equations** associated with the total energy I, in the sense that any smooth enough **minimizer** $\boldsymbol{\varphi}$ of the total energy I over the *set of admissible solutions* $\boldsymbol{\Phi}$, i.e., any $\boldsymbol{\varphi} \in \boldsymbol{\Phi}$ that satisfies $I(\boldsymbol{\varphi}) = \inf_{\boldsymbol{\psi} \in \boldsymbol{\Phi}} I(\boldsymbol{\psi})$, is a solution of this boundary value problem.

Remarks. (1) In this framework, the orientation-preserving condition $\det \boldsymbol{\nabla}\boldsymbol{\varphi} > 0$ that a genuine deformation should satisfy in Ω is easily taken care of, since the set

$$\{\boldsymbol{\chi} \in \mathscr{C}^1(\bar{\Omega}; \mathbb{R}^3); \det(\boldsymbol{\nabla}(\boldsymbol{\varphi}_0 + \boldsymbol{\chi})) > 0 \text{ in } \bar{\Omega}, \ \boldsymbol{\chi} = \boldsymbol{o} \text{ on } \Gamma_0\}$$

is an open subset of the vector space

$$\{\chi \in \mathscr{C}^1(\bar{\Omega}; \mathbb{R}^3);\ \chi = o \text{ on } \Gamma_0\}$$

(recall that the differentiability and the stationarity properties hold for functions defined on open sets; cf. Sect. 1.2).

(2) All the above considerations can be equivalently expressed in terms of the displacement field rather than in terms of the deformation field.

(3) There is an abundant literature on the principle of virtual work, which corresponds to the minimization of the associated energy, and on various "*complementary*" *variational principles*, which correspond to the maximization of an associated "*complementary energy*". See e.g. Bufler [1983], Bielski & Telega [1986], de Campos & Oden [1984], Gałka & Telega [1982], Guo [1980], Labisch [1982], Oden & Reddy [1983], Reissner [1984, 1986], Valid [1977], Washizu [1975]. ■

Let us now examine how the various considerations made in Chapter 3 about a *constitutive equation* "translate" in terms of the *stored energy function* of a hyperelastic material. In the absence of any particular assumption on the stored energy function, there is evidently no reason to expect the associated constitutive equation to satisfy the axiom of material frame-indifference, or the corresponding material to be isotropic. Likewise, the other part of the equilibrium equations (the symmetry of the second Piola–Kirchhoff stress tensor) is yet to be taken into account.

4.2. MATERIAL FRAME-INDIFFERENCE FOR HYPERELASTIC MATERIALS

By a natural extension of the definition given in Sect. 3.3, we shall say that *a stored energy function satisfies the axiom of material frame-indifference*, or simply is **frame-indifferent**, if the response function $\hat{T}^D$ for the Cauchy stress is itself frame-indifferent. The following result, which gives two necessary and sufficient conditions for this property to hold, should be compared with Theorem 3.3-1.

Theorem 4.2-1. *The stored energy function* $\hat{W}: \bar{\Omega} \times \mathbb{M}^3_+ \to \mathbb{R}$ *of a hyperelastic material is frame-indifferent if and only if at all points* $x \in \bar{\Omega}$:

$$\boxed{\hat{W}(x, QF) = \hat{W}(x, F) \quad \text{for all } F \in \mathbb{M}^3_+,\ Q \in \mathbb{O}^3_+\ ;}$$

or equivalently, if and only if there exists a function $\tilde{W}: \bar{\Omega} \times \mathbb{S}^3_> \to \mathbb{R}$ *such that*

$$\boxed{\hat{W}(x, F) = \tilde{W}(x, F^{\mathrm{T}}F) \quad \text{for all } F \in \mathbb{M}^3_+ .}$$

Proof. For notational brevity the dependence on $x \in \bar{\Omega}$ is dropped in the proof. We saw in Theorem 3.3-1 that the response function $\hat{T}^{\mathrm{D}}$ for the Cauchy stress is frame-indifferent if and only if

$$\hat{T}^{\mathrm{D}}(QF) = Q\hat{T}^{\mathrm{D}}(F)Q^{\mathrm{T}} \quad \text{for all } F \in \mathbb{M}^3_+, Q \in \mathbb{O}^3_+ .$$

Since $\hat{T}(F) = (\det F)\hat{T}^{\mathrm{D}}(F)F^{-\mathrm{T}}$, we deduce that the response function $\hat{T}$ for the first Piola-Kirchhoff sttess is frame-indifferent if and only if

$$\hat{T}(QF) = Q\hat{T}(F) \quad \text{for all } F \in \mathbb{M}^3_+, Q \in \mathbb{O}^3_+ ,$$

i.e., if and only if

$$\frac{\partial \hat{W}}{\partial F}(F) = Q^{\mathrm{T}} \frac{\partial \hat{W}}{\partial F}(QF) \quad \text{for all } F \in \mathbb{M}^3_+, Q \in \mathbb{O}^3_+ .$$

Let us compute the derivative of the mapping

$$\hat{W}_Q : F \in \mathbb{M}^3_+ \to \hat{W}_Q(F) := \hat{W}(QF)$$

for a fixed $Q \in \mathbb{O}^3_+$. We have

$$\begin{aligned}\hat{W}_Q(F + G) &= \hat{W}(QF + QG) = \hat{W}(QF) + \frac{\partial \hat{W}}{\partial F}(QF) : QG + o(QG) \\ &= \hat{W}_Q(F) + Q^{\mathrm{T}} \frac{\partial \hat{W}}{\partial F}(QF) : G + o(G) ,\end{aligned}$$

since $A : BC = B^{\mathrm{T}}A : C$ for arbitrary matrices A, B, C and $o(QG) = o(G)$ for a fixed Q. Thus

$$\frac{\partial \hat{W}_Q}{\partial F}(F) = Q^{\mathrm{T}} \frac{\partial \hat{W}}{\partial F}(QF) .$$

The conjunction of the above relations shows that the stored energy function is frame-indifferent if and only if

$$\frac{\partial}{\partial F}(\hat{W}(F) - \hat{W}_Q(F)) = \mathbf{0} \quad \text{for all } F \in \mathbb{M}^3_+, Q \in \mathbb{O}^3_+ .$$

If $\hat{W}(F) = \hat{W}(QF)(=\hat{W}_Q(F))$ for all $F \in \mathbb{M}^3_+$, $Q \in \mathbb{O}^3_+$, then clearly the derivative $(\partial/\partial F)(\hat{W}(F) - \hat{W}_Q(F))$ vanishes. The converse is more subtle to prove. The set $\mathbb{M}^3_+$ being connected (Exercise 4.2), we deduce from the relation $(\partial/\partial F)(\hat{W}(F) - \hat{W}_Q(F)) = \boldsymbol{0}$ that, for each $Q \in \mathbb{O}^3_+$, the difference $\{\hat{W}(QF) - \hat{W}(F)\}$ is a constant with respect to $F \in \mathbb{M}^3_+$; hence there exists a mapping $C: \mathbb{O}^3_+ \to \mathbb{R}$ such that

$$\hat{W}(QF) = \hat{W}(F) + C(Q) \quad \text{for all } F \in \mathbb{M}^3_+, Q \in \mathbb{O}^3_+ .$$

Letting $F = Q^r$, $r \geq 0$, in the above relation, we find that

$$\hat{W}(Q^p) = \hat{W}(I) + pC(Q) \quad \text{for all } Q \in \mathbb{O}^3_+, p \geq 1 ,$$

and thus

$$|\hat{W}(Q^p)| \geq p|C(Q)| - |\hat{W}(I)| .$$

If $C(Q) \neq 0$, then $\lim_{p\to+\infty} \hat{W}(Q^p) = +\infty$. The set $\mathbb{O}^3_+$ being compact (the relation $\|Q\| = \{Q:Q\}^{1/2} = 1$ satisfied by all $Q \in \mathbb{O}^3$ shows that any infinite sequence of matrices in $\mathbb{O}^3_+$ contains a subsequence converging in $\mathbb{O}^3_+$), and the function $\hat{W}$ being continuous (the function $\hat{W}$ is differentiable with respect to its argument F, by definition), we are thus led to the conclusion that $C(Q) = 0$ (the same conclusion can be reached by group theoretic methods; cf. Exercise 4.3).

To prove the second equivalence, let $F \in \mathbb{M}^3_+$ be given. By the polar factorization theorem (Theorem 3.2-2),

$$F = RU, \quad \text{with } R \in \mathbb{O}^3_+, U = (F^{\mathrm{T}}F)^{1/2} \in \mathbb{S}^3_> ,$$

and thus the first equivalence shows that

$$\hat{W}(F) = \hat{W}(RU) = \hat{W}(U) = \tilde{W}(F^{\mathrm{T}}F) ,$$

with

$$\tilde{W}(C) := \hat{W}(C^{1/2}) \quad \text{for all } C \in \mathbb{S}^3_> .$$

If conversely $\hat{W}(F) = \tilde{W}(F^{\mathrm{T}}F)$, then

$$\hat{W}(QF) = \tilde{W}(F^T Q^T Q F) = \tilde{W}(F^T F) = \hat{W}(F)$$
$$\text{for all } Q \in \mathbb{O}^3_+,\ F \in \mathbb{M}^3_+ ,$$

and the assertion again follows from the first equivalence. ■

Remarks. (1) The notation $(\partial \hat{W}/\partial F)(QF)$ used in the proof stands for the matrix $\partial \hat{W}/\partial F$ computed at the matrix QF; it does *not* represent the derivative of the mapping $\hat{W}_Q$.

(2) We shall see later (Theorem 4.8-1) why the requirement that $\hat{W}(QF) = \hat{W}(F)$ for all $F \in \mathbb{M}^3_+$, $Q \in \mathbb{O}^3_+$ precludes the *convexity* of the mapping $F \to \hat{W}(F)$. This seemingly innocent consequence is at the origin of a major mathematical difficulty, which has led John Ball to introduce a weaker condition of *polyconvexity* (Sect. 4.9), an essential assumption in the existence theory of Chapter 7.

(3) Some authors say that a function $\hat{W}$ that satisfies the property established in the theorem, viz., $\hat{W}(QF) = \hat{W}(F)$ for all $F \in \mathbb{M}^3_+$, $Q \in \mathbb{O}^3_+$, is *objective*. ■

Let us also call **stored energy function** the function $\tilde{W}$ found in Theorem 4.2-1. We next show that, just as the *first* Piola–Kirchhoff stress tensor is the gradient of the "first" stored energy function $\hat{W}$, the *second* Piola–Kirchhoff stress tensor is similarly related to the "second" stored energy function $\tilde{W}$, in that it is simply twice the gradient of the function $\tilde{W}$. At the same time, this result provides another useful means for recognizing that a material is hyperelastic (cf. Theorem 4.4-3, where it is applied to a St Venant–Kirchhoff material).

As a preparation, let us first observe that the stored energy function $\tilde{W}: \bar{\Omega} \times \mathbb{S}^3_> \to \mathbb{R}$ satisfies at each point $x \in \bar{\Omega}$:

$$\tilde{W}(x, C) = \hat{W}(x, C^{1/2}) \quad \text{for all } C \in \mathbb{S}^3_> ,$$

so that *it is differentiable on the open subset* $\mathbb{S}^3_>$ *of the vector space* $\mathbb{S}^3$. To see this, consider the mapping

$$G: C \in \mathbb{S}^3_> \to G(C) = C^2 \in \mathbb{S}^3 .$$

Since for each $C = \mathbb{S}^3_>$, the equation $G(D) - C = O$ has the unique solution $D = C^{1/2}$, and since the mapping G is differentiable, the inverse mapping $F: C \in \mathbb{S}^3_> \to F(C) = C^{1/2}$ is also differentiable by the local inversion theorem (Theorem 1.2-4). Notice next that at each point $x \in \bar{\Omega}$, the derivative $(\partial \tilde{W}/\partial C)(x, C)$ may be always assumed to be a *symmetric* tensor. Otherwise it suffices to compute the derivative of the mapping

$$C \in \mathbb{M}^3 \to \tilde{W}(x, \tfrac{1}{2}(C + C^T)),$$

which is clearly equal to the mapping $\tilde{W}(x, \cdot)$ on the subset $\mathbb{S}^3_>$ of $\mathbb{M}^3$. For example, if $\tilde{W}(C) = C_{12} - 4C_{21}^2$, write $\tilde{W}(C)$ as

$$\tilde{W}(C) = \tfrac{1}{2}(C_{12} + C_{21}) - (C_{12} + C_{21})^2 .$$

A detailed discussion on this question is given in Cohen & Wang [1984].

Theorem 4.2-2. *Given a hyperelastic material whose stored energy function $\hat{W} : \bar{\Omega} \times \mathbb{M}^3 \to \mathbb{R}$ is frame-indifferent, let the function $\tilde{W} : \bar{\Omega} \times \mathbb{S}^3_> \to \mathbb{R}$ be defined at each point $x \in \bar{\Omega}$ by:*

$$\tilde{W}(x, C) = \hat{W}(x, C^{1/2}) \quad \text{for all } C \in \mathbb{S}^3_> ,$$

and assume without loss of generality that the derivative $(\partial \tilde{W}/\partial C)(x, C)$ is a symmetric tensor. Then the response function for the second Piola–Kirchhoff stress is given by

$$\boxed{\hat{\Sigma}(x, F) = \tilde{\Sigma}(x, C) = 2 \frac{\partial \tilde{W}}{\partial C}(x, C), \quad C = F^T F , \quad \text{for all } F \in \mathbb{M}^3_+ ,}$$

or equivalently by

$$\boxed{\begin{gathered}\hat{\Sigma}(x, F) = \check{\Sigma}(x, E) = \frac{\partial \check{W}}{\partial E}(x, E) \\ \text{with } \check{W}(x, E) = \tilde{W}(x, C), \quad I + 2E = C = F^T F \quad \text{for all } F \in \mathbb{M}^3_+ .\end{gathered}}$$

Conversely, an elastic material whose response $\hat{\Sigma}$ is of the form

$$\hat{\Sigma}(x, F) = 2 \frac{\partial \tilde{W}}{\partial C}(x, F^T F) \quad \text{for all } F \in \mathbb{M}^3_+ , \quad \text{with } \tilde{W} : \bar{\Omega} \times \mathbb{S}^3_> \to \mathbb{R} ,$$

is hyperelastic with a stored energy function given by

$$\hat{W}(x, F) = \tilde{W}(x, F^{\mathrm{T}}F) \quad \text{for all } F \in M^3_+ .$$

Proof. The dependence on $x \in \bar{\Omega}$ is dropped in the proof. Since

$$\hat{\Sigma}(F) = F^{-1}\hat{T}(F) = F^{-1}\frac{\partial \hat{W}}{\partial F}(F) \text{ and } \hat{W}(F) = \tilde{W}(F^{\mathrm{T}}F)$$

for all $F \in \mathbb{M}^3_+$, we need to compute the gradient $\partial \hat{W}/\partial F$ in terms of the gradient $\partial \tilde{W}/\partial C$. Since the mapping $F \in \mathbb{M}^3_+ \rightarrow F^{\mathrm{T}}F \in \mathbb{S}^3_>$ is continuous, the matrix $(F+G)^{\mathrm{T}}(F+G)$ is also in the set $\mathbb{S}^3_>$ for sufficiently small $\|G\|$ if the matrix $(F+G)$ is in $\mathbb{M}^3_+$. For such matrices G, we have

$$\begin{aligned}\hat{W}(F+G) - \hat{W}(F) &= \tilde{W}(F^{\mathrm{T}}F + G^{\mathrm{T}}F + F^{\mathrm{T}}G + G^{\mathrm{T}}G) - \tilde{W}(F^{\mathrm{T}}F) \\ &= \frac{\partial \tilde{W}}{\partial C}(F^{\mathrm{T}}F) : (G^{\mathrm{T}}F + F^{\mathrm{T}}G) + o(G) \\ &= F\left(\left\{\frac{\partial \tilde{W}}{\partial C}(F^{\mathrm{T}}F)\right\}^{\mathrm{T}} + \frac{\partial \tilde{W}}{\partial C}(F^{\mathrm{T}}F)\right) : G + o(G),\end{aligned}$$

using the relations $A : BC = CA^{\mathrm{T}} : B^{\mathrm{T}} = B^{\mathrm{T}}A : C$ valid for arbitrary matrices A, B, C. The above relation implies that

$$\frac{\partial \hat{W}}{\partial F}(F) = \left\{\frac{\partial \tilde{W}}{\partial C}(F^{\mathrm{T}}F)\right\}^{\mathrm{T}} + \frac{\partial \tilde{W}}{\partial C}(F^{\mathrm{T}}F) = 2\frac{\partial \tilde{W}}{\partial C}(F^{\mathrm{T}}F),$$

since it is assumed that the matrix $\partial \tilde{W}/\partial C$ is symmetric, and all the conclusions of the theorem follow. ∎

As a first consequence of this theorem, it should be noted that *the second Piola–Kirchhoff stress tensor associated with a hyperelastic material satisfying the axiom of material frame-indifference is automatically symmetric*, i.e., *the remaining equations of equilibrium are automatically satisfied by such materials.*

As a second consequence, *the second Piola–Kirchhoff stress tensor is a function of the matrix $F^{\mathrm{T}}F$ only*, but as we already proved in Theorem 3.3-1, this is true even without the assumption of hyperelasticity.

4.3. ISOTROPIC HYPERELASTIC MATERIALS

By analogy with the definition given in Sect. 3.4, we shall say that *a stored energy function is* **isotropic at a point** x of the reference configuration $\bar{\Omega}$ if the corresponding response function $\hat{T}^{\mathrm{D}}$ is isotropic at x, i.e., if it satisfies

$$\hat{T}^{\mathrm{D}}(x, F) = \hat{T}^{\mathrm{D}}(x, FQ) \quad \text{for all } F \in \mathbb{M}^3_+, Q \in \mathbb{O}^3_+ .$$

We then have the following necessary and sufficient condition for isotropy, which should be compared with Theorem 3.4-1:

Theorem 4.3-1. *The stored energy function* $\hat{W} : \bar{\Omega} \times \mathbb{M}^3_+ \to \mathbb{R}$ *of a hyperelastic material is isotropic at* x *if and only if*

$$\boxed{\hat{W}(x, F) = \hat{W}(x, FQ) \quad \text{for all } F \in \mathbb{M}^3_+, Q \in \mathbb{O}^3_+ .}$$

Proof. The dependence on $x \in \bar{\Omega}$ is dropped in the proof. Since $\hat{T}(F) = (\det F)\hat{T}^{\mathrm{D}}(F)F^{-\mathrm{T}}$, the response function $\hat{T}$ is isotropic if and only if

$$\hat{T}(FQ) = \hat{T}(F)Q \quad \text{for all } F \in \mathbb{M}^3_+, Q \in \mathbb{O}^3_+ ,$$

i.e., if and only if

$$\frac{\partial \hat{W}}{\partial F}(F) = \frac{\partial \hat{W}}{\partial F}(FQ)Q^{\mathrm{T}} \quad \text{for all } F \in \mathbb{M}^3_+, Q \in \mathbb{O}^3_+ .$$

The rest of the proof proceeds along exactly the same lines as the proof of Theorem 4.2-1; it relies on the fact that the derivative of the mapping

$$\hat{W}^Q : F \in \mathbb{M}^3_+ \to \hat{W}^Q(F) := \hat{W}(FQ)$$

is given by

$$\frac{\partial \hat{W}^Q}{\partial F}(F) = \frac{\partial \hat{W}}{\partial F}(FQ)Q^{\mathrm{T}} . \qquad \blacksquare$$

As was already the case for constitutive equations, material frame-indifference thus involves multiplication *on the left* by matrices in $\mathbb{O}^3_+$, while isotropy involves multiplication *on the right* by such matrices. Interesting complements on material frame-indifference and isotropy for hyperelastic material, including *differential calculus in the group* $\mathbb{O}^3_+$, are given in Moreau [1979].

4.4. THE STORED ENERGY FUNCTION OF AN ISOTROPIC HYPERELASTIC MATERIAL

By simultaneously taking into account the axiom of material frame-indifference and the property of isotropy, we obtain the following result, which should be compared with Theorem 3.6-2. We recall that $\boldsymbol{\iota}_A$ denotes the triple formed by the three principal invariants of a matrix $\boldsymbol{A}$, that $\boldsymbol{\iota}_{F^{\mathrm{T}}F} = \boldsymbol{\iota}_{FF^{\mathrm{T}}}$, and that

$$\boldsymbol{\iota}(\mathbb{S}^3_>) = \{\boldsymbol{\iota}_A \in \mathbb{R}^3; \boldsymbol{A} \in \mathbb{S}^3_>\} \subset \,]0, +\infty[^3 .$$

Theorem 4.4-1. *The stored energy function* $\hat{W} : \bar{\Omega} \times \mathbb{M}^3_+ \to \mathbb{R}$ *of a hyperelastic material is frame-indifferent and is isotropic at x if and only if there exists a function* $\dot{W}(x, \cdot) : \boldsymbol{\iota}(\mathbb{S}^3_>) \to \mathbb{R}$ *such that*

$$\boxed{\hat{W}(x, \boldsymbol{F}) = \dot{W}(x, \boldsymbol{\iota}_{F^{\mathrm{T}}F}) = \dot{W}(x, \boldsymbol{\iota}_{FF^{\mathrm{T}}}) \quad \text{for all } \boldsymbol{F} \in \mathbb{M}^3_+ .}$$

Proof. The dependence on $x \in \bar{\Omega}$ is dropped in the proof. Assuming material frame-indifference and isotropy at x, the characterizations established in Theorems 4.2-1 and 4.3-1 imply that

$$\tilde{W}(\boldsymbol{F}^{\mathrm{T}}\boldsymbol{F}) = \hat{W}(\boldsymbol{F}) = \hat{W}(\boldsymbol{F}\boldsymbol{Q}) = \tilde{W}(\boldsymbol{Q}^{\mathrm{T}}\boldsymbol{F}^{\mathrm{T}}\boldsymbol{F}\boldsymbol{Q}) \qquad \text{for all } \boldsymbol{F} \in \mathbb{M}^3_+, \boldsymbol{Q} \in \mathbb{O}^3_+ .$$

Since for any matrix $\boldsymbol{C} \in \mathbb{S}^3_>$, the matrix $\boldsymbol{F} := \boldsymbol{C}^{1/2} \in \mathbb{S}^3_> \subset \mathbb{M}^3_+$ satisfies $\boldsymbol{F}^{\mathrm{T}}\boldsymbol{F} = \boldsymbol{C}$, we deduce that

$$\tilde{W}(\boldsymbol{C}) = \tilde{W}(\boldsymbol{Q}^{\mathrm{T}}\boldsymbol{C}\boldsymbol{Q}) \quad \text{for all } \boldsymbol{C} \in \mathbb{S}^3_>, \boldsymbol{Q} \in \mathbb{O}^3_+ .$$

The same argument as in part (v) of the proof of Theorem 3.6-1 then shows that the function $\tilde{W}$ is in fact only a function of the principal invariants of the matrix $C = F^{\mathrm{T}}F$. If conversely $\hat{W}(F) = \dot{W}(\iota_{F^{\mathrm{T}}F})$, then

$$\hat{W}(F) = \hat{W}(QF) = \hat{W}(FQ) \quad \text{for all } F \in \mathbb{M}^3_+,\ Q \in \mathbb{O}^3_+ ,$$

since the principal invariants of the matrices $(QF)^{\mathrm{T}}QF = F^{\mathrm{T}}F$ and $(FQ)^{\mathrm{T}}FQ = Q^{\mathrm{T}}F^{\mathrm{T}}FQ$ are the same as those of the matrix $F^{\mathrm{T}}F$. ■

Given a stored energy function expressed in terms of the principal invariants of the matrix $F^{\mathrm{T}}F$, useful forms of the associated constitutive equation are given in the following result, which should also be compared with Theorem 3.6-2.

Theorem 4.4-2. *Assume that at a point $x \in \bar{\Omega}$, the stored energy function $\hat{W}$ is of the form*

$$\hat{W}(x, F) = \dot{W}(x, \iota_{F^{\mathrm{T}}F}) , \quad F \in \mathbb{M}^3_+ ,$$

and that the function $\dot{W}(x, \cdot): \iota(\mathbb{S}^3_>) \to \mathbb{R}$ is differentiable at a point $\iota_{F^{\mathrm{T}}F}$, $F \in \mathbb{M}^3_+$. Then the associated response functions $\hat{T}$ and $\tilde{\Sigma}$ are given by

$$\tfrac{1}{2}\hat{T}(x, F) = \frac{\partial \dot{W}}{\partial \iota_1} F + \frac{\partial \dot{W}}{\partial \iota_2} (\iota_1 I - FF^{\mathrm{T}})F + \frac{\partial \dot{W}}{\partial \iota_3} \iota_3 F^{-\mathrm{T}} ,$$

$$\tfrac{1}{2}\tilde{\Sigma}(x, C) = \frac{\partial \dot{W}}{\partial \iota_1} I + \frac{\partial \dot{W}}{\partial \iota_2} (\iota_1 I - C) + \frac{\partial \dot{W}}{\partial \iota_3} \iota_3 C^{-1}$$

$$= \left\{\frac{\partial \dot{W}}{\partial \iota_1} + \frac{\partial \dot{W}}{\partial \iota_2} \iota_1 + \frac{\partial \dot{W}}{\partial \iota_3} \iota_2\right\} I - \left\{\frac{\partial \dot{W}}{\partial \iota_2} + \frac{\partial \dot{W}}{\partial \iota_3} \iota_1\right\} C + \frac{\partial \dot{W}}{\partial \iota_3} C^2 ,$$

where

$$\frac{\partial \dot{W}}{\partial \iota_k} := \frac{\partial \dot{W}}{\partial \iota_k}(x, \iota_C) , \quad \iota_k = \iota_k(C),\ C = F^{\mathrm{T}}F .$$

Proof. We drop the dependence on $x \in \bar{\Omega}$ in the proof. Considering the function $\hat{W}$ as a composite mapping, we obtain by the chain rule:

$$\frac{\partial \hat{W}}{\partial F}(F) : G = \hat{W}'(F)G = \frac{\partial \dot{W}}{\partial \iota_k}(\iota_C)\iota_k'(C)\Gamma'(F)G$$

$$\text{for all } F \in \mathbb{M}^3_+,\ G \in \mathbb{M}^3 ,$$

where the derivative $\boldsymbol{\Gamma}'$ of the function

$$\boldsymbol{\Gamma}: \boldsymbol{F} \in \mathbb{M}^3_+ \to \boldsymbol{\Gamma}(\boldsymbol{F}) := \boldsymbol{F}^{\mathrm{T}}\boldsymbol{F} \in \mathbb{S}^3_>,$$

is given by (Sect. 1.2)

$$\boldsymbol{\Gamma}'(\boldsymbol{F})\boldsymbol{G} = \boldsymbol{F}^{\mathrm{T}}\boldsymbol{G} + \boldsymbol{G}^{\mathrm{T}}\boldsymbol{F}.$$

In Sect. 1.2, we also showed that the Gâteaux derivatives of the principal invariants are given by

$$\iota_1'(\boldsymbol{C})\boldsymbol{D} = \operatorname{tr} \boldsymbol{D},$$

$$\iota_2'(\boldsymbol{C})\boldsymbol{D} = \det \boldsymbol{C} \operatorname{tr}(\boldsymbol{C}^{-1}\boldsymbol{D}) \operatorname{tr} \boldsymbol{C}^{-1} - \det \boldsymbol{C} \operatorname{tr}(\boldsymbol{C}^{-1}\boldsymbol{D}\boldsymbol{C}^{-1}),$$

$$\iota_3'(\boldsymbol{C})\boldsymbol{D} = \det \boldsymbol{C} \operatorname{tr}(\boldsymbol{C}^{-1}\boldsymbol{D}), \quad \text{for all } \boldsymbol{C} \in \mathbb{S}^3_>, \boldsymbol{D} \in \mathbb{S}^3.$$

Consequently,

$$\iota_1'(\boldsymbol{C})\boldsymbol{\Gamma}'(\boldsymbol{F})\boldsymbol{G} = \operatorname{tr}(\boldsymbol{F}^{\mathrm{T}}\boldsymbol{G} + \boldsymbol{G}^{\mathrm{T}}\boldsymbol{F}) = 2\boldsymbol{F} : \boldsymbol{G},$$

$$\begin{aligned}\iota_2'(\boldsymbol{C})\boldsymbol{\Gamma}'(\boldsymbol{F})\boldsymbol{G} &= \det \boldsymbol{C} \operatorname{tr}(\boldsymbol{C}^{-1}(\boldsymbol{F}^{\mathrm{T}}\boldsymbol{G} + \boldsymbol{G}^{\mathrm{T}}\boldsymbol{F})) \operatorname{tr} \boldsymbol{C}^{-1} \\ &\quad - \det \boldsymbol{C} \operatorname{tr}(\boldsymbol{C}^{-1}(\boldsymbol{F}^{\mathrm{T}}\boldsymbol{G} + \boldsymbol{G}^{\mathrm{T}}\boldsymbol{F})\boldsymbol{C}^{-1}) \\ &= 2 \det \boldsymbol{C}\{(\operatorname{tr} \boldsymbol{C}^{-1})\boldsymbol{F}^{-\mathrm{T}} - \boldsymbol{F}^{-\mathrm{T}}\boldsymbol{F}^{-1}\boldsymbol{F}^{-\mathrm{T}}\} : \boldsymbol{G},\end{aligned}$$

$$\iota_3'(\boldsymbol{C})\boldsymbol{\Gamma}'(\boldsymbol{F})\boldsymbol{G} = \det \boldsymbol{C} \operatorname{tr}(\boldsymbol{C}^{-1}(\boldsymbol{F}^{\mathrm{T}}\boldsymbol{G} + \boldsymbol{G}^{\mathrm{T}}\boldsymbol{F})) = 2(\det \boldsymbol{C})\boldsymbol{F}^{-\mathrm{T}} : \boldsymbol{G}.$$

In order to obtain the expression of the response function $\hat{\boldsymbol{T}}(\boldsymbol{F}) = (\partial \hat{W}/\partial \boldsymbol{F})(\boldsymbol{F})$, it remains to transform the above expression of $\iota_2'(\boldsymbol{C})\boldsymbol{\Gamma}'(\boldsymbol{F})\boldsymbol{G}$ by the Cayley–Hamilton theorem; this yields

$$\begin{aligned}\det \boldsymbol{C}((\operatorname{tr} \boldsymbol{C}^{-1})\boldsymbol{F}^{-\mathrm{T}} - \boldsymbol{F}^{-\mathrm{T}}\boldsymbol{F}^{-1}\boldsymbol{F}^{-\mathrm{T}}) &= \boldsymbol{F}^{-\mathrm{T}}(\iota_2(\boldsymbol{C})\boldsymbol{I} - \iota_3(\boldsymbol{C})\boldsymbol{C}^{-1}) \\ &= \boldsymbol{F}^{-\mathrm{T}}(\iota_1(\boldsymbol{C})\boldsymbol{C} - \boldsymbol{C}^2) \\ &= \iota_1(\boldsymbol{C})\boldsymbol{F} - \boldsymbol{F}\boldsymbol{F}^{\mathrm{T}}\boldsymbol{F}.\end{aligned}$$

The first expression of the response function $\tilde{\boldsymbol{\Sigma}}$ is an immediate consequence of the relation $\tilde{\boldsymbol{\Sigma}}(\boldsymbol{F}^{\mathrm{T}}\boldsymbol{F}) = \boldsymbol{F}^{-1}\hat{\boldsymbol{T}}(\boldsymbol{F})$, and the second one follows by another application of the Cayley–Hamilton theorem. ■

Remark. In order to avoid any difficulty related to the differentiability of the function $\dot{W}(x, \cdot)$, it suffices to assume that its domain of definition is any open subset of $\mathbb{R}^3$ containing the set $\boldsymbol{\iota}(\mathbb{S}^3_>)$ (which is not open; cf. Exercise 3.10). The study of the relation between the differentiability of the functions $\hat{W}$ and $\dot{W}$ is the object of Exercise 4.4. ■

As a first illustration of the above results, we consider a familiar example.

Theorem 4.4-3. *A St Venant–Kirchhoff material, whose response function is given by*

$$\check{\boldsymbol{\Sigma}}(\boldsymbol{E}) = \lambda(\operatorname{tr}\boldsymbol{E})\boldsymbol{I} + 2\mu\boldsymbol{E} = \left\{\frac{\lambda}{2}(\iota_1 - 3) - \mu\right\}\boldsymbol{I} + \mu\boldsymbol{C} = \tilde{\boldsymbol{\Sigma}}(\boldsymbol{C}),$$

$$\boldsymbol{C} = \boldsymbol{I} + 2\boldsymbol{E},$$

is hyperelastic, with a stored energy function given by

$$\boxed{\check{W}(\boldsymbol{E}) = \frac{\lambda}{2}(\operatorname{tr}\boldsymbol{E})^2 + \mu \operatorname{tr}\boldsymbol{E}^2 .}$$

One also has

$$\check{W}(\boldsymbol{E}) = \mu(\iota_1 - 3) + \frac{\lambda + 2\mu}{8}(\iota_1 - 3)^2 - \frac{\mu}{3}(\iota_2 - 3) = \dot{W}(\boldsymbol{\iota}_C),$$

where $\iota_1 = \operatorname{tr}\boldsymbol{C}$; $\iota_2 = \det\boldsymbol{C}\operatorname{tr}\boldsymbol{C}^{-1}$.

Proof. By Theorem 4.2-2, it suffices to verify that

$$\check{\boldsymbol{\Sigma}}(\boldsymbol{E}) = \frac{\partial\check{W}}{\partial\boldsymbol{E}}(\boldsymbol{E}).$$

Using the expression given in the theorem for $\check{W}(\boldsymbol{E})$, we find by an easy computation:

$$\begin{aligned}\check{W}(\boldsymbol{E}+\boldsymbol{H}) &= \check{W}(\boldsymbol{E}) + \lambda(\operatorname{tr}\boldsymbol{E})(\operatorname{tr}\boldsymbol{H}) + 2\mu\operatorname{tr}(\boldsymbol{E}\boldsymbol{H}) + O(\|\boldsymbol{H}\|^2)\\ &= \check{W}(\boldsymbol{E}) + (\lambda\operatorname{tr}\boldsymbol{E} + 2\mu\boldsymbol{E}):\boldsymbol{H} + O(\|\boldsymbol{H}\|^2).\end{aligned}$$

Alternatively, we could use the expression $\dot{W}(\iota_C)$ of the stored energy function in terms of the principal invariants of the matrix C and Theorem 4.4-2. Another straightforward computation yields in this fashion the expression given in the theorem for $\tilde{\Sigma}(C)$. ∎

Observe in passing that this result combined with Theorem 4.4-1 provides another proof that a St Venant–Kirchhoff material satisfies the axiom of material frame-indifference and is isotropic. Also notice that the third invariant $\iota_3 = \det C$ does not appear in the stored energy function of a St Venant–Kirchhoff material.

4.5. THE STORED ENERGY FUNCTION NEAR A NATURAL STATE

We recall that near a natural state the constitutive equation of a homogeneous, isotropic, elastic material is necessarily of the form (Theorem 3.8-1):

$$\Sigma = \check{\Sigma}(E) = \lambda(\operatorname{tr} E)I + 2\mu E + o(E),$$

where λ and μ are the Lamé constants of the material. Paralleling this result, we now prove:

Theorem 4.5-1. *Let there be given a homogeneous, isotropic, hyperelastic material, whose reference configuration is a natural state, so that the stored energy function is of the form*

$$\hat{W}(F) = \dot{W}(\iota_{F^TF}) = \check{W}(E), \quad F^TF = I + 2E, F \in \mathbb{M}^3_+ .$$

Then, if the function $\dot{W} : \iota(\mathbb{S}^3_>) \to \mathbb{R}$ *is twice differentiable at the point* ι_I,

$$\boxed{\check{W}(E) = \frac{\lambda}{2}(\operatorname{tr} E)^2 + \mu \operatorname{tr} E^2 + o(\|E\|^2), \quad F^TF = I + 2E .}$$

Proof. Since the function $\dot{W}$ is twice differentiable at the point ι_I, the functions $(\partial \dot{W}/\partial \iota_k)(\iota_C)$ are differentiable at the point $C = I$, and by Theorem 4-4.2, the differentiability assumptions of Theorem 3.8-1 are satisfied. Consequently, we must have

$$\check{\Sigma}(E) = \frac{\partial \check{W}}{\partial E}(E) = \lambda(\operatorname{tr} E)I + 2\mu E + o(E).$$

Define the function $\check{\Delta}$ by

$$\check{\Delta}(E) = \check{W}(E) - \left\{\frac{\lambda}{2}(\operatorname{tr} E)^2 + \mu \operatorname{tr} E^2\right\}, \quad I + 2E \in \mathbb{S}^3_>.$$

By Theorem 4.4-3,

$$\frac{\partial \check{\Delta}}{\partial E}(E) = \frac{\partial \check{W}}{\partial E}(E) - \{\lambda(\operatorname{tr} E)I + 2\mu \operatorname{tr} E\},$$

so that the function $\check{\Delta}$ satisfies

$$\frac{\partial \check{\Delta}}{\partial E}(E) = o(E).$$

Since the function $\dot{W}$ is continuously differentiable in a neighborhood of the point ι_I (it is twice differentiable at this point by assumption), the function $\check{\Delta}$ is continuously differentiable in a neighborhood of $E = 0$. Thus if the norm $\|E\|$ is sufficiently small, we can write

$$\check{\Delta}(E) = \int_0^1 \frac{\partial \check{\Delta}}{\partial E}(tE) : E \, dt = O(\|E\|^2),$$

by an application of Taylor's formula with integral remainder (Theorem 1.3-3), and the proof is complete. ∎

Remark. With further differentiability assumptions, the remainder can be shown to be of the form $O(\|E\|^3)$, and the terms of the third order can be explicitly computed (Exercise 4.8). ∎

4.6. BEHAVIOR OF THE STORED ENERGY FUNCTION FOR LARGE STRAINS

It is desirable that constitutive equations reflect in some fashion the intuitive idea that "infinite stress must accompany extreme strains" (Antman [1983]), as suggested by the most immediate physical evidence. Such a property is not easily expressed in terms of a response function, as shown notably in the pioneering works of S.S. Antman (Antman [1970, 1983]). In case of hyperelastic materials, it corresponds to the requirement that the stored energy function $\hat{W}$ approach $+\infty$ if any one of the eigenvalues $\lambda_i(\boldsymbol{C})$ of the matrix $\boldsymbol{C} = \boldsymbol{F}^{\mathrm{T}}\boldsymbol{F}$ approaches 0 or $+\infty$ (a possible measure of "extreme" strains). If this is the case and if the function $\hat{W}$ is sufficiently well-behaved, we can expect the norm

$$\|\hat{\boldsymbol{T}}(x, \boldsymbol{F})\| = \left\|\frac{\partial \hat{W}}{\partial \boldsymbol{F}}(x, \boldsymbol{F})\right\|$$

to approach $+\infty$ (a possible measure of "infinite" stress) as a consequence of the mean value theorem (Theorem 1.2-2). Keeping two eigenvalues $\lambda_{i+1}(\boldsymbol{C})$ and $\lambda_{i+2}(\boldsymbol{C})$ in a compact interval of $]0, +\infty[$, we have the following equivalences:

$$\begin{aligned}
&\lambda_i(\boldsymbol{C}) \to 0^+ \Leftrightarrow \det \boldsymbol{F} \to 0^+ ,\\
&\lambda_i(\boldsymbol{C}) \to +\infty \Leftrightarrow \|\boldsymbol{F}\| \to +\infty ,\\
&\lambda_i(\boldsymbol{C}) \to +\infty \Leftrightarrow \|\mathbf{Cof}\, \boldsymbol{F}\| \to +\infty ,\\
&\lambda_i(\boldsymbol{C}) \to +\infty \Leftrightarrow \det \boldsymbol{F} \to +\infty .
\end{aligned}$$

We are thus led to the following *assumptions* governing the **behavior for large strains** of the stored energy function at each point $x \in \bar{\Omega}$:

$$\boxed{\begin{aligned}
&\hat{W}(x, \boldsymbol{F}) \to +\infty \text{ as } \det \boldsymbol{F} \to 0^+, \boldsymbol{F} \in \mathbb{M}^3_+ ,\\
&\hat{W}(x, \boldsymbol{F}) \to +\infty \text{ as } \{\|\boldsymbol{F}\| + \|\mathbf{Cof}\, \boldsymbol{F}\| + (\det \boldsymbol{F})\} \to +\infty, \boldsymbol{F} \in \mathbb{M}^3_+ .
\end{aligned}}$$

The first assumption is called the **behavior as** $\det \boldsymbol{F} \to 0^+$.

A sharper version of the second assumption takes the form of the following **coerciveness inequality**, which will prove to be an essential tool in the existence theorems of Chapter 7: There exist constants α, β, p, q, r such that

$$\boxed{\begin{cases}
\alpha > 0, p > 0, q > 0, r > 0 ,\\
\hat{W}(x, \boldsymbol{F}) \geq \alpha\{\|\boldsymbol{F}\|^p + \|\mathbf{Cof}\, \boldsymbol{F}\|^q + (\det \boldsymbol{F})^r\} + \beta\\
\quad \text{for all } (x, \boldsymbol{F}) \in \bar{\Omega} \times \mathbb{M}^3_+ .
\end{cases}}$$

There exists convincing evidence that the rate of growth of the stored energy function as reflected by the above coerciveness inequality is a measure of the "strength" of the material, which should be able to "sustain" large enough strains. For example, Antman [1970] has shown that if the stored energy function does not grow sufficiently rapidly with $\|\boldsymbol{F}\|$, the mathematical model for the inflation of a membrane under the influence of a boundary condition of pressure may have no solution for large values of the pressure. In the same spirit, Simpson & Spector [1984a] have shown that growth conditions are needed in order that a specific class of problems possess "natural" solutions. We shall also see in Chapter 7 that the exponents p, q, r should indeed satisfy sharper inequalities in order to produce useful existence results, typically such as

$$p \geq 2,\ q \geq \frac{p}{p-1},\ r > 1 .$$

The behavior of the stored energy function for "small" values of $\det \boldsymbol{F}$ as expressed by the first assumption gives mathematical substance to the intuitive idea that "an infinite energy should be required in order to annihilate volumes" (see Exercise 4.9). Of course, if the stored energy function does not depend explicitly on $\det \boldsymbol{F}$, the first assumption becomes "void", and there is no term such as $(\det \boldsymbol{F})^r$, $r > 0$, in the coerciveness inequality. This is the case of a St Venant–Kirchhoff material, for which it was already noted in Sect. 3.9 that it should not be expected to behave well for large strains, since it is only modeled according to the behavior for "small" values of $\|\boldsymbol{E}\|$.

To conclude, we note that the question of deciding whether an actual material satisfies either one of the above assumptions is of a philosophical nature: By contrast with the behavior for small strains, which can be verified experimentally, any behavior for large strains (understood in its mathematical sense) is essentially a *mathematical* assumption, since numerical information can be gathered only for values of $\|\boldsymbol{F}\|$, $\|\mathbf{Cof}\, \boldsymbol{F}\|$, $\det \boldsymbol{F}$, $(\det \boldsymbol{F})^{-1}$ that lie in compact intervals.

*4.7. CONVEX SETS AND CONVEX FUNCTIONS

Since from now on we will frequently need the notion of *convexity* for *sets* and *functions*, we give a brief review of the main definitions and of some of the basic theorems involving this notion. For general information, see Rockafellar [1970], Ekeland & Temam [1974], Roberts &

Varberg [1973], Ioffe & Tikhomirov [1974], Lay [1982], van Tiel [1984].

A subset of a vector space is **convex** if, whenever it contains two points u and v, it contains the closed segment $[u, v]$. The following basic properties of convex subsets in a normed vector space are proved e.g. in Schwartz [1970, pp. 258 ff].

Theorem 4.7-1. *Let U be a convex subset of a normed vector space. Then both sets $\bar{U}$ and int U are also convex. Besides,*

$$u \in \operatorname{int} U \text{ and } v \in \bar{U} \Rightarrow [u, v[\subset \operatorname{int} U ,$$

so that $\overline{\operatorname{int} U} = \bar{U}$ *whenever* $\operatorname{int} U \neq \emptyset$. ■

In a normed vector space V, a **hyperplane** is a subset of V of the form (V' denotes the dual space of V):

$$P(L; \alpha) = \{v \in V; L(v) \leqslant \alpha\}, \quad L \in (V' - \{0\}), \alpha \in \mathbb{R},$$

and a **closed half-space** is a subset of V of the form

$$H(L; \alpha) = \{v \in V; L(v) \leqslant \alpha\}, \quad L \in (V' - \{0\}), \alpha \in \mathbb{R};$$

note that $P(L; \alpha) = \partial H(L; \alpha)$. There are remarkable links between closed convex subsets and closed half spaces, which are themselves particular closed convex sets (for a proof, see e.g. Brezis [1983, p. 7]):

Theorem 4.7-2. (a) *Let U be a nonempty closed convex subset of a normed vector space V, and let $w \notin U$. Then there exist $L \in (V' - \{0\})$, $\alpha \in \mathbb{R}$, and $\varepsilon > 0$, such that*

$$L(w) \geqslant \alpha + \varepsilon, \text{ and } L(v) \leqslant \alpha - \varepsilon \quad \text{for all } v \in U .$$

(b) *A nonempty closed convex subset strictly contained in a normed vector space is contained in at least one closed half space, and it is the intersection of all the closed half spaces that contain it.* ■

Let U be a nonempty subset of a vector space V. The **convex hull** co U of U is the intersection of all the convex subsets of V that contain U, or equivalently it is the smallest convex subset of V that contains U. It is also the set formed by all *convex combinations* of elements of U, i.e., elements of V of the form

$$v = \sum_{p=1}^{N} \mu_p v_p,\ v_p \in U,\ \sum_{p=1}^{N} \mu_p = 1,\ \mu_p \geq 0\,,\quad N \text{ arbitrary}\,.$$

For instance, let there be given $(n+1)$ points $a_j = (a_{ij})_{i=1}^n \in \mathbb{R}^n$, $1 \leq j \leq n+1$, that are affinely independent, in the sense that they are not contained in a hyperplane of $\mathbb{R}^n$; equivalently, the matrix (a_{ij}), where $a_{n+1,j} = 1$, $1 \leq j \leq n+1$, is invertible. Then the convex hull of the set $\bigcup_{j=1}^{n+1} \{a_j\}$:

$$T = \left\{ v \in \mathbb{R}^n\ ;\ v = \sum_{j=1}^{n+1} \lambda_j a_j,\ \sum_{j=1}^{n+1} \lambda_j = 1,\ \lambda_j \geq 0 \right\},$$

is called an *n-simplex*, and the points a_j are called its *vertices* (a 2-simplex is a triangle, and a 3-simplex is a tetrahedron). Note that T has a nonempty interior (this property is notably used in the proof of Theorem 4.7-4) and that T is compact (this is a special case of property (c) of Theorem 4.7-3).

Let U be a subset of a normed vector space V. The **closed convex hull** $\overline{\text{co}}\, U$ of U is the intersection of all the closed convex subsets of V that contain U, or equivalently, it is the smallest closed convex subset of V that contains U. The following theorem gives useful properties of the closed convex hull (for proofs, see e.g. Schwartz [1970, p. 260], Cheney [1966, p. 18]).

Theorem 4.7-3. (a) *Let U be a nonempty subset of a normed vector space. Then*

$$\text{co}\, \bar{U} \subset \overline{\text{co}}\, U = \{\text{co}\, U\}^-\,.$$

(b) *The closed convex hull of U is the intersection of all the closed half spaces that contain U.*

(c) *The convex hull of a compact subset U of a finite dimensional space is compact. Hence $\text{co}\, \bar{U} = \overline{\text{co}}\, U$ in this case.* ∎

Remark. Consider the closed subset

$$U = \left\{ v \in \mathbb{R}^2;\ v_2 \geq \frac{1}{1+v_1^2} \right\}$$

of $\mathbb{R}^2$. Then

$$\operatorname{co} \bar{U} = \operatorname{co} U = \{v \in \mathbb{R}^2;\ v_2 > 0\} \subsetneqq \overline{\operatorname{co}}\, U = \{v \in \mathbb{R}^2;\ v_2 \geq 0\}\ . \qquad \blacksquare$$

We now identify two convex hulls of particular interest in elasticity.

Theorem 4.7-4. *Let as usual* $\mathbb{M}^3_+ = \{\boldsymbol{F} \in \mathbb{M}^3;\ \det \boldsymbol{F} > 0\}$. *Then,*

$$\operatorname{co} \mathbb{M}^3_+ = \mathbb{M}^3\ ,$$

$$\operatorname{co}\{(\boldsymbol{F}, \mathbf{Cof}\, \boldsymbol{F}, \det \boldsymbol{F}) \in \mathbb{M}^3 \times \mathbb{M}^3 \times \mathbb{R};\ \boldsymbol{F} \in \mathbb{M}^3_+\}$$
$$= \mathbb{M}^3 \times \mathbb{M}^3 \times\,]0, +\infty[\ .$$

Proof. The matrix $-\boldsymbol{I}$ belongs to the set $\operatorname{co} \mathbb{M}^3_+$ since (for instance)

$$-\boldsymbol{I} = \tfrac{1}{2}\,\mathbf{Diag}(-3, 1, -1) + \tfrac{1}{2}\,\mathbf{Diag}(1, -3, -1)\ .$$

Any matrix $\boldsymbol{F} \in \mathbb{M}^3$ can be written as

$$\boldsymbol{F} = \tfrac{1}{2}(\lambda \boldsymbol{I} + 2\boldsymbol{F}) + \tfrac{1}{2}(-\lambda \boldsymbol{I}) \quad \text{for all } \lambda \in \mathbb{R}\ .$$

Since $\det(\lambda \boldsymbol{I} + 2\boldsymbol{F})$ is a polynomial of degree 3 in λ whose monomial of highest degree is λ^3, there exists $\lambda > 0$ such that $(\lambda \boldsymbol{I} + 2\boldsymbol{F}) \in \mathbb{M}^3_+$. This shows that any $\boldsymbol{F} \in \mathbb{M}^3$ can be written as a convex combination of two matrices in $\mathbb{M}^3_+$. Hence $\operatorname{co} \mathbb{M}^3_+ = \mathbb{M}^3$.

Next, let

$$\mathbb{U} := \{(\boldsymbol{F}, \mathbf{Cof}\, \boldsymbol{F}, \det \boldsymbol{F}) \in \mathbb{M}^3 \times \mathbb{M}^3 \times \mathbb{R};\ \boldsymbol{F} \in \mathbb{M}^3_+\}\ .$$

We first show that $\overline{\operatorname{co}}\, \mathbb{U} = \mathbb{M}^3 \times \mathbb{M}^3 \times [0, +\infty[$. Since the set $\mathbb{M}^3 \times \mathbb{M}^3 \times [0, +\infty[$ is closed and convex, it contains the set $\overline{\operatorname{co}}\, \mathbb{U}$. To prove the other inclusion, it suffices by Theorem 4.7-3(b) to prove that whenever a closed half-space in the space $\mathbb{M}^3 \times \mathbb{M}^3 \times \mathbb{R}$ contains the set $\mathbb{U}$, it contains the set $\mathbb{M}^3 \times \mathbb{M}^3 \times [0, +\infty[$. A closed half-space in the space $\mathbb{M}^3 \times \mathbb{M}^3 \times \mathbb{R}$ is of the form

$$H(\boldsymbol{G}, \boldsymbol{K}, \varepsilon; \alpha) := \{(\boldsymbol{F}, \boldsymbol{H}, \delta) \in \mathbb{M}^3 \times \mathbb{M}^3 \times \mathbb{R};\ \boldsymbol{G} : \boldsymbol{F} + \boldsymbol{K} : \boldsymbol{H} + \varepsilon\delta \leq \alpha\}\ ,$$

for some fixed $\boldsymbol{G}, \boldsymbol{K} \in \mathbb{M}^3$ and $\varepsilon, \alpha \in \mathbb{R}$ with $(\boldsymbol{G}, \boldsymbol{K}, \varepsilon) \neq (\boldsymbol{0}, \boldsymbol{0}, 0)$. Hence

$$\mathbb{M}^3 \times \mathbb{M}^3 \times [0, +\infty[\ \subset H(\boldsymbol{G}, \boldsymbol{K}, \varepsilon; \alpha) \Leftrightarrow \boldsymbol{G} = \boldsymbol{0}, \boldsymbol{K} = \boldsymbol{0}\ ,$$
$$\varepsilon < 0, \text{ and } \alpha \geq 0\ .$$

We must therefore show that, if a closed half-space $H(\boldsymbol{G}, \boldsymbol{K}, \varepsilon; \alpha)$ contains $\mathbb{U}$, i.e., if

$$\boldsymbol{G}:\boldsymbol{F} + \boldsymbol{K}:\mathbf{Cof}\,\boldsymbol{F} + \varepsilon \det \boldsymbol{F} \leqslant \alpha \quad \text{for all } \boldsymbol{F} \in \mathbb{M}^3_+ ,$$

then we necessarily have $\boldsymbol{G} = \boldsymbol{0}$, $\boldsymbol{K} = \boldsymbol{0}$, $\varepsilon < 0$, and $\alpha \geqslant 0$. Using the singular value decomposition of the matrix $\boldsymbol{K}$ (Theorem 3.2-3), we can find two matrices $\boldsymbol{U}$, $\boldsymbol{V}$ such that

$$\boldsymbol{U}, \boldsymbol{V} \in \mathbb{O}^3_+, \boldsymbol{K}^{\mathrm{T}} = \boldsymbol{U}^{\mathrm{T}}\boldsymbol{D}\boldsymbol{V}, \boldsymbol{D} = \mathbf{Diag}\, d_i, |d_i| = v_i(\boldsymbol{K}) ,$$

so that (note that $\mathbf{Cof}\,\boldsymbol{U} = \boldsymbol{U}$ if $\boldsymbol{U} \in \mathbb{O}^3_+$):

$$\boldsymbol{G}:\boldsymbol{F} + \boldsymbol{K}:\mathbf{Cof}\,\boldsymbol{F} + \varepsilon \det \boldsymbol{F} = (\boldsymbol{V}\boldsymbol{G}\boldsymbol{U}^{\mathrm{T}}):(\boldsymbol{V}\boldsymbol{F}\boldsymbol{U}^{\mathrm{T}}) + \boldsymbol{D}:\mathbf{Cof}(\boldsymbol{V}\boldsymbol{F}\boldsymbol{U}^{\mathrm{T}}) + \varepsilon \det(\boldsymbol{V}\boldsymbol{F}\boldsymbol{U}^{\mathrm{T}}) .$$

Let $\boldsymbol{V}\boldsymbol{G}\boldsymbol{U}^{\mathrm{T}} = (\tilde{g}_{ij})$. The particular choice $\boldsymbol{V}\boldsymbol{F}\boldsymbol{U}^{\mathrm{T}} = \mathbf{Diag}(\lambda^{-1}, \lambda^{1/2}, \lambda^{1/2})$, $\lambda > 0$, shows that we must have

$$\lambda^{-1}\tilde{g}_{11} + \lambda^{1/2}(\tilde{g}_{22} + \tilde{g}_{33}) + \lambda d_1 + \lambda^{-1/2}(d_2 + d_3) + \varepsilon \leqslant \alpha \quad \text{for all } \lambda > 0 ,$$

and thus $\tilde{g}_{11} \leqslant 0$ (let $\lambda \to 0^+$) and $d_1 \leqslant 0$ (let $\lambda \to +\infty$). The particular choice $\boldsymbol{V}\boldsymbol{F}\boldsymbol{U}^{\mathrm{T}} = \mathbf{Diag}(-\lambda^{-1}, -\lambda^{1/2}, \lambda^{1/2})$, $\lambda > 0$, next shows that we must have

$$-\lambda^{-1}\tilde{g}_{11} + \lambda^{1/2}(-\tilde{g}_{22} + \tilde{g}_{33}) - \lambda d_1 + \lambda^{-1/2}(-d_2 + d_3) + \varepsilon \leqslant \alpha \quad \text{for all } \lambda > 0 ,$$

and thus $\tilde{g}_{11} \geqslant 0$ (let $\lambda \to 0^+$) and $d_1 \geqslant 0$ (let $\lambda \to +\infty$). Continuing in this manner, we obtain $\tilde{g}_{ii} = 0$ and $\boldsymbol{D} = \boldsymbol{0}$; hence $\boldsymbol{K} = \boldsymbol{O}$.

We can find two matrices $\boldsymbol{P}$, $\boldsymbol{Q}$ such that

$$\boldsymbol{P}, \boldsymbol{Q} \in \mathbb{O}^3_+, \boldsymbol{G}^{\mathrm{T}} = \boldsymbol{P}^{\mathrm{T}}\boldsymbol{\Delta}\boldsymbol{Q}, \boldsymbol{\Delta} = \mathbf{Diag}\, \delta_i, \quad |\delta_i| = v_i(\boldsymbol{G}) ,$$

so that

$$\boldsymbol{G}:\boldsymbol{F} + \varepsilon \det \boldsymbol{F} = \boldsymbol{\Delta}:(\boldsymbol{Q}\boldsymbol{F}\boldsymbol{P}^{\mathrm{T}}) + \varepsilon \det(\boldsymbol{Q}\boldsymbol{F}\boldsymbol{P}^{\mathrm{T}}) ,$$

and the relations $\tilde{g}_{ii} = 0$ reduce in this case to $\boldsymbol{\Delta} = \boldsymbol{O}$; hence $\boldsymbol{G} = \boldsymbol{O}$. It then follows that $\varepsilon < 0$ (recall that $(\boldsymbol{G}, \boldsymbol{H}, \varepsilon) \neq (\boldsymbol{0}, \boldsymbol{0}, 0)$) and $\alpha \geq 0$. We have thus proved that $\overline{\text{co}}\, \mathbb{U} = \mathbb{M}^3 \times \mathbb{M}^3 \times [0, +\infty[$.

Since the set $\mathbb{U}$ is contained in the convex set $\mathbb{M}^3 \times \mathbb{M}^3 \times]0, +\infty[$, it follows by definition of the convex hull that

$$\text{co}\, \mathbb{U} \subset \mathbb{M}^3 \times \mathbb{M}^3 \times]0, \infty[\,.$$

In order to establish the other inclusion, we first observe that $\text{int co}\, \mathbb{U} \neq \emptyset$, for either $\text{co}\, \mathbb{U}$ is contained in a hyperplane $\mathbb{H}$, but since $\mathbb{H}$ is closed and convex, this would imply $\overline{\text{co}}\, \mathbb{U} \subset \mathbb{H}$, in contradiction with the relation $\overline{\text{co}}\, \mathbb{U} = \mathbb{M}^3 \times \mathbb{M}^3 \times [0, +\infty[$; or $\text{co}\, \mathbb{U}$ contains at least $(n + 1)$ affinely independent points ($n = 19$), which are thus the vertices of an n-simplex $\mathbb{S}$. Since an n-simplex is convex, $\mathbb{S}$ is contained in $\text{co}\, \mathbb{U}$, and since an n-simplex has a nonempty interior,

$$\emptyset \neq \text{int}\, \mathbb{S} \subset \text{int co}\, \mathbb{U}\,.$$

Next, let $(\boldsymbol{F}, \boldsymbol{G}, \delta) \in \mathbb{M}^3 \times \mathbb{M}^3 \times]0, +\infty[$ and $(\boldsymbol{F}_0, \boldsymbol{G}_0, \delta_0) \in \text{int co}\, \mathbb{U}$ be given, so that $\delta_0 > 0$ in particular. There exists t such that

$$0 < t < 1 \text{ and } (1 - t)\delta_0 \leq \delta\,.$$

Let

$$\boldsymbol{F}_1 = \frac{1}{t}(\boldsymbol{F} - (1 - t)\boldsymbol{F}_0),\ \boldsymbol{G}_1 = \frac{1}{t}(\boldsymbol{G} - (1 - t)\boldsymbol{G}_0)\,,$$

$$\delta_1 = \frac{1}{t}(\delta - (1 - t)\delta_0)\,.$$

Then

$$(\boldsymbol{F}_1, \boldsymbol{G}_1, \delta_1) \in \mathbb{M}_3 \times \mathbb{M}_3 \times [0, +\infty[= \overline{\text{co}}\, \mathbb{U}\,,$$

$$(\boldsymbol{F}, \boldsymbol{G}, \delta) = t(\boldsymbol{F}_1, \boldsymbol{G}_1, \delta_1) + (1 - t)(\boldsymbol{F}_0, \boldsymbol{G}_0, \delta_0)$$
$$\in](\boldsymbol{F}_0, \boldsymbol{G}_0, \delta_0), (\boldsymbol{F}_1, \boldsymbol{G}_1, \delta_1)[\,,$$

so that, combining with Theorem 4.7-1, we obtain the desired implication:

$$(\boldsymbol{F}, \boldsymbol{G}, \delta) \in \mathbb{M}^3 \times \mathbb{M}^3 \times]0, +\infty] \Rightarrow (\boldsymbol{F}, \boldsymbol{G}, \delta) \in \text{int co}\, \mathbb{U} \subset \text{co}\, \mathbb{U}\,,$$

and the proof is complete. ■

An immediate consequence of the relation $\mathrm{co}\,\mathbb{M}^3_+ = \mathbb{M}^3$ is that *the set* $\mathbb{M}^3_+$ *is not convex* (this can also be proved directly; cf. Exercise 4.2). The identification of the set $\mathrm{co}\,\mathbb{U}$ is due to Ball [1977, Theorem 4.3], whose original proof, similar in its principle to the one given here, is suggested in Exercise 4.11. For yet another proof, see Exercise 4.12.

We now list various results relating *convexity* and *minima of functions* (for proofs, see e.g. Ciarlet [1985, p. 151 ff]). The necessary condition $J'(u) = 0$ that a point u be a local minimum of a function J defined on an *open* set containing u, admits the following extension for functions defined over a *convex* set (the converse holds for convex functions; cf. Theorem 4.7-8).

Theorem 4.7-5 (necessary condition for a local minimum). *Let $J : U \to \mathbb{R}$ be a function defined on a convex subset U of a normed vector space. If a point $u \in U$ is a local minimum of the function $J : U \to \mathbb{R}$ and if the function J is differentiable at u, then*

$$J'(u)(v - u) \geq 0 \quad \text{for all } v \in U . \quad ■$$

The above inequalities are called the *Euler inequalities*.

Remark. According to the definition of differentiable functions (Sect. 1.2), the function J should be defined in a neighborhood of each point where it is assumed to possess a derivative. For simplicity, this kind of hypothesis is not explicitly stated in the above theorem, nor in the following ones. ■

A function $J : U \subset V \to \mathbb{R}$ defined on a *convex* subset U of a vector space V is **convex** on U if

$$u, v \in U \text{ and } \theta \in [0, 1] \Rightarrow J(\theta u + (1-\theta)v) \leq \theta J(u) + (1-\theta)J(v) ,$$

and **strictly convex** on U if

$$u, v \in U,\ u \neq v, \text{ and } \theta \in \,]0, 1[\, \Rightarrow J(\theta u + (1-\theta)v) < \theta J(u) + (1-\theta)J(v) .$$

The next two theorems give useful relations between convexity and derivability:

Theorem 4.7-6 (convexity and the derivative). *Let $J : U \to \mathbb{R}$ be a function defined and differentiable over a convex subset U of a normed vector space.*

(a) *The function J is convex on U if and only if*

$$J(v) \geqslant J(u) + J'(u)(v-u) \quad \text{for all } u, v \in U .$$

(b) *The function J is strictly convex on U if and only if*

$$J(v) > J(u) + J'(u)(v-u) \quad \text{for all } u, v \in U, u \neq v . \qquad \blacksquare$$

The geometrical interpretation of these inequalities is clear (Fig. 4.7-1); they simply express that the function is "everywhere above its tangent plane".

Theorem 4.7-7 (convexity and the second derivative). *Let $J : U \to \mathbb{R}$ be a function defined and twice differentiable over a convex subset U of a normed vector space.*

(a) *The function J is convex on U if and only if*

$$J''(u)(v-u, v-u) \geqslant 0 \quad \text{for all } u, v \in U .$$

(b) *If*

$$J''(u)(v-u, v-u) > 0 \quad \text{for all } u, v \in U, u \neq v ,$$

the function J is strictly convex on U. $\blacksquare$

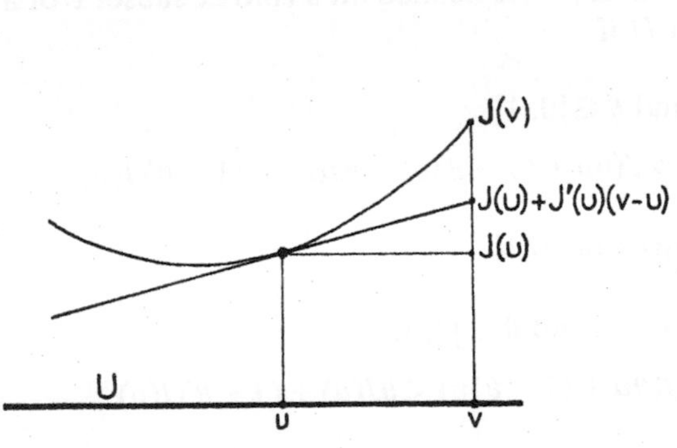

Fig. 4.7-1. The geometrical characterization $J(u) + J'(u)(v-u) \leqslant J(v)$ of convex functions.

Remarks. (1) There is no converse *in general* to condition (b). Consider for instance the strictly convex function $v \in \mathbb{R} \to J(v) = v^4$ for which $J''(0) = 0$.

(2) Theorem 4.7-7 should be compared with Theorems 1.3-1 and 1.3-2. ■

As a first application of Theorem 4.7-7, we infer that *the function*

$$g : F \in \mathbb{M}^n \to \|F\|^2 = \operatorname{tr} F^{\mathrm{T}} F$$

is strictly convex, since its second derivative satisfies

$$g''(F)(G, G) = 2\|G\|^2 \geq 0 \quad \text{for all } F, G \in \mathbb{M}^n .$$

Let $J : U \to \mathbb{R}$ be a function defined on an arbitrary set U. A point $u \in U$ is a **minimum** *of J on U* if $J(u) \leq J(v)$ for all $v \in U$; it is a **strict minimum** *of J on U* if $J(u) < J(v)$ for all $v \in U$, $v \neq u$. The following simple properties of minima of convex functions are constantly used:

Theorem 4.7-8 (minima of convex functions). *Let $J : U \to \mathbb{R}$ be a convex function defined on a convex subset U of a normed vector space.*

(a) *Any local minimum of J on U is a minimum.*

(b) *If J is strictly convex, it has at most one minimum on U, and it is a strict minimum.*

(c) *Let J be differentiable at a point $u \in U$. A point u is a minimum of J on U if and only if*

$$J'(u)(v - u) \geq 0 \quad \text{for all } v \in U .$$

(d) *If the set U is open, a point u is a minimum of J on U if and only if $J'(u) = 0$.* ■

We next extend in two directions the definition of a convex function. First, we define convexity for functions that are not necessarily defined on a convex set: Let U be an arbitrary nonempty subset of a vector space V and let co U denote its convex hull. A function $J^* : U \to \mathbb{R}$ is **convex** if there exists a convex function (in the sense understood so far) $J : \text{co } U \to \mathbb{R}$ such that $J^*(v) = J(v)$ for all $v \in U$. Figure 4.7-2 shows an instance of a function that is convex on a nonconvex subset U of $\mathbb{R}$, and an instance of a function that is not convex on the same subset. A

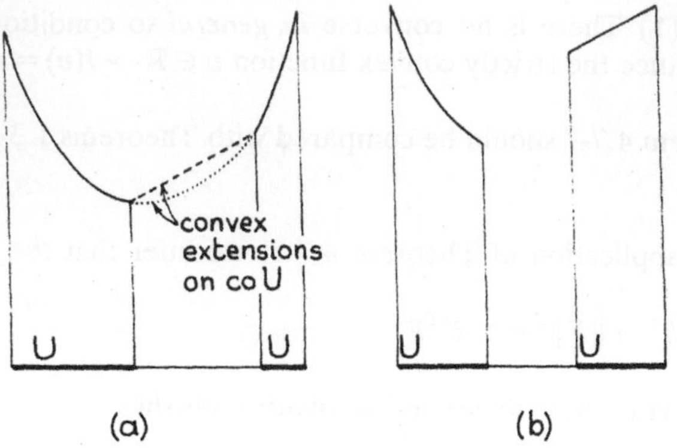

Fig. 4.7-2. A convex function (a) and a nonconvex function (b) over a nonconvex set U.

necessary and sufficient condition for convexity of functions defined on nonconvex sets is given in Exercise 4.13.

We next recall that the *set of extended real numbers* is the set $\{-\infty\} \cup \mathbb{R} \cup \{+\infty\}$, equipped with the natural operations and ordering it inherits from the set $\mathbb{R}$, with particular rules concerning the symbols $-\infty$ and $+\infty$ (for details, see for example Bourbaki [1966, Ch. IV, §4], Taylor [1965, p. 29, p. 178]). Then the notion of convexity can be extended to functions with values in the set $\mathbb{R} \cup \{+\infty\}$ (the value $-\infty$ has to be excluded in order to avoid pathological situations; see for instance the discussion in Ekeland & Temam [1974, pp. 8–10]), as follows:

Let U be a convex subset of a vector space V. A function $J: U \to \mathbb{R} \cup \{+\infty\}$ is **convex** if

$$J(\lambda u + (1-\lambda)v) \leq \lambda J(u) + (1-\lambda)J(v) \quad \text{for all } u, v \in U,\ \lambda \in [0,1]\,.$$

Notice that, since the value $-\infty$ is excluded, the right-hand side of the above inequality is a well defined number in the set $\mathbb{R} \cup \{+\infty\}$. The interest of allowing the value $+\infty$ lies in the observation that convex functions with values in the set $\mathbb{R} \cup \{+\infty\}$ can be always assumed to be defined over the *whole* space. More precisely:

Theorem 4.7-9. *Let U be a subset of a vector space V, and let $J: U \to \mathbb{R}$ be a real-valued function. Then the function $\tilde{J}: V \to \mathbb{R} \cup \{+\infty\}$ defined by*

$$\tilde{J}: v \in V \to \begin{cases} J(v) & \text{if } v \in U\,, \\ +\infty & \text{if } v \notin U\,, \end{cases}$$

is convex if and only if the set U is convex and the function $J: V \to \mathbb{R}$ is convex. ■

To conclude, we state three other important properties of convex functions with values in the set $\mathbb{R} \cup \{+\infty\}$. The first two are elementary. The third one is proved e.g. in Ekeland & Temam [1974, p. 12].

Theorem 4.7-10. (a) *Let V be a vector space. A function $J: V \to \mathbb{R} \cup \{+\infty\}$ is convex if and only if its* **epigraph**

$$\operatorname{epi} J := \{(v, \alpha) \in V \times \mathbb{R};\ J(v) \leq \alpha\}$$

is a convex subset of the space $V \times \mathbb{R}$ (Fig. 4.7-3).

(b) *Let V be a vector space, and let $(J_\lambda : V \to \mathbb{R} \cup \{+\infty\})_{\lambda \in \Lambda}$ be any family of convex functions. Then the function*

$$J = \sup_{\lambda \in \Lambda} J_\lambda$$

is also convex.

(c) *Let V be a finite-dimensional space. A convex function $J: V \to \mathbb{R} \cup \{+\infty\}$ is continuous on the interior of the set $\{v \in V;\ J(v) < +\infty\}$.* ■

As shown in Fig. 4.7-3, a convex function J may be discontinuous on the boundary of the set $\{v \in V;\ J(v) < +\infty\}$. The assumption that the

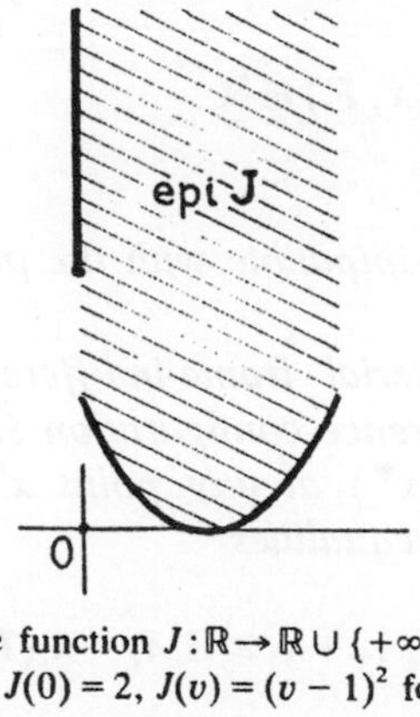

Fig. 4.7-3. The epigraph of the function $J: \mathbb{R} \to \mathbb{R} \cup \{+\infty\}$ defined by $J(v) = +\infty$ for $v < 0$ and $2 \leq v$, $J(0) = 2$, $J(v) = (v-1)^2$ for $0 < v < 2$.

space V be finite-dimensional is essential in (c). For instance the convex functional $J(p)=p(2)$ is discontinuous on the space of all polynomials $p:\mathbb{R}\to\mathbb{R}$ normed by $\|p\|=\sup_{0\leq x\leq 1}|p(x)|$; to see this, consider the polynomials $p_n: x\in\mathbb{R}\to x^n/n$, for which $p_n\to 0$ and $J(p_n)\to+\infty$. For the extension of property (c) to infinite-dimensional spaces, see Ekeland & Temam [1974, p. 13].

In Chapter 7, we shall study the relation between convexity and the important notion of lower semi-continuous functionals, notably in connection with the existence of minimizers for such functionals.

4.8. NONCONVEXITY OF THE STORED ENERGY FUNCTION

Convexity of the stored energy function with respect to its argument $F\in\mathbb{M}^3_+$ at each point $x\in\bar{\Omega}$ would render the mathematical analysis of the associated minimization problem straightforward (Theorem 7.3-2). If this property were verified, simple convex functions such as $F\to\operatorname{tr}F^TF=\|F\|^2$ (Theorem 4.7-7), or more generally $F\to\operatorname{tr}(F^TF)^{\gamma/2}$, $\gamma\geq 1$ (Theorem 4.9-2), would be good candidates for constructing stored energy functions. Nevertheless *such seemingly innocuous examples must be ruled out, on the ground that they contradict the most immediate physical experience*, as we now show. To begin with, let us derive *mathematical* consequences that this property would imply (the definition of convex functions over nonconvex sets has been given in the previous section).

Theorem 4.8-1. *Let $x\in\bar{\Omega}$ be such that the function*

$$\hat{W}(x,\cdot):F\in\mathbb{M}^3_+\to\hat{W}(x,F)\in\mathbb{R}$$

is convex. Then:

(a) *This property is incompatible with the property:* $\hat{W}(x,F)\to+\infty$ *as* $\det F\to 0^+$, $F\in\mathbb{M}^3_+$.

(b) *The axiom of material frame-indifference implies that for any deformation φ of the reference configuration $\bar{\Omega}$, the eigenvalues τ_i of the Cauchy stress tensor $T^\varphi(x^\varphi)$ at any point $x^\varphi=\varphi(x)$ of the deformed configuration satisfy the inequalities*

$$\tau_1+\tau_2\geq 0,\ \tau_2+\tau_3\geq 0,\ \tau_3+\tau_1\geq 0,\ \tau_i=\lambda_i(T^\varphi(x^\varphi))\,.$$

Proof. (i) Since the set $\mathbb{M}^3_+$ is *not* convex (Theorem 4.7-4), there exist a number μ_0 and matrices $\boldsymbol{F}_0$, $\boldsymbol{F}_1$ such that

$$\mu_0 \in]0, 1[, \boldsymbol{F}_0 \in \mathbb{M}^3_+, \boldsymbol{G}_0 \in \mathbb{M}^3_+, \{(1-\mu_0)\boldsymbol{F}_0 + \mu_0 \boldsymbol{G}_0\} \not\in \mathbb{M}^3_+ .$$

Let us denote by $\hat{W}^*(x, \cdot): \mathbb{M}^3 \to \mathbb{R} \cup \{+\infty\}$ any convex extension of the function $\hat{W}(x, \cdot): \mathbb{M}^3_+ \to \mathbb{R}$ to the set $\operatorname{co} \mathbb{M}^3_+ = \mathbb{M}^3$ (Theorem 4.7-4), and consider the convex function

$$\omega : \lambda \in [0, 1] \to \omega(\lambda) := \hat{W}^*(x, (1-\lambda)\boldsymbol{F}_0 + \lambda \boldsymbol{G}_0) \in \mathbb{R} \cup \{+\infty\} .$$

On the one hand, we must have

$$\sup_{0 \leq \lambda \leq 1} \omega(\lambda) \leq \max\{\omega(0), \omega(1)\} < +\infty .$$

On the other hand, there exists $\lambda_0 \in]0, \mu_0]$ such that

$$\det\{(1-\lambda)\boldsymbol{F}_0 + \lambda \boldsymbol{G}_0\} > 0 \quad \text{for } 0 \leq \lambda < \lambda_0,$$
$$\det\{(1-\lambda_0)\boldsymbol{F}_0 + \lambda_0 \boldsymbol{G}_0\} = 0$$

(the function $\lambda \to \det\{(1-\lambda)\boldsymbol{F}_0 + \lambda \boldsymbol{G}_0\}$ is a polynomial of degree ≤ 3); hence we must have $\lim_{\lambda \to \lambda_0^-} \omega(\lambda) = +\infty$, which is impossible.

(ii) We have seen (Theorem 4.2-1) that for hyperelastic materials, the axiom of material frame-indifference is equivalent to the relations

$$\hat{W}(x, \boldsymbol{QF}) = \hat{W}(x, \boldsymbol{F}) \quad \text{for all } \boldsymbol{Q} \in \mathbb{O}^3_+, \boldsymbol{F} \in \mathbb{M}^3_+ .$$

Assume that the function $\hat{W}(x, \cdot): \mathbb{M}^3_+ \to \mathbb{R}$ is convex. Given arbitrary matrices $\boldsymbol{Q} \in \mathbb{O}^3_+$ and $\boldsymbol{F} \in \mathbb{M}^3_+$, the criterion for convexity given in Theorem 4.7-6 shows that

$$\hat{W}(x, \boldsymbol{F}) + \frac{\partial \hat{W}}{\partial \boldsymbol{F}}(x, \boldsymbol{F}) : (\boldsymbol{QF} - \boldsymbol{F}) \leq \hat{W}(x, \boldsymbol{QF}) .$$

Observe that it is not necessary to replace the function $\hat{W}(x, \cdot)$ in this inequality by its extension to the set $\mathbb{M}^3$ since both arguments $\boldsymbol{F}$ and $\boldsymbol{QF}$ are in $\mathbb{M}^3_+$; observe also that the point $\boldsymbol{F}$ belongs to the open set $\mathbb{M}^3_+$, on which the function $\hat{W}(x, \cdot)$ is differentiable by assumption. Since $\hat{W}(x, \boldsymbol{QF}) = \hat{W}(x, \boldsymbol{F})$, we must therefore have

$$\frac{\partial \hat{W}}{\partial F}(x, F):(QF - F) \leq 0 \quad \text{for all } Q \in \mathbb{O}^3_+,\ F \in \mathbb{M}^3_+ .$$

Using the relation $A:BC = AC^{\mathrm{T}}:B$ and the inequality $\det F > 0$, we must also have

$$(\det F)^{-1} \frac{\partial \hat{W}}{\partial F}(x, F)F^{\mathrm{T}}:(Q - I) \leq 0 \quad \text{for all } Q \in \mathbb{O}^3_+,\ F \in \mathbb{M}^3_+ .$$

The particular choice $F = \nabla\varphi(x)$ in the last inequality implies that the Cauchy stress tensor at the point $x^\varphi = \varphi(x)$ (Sect. 2.5):

$$T^\varphi(x^\varphi) = (\det \nabla\varphi(x))^{-1} T(x)\nabla\varphi(x)^{\mathrm{T}}, \text{ with } T(x) = \frac{\partial \hat{W}}{\partial F}(x, \nabla\varphi(x)),$$

satisfies

$$T^\varphi(x^\varphi):(Q - I) \leq 0 \quad \text{for all } Q \in \mathbb{O}^3_+,$$

so that it remains to characterize those tensors T that satisfy

$$T \in \mathbb{S}^3 \text{ and } T:(Q - I) \leq 0 \quad \text{for all } Q \in \mathbb{O}^3_+ .$$

Given such a tensor T, let P be such that

$$P \in \mathbb{O}^3_+,\ T = P^{\mathrm{T}}DP \quad \text{with } D = \mathbf{Diag}\,\tau_i$$

(we may always assume that an orthogonal matrix that diagonalizes a symmetric matrix has a determinant >0; otherwise, we replace one of its column vectors by its opposite). By virtue of the relations $AB:C = A:CB^{\mathrm{T}} = B:A^{\mathrm{T}}C$, we also have

$$T:(Q - I) = P^{\mathrm{T}}DP:Q - P^{\mathrm{T}}DP:I = D:(PQP^{\mathrm{T}} - I) .$$

Thus it suffices to characterize those *diagonal* matrices $\mathbf{Diag}\,\tau_i$ that satisfy

$$\mathbf{Diag}\,\tau_i:(R - I) \leq 0 \quad \text{for all } R \in \mathbb{O}^3_+ .$$

The particular choices $R = \mathbf{Diag}(1, -1, -1)$, $R = \mathbf{Diag}(-1, 1, -1)$, $R = \mathbf{Diag}(-1, -1, 1)$ show that $\mathbb{T} \subset \mathbb{T}'$, where

$$\mathbb{T} := \{\mathbf{Diag}\,\tau_i \in \mathbb{M}^3;\ \mathbf{Diag}\,\tau_i : (\boldsymbol{R} - \boldsymbol{I}) \leqslant 0 \quad \text{for all } \boldsymbol{R} \in \mathbb{O}^3_+\}\,,$$

$$\mathbb{T}' := \{\mathbf{Diag}\,\tau_i \in \mathbb{M}^3;\ \tau_1 + \tau_2 \geqslant 0,\ \tau_2 + \tau_3 \geqslant 0,\ \tau_3 + \tau_1 \geqslant 0\}\,.$$

In order to show that the opposite inclusion also holds, we first observe that since both $\mathbb{T}$ and $\mathbb{T}'$ are cones with vertices at the origin ($\lambda\boldsymbol{D} \in \mathbb{T}$ for all $\lambda \geqslant 0$ if $\boldsymbol{D} \in \mathbb{T}$), it suffices to prove that

$$\mathbb{T}' \cap \mathbb{P} \subset \mathbb{T} \cap \mathbb{P}\,,$$

where $\mathbb{P}$ is the plane defined by

$$\mathbb{P} := \{\mathbf{Diag}\,\tau_i \in \mathbb{M}^3;\ \tau_1 + \tau_2 + \tau_3 = 1\}\,.$$

Since each matrix in the triangle $\mathbb{T}' \cap \mathbb{P}$ can be written as a convex combination of its three vertices $\mathbf{Diag}(-1, 1, 1)$, $\mathbf{Diag}(1, -1, 1)$, $\mathbf{Diag}(1, 1, -1)$, and since both sets $\mathbb{T}' \cap \mathbb{P}$ and $\mathbb{T} \cap \mathbb{P}$ are convex, it suffices to show that the set $\mathbb{T} \cap \mathbb{P}$ contains these three vertices. Let $\boldsymbol{S}$ be any one of these three diagonal matrices. Since

$$-\boldsymbol{S} \in \mathbb{O}^3_+ \text{ and } (-\boldsymbol{S}) : (\boldsymbol{R} - \boldsymbol{I}) = \operatorname{tr}(-\boldsymbol{S}\boldsymbol{R}) + 1\,,$$

our problem is reduced to showing that

$$\operatorname{tr} \boldsymbol{Q} \geqslant -1 \quad \text{for all } \boldsymbol{Q} \in \mathbb{O}^3_+\,.$$

But the trace is an invariant, so that there is no loss of generality in considering orthogonal matrices of the particular form

$$\boldsymbol{Q} = \begin{pmatrix} 1 & 0 & 0 \\ 0 & \cos\theta & -\sin\theta \\ 0 & \sin\theta & \cos\theta \end{pmatrix},$$

which clearly satisfy the inequality $\operatorname{tr} \boldsymbol{Q} \geqslant -1$. Hence the proof is complete (another proof is suggested in Exercise 4.14). ■

The first implication proved in Theorem 4.8-1 rules out the convexity of those stored energy functions that have an explicit dependence on det $\boldsymbol{F}$, since we have shown in Section 4.6 that there exists in this case strong evidence in favor of the limit behavior $\lim_{\det \boldsymbol{F} \to 0^+} \hat{W}(x, \boldsymbol{F}) = +\infty$.

The second implication leads to even more serious impossibilities, since the eigenvalues $\tau_i = \lambda_i(T^\varphi(x^\varphi))$ of the Cauchy stress tensor cannot be expected to satisfy the inequalities $\tau_i + \tau_{i+1} \geqslant 0$ at *all* points x^φ in *all* deformed configurations: Consider for example a sphere subjected to a uniform pressure, in which case $T^\varphi(x^\varphi) = -\pi I$, $\pi > 0$ (Fig. 3.8-2). The inequalities $\tau_i + \tau_{i+1} \geqslant 0$ have been first derived by Coleman & Noll [1959] (see also Truesdell & Noll [1965, p. 163]). That convexity of the stored energy function is incompatible with its behavior as $\det F \to 0^+$ was first observed by Antman [1970] (see also Antman [1983]).

Remark. If in addition the stored energy function were assumed to be *strictly* convex, the associated energy would have at most one stationary point in the situation considered in Theorem 4.1-1 (Exercise 4.15), in contradiction with the counter-examples to uniqueness described in Sect. 5.8. For complements, see Hill [1957, 1968, 1970], Rivlin [1973], Sidoroff [1974]. ■

4.9. JOHN BALL'S POLYCONVEX STORED ENERGY FUNCTIONS

John Ball has made the crucial observation that, if the convexity of the stored energy function $\hat{W}(x, F)$ must be ruled out, it can be replaced by the *weaker* requirement that the function $\mathbb{W}^*(x, \cdot)$ defined by $\mathbb{W}^*(x, F, \operatorname{Cof} F, \det F) := \hat{W}(x, F)$ for all $F \in \mathbb{M}^3_+$ be convex on the set $\{(F, \operatorname{Cof} F, \det F) \in \mathbb{M}^3 \times \mathbb{M}^3 \times \mathbb{R},\ F \in \mathbb{M}^3_+\}$. Contrary to convexity (Theorem 4.8-1), such an assumption does not conflict with any physical requirement and indeed, it is satisfied by realistic models (Theorems 4.9-2 and 4.10-1). Above all, its paramount virtue lies in the powerful *existence results* proved in Chapter 7.

More specifically, the following general definition has been proposed by John Ball (cf. Ball [1977, p. 359]): A function $\hat{W} : \mathbb{F} \to \mathbb{R}$ defined on an arbitrary subset $\mathbb{F}$ of the set $\mathbb{M}^3$ is *polyconvex* if there exists a convex function $\mathbb{W}^* : \mathbb{U} \to \mathbb{R}$, where

$$\mathbb{U} := \{(F, \operatorname{Cof} F, \det F) \in \mathbb{M}^3 \times \mathbb{M}^3 \times \mathbb{R};\ F \in \mathbb{F}\},$$

such that

$$\hat{W}(F) = \mathbb{W}^*(F, \operatorname{Cof} F, \det F) \quad \text{for all } F \in \mathbb{F},$$

or equivalently (according to the definition of convex functions over nonconvex sets; cf. Sect. 4.7) if there exists a convex function $\mathbb{W}: \text{co}\, \mathbb{U} \rightarrow \mathbb{R}$ such that

$$\hat{W}(\boldsymbol{F}) = \mathbb{W}(\boldsymbol{F}, \mathbf{Cof}\, \boldsymbol{F}, \det \boldsymbol{F}) \quad \text{for all } \boldsymbol{F} \in \mathbb{F}\, .$$

Since (Theorem 4.7-4)

$$\begin{aligned}\text{co}\{(\boldsymbol{F}, \mathbf{Cof}\, \boldsymbol{F}, \det \boldsymbol{F}) \in \mathbb{M}^3 \times \mathbb{M}^3 \times \mathbb{R};\, \boldsymbol{F} \in \mathbb{M}^3_+\} \\ = \mathbb{M}^3 \times \mathbb{M}^3 \times]0, +\infty[\, ,\end{aligned}$$

this definition admits the following special case, which will be of essential interest to us: *A stored energy function* $\hat{W}: \bar{\Omega} \times \mathbb{M}^3_+ \rightarrow \mathbb{R}$ *is* **polyconvex** *if for each* $x \in \bar{\Omega}$, *there exists a convex function*

$$\mathbb{W}(x, \cdot): \mathbb{M}^3 \times \mathbb{M}^3 \times]0, +\infty[\rightarrow \mathbb{R}$$

such that

$$\boxed{\hat{W}(x, \boldsymbol{F}) = \mathbb{W}(x, \boldsymbol{F}, \mathbf{Cof}\, \boldsymbol{F}, \det \boldsymbol{F}) \quad \text{for all } \boldsymbol{F} \in \mathbb{M}^3_+\, .}$$

Remarks. (1) Since the convex set $\mathbb{M}^3 \times \mathbb{M}^3 \times]0, +\infty[$ is open in $\mathbb{M}^3 \times \mathbb{M}^3 \times \mathbb{R}$, the convex function $\mathbb{W}(x, \cdot): \mathbb{M}^3 \times \mathbb{M}^3 \times]0, +\infty[\rightarrow \mathbb{R}$ is continuous (Theorem 4.7-10). On the other hand, the function

$$\bar{\mathbb{W}}: (x, \boldsymbol{F}, \boldsymbol{H}, \delta) \in \mathbb{M}^3 \times \mathbb{M}^3 \times \mathbb{R} \rightarrow \begin{cases} \mathbb{W}(x, \boldsymbol{F}, \boldsymbol{H}, \delta) \text{ if } \delta > 0\, , \\ +\infty \text{ if } \delta \leq 0\, , \end{cases}$$

with values in the set $\mathbb{R} \cup \{+\infty\}$, is convex (Theorem 4.7-9), but it need not be continuous for $\delta = 0$, unless the function $\mathbb{W}$ is such that

$$\left.\begin{aligned}(\boldsymbol{F}_k, \boldsymbol{H}_k) \rightarrow (\boldsymbol{F}, \boldsymbol{H}) \text{ in } \mathbb{M}^3 \times \mathbb{M}^3 \\ \delta_k \rightarrow 0^+\end{aligned}\right\} \Rightarrow \mathbb{W}(x, \boldsymbol{F}_k, \boldsymbol{H}_k, \delta_k) \rightarrow +\infty\, .$$

Observe, however, that this property implies, but is *not* necessarily implied by, the property $\hat{W}(x, \boldsymbol{F}) \rightarrow +\infty$ as $\det \boldsymbol{F} \rightarrow 0^+$ (Exercise 7.11).

(2) A definition of *two-dimensional polyconvex stored energy functions* can be derived by a "limit" analysis for von Kármán plates; cf. Ciarlet & Quintela-Estevez [1987]. ■

As a first indication of the merits of polyconvexity, consider the function

$$\hat{W}: F \in \mathbb{M}^3_+ \rightarrow \hat{W}(F) = \operatorname{tr} \mathbf{Cof}\, F^{\mathrm{T}} F = \|\mathbf{Cof}\, F\|^2 .$$

This is not a convex function: Consider for example the matrices

$$F_\lambda = \lambda\, \mathbf{Diag}(2, 1, 1) + (1 - \lambda)\, \mathbf{Diag}(1, 2, 1) \in \mathbb{M}^3_+, \quad 0 \leqslant \lambda \leqslant 1 ,$$

for which the associated function

$$\lambda \in [0, 1] \rightarrow \operatorname{tr} \mathbf{Cof}\, F_\lambda^{\mathrm{T}} F_\lambda = 9 + 2\lambda - \lambda^2 - 2\lambda^3 + \lambda^4$$

is not convex in a neighborhood of 0. It is, however, polyconvex since the function

$$\mathbb{W}: H \in \mathbb{M}^3 \rightarrow \mathbb{W}(H) = \operatorname{tr} H^{\mathrm{T}} H = \|H\|^2$$

is convex (Theorem 4.7-7). Consider likewise the function

$$\hat{W}: F \in \mathbb{M}^3_+ \rightarrow \hat{W}(F) = \det F .$$

This is not a convex function: Consider again the above matrices F_λ, $0 \leqslant \lambda \leqslant 1$, for which the associated function

$$\lambda \in [0, 1] \rightarrow \det F_\lambda = 2 + \lambda - \lambda^2$$

is not convex. It is however polyconvex since the function $\mathbb{W}(\delta) = \delta$ is convex.

Combining these examples, we conclude that a stored energy function of the form

$$\hat{W}: F \in \mathbb{M}^3_+ \rightarrow \hat{W}(F) = a\|F\|^2 + b\|\mathbf{Cof}\, F\|^2 + \Gamma(\det F)$$

where $a > 0$, $b > 0$, and $\Gamma :]0, +\infty[\rightarrow \mathbb{R}$ is a convex function, is polyconvex, since the function $\mathbb{W}$ defined by

$$\mathbb{W}(x, F, H, \delta) = a\|F\|^2 + b\|H\|^2 + \Gamma(\delta) ,$$

is convex on $\mathbb{M}^3 \times \mathbb{M}^3 \times]0, +\infty[$. *Our objective is to define an important class of polyconvex stored energy functions that generalize the above example.*

As a preparation, we establish a general criterion of convexity for functions of matrices (various complements to this result are given in Exercises 4.16, 4.17, 4.18, 4.19). We recall that the *singular values* $v_i(\boldsymbol{F})$ of a matrix $\boldsymbol{F} \in \mathbb{M}^3$ are the square roots of the eigenvalues of the positive semi-definite matrix $\boldsymbol{F}^T\boldsymbol{F}$ (Sect. 3.2). The following result is due to Thompson & Freede [1971]; the proof given here is due to Ball [1977, Theorem 5.1].

Theorem 4.9-1. *Let there be given a function* $\Phi : [0, +\infty[^n \to \mathbb{R}$ *that is symmetric, convex, and nondecreasing in each variable. Then the function*

$$W : \boldsymbol{F} \in \mathbb{M}^n \to W(\boldsymbol{F}) = \Phi(v_1(\boldsymbol{F}), v_2(\boldsymbol{F}), \ldots, v_n(\boldsymbol{F}))$$

is convex.

Proof. Let us denote by $\mathfrak{S}_n$ the permutation group of the set $\{1, 2, \ldots, n\}$. The assumed symmetry of the function Φ means that

$$\Phi(v_1, v_2, \ldots, v_n) = \Phi(v_{\sigma(1)}, v_{\sigma(2)}, \ldots, v_{\sigma(n)}) \quad \text{for all } \sigma \in \mathfrak{S}_n .$$

(i) Given two arbitrary matrices $\boldsymbol{F}, \boldsymbol{G} \in \mathbb{M}^n$ and $\lambda \in]0, 1[$, we must show that

$$W(\lambda \boldsymbol{F} + (1 - \lambda)\boldsymbol{G}) \leq \lambda W(\boldsymbol{F}) + (1 - \lambda)W(\boldsymbol{G}) ,$$

or equivalently, that

$$\Phi(\boldsymbol{a}) \leq \lambda \Phi(\boldsymbol{u}) + (1 - \lambda)\Phi(\boldsymbol{v}) ,$$

where

$$a_1 \geq a_2 \geq \cdots \geq a_n \geq 0, \; u_1 \geq u_2 \geq \cdots \geq u_n \geq 0, \; v_1 \geq v_2 \geq \cdots \geq v_n \geq 0$$

denote the singular values of the matrices $\lambda \boldsymbol{F} + (1 - \lambda)\boldsymbol{G}$, $\boldsymbol{F}$, $\boldsymbol{G}$, respectively, and $\boldsymbol{a} = (a_i)$, $\boldsymbol{u} = (u_i)$, $\boldsymbol{v} = (v_i)$. Letting (cf. Fig. 4.9-1 for $n = 2$)

$$\boldsymbol{c} = \lambda \boldsymbol{u} + (1 - \lambda)\boldsymbol{v} = (c_i) ,$$

$$\boldsymbol{c}^0 = \boldsymbol{o} = (c_i^0), \; \boldsymbol{c}^l = (c_1, c_2, \ldots, c_l, 0, 0, \ldots, 0) = (c_i^l) ,$$

$$\boldsymbol{c}_\sigma^l = (c_{\sigma(i)}^l), \; 0 \leq l \leq n, \; \sigma \in \mathfrak{S}_n ,$$

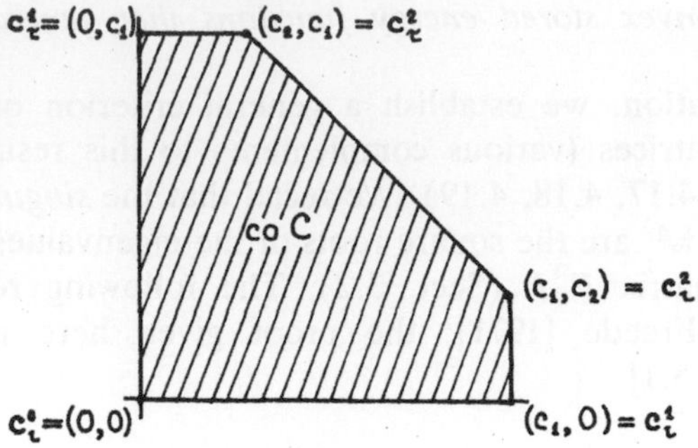

Fig. 4.9-1. The convex hull of the finite set $C = \{c_\sigma^l;\ l = 0, 1, 2,\ \sigma \in \mathfrak{S}_2\}$, $\mathfrak{S}_2 = \{\iota, \tau\}$.

we shall show (part (iii) of the proof) that the vector $a = (a_i)$ can be written as a convex combination of the vectors c_σ^l, i.e., in the form

$$\begin{cases} a = \sum_{l=0}^{n} \sum_{\sigma \in \mathfrak{S}_n} \mu_\sigma^l c_\sigma^l\,, \\ 1 = \sum_{l=0}^{n} \sum_{\sigma \in \mathfrak{S}_n} \mu_\sigma^l\,, \quad \mu_\sigma^l \geq 0\,. \end{cases}$$

Then we claim that this implies the convexity of the function W: Since the function Φ is convex,

$$\Phi(a) \leq \sum_{l=0}^{n} \sum_{\sigma \in \mathfrak{S}_n} \mu_\sigma^l \Phi(c_\sigma^l)\,;$$

since it is symmetric and nondecreasing in each variable,

$$\Phi(c_0^l) = \Phi(c^l) \geq \Phi(c)\,;$$

and thus, using again the convexity of the function Φ,

$$\Phi(a) \leq \Phi(c) = \Phi(\lambda u + (1-\lambda)v) \leq \lambda\Phi(u) + (1-\lambda)\Phi(v)\,.$$

(ii) As an intermediary result, we show that, *given any numbers* $r_1 \geq r_2 \geq \cdots \geq r_n \geq 0$, *the particular function*

$$F \in \mathbb{M}^3 \to \theta_r(F) := \sum_{i=1}^{n} r_i v_i(F)\,,$$

where the singular values of $\boldsymbol{F}$ *are ordered as*

$$v_1(\boldsymbol{F}) \geqslant v_2(\boldsymbol{F}) \geqslant \cdots \geqslant v_n(\boldsymbol{F}) \geqslant 0\,,$$

is convex. To this end, we shall establish that

$$\theta_r(\boldsymbol{F}) = \max_{\boldsymbol{Q},\boldsymbol{R}\in\mathbb{O}^n} \operatorname{tr}(\boldsymbol{F}\boldsymbol{Q}\boldsymbol{D}_r\boldsymbol{R})\,, \quad \boldsymbol{D}_r = \mathbf{Diag}\, r_i\,.$$

Since each function $\boldsymbol{F} \in \mathbb{M}^n \to \operatorname{tr}(\boldsymbol{F}\boldsymbol{Q}\boldsymbol{D}_r\boldsymbol{R})$ is linear and hence convex, and since a supremum of convex functions is convex (Theorem 4.7-10), this will prove the assertion. First, we have by Theorem 3.2-4,

$$\operatorname{tr}(\boldsymbol{F}\boldsymbol{Q}\boldsymbol{D}_r\boldsymbol{R}) \leqslant \sum_{i=1}^n v_i(\boldsymbol{F}\boldsymbol{Q})v_i(\boldsymbol{D}_r\boldsymbol{R}) = \sum_{i=1}^n r_i v_i(\boldsymbol{F}) \quad \text{for all } \boldsymbol{Q}, \boldsymbol{R} \in \mathbb{O}^n\,.$$

Secondly, using the singular value decomposition theorem (Theorem 3.2-3), we can write the matrix $\boldsymbol{F}$ as

$$\boldsymbol{F} = \boldsymbol{S}\boldsymbol{D}_v\boldsymbol{T} \text{ with } \boldsymbol{D}_v = \mathbf{Diag}\, v_i(\boldsymbol{F}), \quad \boldsymbol{S}, \boldsymbol{T} \in \mathbb{O}^n\,,$$

so that the particular choices $\boldsymbol{Q} = \boldsymbol{T}^{-1}$ and $\boldsymbol{R} = \boldsymbol{S}^{-1}$ imply

$$\max_{\boldsymbol{Q},\boldsymbol{R}\in\mathbb{O}^n} \operatorname{tr}(\boldsymbol{F}\boldsymbol{Q}\boldsymbol{D}_r\boldsymbol{R}) \geqslant \operatorname{tr}(\boldsymbol{S}\boldsymbol{D}_v\boldsymbol{D}_r\boldsymbol{S}^{-1}) = \sum_{i=1}^n r_i v_i(\boldsymbol{F})\,.$$

(iii) It remains to show that the vector $\boldsymbol{a}$ can be written as a convex combination of the vectors $\boldsymbol{c}_\sigma^l$, $0 \leqslant l \leqslant n$, $\sigma \in \mathfrak{S}_n$, defined in part (i), i.e., that $\boldsymbol{a}$ belongs to the convex hull of the set (Fig. 4.9-1)

$$C := \{\boldsymbol{c}_\sigma^l \in \mathbb{R}^n; 0 \leqslant l \leqslant n, \sigma \in \mathfrak{S}_n\}\,.$$

Since the set C is finite, its convex hull co C is closed and thus it is the intersection of all the closed half-spaces $\{\boldsymbol{x} \in \mathbb{R}^n; \boldsymbol{x} \cdot \boldsymbol{d} \leqslant \delta\}$ that contain C (Theorem 4.7-2). In other words, we must prove the implication:

$$\boldsymbol{c}_\sigma^l \cdot \boldsymbol{d} \leqslant \delta, 0 \leqslant l \leqslant n, \sigma \in \mathfrak{S}_n \Rightarrow \boldsymbol{a} \cdot \boldsymbol{d} \leqslant \delta\,.$$

In order to relate the vectors $\boldsymbol{a}$ and $\boldsymbol{c}$ (and hence the vectors $\boldsymbol{c}_\sigma^l$), we use the result of part (ii): Given any numbers $r_1 \geqslant r_2 \geqslant \cdots \geqslant r_n \geqslant 0$, we have shown that

$$\theta_r(\lambda \boldsymbol{F} + (1-\lambda)\boldsymbol{G}) \leqslant \lambda\theta_r(\boldsymbol{F}) + (1-\lambda)\theta_r(\boldsymbol{G})\,,$$

i.e., that

$$\sum_{i=1}^{n} r_i a_i \leq \lambda \sum_{i=1}^{n} r_i u_i + (1-\lambda) \sum_{i=1}^{n} r_i v_i = \sum_{i=1}^{n} r_i c_i .$$

Consequently, the vectors $\boldsymbol{a}$ and $\boldsymbol{c}$ satisfy

$$\sum_{i=1}^{n} r_i a_i \leq \sum_{i=1}^{n} r_i c_i \quad \text{for all } r_1 \geq \cdots \geq r_n \geq 0 .$$

Let then $\boldsymbol{d} \in \mathbb{R}^n$ and $\delta \in \mathbb{R}$ be such that $\boldsymbol{c}_\sigma^l \cdot \boldsymbol{d} \leq \delta$, $0 \leq l \leq n$, $\sigma \in \mathfrak{S}_n$. Since $\boldsymbol{c}_\sigma^0 = 0$, we infer that $\delta \geq 0$. If all the components d_i are <0, the inequality $\boldsymbol{a} \cdot \boldsymbol{d} \leq \delta$ is surely satisfied, so that it suffices to consider the case where at least one component d_i is ≥ 0. Let $\sigma \in \mathfrak{S}_n$ be such that

$$d_{\sigma^{-1}(1)} \geq \cdots \geq d_{\sigma^{-1}(l)} \geq 0 > d_{\sigma^{-1}(l+1)} \geq \cdots \geq d_{\sigma^{-1}(n)} ,$$

with $1 \leq l \leq n$ (if $l = n$, there is no term to the right of 0). We can write

$$\boldsymbol{a} \cdot \boldsymbol{d} = \sum_{i=1}^{n} a_{\sigma^{-1}(i)} d_{\sigma^{-1}(i)} \leq \sum_{i=1}^{n} a_i d_{\sigma^{-1}(i)}$$

$$\leq \sum_{i=1}^{l} a_i d_{\sigma^{-1}(i)} \leq \sum_{i=1}^{l} c_i d_{\sigma^{-1}(i)} = \boldsymbol{c}_\sigma^l \cdot \boldsymbol{d} \leq \delta ,$$

by using the inequality $\Sigma_{i=1}^n r_i a_i \leq \Sigma_{i=1}^n r_i c_i$ with the particular choices

$$r_i = d_{\sigma^{-1}(i)} \quad \text{for } 1 \leq i \leq l,\ r_i = 0 \text{ for } l+1 \leq i \leq n ,$$

and the proof is complete. ∎

Remark. In the above proof, we have used the implication

$$\left.\begin{array}{l} \alpha_1 \geq \alpha_2 \geq \cdots \geq \alpha_n \\ \beta_1 \geq \beta_2 \geq \cdots \geq \beta_n \end{array}\right\} \Rightarrow \sum_{i=1}^{n} \alpha_i \beta_i \geq \sum_{i=1}^{n} \alpha_i \beta_{\sigma(i)} \quad \text{for all } \sigma \in \mathfrak{S}_n ,$$

which can be easily established by induction on n. ∎

We now define a general class of polyconvex stored energy functions (for notational brevity, we only consider the case of homogeneous materials; the extension to nonhomogeneous materials is straightforward).

Theorem 4.9-2. *Let $\hat{W}$ be a stored energy function of the form*

$$\hat{F} \in \mathbb{M}^3_+ \to \hat{W}(F) = \sum_{i=1}^{M} a_i(v_1^{\gamma_i} + v_2^{\gamma_i} + v_3^{\gamma_i})$$
$$+ \sum_{j=1}^{N} b_j((v_2 v_3)^{\delta_j} + (v_3 v_1)^{\delta_j} + (v_1 v_2)^{\delta_j}) + \Gamma(\det F),$$

where:

$$\begin{cases} v_i = v_i(F) \text{ are the singular values of } F; \\ a_i > 0 \text{ and } \gamma_i \geq 1,\ 1 \leq i \leq M;\ b_j > 0 \text{ and } \delta_j \geq 1,\ 1 \leq j \leq N; \\ \Gamma :]0, +\infty[\to \mathbb{R} \text{ is a convex function.} \end{cases}$$

Then the function $\hat{W}$ is polyconvex, and it satisfies a coerciveness inequality of the form

$$\hat{W}(F) \geq \alpha\{\|F\|^p + \|\mathbf{Cof}\, F\|^q\} + \Gamma(\det F) \quad \text{for all } F \in \mathbb{M}^3_+,$$
$$\text{with } \alpha > 0,\ p = \max_i \gamma_i,\ q = \max_j \delta_j.$$

Proof. (i) The particular function

$$\Phi_\gamma(v_1, v_2, v_3) := v_1^\gamma + v_2^\gamma + v_3^\gamma, \quad \gamma \geq 1,$$

satisfies all the assumptions of Theorem 4.9-1. First, it is symmetric and nondecreasing in each variable. Secondly, it is convex on the set $[0, +\infty[^3$ since it is convex on the set $]0, +\infty[^3$ (the matrix $\gamma(\gamma-1)\,\mathbf{Diag}\, v_i^{\gamma-2}$ representing its second derivative is semi-positive definite for $\gamma \geq 1$; cf. Theorem 4.7-7) and continuous on the set $[0, +\infty[^3$. Hence the function Φ_γ is convex for $\gamma \geq 1$.

(ii) Since the singular values of the matrix $\mathbf{Cof}\, F$ are precisely $v_2 v_3$, $v_3 v_1$, $v_1 v_2$ (Theorem 1.1-1), the function $\hat{W}$ takes the form

$$\hat{W}(F) = A(F) + B(\mathbf{Cof}\, F) + \Gamma(\det F) := \mathbb{W}(F, \mathbf{Cof}\, F, \det F),$$

where the functions $A : \mathbb{M}^3 \to \mathbb{R}$ and $B : \mathbb{M}^3 \to \mathbb{R}$ are convex by (i), and the function $\Gamma :]0, +\infty[\to \mathbb{R}$ is convex by assumption. Therefore the function $\mathbb{W} : \mathbb{M}^3 \times \mathbb{M}^3 \times]0, +\infty[\to \mathbb{R}$ is convex and thus the function $\hat{W}$ is polyconvex.

(iii) Let us finally establish the coerciveness inequality. Since for $\gamma \geqslant 1$, the mapping

$$(v_i) \in \mathbb{R}^3 \rightarrow (|v_1|^\gamma + |v_2|^\gamma + |v_3|^\gamma)^{1/\gamma},$$

is a norm on the space $\mathbb{R}^3$, and since all norms are equivalent on a finite-dimensional space, there exists for each $\gamma \geqslant 1$ a constant c_γ such that

$$(|v_1|^\gamma + |v_2|^\gamma + |v_3|^\gamma) \geqslant c_\gamma(|v_1|^2 + |v_2|^2 + |v_3|^2)^{\gamma/2}.$$

We thus have, with $v_i = v_i(\boldsymbol{F})$,

$$v_1^\gamma + v_2^\gamma + v_3^\gamma \geqslant c_\gamma(v_1^2 + v_2^2 + v_3^2)^{\gamma/2} = c_\gamma(\operatorname{tr} \boldsymbol{F}^{\mathrm{T}}\boldsymbol{F})^{\gamma/2} = c_\gamma \|\boldsymbol{F}\|^\gamma,$$

and the conclusion follows. ∎

Remarks. (1) While demonstrating the convexity of the function $\boldsymbol{F} \in \mathbb{M}^3 \rightarrow (v_1^\gamma + v_2^\gamma + v_3^\gamma)$ is easy for $\gamma = 2$ (as we have already seen, it is an immediate consequence of Theorem 4.7-7), the proof is surprisingly difficult for $\gamma \neq 2$. By contrast, the convexity of the function $\boldsymbol{F} \in \mathbb{M}^3 \rightarrow \|\boldsymbol{F}\|^\gamma$, $\gamma \geqslant 1$, is easily established (Exercise 4.20).

(2) In order to establish the convexity of a function defined on the set $\mathbb{M}_+^3$, it is by definition necessary to find an appropriate extension that is convex on the set $\mathbb{M}^3 = \operatorname{co} \mathbb{M}_+^3$, and this may not be easy in general (Exercise 4.13). By chance, this is immediate in the case of the function $(v_1^\gamma + v_2^\gamma + v_3^\gamma)$. ∎

The dependence on the matrix $\boldsymbol{F}$ itself can be displayed more explicitly in the stored energy function of Theorem 4.9-2, as follows. For any $\boldsymbol{F} \in \mathbb{M}_+^3$, the matrix $(\boldsymbol{F}^{\mathrm{T}}\boldsymbol{F})^{1/2}$ is uniquely defined and it can be written as

$$(\boldsymbol{F}^{\mathrm{T}}\boldsymbol{F})^{1/2} = \boldsymbol{P}\, \mathbf{Diag}\, v_i \boldsymbol{P}^{\mathrm{T}},\ \boldsymbol{P} \in \mathbb{O}^3,\ v_i = v_i(\boldsymbol{F}) = \{\lambda_i(\boldsymbol{F}^{\mathrm{T}}\boldsymbol{F})\}^{1/2}.$$

For any $\delta > 0$, we define the symmetric matrix

$$(\boldsymbol{F}^{\mathrm{T}}\boldsymbol{F})^{\delta/2} := \boldsymbol{P}\, \mathbf{Diag}\, v_i^\delta \boldsymbol{P}^{\mathrm{T}},$$

which can be shown to be independent of the particular orthogonal matrix $\boldsymbol{P}$ that diagonalizes the matrix $\boldsymbol{F}^{\mathrm{T}}\boldsymbol{F}$ (Exercise 4.21). Using this definition, we can write

$$v_1^\gamma + v_2^\gamma + v_3^\gamma = \operatorname{tr}(\boldsymbol{F}^{\mathrm{T}}\boldsymbol{F})^{\gamma/2},$$
$$(v_2 v_3)^\delta + (v_3 v_1)^\delta + (v_1 v_2)^\delta = \operatorname{tr}(\mathbf{Cof}\,\boldsymbol{F}^{\mathrm{T}}\boldsymbol{F})^{\delta/2},$$

and the stored energy function of Theorem 4.9-2 becomes

$$\hat{W}(\boldsymbol{F}) = \sum_{i=1}^{M} a_i \operatorname{tr}(\boldsymbol{F}^{\mathrm{T}}\boldsymbol{F})^{\gamma_i/2} + \sum_{j=1}^{N} \operatorname{tr}(\mathbf{Cof}\,\boldsymbol{F}^{\mathrm{T}}\boldsymbol{F})^{\delta_j/2} + \Gamma(\det \boldsymbol{F}).$$

A hyperelastic material whose stored energy function is of the above form and satisfies the assumptions of Theorem 4.9-2 together with the additional property $\lim_{\delta \to 0^+} \Gamma(\delta) = +\infty$ is called an **Ogden material**. Such materials, which were proposed by Ogden [1972b], are very important from both practical and theoretical viewpoints.

To conclude this chapter, we construct specific examples of Ogden's materials.

4.10. EXAMPLES OF OGDEN'S AND OTHER HYPERELASTIC MATERIALS

We have seen in Theorem 4.4-3 that a St Venant–Kirchhoff material is hyperelastic, and that its stored energy function is of the form

$$\hat{W}(\boldsymbol{F}) = \check{W}(\boldsymbol{E}) = \frac{\lambda}{2}(\operatorname{tr}\boldsymbol{E})^2 + \mu \operatorname{tr}\boldsymbol{E}^2, \quad \boldsymbol{I} + 2\boldsymbol{E} = \boldsymbol{F}^{\mathrm{T}}\boldsymbol{F},$$

or equivalently,

$$\begin{aligned}
\hat{W}(\boldsymbol{F}) &= -\left(\frac{3\lambda + 2\mu}{4}\right)(v_1^2 + v_2^2 + v_3^2) + \left(\frac{\lambda + 2\mu}{8}\right)(v_1^4 + v_2^4 + v_3^4) \\
&\quad + \frac{\lambda}{4}(v_2^2 v_3^2 + v_3^2 v_1^2 + v_1^2 v_2^2) + \left(\frac{6\mu + 9\lambda}{8}\right), \quad v_i = v_i(\boldsymbol{F}), \\
&= -\left(\frac{3\lambda + 2\mu}{4}\right)\operatorname{tr}\boldsymbol{C} + \left(\frac{\lambda + 2\mu}{8}\right)\operatorname{tr}\boldsymbol{C}^2 + \frac{\lambda}{4}\operatorname{tr}\mathbf{Cof}\,\boldsymbol{C} \\
&\quad + \left(\frac{6\mu + 9\lambda}{8}\right), \quad \boldsymbol{C} = \boldsymbol{F}^{\mathrm{T}}\boldsymbol{F}.
\end{aligned}$$

While the second expression resembles that of an Ogden material as given in Theorem 4.9-2, *we now show that the stored energy function of a St Venant–Kirchhoff material is not polyconvex*, basically because there is a minus sign in front of the function $(v_1^2 + v_2^2 + v_3^2)$. But this observation is not a proof, since the above representation as a function of $(\boldsymbol{F}, \mathbf{Cof}\,\boldsymbol{F}, \det \boldsymbol{F})$ is not unique (the lack of an additive term of the form $\Gamma(\det \boldsymbol{F})$, where Γ is a convex function, is irrelevant here; if a stored energy is polyconvex, it remains so if such a term is added). The ingenious proof given here is due to Raoult [1986].

Theorem 4.10-1. *A stored energy function of the form*

$$\hat{W}(\boldsymbol{F}) = a_1 \operatorname{tr} \boldsymbol{C} + a_2 \operatorname{tr} \boldsymbol{C}^2 + b \operatorname{tr} \mathbf{Cof}\,\boldsymbol{C}, \quad \boldsymbol{C} = \boldsymbol{F}^{\mathrm{T}}\mathbf{F}, \quad \text{with } a_2 > 0,\ b > 0,$$

is not polyconvex if $a_1 < 0$.

Proof. For each $\varepsilon > 0$, the matrices

$$\boldsymbol{F}_\varepsilon := \varepsilon \boldsymbol{I}, \quad \boldsymbol{G}_\varepsilon := \varepsilon\, \mathbf{Diag}(1, 1, 3),$$

belong to the set $\mathbb{M}^3_+$, and they satisfy

$$\mathbf{Cof}\,\tfrac{1}{2}(\boldsymbol{F}_\varepsilon + \boldsymbol{G}_\varepsilon) = \tfrac{1}{2}(\mathbf{Cof}\,\boldsymbol{F}_\varepsilon + \mathbf{Cof}\,\boldsymbol{G}_\varepsilon),$$
$$\det \tfrac{1}{2}(\boldsymbol{F}_\varepsilon + \boldsymbol{G}_\varepsilon) = \tfrac{1}{2}(\det \boldsymbol{F}_\varepsilon + \det \boldsymbol{G}_\varepsilon).$$

Let $\hat{W} : \mathbb{M}^3_+ \to \mathbb{R}$ be a polyconvex stored energy function. By definition, there exists a convex function $\mathbb{W} : \mathbb{M}^3 \times \mathbb{M}^3 \times]0, +\infty[\to \mathbb{R}$ such that

$$\begin{aligned} \hat{W}(\tfrac{1}{2}(\boldsymbol{F}_\varepsilon + \boldsymbol{G}_\varepsilon)) &= \mathbb{W}(\tfrac{1}{2}(\boldsymbol{F}_\varepsilon + \boldsymbol{G}_\varepsilon), \mathbf{Cof}\,\tfrac{1}{2}(\boldsymbol{F}_\varepsilon + \boldsymbol{G}_\varepsilon), \det \tfrac{1}{2}(\boldsymbol{F}_\varepsilon + \boldsymbol{G}_\varepsilon)) \\ &= \mathbb{W}(\tfrac{1}{2}(\boldsymbol{F}_\varepsilon + \boldsymbol{G}_\varepsilon), \tfrac{1}{2}(\mathbf{Cof}\,\boldsymbol{F}_\varepsilon + \mathbf{Cof}\,\boldsymbol{G}_\varepsilon), \tfrac{1}{2}(\det \boldsymbol{F}_\varepsilon + \det \boldsymbol{G}_\varepsilon)) \\ &\leq \tfrac{1}{2}\mathbb{W}(\boldsymbol{F}_\varepsilon, \mathbf{Cof}\,\boldsymbol{F}_\varepsilon, \det \boldsymbol{F}_\varepsilon) + \tfrac{1}{2}\mathbb{W}(\boldsymbol{G}_\varepsilon, \mathbf{Cof}\,\boldsymbol{G}_\varepsilon, \det \boldsymbol{G}_\varepsilon) \\ &= \tfrac{1}{2}(\hat{W}(\boldsymbol{F}_\varepsilon) + \hat{W}(\boldsymbol{G}_\varepsilon)). \end{aligned}$$

For a stored energy function of the form

$$\hat{W}(\boldsymbol{F}) = a_1 \operatorname{tr} \boldsymbol{C} + a_2 \operatorname{tr} \boldsymbol{C}^2 + b \operatorname{tr} \mathbf{Cof}\,\boldsymbol{C}, \quad \boldsymbol{C} = \boldsymbol{F}^{\mathrm{T}}\boldsymbol{F},$$

the inequality $\hat{W}(\frac{1}{2}(F_\varepsilon + G_\varepsilon)) \leq \frac{1}{2}(\hat{W}(F_\varepsilon) + \hat{W}(G_\varepsilon))$ is equivalent to the inequality

$$0 \leq a_1\varepsilon^2 + (25a_2 + 2b)\varepsilon^4 ,$$

which does not hold for ε small enough if a_1 is <0. ■

On the other hand, the stored energy function of a St Venant–Kirchhoff material is the simplest one that agrees with the expansion (Theorem 4.5-1)

$$\check{W}(E) = \frac{\lambda}{2}(\mathrm{tr}\, E)^2 + \mu\, \mathrm{tr}\, E^2 + o(\|E\|^2)$$

of the stored energy function of a homogeneous, isotropic, hyperelastic material near a natural state. We now construct a family of *Ogden materials* whose stored energy functions reconcile these two objectives (agreement with the above expansion for arbitrary Lamé constants $\lambda > 0$ and $\mu > 0$, and polyconvexity), and yet retain a remarkably simple expression. The following result is due to Ciarlet & Geymonat [1982].

Theorem 4.10-2. *Let $\lambda > 0$ and $\mu > 0$ be two given Lamé constants. There exist polyconvex stored energy functions of the form*

$$F \in \mathbb{M}^3_+ \to \hat{W}(F) = a\|F\|^2 + b\|\mathbf{Cof}\, F\|^2 + \Gamma(\det F) + e ,$$

with

$$a > 0,\ b > 0,\ \Gamma(\delta) = c\delta^2 - d\,\mathrm{Log}\,\delta,\ c > 0,\ d > 0,\ e \in \mathbb{R} ,$$

that satisfy

$$\hat{W}(F) = \check{W}(E) = \frac{\lambda}{2}(\mathrm{tr}\, E)^2 + \mu\, \mathrm{tr}\, E^2 + O(\|E\|^3) ,$$

$$I + 2E = F^{\mathrm{T}}F .$$

A stored energy function of this form satisfies the coerciveness inequality:

$$\hat{W}(F) \geq \alpha(\|F\|^2 + \|\mathrm{Cof}\, F\|^2 + (\det F)^2) + \beta\,, \quad \alpha > 0\,.$$

Proof. The following relations hold:

$$\|F\|^2 = \mathrm{tr}\, F^{\mathrm{T}}F = \mathrm{tr}(I + 2E) = 3 + 2\,\mathrm{tr}\, E\,,$$

$$\begin{aligned}\|\mathrm{Cof}\, F\|^2 &= \mathrm{tr}\, \mathrm{Cof}\, F^{\mathrm{T}}F = \tfrac{1}{2}(\mathrm{tr}\, F^{\mathrm{T}}F)^2 - \tfrac{1}{2}\,\mathrm{tr}(F^{\mathrm{T}}F)^2 \\ &= \tfrac{1}{2}\{\mathrm{tr}(I + 2E)\}^2 - \tfrac{1}{2}\,\mathrm{tr}(I + 2E)^2 \\ &= 3 + 4\,\mathrm{tr}\, E + 2(\mathrm{tr}\, E)^2 - 2\,\mathrm{tr}\, E^2\,,\end{aligned}$$

$$\begin{aligned}\det F^{\mathrm{T}}F &= \tfrac{1}{6}\{\mathrm{tr}\, F^{\mathrm{T}}F\}^3 - \tfrac{1}{2}\{\mathrm{tr}\, F^{\mathrm{T}}F\}\{\mathrm{tr}(F^{\mathrm{T}}F)^2\} + \tfrac{1}{3}\,\mathrm{tr}(F^{\mathrm{T}}F)^3 \\ &= 1 + 2\,\mathrm{tr}\, E + 2(\mathrm{tr}\, E)^2 - 2\,\mathrm{tr}\, E^2 + O(\|E\|^3)\,,\end{aligned}$$

$$\begin{aligned}\Gamma(\det F) &= \Gamma(\{\det F^{\mathrm{T}}F\}^{1/2}) \\ &= \Gamma(1 + \mathrm{tr}\, E + \tfrac{1}{2}(\mathrm{tr}\, E)^2 - \mathrm{tr}\, E^2 + O(\|E\|^3)) \\ &= \Gamma(1) + \Gamma'(1)\{\mathrm{tr}\, E + \tfrac{1}{2}(\mathrm{tr}\, E)^2 - \mathrm{tr}\, E^2\} \\ &\qquad + \tfrac{1}{2}\Gamma''(1)(\mathrm{tr}\, E)^2 + O(\|E\|^3)\,,\end{aligned}$$

where the form of the function $\Gamma :]0, +\infty[\to \mathbb{R}$ is for the time being left unspecified; the only requirement is that Γ be convex and twice differentiable at the point 1. In order that

$$a\|F\|^2 + b\|\mathrm{Cof}\, F\|^2 + \Gamma(\det F) + e = \frac{\lambda}{2}(\mathrm{tr}\, E)^2 + \mu\,\mathrm{tr}\, E^2 + O(\|E\|^3)\,,$$

we must have:

$$\begin{cases} 3a + 3b + \Gamma(1) + e = 0\,, \\ 2a + 4b + \Gamma'(1) = 0\,, \\ 2b + \tfrac{1}{2}\Gamma'(1) + \tfrac{1}{2}\Gamma''(1) = \dfrac{\lambda}{2}\,, \\ -2b - \Gamma'(1) = \mu\,. \end{cases}$$

We verify that these equations can be solved in such a way that

$$a > 0,\ b > 0,\ \Gamma''(1) \geq 0\,.$$

The last two equations impose that

$$\Gamma''(1) = (\lambda + 2\mu) + \Gamma'(1)\,,$$

and the inequalities $a > 0$ and $b > 0$ are respectively equivalent to $\Gamma'(1) > -2\mu$ and $\Gamma'(1) < -\mu$. Thus any point $(\Gamma'(1), \Gamma''(1))$ of the open segment with end-points $(-2\mu, \lambda)$ and $(-\mu, \lambda + \mu)$ satisfies all the requirements set so far (cf. Fig. 4.10-1). It remains to show that we can find $c > 0$ and $d > 0$ such that the pair formed by the derivatives $\Gamma'(1)$ and $\Gamma''(1)$ of the function

$$\Gamma : \delta > 0 \to \Gamma(\delta) = c\delta^2 - d \operatorname{Log} \delta$$

belongs to this open segment. Since $\Gamma''(\delta) = 2c + d/\delta^2 > 0$ for all $\delta > 0$,

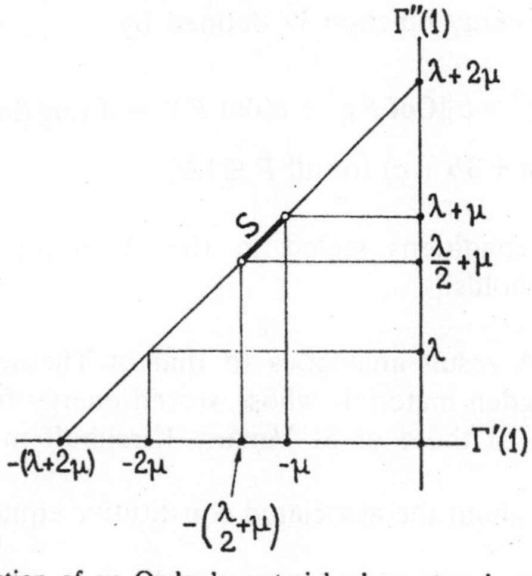

Fig. 4.10-1. Construction of an Ogden's material whose stored energy function $\hat{W}(\boldsymbol{F}) = a\|\boldsymbol{F}\|^2 + b\|\mathbf{Cof}\,\boldsymbol{F}\|^2 + \Gamma(\det \boldsymbol{F})$ satisfies

$$\hat{W}(\boldsymbol{F}) = \check{W}(\boldsymbol{E}) = \frac{\lambda}{2}(\operatorname{tr} \boldsymbol{E})^2 + \mu \operatorname{tr} \boldsymbol{E}^2 + O(\|\boldsymbol{E}\|^3), \quad \boldsymbol{F}^T\boldsymbol{F} = \boldsymbol{I} + 2\boldsymbol{E}\,,$$

near a natural state, where λ and μ are arbitrary Lamé constants. The set S is formed by the admissible pairs $(\Gamma'(1), \Gamma''(1))$.

we first remark that the function Γ is convex. Since $\Gamma'(1) = 2c - d$ and $\Gamma''(1) = 2c + d$, the inequalities $c > 0$ and $d > 0$ are equivalent to the inequality

$$\Gamma''(1) > |\Gamma'(1)| ,$$

whose effect is to reduce the set of admissible pairs $(\Gamma'(1), \Gamma''(1))$ to the open segment (Fig. 4.10-1)

$$S = \Big]\Big(-\frac{\lambda}{2} - \mu, \frac{\lambda}{2} + \mu\Big), (-\mu, \lambda + \mu)\Big[.$$

To sum up, consider any point $(\Gamma'(1), \Gamma''(1))$ of the open segment S, and let

$$a = \mu + \tfrac{1}{2}\Gamma'(1) , \quad b = -\frac{\mu}{2} - \frac{1}{2}\Gamma'(1) ,$$

$$c = \tfrac{1}{4}(\Gamma'(1) + \Gamma''(1)) , \quad d = \tfrac{1}{2}(\Gamma''(1) - \Gamma'(1)) .$$

Then the stored energy function $\hat{W}$ defined by

$$\hat{W}(\boldsymbol{F}) = a\|\boldsymbol{F}\|^2 + b\|\mathbf{Cof}\,\boldsymbol{F}\|^2 + c(\det \boldsymbol{F})^2 - d \operatorname{Log} \det \boldsymbol{F}$$
$$- (3a + 3b + c) \text{ for all } \boldsymbol{F} \in \mathbb{M}^3_+$$

satisfies all the conditions stated in the theorem; the coerciveness inequality clearly holds. ■

Remarks. (1) A result analogous to that of Theorem 4.10-2 can be established for Ogden materials whose stored energy functions are even more reminiscent of those of St Venant–Kirchhoff materials (Exercise 4.23).

(2) Indications about the associated constitutive equations are given in Exercise 4.23. ■

By Theorem 4.10-2, it is thus always possible to adjust the stored energy function of an Ogden's material so that it models a given hyperelastic material with known Lamé constants, themselves determined by the experiments described in Sect. 3.8.

To conclude this chapter, we list the stored energy functions encountered so far, and we also indicate closely related examples:

(i) *Ogden's materials* (Ogden [1972b]):

$$\begin{aligned}\hat{W}(\boldsymbol{F}) &= \sum_{i=1}^{M} a_i(v_1^{\gamma_i} + v_2^{\gamma_i} + v_3^{\gamma_i}) + \sum_{j=1}^{N} b_j((v_2v_3)^{\delta_j} + (v_3v_1)^{\delta_j} + (v_1v_2)^{\delta_j}) \\ &\quad + \Gamma(\det \boldsymbol{F}) \\ &= \sum_{i=1}^{M} a_i \operatorname{tr} \boldsymbol{C}^{\gamma_i/2} + \sum_{j=1}^{N} b_j \operatorname{tr}(\mathbf{Cof}\, \boldsymbol{C})^{\delta_j/2} + \Gamma(\det \boldsymbol{F}),\ \boldsymbol{C} = \boldsymbol{F}^{\mathrm{T}}\boldsymbol{F}\,,\end{aligned}$$

where $v_i = \{\lambda_i(\boldsymbol{C})\}^{1/2}$, $a_i > 0$, $\gamma_i \geq 1$, $b_j > 0$, $\delta_j \geq 1$, and $\Gamma :]0, +\infty[\to \mathbb{R}$ is a convex function satisfying $\Gamma(\delta) \to +\infty$ as $\delta \to 0^+$ and subjected to suitable growth conditions as $\delta \to +\infty$; such materials are discussed at length in Sect. 4.9. Notice that in the literature, the normalizing constant $3 = \operatorname{tr} \boldsymbol{I}$ is often introduced, as in

$$\hat{W}(\boldsymbol{F}) = \sum_{i=1}^{M} a_i\{\operatorname{tr} \boldsymbol{C}^{\gamma_i/2} - 3\} + \sum_{j=1}^{N} b_j\left\{\operatorname{tr}(\mathbf{Cof}\, \boldsymbol{C})^{\delta_j/2} - 3\right\} + \Gamma(\det \boldsymbol{F})\,,$$

in order that the first terms vanish when $\boldsymbol{F}^{\mathrm{T}}\boldsymbol{F} = \boldsymbol{I}$.

(ii) *Compressible neo-Hookean materials* (Blatz [1971]):

$$\hat{W}(\boldsymbol{F}) = a\|\boldsymbol{F}\|^2 + \Gamma(\det \boldsymbol{F})\,, \quad a > 0\,.$$

We recall that $\|\boldsymbol{F}\|^2 = \operatorname{tr} \boldsymbol{C}$.

(iii) *Compressible Mooney–Rivlin materials* (Ciarlet & Geymonat [1982]; cf. also Theorem 4.10-2):

$$\hat{W}(\boldsymbol{F}) = a\|\boldsymbol{F}\|^2 + b\|\mathbf{Cof}\, \boldsymbol{F}\|^2 + \Gamma(\det \boldsymbol{F})\,,$$

with $a > 0$, $b > 0$ and $\Gamma(\delta) = c\delta^2 - d \operatorname{Log} \delta$, $c > 0$, $d > 0$. Examples (ii) and (iii) are named after similar stored energy functions, but without the term $\Gamma(\det \boldsymbol{F})$, that are used for modeling *incompressible neo-Hookean*, and *incompressible Mooney–Rivlin*, materials, respectively. Example (iii) can be generalized as (Ciarlet & Geymonat [1982]; cf. also Exercise 4.23):

$$\hat{W}(\boldsymbol{F}) = a_1 \operatorname{tr} \boldsymbol{C} + a_2 \operatorname{tr} \boldsymbol{C}^2 + b \operatorname{tr} \mathbf{Cof}\, \boldsymbol{C} + \Gamma(\det \boldsymbol{F}),$$

with $a_1 > 0$, $a_2 > 0$, $b > 0$, and $\Gamma(\delta) = c\delta^2 - d \operatorname{Log} \delta$, $c > 0$, $d > 0$.

(iv) The stored energy function

$$\hat{W}(\boldsymbol{F}) = \tfrac{1}{2}\|\boldsymbol{F}\|^2 + \frac{1}{\sigma}(\det \boldsymbol{F})^{-\sigma}, \quad \sigma > 0,$$

is used by Burgess & Levinson [1972], Simpson & Spector [1984b]. It is a special case of the stored energy functions

$$\hat{W}(\boldsymbol{F}) = a_0(\operatorname{tr} \boldsymbol{C}^{1/2})^{\rho} + a_1(\operatorname{tr} \boldsymbol{C})^{\rho} + c(\det \boldsymbol{F})^{-\sigma},$$

with $a_0 > 0$, $a_1 > 0$, $\rho > 0$, $\sigma > 0$, introduced by Antman [1979] for studying the eversion of spherical shells.

(v) *St Venant–Kirchhoff materials* (cf. Sect. 3.9, Theorems 4.4-3 and 4.10-1):

$$\hat{W}(\boldsymbol{F}) = -\left(\frac{3\lambda + 2\mu}{4}\right) \operatorname{tr} \boldsymbol{C} + \left(\frac{\lambda + 2\mu}{8}\right) \operatorname{tr} \boldsymbol{C}^2 + \frac{\lambda}{4} \operatorname{tr} \mathbf{Cof}\, \boldsymbol{C} + \left(\frac{6\mu + 9\lambda}{8}\right),$$

$$= \frac{\lambda}{2}(\operatorname{tr} \boldsymbol{E})^2 + \mu \operatorname{tr} \boldsymbol{E}^2, \quad \boldsymbol{F}^{\mathrm{T}}\boldsymbol{F} = \boldsymbol{I} + 2\boldsymbol{E},$$

where $\lambda > 0$ and $\mu > 0$ are the Lamé constants of the material.

(vi) *Hadamard–Green material* (Simpson & Spector [1984a]; see also Exercise 5.18):

$$\hat{W}(\boldsymbol{F}) = \frac{\alpha}{2}\|\boldsymbol{F}\|^2 + \frac{\beta}{4}\{\|\boldsymbol{F}\|^4 - \|\boldsymbol{F}\boldsymbol{F}^{\mathrm{T}}\|^2\} + \Gamma(\det \boldsymbol{F}),$$

with $\alpha > 0$, $\beta > 0$.

For further examples of stored energy functions of *compressible* elastic materials, see Blatz & Ko [1962], Ogden [1970, 1976, 1984], Knowles & Sternberg [1973, 1975], Flory & Tatara [1975], Fong & Penn [1975], Peng & Landel [1975], Charrier, Dacorogna, Hanouzet & Laborde [1985], and the survey of Davet [1985]; for *incompressible* materials, see Ogden [1972a, 1984], Treloar [1975], Storåkers [1979], Davet [1985]. In these references, all indications are given regarding the *adjustment to experiments* of a stored energy function given *a priori* in a specific form.

EXERCISES

4.1. Let there be given a stored energy function $\hat{W}$ with the property that the mapping $\partial\hat{W}/\partial F : \bar{\Omega} \times \mathbb{M}^3_+ \to \mathbb{M}^3$ is continuous and satisfies the following condition: For any $r > 0$, there exists $l(r)$ such that

$$\left|\frac{\partial\hat{W}}{\partial F}(x, F) - \frac{\partial\hat{W}}{\partial F}(x, G)\right| \leq l(r)\|F - G\|$$

for all $x \in \bar{\Omega}$ and all $F, G \in \mathbb{M}^3_+$ satisfying $\|F\| \leq r$ and $\|G\| \leq r$. Show that the mapping:

$$W : \psi \in \mathscr{C}^1(\bar{\Omega}; \mathbb{R}^3) \to W(\psi) = \int_\Omega \hat{W}(x, \nabla\psi(x))\, dx$$

is differentiable, with

$$W'(\psi)\theta = \int_\Omega \frac{\partial\hat{W}}{\partial F}(x, \nabla\psi(x)) : \nabla\theta(x)\, dx\,.$$

The space $\mathscr{C}^1(\bar{\Omega}; \mathbb{R}^3)$ is equipped with its natural norm:

$$\|\psi\|_{\mathscr{C}^1(\bar{\Omega};\mathbb{R}^3)} = \max_{x\in\bar{\Omega}} |\psi(x)| + \max_{x\in\bar{\Omega}} \|\nabla\psi(x)\|\,.$$

4.2. (1) Show that the subset $\mathbb{M}^3_+$ of $\mathbb{M}^3$ is connected.

(2) Show directly that the set $\mathbb{M}^3_+$ is not a convex subset of $\mathbb{M}^3$, i.e., without resorting to Theorem 4.7-4.

4.3. Let $C : \mathbb{O}^3_+ \to \mathbb{R}$ be the mapping appearing in the relation $\hat{W}(QF) = \hat{W}(F) + C(Q)$ for all $F \in \mathbb{M}^3_+$, $Q \in \mathbb{O}^3_+$, found in the proof of Theorem 4.2-1.

(1) Show that the mapping C is a group homomorphism between $\mathbb{O}^3_+$ and $\mathbb{R}$, in the sense that

$$C(PQ) = C(P) + C(Q) \quad \text{for all } P, Q \in \mathbb{O}^3_+\,.$$

(2) Without assuming that the mapping C is continuous, show that the above relation implies that $C(Q) = 0$ for all $Q \in \mathbb{O}^3_+$.

4.4. Consider a homogeneous isotropic elastic material whose stored energy function $\hat{W} : \mathbb{M}^3_+ \to \mathbb{R}$ is of the form (Theorem 4.4-1):

$$\hat{W}(\boldsymbol{F}) = \dot{W}(\boldsymbol{\iota}_{\boldsymbol{F}^{\mathrm{T}}\boldsymbol{F}}) \quad \text{for all } \boldsymbol{F} \in \mathbb{M}^3_+ .$$

Clearly, the function $\hat{W}$ is differentiable if the function $\dot{W}$ is differentiable. Is the converse true?

Remark. The differentiability of the function $\dot{W}$ was an assumption in Theorem 4.4-2.

4.5. Given an elastic material whose constitutive equation is of the form (Theorem 3.6-2):

$$\bar{\boldsymbol{T}}^{\mathrm{D}}(\boldsymbol{B}) = \beta_0(\boldsymbol{\iota}_{\boldsymbol{B}})\boldsymbol{I} + \beta_1(\boldsymbol{\iota}_{\boldsymbol{B}})\boldsymbol{B} + \beta_2(\boldsymbol{\iota}_{\boldsymbol{B}})\boldsymbol{B}^2 ,$$

find necessary and sufficient conditions, in the form of a system of three equations relating the partial derivatives of the functions β_α, that it be hyperelastic.

4.6. Show that the Cauchy stress tensor can be also related to the derivative of an appropriate stored energy function expressed in terms of variables associated with the *deformed* configuration. This observation is due to Doyle & Ericksen [1956] (see also Marsden & Hughes [1983, p. 197] and Simo & Marsden [1984a, 1984b]).

4.7. If a homogeneous elastic body undergoes a *homogeneous deformation* $\boldsymbol{\varphi}: \bar{\Omega} \to \mathbb{R}^3$, i.e., whose deformation gradient is a constant throughout $\bar{\Omega}$, the first equation of equilibrium reduces to $-\mathbf{div}\, \boldsymbol{T} = \boldsymbol{o}$ in Ω. Consequently, such a deformation is necessarily produced by the application of surface forces alone in homogeneous elastic bodies. Remarkably, a converse holds for hyperelastic materials:

Let $\boldsymbol{\varphi}: \bar{\Omega} \to \mathbb{R}^3$ be a smooth enough deformation with the property that it can occur in *any* homogeneous, isotropic, hyperelastic body as the result of the application of surface forces alone (which may depend on the body considered). Then show that the deformation $\boldsymbol{\varphi}$ is homogeneous.

Remark. The above result is known as *Ericksen's theorem*, following Ericksen [1955a]. For other proofs, see Truesdell & Noll [1965, p. 336], Marsden & Hughes [1983], Wang & Truesdell [1973, p. 276], Shield [1971]. A similar result, again due to J.L. Ericksen, holds for incompressible bodies (Ericksen [1954]).

4.8. (1) Let there be given a homogeneous, isotropic, hyperelastic material, whose reference configuration is a natural state. Assuming adequate smoothness, show that

$$\check{W}(\boldsymbol{E}) = \frac{\lambda}{2}(\operatorname{tr}\boldsymbol{E})^2 + \mu \operatorname{tr}\boldsymbol{E}^2 + \alpha_1(\operatorname{tr}\boldsymbol{E})^3 + \alpha_2(\operatorname{tr}\boldsymbol{E})\operatorname{tr}\boldsymbol{E}^2 + \alpha_3 \operatorname{tr}\boldsymbol{E}^3 + o(\|\boldsymbol{E}\|^3),$$

where $\alpha_1, \alpha_2, \alpha_3$ are constants (for further details, see Novozhilov [1953, pp. 117, 124]).

(2) Compute the associated response function $\check{\Sigma}$, and compare with the expansion found in Exercise 3.11 (refer also to the comments given there).

(3) Why does the assumption of hyperelasticity result in a reduction (from 4 to 3) of the number of arbitrary constants found in the third-order term of the expansion?

(4) Compare the differentiability properties at the point $(0\,0, 0)$ of the functions W_1 and W_2 defined by the relations

$$\dot{W}(\boldsymbol{\iota}_{I+2E}) = W_1(\boldsymbol{\iota}_E) = W_2(\operatorname{tr}\boldsymbol{E}, \operatorname{tr}\boldsymbol{E}^2, \operatorname{tr}\boldsymbol{E}^3),$$

with those of the function $\dot{W}$ at the point $\boldsymbol{\iota}_I$.

4.9. Consider a homogeneous, isotropic, hyperelastic material whose reference configuration is the unit ball $\bar{\Omega} = \{x \in \mathbb{R}^3; |x| \leq 1\}$. Assume that it undergoes a deformation of the form $\varphi_\varepsilon = \varepsilon\boldsymbol{id}$, $\varepsilon > 0$, when it is subjected to a *boundary condition of pressure*, with (the notations are those of Sect. 2.7):

$$\boldsymbol{g}(x) = -\pi_\varepsilon(\det \nabla\varphi(x))\nabla\varphi(x)^{-\mathrm{T}}\boldsymbol{n}(x), \quad \pi_\varepsilon > 0,$$

and that for each $\alpha > 0$, the associated total energy

$$I(\psi) = \int_\Omega \hat{W}(\nabla\psi(x))\,\mathrm{d}x - G(\psi)$$

(the function G is given in Theorem 2.7-1) satisfies

$$I(\varphi_\varepsilon) = \inf_{\psi \in \Phi} I(\psi),$$

where (as usual smoothness is left unspecified):

$$\boldsymbol{\Phi} = \left\{ \boldsymbol{\psi} : \bar{\Omega} \to \mathbb{R}^3 ; \det \boldsymbol{\nabla\psi} > 0, \int_\Omega \boldsymbol{\psi}(x)\, \mathrm{d}x = \boldsymbol{o} \right\}.$$

Then show that

$$\lim_{\det \boldsymbol{F} \to 0^+} \hat{W}(\boldsymbol{F}) = +\infty \Rightarrow \lim_{\varepsilon \to 0^+} \pi_\varepsilon = +\infty .$$

In other words, "an infinite pressure is required in order to annihilate a volume".

Remark. For such a "pure traction problem", an additional condition such as $\int_\Omega \boldsymbol{\psi}(x)\, \mathrm{d}x = \boldsymbol{o}$ is needed in the definition of the set $\boldsymbol{\Phi}$ of admissible deformations, in order that the problem be well posed (cf. the discussion given in Sect. 5.1).

4.10. Show that the stored energy function of a St Venant–Kirchhoff material satisfies a coerciveness inequality of the form

$$\hat{W}(\boldsymbol{F}) \geq \alpha(\|\boldsymbol{F}\|^4 + \|\mathbf{Cof}\,\boldsymbol{F}\|^2) + \beta , \quad \alpha > 0 .$$

4.11. (1) Given $\delta \in \mathbb{R}$, let

$$\mathbb{U}_\delta = \{(\boldsymbol{F}, \mathbf{Cof}\,\boldsymbol{F}) \in \mathbb{M}^3 \times \mathbb{M}^3 ; \det \boldsymbol{F} = \delta\} ,$$

and show that $\operatorname{co} \mathbb{U}_\delta = \mathbb{M}^3 \times \mathbb{M}^3$.

(2) Infer from (1) that $\operatorname{co} \mathbb{U} = \mathbb{M}^3 \times \mathbb{M}^3 \times]0, +\infty[$, where

$$\mathbb{U} = \{(\boldsymbol{F}, \mathbf{Cof}\,\boldsymbol{F}, \det \boldsymbol{F}) \in \mathbb{M}^3 \times \mathbb{M}^3 \times \mathbb{R} ; \boldsymbol{F} \in \mathbb{M}^3_+\} .$$

Remark. This is the proof of Theorem 4.7-4 as originally given by Ball [1977, Theorem 4.3].

4.12. The following ingenious proof of Theorem 4.7-4 is due to A. Mielke.

(1) Show that

$$(\boldsymbol{F}, \boldsymbol{H}, \delta) \in \operatorname{co} \mathbb{U} \text{ and } \boldsymbol{G} \in \mathbb{M}^3_+ \Rightarrow (\boldsymbol{GF}, (\mathbf{Cof}\,\boldsymbol{G})\boldsymbol{H}, (\det \boldsymbol{G})\delta) \in \operatorname{co} \mathbb{U} .$$

(2) Show that

$$(I, 0, \delta), (-I, 0, \delta), (0, I, \delta), (0, -I, \delta) \in \operatorname{co} \mathbb{U} \quad \text{for all } \delta > 0 .$$

(3) Show that

$$(F, 0, \delta), (0, H, \delta) \in \operatorname{co} \mathbb{U} \quad \text{for all } F, H \in \mathbb{M}^3, \delta > 0 .$$

Hint: Argue as in the proof of $\operatorname{co} \mathbb{M}^3_+ = \mathbb{M}^3$ given in Theorem 4.7-4.
(4) Conclude that $\operatorname{co} \mathbb{U} = \mathbb{M}^3 \times \mathbb{M}^3 \times]0, +\infty[$.

4.13. (1) Following Busemann, Ewald & Shephard [1963], show that a function $J : U \subset V \to \mathbb{R}$, where U is a nonempty subset of a finite-dimensional vector space V, is convex if and only if there exists an affine function $G : V \to \mathbb{R}$ such that $J(v) \geq G(v)$ for all $v \in V$, and

$$J\Big(\sum_{i=1}^{N} \lambda_i v_i\Big) \leq \sum_{i=1}^{N} \lambda_i J(v_i) ,$$

for all those points $v_i \in U$ and convex combinations $\Sigma_{i=1}^{N} \lambda_i v_i$, $\Sigma_{i=1}^{N} \lambda_i = 1$, $\lambda_i \geq 0$, that belong to the set U.
(2) Show that one possible convex extension $\bar{J} : \operatorname{co} U \to \mathbb{R}$ of the function J is given by

$$\bar{J}(v) = \inf\Big\{ \sum_{i=1}^{N} \lambda_i J(v_i);\ \sum_{i=1}^{N} \lambda_i v_i = v,\ \sum_{i=1}^{N} \lambda_i = 1,\ \lambda_i \geq 0 \Big\} \quad \text{for } v \in \operatorname{co} U .$$

4.14. (1) Using the *Euler angles* (cf. Euler [1758]; see e.g. Synge [1960, p. 18]), find the general form of the elements of a matrix $R \in \mathbb{O}^3_+$. With the expression found for the diagonal elements, give another proof of the equivalence

$$\operatorname{Diag} \tau_i : (R - I) \leq 0 \quad \text{for all } R \in \mathbb{O}^3_+ \Leftrightarrow \tau_i + \tau_{i+1} \geq 0 ,$$
$$i = 1, 2, 3 \ (\text{mod. } 3) ,$$

already established in the proof of Theorem 4.8-1.
(2) Using likewise the Euler Angles, give a new proof for $n = 3$ of the relation

$$|\operatorname{tr} AB| \leq \sum_{i=1}^{n} v_i(A) v_i(B) ,$$

established for arbitrary n in Theorem 3.2-4 and used in the proof of Theorem 4.9-1.

Remark. Other useful representations of a general orthogonal matrix of order 3 are found in Guo [1981].

4.15. (1) Assume that the stored energy function $\hat{W}:\mathbb{M}^3_+ \rightarrow \mathbb{R}$ is strictly convex. Using a formal argument, show that the total energy considered in Theorem 4.1-1 can have at most one stationary point.

(2) Explain why on the other hand the *strict polyconvexity* of the stored energy function is compatible with the existence of séveral distinct solutions to the associated boundary value problem.

4.16. Let $g:\mathbb{R}\rightarrow\mathbb{R}$ be a convex function. Show that the function $A \in \mathbb{S}^n \rightarrow \Sigma_{i=1}^n g(\lambda_i(A))$ is convex (Yang [1980]).

4.17. Let there be given a symmetric function $\Phi:[0, +\infty[^n \rightarrow \mathbb{R}$. Show that the function

$$A \in \mathbb{S}^n_{\geqslant} \rightarrow \Phi(\lambda_1(A), \ldots, \lambda_n(A)),$$

is convex if and only if the function Φ is convex (cf. Ball [1977, Therem 5.1(i)]; see also Marques & Moreau [1982] for a generalization).

4.18. Let there be given a symmetric function $\Phi:[0, +\infty[^n \rightarrow \mathbb{R}$ such that the function

$$F \in \mathbb{M}^n \rightarrow \Phi(v_1(F), \ldots, v_n(F))$$

is convex. Show that the function Φ is convex and nondecreasing in each variable (Ball [1977, Theorem 5.1(ii)]).

Remark. This result is the converse to Theorem 4.9-1.

4.19. Consider a stored energy function of the form

$$\hat{W}:F \in \mathbb{M}^3_+ \rightarrow \hat{W}(F) = \Theta(v_1, v_2, v_3, v_1v_2, v_2v_3, v_3v_1, v_1v_2v_3),$$
$$v_i = v_i(F),$$

where the function $\Theta:([0, +\infty[^6 \times]0, +\infty[) \rightarrow \mathbb{R}$ is convex, nondecreasing in the variables v_i and $v_i v_{i+1}$, and satisfies

$$\Theta(v_1, v_2, v_3, w_1, w_2, w_3, \delta)$$
$$= \Theta(v_{\sigma(1)}, v_{\sigma(2)}, v_{\sigma(3)}, w_{\tau(1)}, w_{\tau(2)}, w_{\tau(3)}, \sigma)$$
for all $\sigma, \tau \in \mathfrak{S}_3$.

Show that the function $\hat{W}$ is polyconvex (Ball [1977, Theorem 5.2]).

Remark: This result contains Theorem 4.9-2 as a special case.

4.20. Let U be a convex subset of a vector space, and let $J : U \to \mathbb{R}$ be convex and $\geqslant 0$. Show that the function $J^p : U \to \mathbb{R}$ is convex for all $p \geqslant 1$. Infer from this result that the function $F \in \mathbb{M}^n \to \|F\|^p$, $p \geqslant 1$, is convex.

4.21. Let K be a symmetric positive definite matrix of order n, diagonalized as

$$K = P(\text{Diag } \lambda_i)P^{\mathrm{T}} = Q(\text{Diag } \lambda_{\sigma(i)})Q^{\mathrm{T}}, \quad \sigma \in \mathfrak{S}^n .$$

Show that, for any $\delta > 0$,

$$P(\text{Diag } \lambda_i^{\delta})P^{\mathrm{T}} = Q(\text{Diag } \lambda_{\sigma(i)}^{\delta})Q^{\mathrm{T}} .$$

Remark. This result justifies the definition of the matrix $K^{\delta} \in \mathbb{S}^n_>$ for any $K \in \mathbb{S}^n_>$, used for representing the stored energy functions of Ogden's materials (Theorem 4.9-2).

4.22. Let $\hat{W}(F) = \operatorname{tr} E^2$, with $I + 2E = F^{\mathrm{T}}F$, for all $F \in \mathbb{M}^3$ (the function $\hat{W}$ may be thought of as the stored energy function of a "limit" St Venant–Kirchhoff material, with $\lambda = 0$, $\mu = 1$).

(1) Show that $\hat{W}(F) \to +\infty$ as $\|F\| \to +\infty$.

(2) Let $\mathbb{V}$ be an arbitrary convex neighborhood of I in $\mathbb{M}^3$. Show that $\hat{W}$ is not convex on $\mathbb{V}$.

(3) Show that there exists a neighborhood $\mathbb{V}$ of I in $\mathbb{M}^3$ such that $\hat{W}$ is polyconvex on $\mathbb{V}$. Why is this property compatible with Theorem 4.10-1?

Remark. For details, see Atteia & Rassouli [1986].

4.23. This exercise is a complement to Theorem 4.10-2. Let $\lambda > 0$ and $\mu > 0$ be two given Lamé constants.

(1) Show that there exist stored energy functions of the form

$$\hat{W}(\boldsymbol{F}) = a_1 \operatorname{tr} \boldsymbol{C} + a_2 \operatorname{tr} \boldsymbol{C}^2 + b \operatorname{tr} \operatorname{Cof} \boldsymbol{C} + \Gamma(\det \boldsymbol{F}) + e\,, \quad \boldsymbol{C} = \boldsymbol{F}^{\mathrm{T}}\boldsymbol{F}\,,$$

with

$$a_1 > 0,\ a_2 > 0,\ b > 0,\ \Gamma(\delta) = c\delta^2 - d \operatorname{Log} \delta,\ c > 0,\ d > 0,\ e \in \mathbb{R}\,,$$

that satisfy

$$\hat{W}(\boldsymbol{F}) = \check{W}(\boldsymbol{E}) = \frac{\lambda}{2} (\operatorname{tr} \boldsymbol{E})^2 + \mu \operatorname{tr} \boldsymbol{E}^2 + O(\|\boldsymbol{E}\|^3)\,, \quad \boldsymbol{I} + 2\boldsymbol{E} = \boldsymbol{F}^{\mathrm{T}}\boldsymbol{F}\,.$$

Hint. The proof closely follows that of Theorem 4.10-2; in particular, the set of admissible pairs $(\Gamma'(1), \Gamma''(1))$ of Fig. 4.10-1 now becomes the interior of a triangle (Ciarlet & Geymonat [1982]).

(2) Write the associated constitutive equation as (Theorem 4.2-2):

$$\check{\boldsymbol{\Sigma}}(\boldsymbol{E}) = \frac{\partial \check{W}}{\partial \boldsymbol{E}}(\boldsymbol{E}) = \lambda(\operatorname{tr} \boldsymbol{E})\boldsymbol{I} + 2\mu \boldsymbol{E} + \check{\boldsymbol{R}}(\boldsymbol{E})\,.$$

Give an explicit form of the remainder $\check{\boldsymbol{R}}(\boldsymbol{E})$ and show directly that $\check{\boldsymbol{R}}(\boldsymbol{E}) = O(\|\boldsymbol{E}\|^2)$, in accordance with Theorem 3.8-1.

CHAPTER 5

THE BOUNDARY VALUE PROBLEMS OF THREE-DIMENSIONAL ELASTICITY

INTRODUCTION

We have encountered so far two types of boundary conditions, the *boundary condition of place*:

$$\varphi(x) = \varphi_0(x), \quad x \in \Gamma_0 \subset \Gamma ,$$

and the *boundary condition of traction*:

$$\hat{T}(x, \nabla\varphi(x))n = \hat{g}(x, \nabla\varphi(x)), \quad x \in \Gamma_1 \subset \Gamma .$$

If only these two occur in a boundary value problem of three-dimensional elasticity, it is called a *displacement–traction problem* (Sect. 5.1). However, other boundary conditions are commonly found in practice, such as *nonlocal boundary conditions*, or boundary conditions that are partly of place and partly of traction (Sect. 5.2). Of particular importance are the *unilateral boundary conditions of place* (Sect. 5.3), which take the form

$$\varphi(x) \in C, \quad x \in \Gamma_2 \subset \Gamma ,$$

where C is an arbitrary closed subset of $\mathbb{R}^3$. When such a condition is imposed to the minimizers of the total energy, it provides a mathematical model of *contact without friction with an obstacle* (Theorem 5.3-1).

In order to be physically acceptable, a deformation φ must be *injective over the open set* Ω (it looses its injectivity on Γ whenever self-contact occurs). For the solution of a *pure displacement problem* ($\varphi = \varphi_0$ on all of Γ), injectivity is usually obtained as a consequence of the following result (Theorem 5.5-2): If a smooth enough mapping $\varphi : \bar{\Omega} \rightarrow \mathbb{R}^3$, where Ω is a bounded open connected subset of $\mathbb{R}^3$, is orientation-preserving ($\det \nabla\varphi > 0$ in $\bar{\Omega}$) and is equal to an injective mapping $\varphi_0 : \bar{\Omega} \rightarrow \mathbb{R}^3$ on the boundary of the set Ω, then it is also injective on $\bar{\Omega}$.

However, this result cannot be applied to the more realistic situations where the deformation is only prescribed on a portion of Γ (or is even nowhere prescribed on Γ). In order to handle these cases, we propose a different approach (Sect. 5.6), which consists in imposing, together with the orientation preserving condition $\det \nabla\varphi > 0$ in $\bar{\Omega}$, the *injectivity condition*

$$\int_\Omega \det \nabla\varphi(x)\,\mathrm{d}x \leq \operatorname{vol} \varphi(\Omega)$$

to the minimizers $\varphi : \bar{\Omega} \to \mathbb{R}^3$ of the total energy. Then the associated minimization problem is a mathematical model of *self-contact without friction* and of *noninterpenetration of matter* (Theorem 5.6-3).

A *set of admissible deformations* is thus naturally attached to a particular boundary value problem of three-dimensional elasticity: It consists of all smooth enough mappings $\varphi : \bar{\Omega} \to \mathbb{R}^3$ that satisfy all the *geometrical constraints* that we wish to impose, such as the orientation-preserving condition, the injectivity condition, a (possibly unilateral) boundary condition of place, etc. (Sect. 5.7).

We next take a closer look at the *nonlinear character* inherent to the boundary value problem of three-dimensional elasticity, as evidenced experimentally by the *nonuniqueness* of solutions in various physical situations (Sect. 5.8), and mathematically, by the *quasilinearity* of the equations of equilibrium and by the nonlinear conditions that are imposed on the admissible deformations (Sect. 5.9). Finally, we review (Sect. 5.10) various *constitutive assumptions*, such as isotropy, the polyconvexity and the behavior of the stored energy function for small and large strains, etc. These mathematical assumptions play a crucial rôle in the existence theories developed in the next chapters.

5.1. DISPLACEMENT–TRACTION PROBLEMS

Assembling the various notions found in the previous chapters, we are in a position to describe the simplest *boundary value problems of three-dimensional elasticity*. We consider *isotropic*, *compressible*, *elastic materials*, which may be nonhomogeneous. We are given:

– a domain Ω in $\mathbb{R}^3$, i.e., an open, bounded, and connected subset whose boundary Γ is Lipschitz-continuous, and disjoint relatively open subsets Γ_0 and Γ_1 of Γ such that $\mathrm{d}a\text{-meas}\{\Gamma - (\Gamma_0 \cup \Gamma_1)\} = 0$;

– a tensor-valued function $\hat{T} : \bar{\Omega} \times \mathbb{M}^3_+ \to \mathbb{M}^3$, which is the response function for the first Piola–Kirchhoff stress, or equivalently, a symmetric tensor-valued function $\hat{\boldsymbol{\Sigma}}: \bar{\Omega} \times \mathbb{M}^3_+ \to \mathbb{S}^3$, which is the response function for the second Piola–Kirchhoff stress, the two functions being related by $\hat{\boldsymbol{T}}(x, \boldsymbol{F}) = \boldsymbol{F}\hat{\boldsymbol{\Sigma}}(x, \boldsymbol{F})$ for all $x \in \bar{\Omega}$, $\boldsymbol{F} \in \mathbb{M}^3_+$ (Sects. 2.5 and 3.1);

– a vector-valued function $\hat{\boldsymbol{f}}: \bar{\Omega} \times \mathbb{R}^3 \to \mathbb{R}^3$, which measures the density of the applied body force per unit volume in the reference configuration (Sects. 2.6 and 2.7);

– a vector-valued function $\hat{\boldsymbol{g}}: \Gamma_1 \times \mathbb{M}^3_+ \to \mathbb{R}^3$, which measures the density of the applied surface force per unit area in the reference configuration (Sects. 2.6 and 2.7).

We are seeking a deformation $\boldsymbol{\varphi}: \bar{\Omega} \to \mathbb{R}^3$ that solves the following boundary value problem:

$$\left.\begin{aligned} -\mathbf{div}\, \hat{\boldsymbol{T}}(x, \boldsymbol{\nabla\varphi}(x)) &= \hat{\boldsymbol{f}}(x, \boldsymbol{\varphi}(x)),\ x \in \Omega, \\ \boldsymbol{\varphi}(x) &= \boldsymbol{\varphi}_0(x),\ x \in \Gamma_0, \\ \hat{\boldsymbol{T}}(x, \boldsymbol{\nabla\varphi}(x))\boldsymbol{n} &= \hat{\boldsymbol{g}}(x, \boldsymbol{\nabla\varphi}(x)),\ x \in \Gamma_1 \end{aligned}\right\} \text{ or, in short, } \left\{\begin{aligned} -\mathbf{div}\, \hat{\boldsymbol{T}}(\boldsymbol{\nabla\varphi}) &= \hat{\boldsymbol{f}}(\boldsymbol{\varphi}) \text{ in } \Omega, \\ \boldsymbol{\varphi} &= \boldsymbol{\varphi}_0 \text{ on } \Gamma_0, \\ \hat{\boldsymbol{T}}(\boldsymbol{\nabla\varphi})\boldsymbol{n} &= \hat{\boldsymbol{g}}(\boldsymbol{\nabla\varphi}) \text{ on } \Gamma_1, \end{aligned}\right.$$

or equivalently:

$$\left.\begin{aligned} -\mathbf{div}\, \boldsymbol{\nabla\varphi}(x)\hat{\boldsymbol{\Sigma}}(x, \boldsymbol{\nabla\varphi}(x)) &= \hat{\boldsymbol{f}}(x, \boldsymbol{\varphi}(x)),\ x \in \Omega, \\ \boldsymbol{\varphi}(x) &= \boldsymbol{\varphi}_0(x),\ x \in \Gamma_0, \\ \boldsymbol{\nabla\varphi}(x)\hat{\boldsymbol{\Sigma}}(x, \boldsymbol{\nabla\varphi}(x))\boldsymbol{n} &= \boldsymbol{g}(x, \boldsymbol{\nabla\varphi}(x)),\ x \in \Gamma_1 \end{aligned}\right\} \text{ or, in short, } \left\{\begin{aligned} -\mathbf{div}\, \boldsymbol{\nabla\varphi}\hat{\boldsymbol{\Sigma}}(\boldsymbol{\nabla\varphi}) &= \hat{\boldsymbol{f}}(\boldsymbol{\varphi}) \text{ in } \Omega, \\ \boldsymbol{\varphi} &= \boldsymbol{\varphi}_0 \text{ on } \Gamma_0, \\ \boldsymbol{\nabla\varphi}\hat{\boldsymbol{\Sigma}}(\boldsymbol{\nabla\varphi})\boldsymbol{n} &= \hat{\boldsymbol{g}}(\boldsymbol{\nabla\varphi}) \text{ on } \Gamma_1. \end{aligned}\right.$$

Remarks: (1) The assumption that the sets Γ_0 and Γ_1 are relatively open in Γ is a mere mathematical convenience (see for instance the proof of Theorem 5.3-1). As a consequence, we possibly ignore a subset of zero da-measure of the boundary Γ.

(2) The specific dependences of the applied forces on the deformation, viz., $\boldsymbol{f}(x) = \hat{\boldsymbol{f}}(x, \boldsymbol{\varphi}(x))$, $x \in \Omega$, and $\boldsymbol{g}(x) = \hat{\boldsymbol{g}}(x, \boldsymbol{\nabla\varphi}(x))$, $x \in \Gamma_1$, do not necessarily cover all possible cases. They simply correspond to the examples considered so far.

(3) In some instances (as in Chapter 6), it is more convenient to consider the *displacement* $\boldsymbol{u}: \bar{\Omega} \to \mathbb{R}^3$ as the unknown. It is then immediate to express the above boundary value problem in terms of this unknown (recall that $\boldsymbol{\varphi} = \boldsymbol{id} + \boldsymbol{u}$, so that $\boldsymbol{\nabla\varphi} = \boldsymbol{I} + \boldsymbol{\nabla u}$). ■

The conditions that the unknown $\boldsymbol{\varphi}$ be orientation-preserving and injective, except possibly on Γ, are not included in the formulation as a boundary value problem, for they are usually verified *a posteriori* in this case (Sects. 5.5 and 6.9). By contrast, the same conditions can be easily taken into account in the formulation as a minimization problem when the material is hyperelastic (Sect. 5.6).

We recall that the equation $-\mathbf{div}\,\hat{T}(\nabla\varphi)=\hat{f}(\varphi)$ in Ω and the *boundary condition of traction* $\hat{T}(\nabla\varphi)n=\hat{g}(\nabla\varphi)$ on Γ_1 are parts of the *equations of equilibrium in the reference configuration* (cf. Sect. 2.6; the remaining equation of equilibrium expresses the symmetry of the second Piola–Kirchhoff stress tensor), and that "$\varphi=\varphi_0$ on Γ_0" is a *boundary condition of place*.

The above boundary value problem is called a **pure displacement problem** if $\Gamma_1=\emptyset$, a **displacement–traction problem** if $\text{area}(\Gamma_0)>0$ and $\text{area}(\Gamma_1)>0$, and a **pure traction problem** if $\Gamma_0=\emptyset$.

If the material is hyperelastic (Sect. 4.1), i.e., if

$$\hat{T}(x,F)=\frac{\partial\hat{W}}{\partial F}(x,F)\quad\text{for all } x\in\bar{\Omega},\ F\in\mathbb{M}^3_+\,,$$

and if the applied forces are conservative (Sect. 2.7), any one of these problems is formally equivalent to finding the stationary points of the total energy (Theorem 4.1-1):

$$I(\psi)=\int_\Omega \hat{W}(x,\nabla\psi(x))\,dx \\ -\Big\{\int_\Omega \hat{F}(x,\psi(x))\,dx+\int_{\Gamma_1}\hat{G}(x,\psi(x),\nabla\psi(x))\,da\Big\}$$

when the admissible mappings φ vary in the set

$$\Phi=\{\varphi:\bar{\Omega}\to\mathbb{R}^3;\ \det\nabla\varphi>0 \text{ in } \bar{\Omega},\ \varphi=\varphi_0 \text{ on } \Gamma_0\}\,.$$

When the applied forces are conservative, the potential $\hat{F}:\bar{\Omega}\times\mathbb{R}^3\to\mathbb{R}$ of the applied body force and the potential $\hat{G}:\Gamma_1\times\mathbb{R}^3\times\mathbb{M}^3_+\to\mathbb{R}$ of the applied surface force are respectively determined by the densities $\hat{f}$ and $\hat{g}$ (Sect. 2.7).

As we already pointed out in Sect. 4.1, we shall look for particular stationary points of the energy, namely its *minimizers* over the set Φ, i.e., those mappings φ that satisfy

$$\boxed{\varphi \in \Phi \text{ and } I(\varphi) = \inf_{\psi \in \Phi} I(\psi)\,.}$$

The simplest conservative applied forces are the *dead loads*, i.e., those applied forces whose associated densities in the reference configuration are independent of the deformation. A realistic example of an applied surface force that is not a dead load is given by a *pressure load*, whose associated *boundary condition of pressure* takes the form (Sect. 2.7)

$$\hat{T}(\nabla\varphi)n = -\pi(\det \nabla\varphi)\nabla\varphi^{-T}n \text{ on } \Gamma_1\,,$$

where π is a given real number (the corresponding boundary condition in the deformed configuration is $T^{\varphi}n^{\varphi} = -\pi n^{\varphi}$ on $\Gamma_1^{\varphi} = \varphi(\Gamma_1)$). If the condition of traction is of this form, the associated boundary value problem is called a **displacement–pressure problem** if area $(\Gamma_0) > 0$ and area$(\Gamma_1) > 0$, and a **pure pressure problem** if $\Gamma_0 = \emptyset$. Since a pressure load is conservative (Theorem 2.7-1), either one of the pressure boundary value problems is again formally equivalent to finding the stationary points of the energy if the material is hyperelastic.

In the remainder of the book, we shall mostly confine our attention to the case where all the applied forces are dead loads, leaving as exercises the case of more general loads. Accordingly, we make the convention that, *unless explicit stated otherwise, the pure displacement, displacement–traction, and pure traction, problems subsequently encountered correspond to applied body and surface forces that are all dead loads.*

Even though we shall spend a great deal of attention to the mathematical analysis of the *pure displacement problem*, notably in Chapter 6, it should be kept in mind that this is not a commonly encountered case, although it is a realistic example: When the deformation is imposed along the entire boundary, the body under study is completely included in, and glued to, a larger structure. At the other end of the possibilities, genuine *pure traction* or *pure pressure problems* also correspond to actual situations (consider a submarine vessel, or a soap bubble, that move with a constant velocity). *Displacement–traction problems* are more common in applications, although they are far from covering all remaining possibilities, as we shall indicate in the next sections.

The *pure traction problem*

$$\boxed{\begin{aligned} -\mathbf{div}\, \hat{T}(\nabla\varphi) &= f \text{ in } \Omega,\\ \hat{T}(\nabla\varphi)n &= g \text{ on } \Gamma, \end{aligned}}$$

possesses special features that render its analysis more delicate. In particular, an application of the divergence theorem for tensor fields shows that

$$\int_{\Omega} \boldsymbol{f}\,\mathrm{d}x + \int_{\Gamma} \boldsymbol{g}\,\mathrm{d}a = \boldsymbol{o}\,,$$

which thus appears as a *compatibility condition* that the applied forces must satisfy in order that the problem have a solution. In view of the relations $\boldsymbol{f}\,\mathrm{d}x = \boldsymbol{f}^{\varphi}\,\mathrm{d}x^{\varphi}$ and $\boldsymbol{g}\,\mathrm{d}a = \boldsymbol{g}^{\varphi}\,\mathrm{d}a^{\varphi}$, this relation is equivalent to

$$\int_{\varphi(\Omega)} \boldsymbol{f}^{\varphi}\,\mathrm{d}x^{\varphi} + \int_{\varphi(\Gamma)} \boldsymbol{g}^{\varphi}\,\mathrm{d}a^{\varphi} = \boldsymbol{o}\,,$$

which is nothing but the *axiom of force balance* (Sect. 2.2) applied to the particular domain $\bar{\Omega}^{\varphi} = \boldsymbol{\varphi}(\bar{\Omega})$ (the application of the axiom of moment balance is discussed in Exercise 5.1).

Remark. In the case of a displacement–traction problem, this condition is satisfied without imposing any *a priori* condition on the forces, because the first Piola–Kirchhoff stress vector $\hat{\boldsymbol{T}}(\nabla\boldsymbol{\varphi})\boldsymbol{n}$ is left unspecified on the portion Γ_0 of the boundary. Any deformation that is solution of the problem thus automatically "adjusts itself" on Γ_0 so as to satisfy the axiom of force balance. ∎

The assumption of hyperelasticity yields another proof of the necessity of this compatibility condition: Let $\boldsymbol{\varphi}$ be an element in the associated set

$$\boldsymbol{\Phi} = \{\boldsymbol{\psi} : \Omega \to \mathbb{R}^3;\ \det \nabla\boldsymbol{\psi} > 0 \text{ in } \Omega\}$$

that satisfies

$$I(\boldsymbol{\varphi}) = \inf_{\boldsymbol{\psi}\in\boldsymbol{\Phi}} I(\boldsymbol{\psi})\,.$$

Since for any vector $\boldsymbol{d} \in \mathbb{R}^3$, the mapping $\boldsymbol{\varphi} + \boldsymbol{d}$ is again in the set $\boldsymbol{\Phi}$, we must have

$$I(\boldsymbol{\varphi}) \leq I(\boldsymbol{\varphi} + \boldsymbol{d}) = I(\boldsymbol{\varphi}) - \left\{\int_{\Omega} \boldsymbol{f}\,\mathrm{d}x + \int_{\Gamma} \boldsymbol{g}\,\mathrm{d}a\right\} \cdot \boldsymbol{d} \quad \text{for all } \boldsymbol{d} \in \mathbb{R}^3\,.$$

Then the particular choice

$$\boldsymbol{d} = \int_{\Omega} \boldsymbol{f}\,\mathrm{d}x + \int_{\Gamma} \boldsymbol{g}\,\mathrm{d}a$$

shows that we must have $\int_\Omega \boldsymbol{f}\, dx + \int_\Gamma \boldsymbol{g}\, da = \boldsymbol{o}$ in order that a minimum exist. In order to remove this *indeterminacy up to rigid translations*, it is customary to impose an additional condition on the mappings $\boldsymbol{\psi} \in \boldsymbol{\Phi}$, such as:

$$\int_\Omega \boldsymbol{\psi}\, dx = \boldsymbol{e},$$

where $\boldsymbol{e}$ is an arbitrary vector in $\mathbb{R}^3$. We shall see in Chapter 7 that such a condition is also needed for proving existence results (other complements are given in Exercise 5.2).

Finally, we wish to emphasize that the boundary condition of place and the boundary condition of traction are only *approximate models* of the "true" boundary conditions: In practice, there always occurs some interaction between an elastic body and its environment (itself usually made up of another elastic material), which is not taken into account by these ideal boundary conditions. In this direction, see Batra [1972], Podio-Guidugli, Vergara-Caffarelli & Virga [1987].

5.2. OTHER EXAMPLES OF BOUNDARY CONDITIONS

The boundary conditions of place, traction, and pressure, far from exhaust all the situations found in practice. Let us illustrate the variety of other possibilities by means of examples.

Following Noll [1978], we consider first a **balloon problem** (Fig. 5.2-1), where the exterior boundary Γ_e^φ of a balloon is subjected to a pressure load with constant pressure π_e independent of the deformation, while the interior boundary Γ_i^φ is subjected to a pressure load with a pressure that is a known function π_i of the enclosed volume $v_i(\boldsymbol{\varphi})$. The corresponding boundary condition on Γ_i:

$$\boldsymbol{T}(\nabla\boldsymbol{\varphi}(x))\boldsymbol{n}_i(x) = -\pi_i(v_i(\boldsymbol{\varphi}))\det \nabla\boldsymbol{\varphi}(x)\nabla\boldsymbol{\varphi}(x)^{-\mathrm{T}}\boldsymbol{n}_i(x), \quad x \in \Gamma_i,$$

is an example of a *nonlocal boundary condition*, in that the expression of its right-hand side at a point $x \in \Gamma_i$ involves values of the deformation at other points (cf. Exercise 5.3 for further considerations).

As a second example, consider a plate whose reference configuration is a rectangular parallelepiped as indicated in Fig. 5.2-2: It is conceivable that, by means of some appropriate mechanical apparatus, a force is exerted in the direction of the vector $\boldsymbol{e}_2$ on the face Γ_0' and that the face

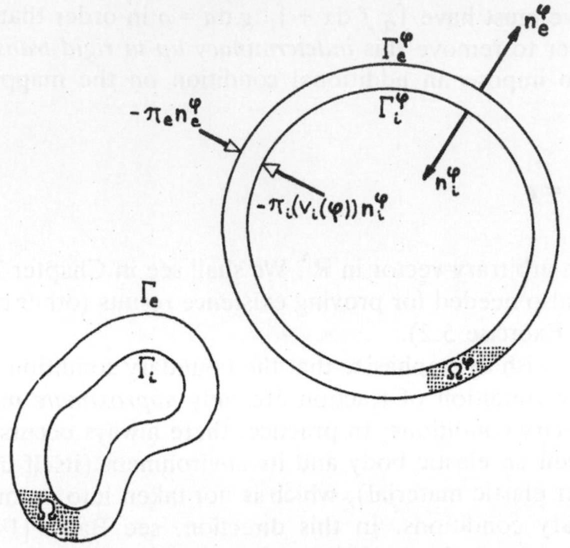

Fig. 5.2-1. The balloon problem: The exterior boundary is subjected to a pressure load independent of the deformation, and the interior boundary is subjected to a pressure load that depends on the enclosed volume. The latter condition provides an example of a nonlocal boundary condition.

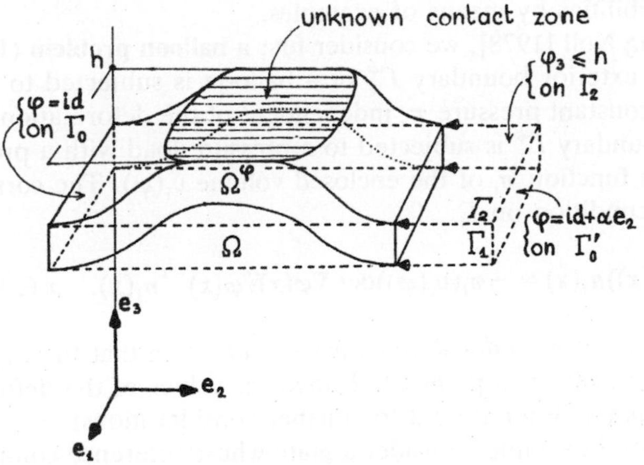

Fig. 5.2-2. A plate problem. The face Γ_0' can only undergo rigid translations parallel to e_2, but of unknown magnitude; the upper face Γ_2 must lie beneath a rigid plane $x_3 = h$.

Γ_0' can only undergo rigid translations in the same direction. The corresponding boundary condition takes the form

$$\varphi = \boldsymbol{id} + \alpha \boldsymbol{e}_2 \text{ on } \Gamma_0', \ \alpha \in \mathbb{R},$$

where the number α is an unknown of the problem.

Other boundary conditions that are very common in practice express that some portions of the boundary of the body are confined in some subsets of $\mathbb{R}^3$. For instance the upper face of the plate considered in Fig. 5.2-2 may be restricted to lie beneath a rigid plane $x_3 = h$. This constraint is expressed by the boundary condition

$$\varphi_3 \leq h \text{ on } \Gamma_2,$$

or equivalently

$$\varphi(\Gamma_2) \subset C, \quad \text{where } C := \{x \in \mathbb{R}^3; x_3 \leq h\}.$$

Notice that *the contact zone* $\{\varphi_3(x) = h; x \in \Gamma_2\}$ *is an unknown of the problem*. Such a boundary condition is an instance of a **unilateral boundary condition of place**, whose general form is

$$\boxed{\varphi(\Gamma_2) \subset C,}$$

where Γ_2 is a portion of the boundary Γ of the reference configuration, and C is some given closed subset (bounded or not) of $\mathbb{R}^3$. Such a condition means that some parts of the boundary of the body should "stay on a given side" of the boundary ∂C of the set C; for this reason, the boundary ∂C is called an **obstacle**. In view of their importance, unilateral boundary conditions of place are discussed separately in the next section (see notably Theorem 5.3-1).

For the time being, we take a closer look at the boundary condition on the face Γ_0'. Let $-\lambda(\text{area } \Gamma_0')$ denote the component along the vector $\boldsymbol{e}_2$ of the resultant of the force applied to the deformed face $\varphi(\Gamma_0')$ (so that the plate is compressed if $\lambda > 0$). Then the application of the axiom of force balance (Sect. 2.2) to the deformed face $\varphi(\Gamma_0')$ implies that

$$\left\{\int_{\varphi(\Gamma_0')} \boldsymbol{T}^{\varphi} \boldsymbol{n}^{\varphi} \, \mathrm{d}a^{\varphi}\right\} \cdot \boldsymbol{e}_2 = -\lambda(\text{area } \Gamma_0').$$

Remark. The attentive reader has immediately noticed that the set $\varphi(\Gamma_0')$ is not a domain since it has an empty interior in $\mathbb{R}^3$; hence the application of the axiom of force balance requires *un acte de foi* in the present circumstances... ■

Combining the above condition with the relations

$$\int_{\varphi(\Gamma_0')} T^{\varphi} n^{\varphi} \, da^{\varphi} = \int_{\Gamma_0'} T n \, da, \; n = e_2 \text{ on } \Gamma_0', \; (Te_2) \cdot e_2 = T_{22},$$

results in another example of a *nonlocal boundary condition*, which takes the form

$$\frac{1}{\text{area } \Gamma_0'} \int_{\Gamma_0'} \hat{T}_{22}(\nabla \varphi) \, da = -\lambda, \quad \lambda \in \mathbb{R} \text{ given, with } da = dy_1 \, dy_3 .$$

Since it is assumed that the face Γ_0' can only undergo rigid transations in the direction of the vector e_2, the available boundary conditions on the face Γ_0' thus read:

$$\begin{cases} \dfrac{1}{\text{area } \Gamma_0'} \displaystyle\int_{\Gamma_0'} \hat{T}_{22}(\nabla \varphi) \, da = -\lambda, & \lambda \in \mathbb{R} \text{ given}, \\ \varphi = id + \alpha e_2, & \alpha \in \mathbb{R} \text{ unknown}. \end{cases}$$

Notice a novelty, in that on the face Γ_0' some information must be *simultaneously* provided about the deformation and about the first Piola–Kirchhoff stress vector Te_2. The way in which both information complement each other is well determined, either by appealing to the axiom of force balance (properly extended) as above, or, as we now show, by verifying the validity of an appropriate *principle of virtual work*, which is in turn equivalent to expressing that a certain functional is stationary. For clarity, we do not take into consideration the boundary condition "$\varphi_3 \leq h$ on Γ_2" at this stage.

Theorem 5.2-1. *The notation is as in Fig. 5.2-2. The boundary value problem*

$$\begin{cases} -\text{div } \hat{T}(\nabla \varphi) = f \text{ in } \Omega, \\ \varphi = id \text{ on } \Gamma_0, \\ \left.\begin{array}{l} \varphi = id + \alpha e_2, \; \alpha \in \mathbb{R} \text{ unknown}, \\ \dfrac{1}{\text{area } \Gamma_0'} \displaystyle\int_{\Gamma_0'} \hat{T}_{22}(\nabla \varphi) \, da = -\lambda, \; \lambda \in \mathbb{R} \text{ given}, \end{array}\right\} \text{ on } \Gamma_0', \\ \hat{T}(\nabla \varphi) n = g \text{ on } \Gamma_1' := \Gamma - \{\Gamma_0 \cup \Gamma_0'\} \end{cases}$$

is formally equivalent to the variational equations

$$\int_\Omega \hat{T}(\nabla\varphi):\nabla\theta\,dx = \int_\Omega f\cdot\theta\,dx - \lambda\int_{\Gamma_0'} e_2\cdot\theta\,da + \int_{\Gamma_1'} g\cdot\theta\,da\,,$$

when the smooth enough maps θ *vary in the space*

$$T_\varphi\Phi := \{\theta:\bar{\Omega}\to\mathbb{R}^3;\ \theta = o \text{ on } \Gamma_0,\ \theta = \beta e_2,\ \beta\in\mathbb{R}, \text{ on } \Gamma_0'\}\,.$$

If the material is hyperelastic ($\hat{T}(F) = (\partial\hat{W}/\partial F)(F)$), *the variational equations are equivalent to the equations*

$$I'(\varphi)\theta = 0 \quad \text{for all } \theta\in T_\varphi\Phi\,,$$

where the functional I is defined for arbitrary deformations $\psi:\bar{\Omega}\to\mathbb{R}^3$ by

$$I(\psi) = \int_\Omega \hat{W}(\nabla\psi)\,dx - \left\{\int_\Omega f\cdot\psi\,dx - \lambda\int_{\Gamma_0'} e_2\cdot\psi\,da + \int_{\Gamma_1'} g\cdot\psi\,da\right\}.$$

Proof. As in Theorems 2.4-1 or 2.6-1, the equivalence with the variational equations relies on the identity

$$\int_\Omega \operatorname{div} T\cdot\theta\,dx = -\int_\Omega T:\nabla\theta\,dx + \int_\Gamma Tn\cdot\theta\,da\,,$$

valid for all smooth enough tensor fields $T:\bar{\Omega}\to\mathbb{M}^3$ and vector fields $\theta:\bar{\Omega}\to\mathbb{R}^3$. It suffices to combine this identity with the relations

$$\int_{\Gamma_0'} Tn\cdot\theta\,da = \left(\int_{\Gamma_0'} T_{22}\,da\right)\beta \quad \text{if } \theta = \beta e_2,\ \beta\in\mathbb{R}, \text{ on } \Gamma_0'\,,$$

and

$$-\lambda\int_{\Gamma_0'} e_2\cdot\theta\,da = -\lambda(\text{area } \Gamma_0')\beta \quad \text{if } \theta = \beta e_2,\ \beta\in\mathbb{R}, \text{ on } \Gamma_0'\,.$$

The equivalence with the stationary character of the functional is straightforward. ∎

Since the present analysis is identical in its principle to that followed in Sections 2.6 and 4.1 for the displacement–traction problem, we are justified in saying that the variational equations and the functional found in Theorem 5.2-1 represent the *principle of virtual work*, and the *total energy* respectively, associated with the particular boundary value problem under consideration. Notice in passing that the space $T_{\varphi}\Phi$ defined in the theorem is nothing but the tangent space (whence the notation) at the point φ of the corresponding *set of admissible solutions*:

$$\Phi = \{\Psi : \bar{\Omega} \to \mathbb{R}^3 ;\ \det \nabla\psi > 0 \text{ in } \Omega,\ \psi = id \text{ on } \Gamma_0,\ \psi = id + \alpha e_2,\\ \alpha \in \mathbb{R}, \text{ on } \Gamma_0'\} .$$

Of course, other "combinations" of boundary conditions involving simultaneously the deformations and the first Piola–Kirchhoff stress vector are possible. In this respect, another example (again borrowed from plate theory) is discussed in Exercise 5.4.

5.3. UNILATERAL BOUNDARY CONDITIONS OF PLACE IN HYPERELASTICITY

We now consider the case where a *unilateral boundary condition of place*: "$\varphi(\Gamma_2) \subset C$", where C is a closed subset of $\mathbb{R}^3$, is imposed on a portion Γ_2 of the boundary of the reference configuration. In order to identify the associated boundary value problem, and especially to decide what kind of complementary boundary condition involving the first Piola–Kirchhoff stress vector should be imposed on Γ_2, we shall resort to a *new approach*: As shown in the next theorem, due to Ciarlet & Nečas [1985], this information can be easily derived from the *a priori* knowledge of the total energy and of the set of admissible solutions, under the assumption that the total energy attains its minimum. This "reverse" approach has the additional advantage of yielding as a by-product the associated principle of virtual work, notably by suggesting the specific forms of the "variations" that should enter this principle (Exercise 5.5). By contrast, the principle of the virtual work and the expression of the total energy have been so far derived from the *a priori* knowledge of the boundary value problem.

Theorem 5.3-1 (displacement–traction problem with a unilateral boundary condition of place). *Let Ω be a domain in $\mathbb{R}^3$ and let $\Gamma_0, \Gamma_1, \Gamma_2$ be disjoint relatively open subsets of $\Gamma = \partial\Omega$ such that* area $\{\Gamma - (\Gamma_0 \cup \Gamma_1 \cup \Gamma_2)\} = 0$ *and* area $\Gamma_2 > 0$. *Let the set of admissible solutions be of the form*

$$\boxed{\Phi = \{\psi : \bar{\Omega} \to \mathbb{R}^3 ;\ \det \nabla\psi > 0 \text{ in } \bar{\Omega},\ \psi = \varphi_0 \text{ on } \Gamma_0,\ \psi(\Gamma_2) \subset C\} ,}$$

where C is a given closed subset of $\mathbb{R}^3$*, and let the total energy be defined by*

$$I(\boldsymbol{\psi}) = \int_\Omega \hat{W}(\nabla\boldsymbol{\psi})\,\mathrm{d}x - \left\{\int_\Omega \boldsymbol{f}\cdot\boldsymbol{\psi}\,\mathrm{d}x + \int_{\Gamma_1} \boldsymbol{g}\cdot\boldsymbol{\psi}\,\mathrm{d}a\right\}.$$

A smooth enough solution $\boldsymbol{\varphi}$ *of the minimization problem:*

$$\boldsymbol{\varphi} \in \boldsymbol{\Phi} \text{ and } I(\boldsymbol{\varphi}) = \inf_{\boldsymbol{\psi}\in\boldsymbol{\Phi}} I(\boldsymbol{\psi}),$$

is, at least formally, a solution of the following boundary value problem:

$$\begin{aligned}
&-\operatorname{div}\hat{\boldsymbol{T}}(\nabla\boldsymbol{\varphi}) = \boldsymbol{f} \text{ in } \Omega, \text{ with } \hat{\boldsymbol{T}}(\boldsymbol{F}) := \frac{\partial\hat{W}}{\partial\boldsymbol{F}}(\boldsymbol{F}) \text{ for } \boldsymbol{F}\in\mathbb{M}^3_+,\\
&\boldsymbol{\varphi} = \boldsymbol{\varphi}_0 \text{ on } \Gamma_0,\\
&\hat{\boldsymbol{T}}(\nabla\boldsymbol{\varphi})\boldsymbol{n} = \boldsymbol{g} \text{ on } \Gamma_1,\\
&\boldsymbol{\varphi}(\Gamma_2) \subset C,\\
&\hat{\boldsymbol{T}}(\nabla\boldsymbol{\varphi}(x))\boldsymbol{n}(x) = \boldsymbol{o} \quad \text{if } x\in\Gamma_2 \text{ and } \boldsymbol{\varphi}(x)\in \operatorname{int} C,\\
&\hat{\boldsymbol{T}}(\nabla\boldsymbol{\varphi}(x))\boldsymbol{n}(x) = \lambda(x)\boldsymbol{n}^{\varphi}(x^{\varphi}) \text{ with } \lambda(x)\le 0 \text{ if } x\in\Gamma_2\\
&\qquad\qquad\text{and } x^{\varphi} = \boldsymbol{\varphi}(x)\in\partial C,
\end{aligned}$$

where $\boldsymbol{n}^{\varphi}$ *denotes the unit outer normal vector along the deformed surface* $\boldsymbol{\varphi}(\Gamma)$.

Proof. We shall repeatedly use the familiar Green formula:

$$\int_\Omega \boldsymbol{T}:\nabla\boldsymbol{\theta}\,\mathrm{d}x = -\int_\Omega \operatorname{div}\boldsymbol{T}\cdot\boldsymbol{\theta}\,\mathrm{d}x + \int_\Gamma \boldsymbol{T}\boldsymbol{n}\cdot\boldsymbol{\theta}\,\mathrm{d}a,$$

valid for all smooth enough tensor fields $\boldsymbol{T}:\bar{\Omega}\to\mathbb{M}^3$ and vector fields $\boldsymbol{\theta}:\Omega\to\mathbb{R}^3$. Assume first that $\boldsymbol{\theta}$ vanishes in a neighborhood of $\Gamma_0\cup\Gamma_2$. There exists $\varepsilon_0=\varepsilon_0(\boldsymbol{\theta})$ such that $\boldsymbol{\varphi}^{\varepsilon} := \boldsymbol{\varphi}+\varepsilon\boldsymbol{\theta}\in\boldsymbol{\Phi}$ for $|\varepsilon|\le\varepsilon_0$. The assumed inequality $I(\boldsymbol{\varphi}^{\varepsilon})\ge I(\boldsymbol{\varphi})$ reads:

$$\int_\Omega \{\hat{W}(\nabla\boldsymbol{\varphi}+\varepsilon\nabla\boldsymbol{\theta}) - \hat{W}(\nabla\boldsymbol{\varphi})\}\,\mathrm{d}x - \varepsilon\left\{\int_\Omega \boldsymbol{f}\cdot\boldsymbol{\theta}\,\mathrm{d}x + \int_{\Gamma_1}\boldsymbol{g}\cdot\boldsymbol{\theta}\,\mathrm{d}a\right\} \ge 0,$$

so that, using the relation $\hat{T}(F) = (\partial \hat{W}/\partial F)(F)$ and Green's formula, we infer by a formal argument that

$$\varepsilon\left\{\int_\Omega (-\mathbf{div}\, \hat{T}(\nabla\varphi) - f)\cdot\theta\, dx + \int_{\Gamma_1} (\hat{T}(\nabla\varphi)n - g)\cdot\theta\, da + O(\varepsilon)\right\} \geq 0 \quad \text{for } |\varepsilon| \leq \varepsilon_0\,,$$

and consequently,

$$\int_\Omega (-\mathbf{div}\, \hat{T}(\nabla\varphi) - f)\cdot\theta\, dx + \int_{\Gamma_1} (\hat{T}(\nabla\varphi)n - g)\cdot\theta\, da = 0\,.$$

By considering first fields θ with support in Ω, we deduce that $-\mathbf{div}\, \hat{T}(\nabla\varphi) = f$ in Ω; by considering next fields θ vanishing in a neighborhood of $\Gamma_0 \cup \Gamma_2$, we deduce that $\hat{T}(\nabla\varphi)n = g$ on Γ_1.

Next, let $x \in \Gamma_2$ be such that $\varphi(x) \in \text{int}\, C$, and let $r > 0$ be such that $B_r(x) \cap \Gamma \subset \Gamma_2$, φ can be extended to $\bar{\Omega} \cap B_r(x)$, and $\varphi(B_r(x)) \subset \text{int}\, C$. Given any smooth field $\theta : \bar{\Omega} \to \mathbb{R}^3$ with $\text{supp}\, \theta \subset B_r(x)$, there exists $\varepsilon_1 = \varepsilon_1(\theta) > 0$ such that $\varphi^\varepsilon := \varphi + \varepsilon\theta \in \Phi$ for $|\varepsilon| \leq \varepsilon_1$. Expressing again that $I(\varphi^\varepsilon) \geq I(\varphi)$ for all such functions φ^ε, and taking into account the previous computations, we obtain the inequality

$$\varepsilon\left\{\int_{\Gamma_2} \hat{T}(\nabla\varphi)n\cdot\theta\, da + O(\varepsilon)\right\} \geq 0 \quad \text{for } |\varepsilon| \leq \varepsilon_1\,,$$

from which we deduce that $\int_{\Gamma_2} \hat{T}(\nabla\varphi)n\cdot\theta\, da = 0$. Since this relation holds for all smooth fields θ with support in the ball $B_r(x)$, we deduce that the boundary condition $\hat{T}(\nabla\varphi)n = o$ holds at those points x of Γ_2 where $\varphi(x) \in \text{int}\, C$.

Finally, let $y \in \Gamma_2$ be such that $\varphi(y) \in \partial C$ and assume that the surfaces ∂C and $\varphi(\Gamma_2)$ have the same tangent space at the point $\varphi(y)$; since our proof is formal, this last assumption is indeed licit if both boundaries of the sets C and $\varphi(\Omega)$ are smooth enough. Let t_1^φ, t_2^φ be a smooth field of linearly independent vectors defined in a neighborhood V^φ of the point $\varphi(y)$, with the following properties:

$$\begin{cases} |t_1^\varphi| = |t_2^\varphi| = 1 \text{ in } V^\varphi\,; \\ t_1^\varphi(z^\varphi) \text{ and } t_2^\varphi(z^\varphi) \text{ span the tangent plane at the} \\ \quad \text{surface } \varphi(\Gamma_2) \text{ at all points } z^\varphi \in \varphi(\Gamma_2) \cap V^\varphi\,, \end{cases}$$

and let $B_\rho(y)$ be an open ball centered at y such that

$$B_\rho(y) \cap \Gamma \subset \Gamma_2 \text{ and } \varphi(B_\rho(y)) \subset V^\varphi .$$

Given any smooth functions $\zeta_1, \zeta_2 : \bar{\Omega} \to \mathbb{R}$ with supports in $B_\rho(y)$, there exists $\varepsilon_2 = \varepsilon_2(\zeta_1, \zeta_2) > 0$ and there exist functions $\lambda_1^\varepsilon, \lambda_2^\varepsilon : \bar{\Omega} \to \mathbb{R}$ with supports in $B_\rho(y)$ such that (no summation with respect to α)

$$\begin{cases} \varphi_\alpha^\varepsilon := \varphi + \varepsilon(\zeta_\alpha t_\alpha^\varphi + \lambda_\alpha^\varepsilon n^\varphi) \circ \varphi \in \Phi & \text{for } |\varepsilon| \leqslant \varepsilon_2,\ \alpha = 1, 2 , \\ \sup_{z \in B_\rho(y)} |\lambda_\alpha^\varepsilon(z)| = O(\varepsilon),\ \alpha = 1, 2 . \end{cases}$$

Proceeding as before, we find

$$\varepsilon \left\{ \int_{\Gamma_2} \zeta_\alpha \{ \hat{T}(\nabla\varphi) n \cdot (t_\alpha^\varphi \circ \varphi) \} \, da + O(\varepsilon) \right\} \geqslant 0 \quad \text{for } |\varepsilon| \leqslant \varepsilon_2 ,$$

and thus $\int_{\Gamma_2} \zeta_\alpha \{ \hat{T}(\nabla\varphi) n \cdot (t_\alpha^\varphi \circ \varphi) \} \, da = 0$ for $\alpha = 1, 2$. Since this relation holds for all smooth functions ζ_α with support in $B_\rho(y)$, we deduce that $\hat{T}(\nabla\varphi) n \cdot (t_\alpha^\varphi \circ \varphi) = 0$ for $\alpha = 1, 2$, i.e., the vector $\hat{T}(\nabla\varphi) n$ must either vanish or be parallel to the vector n^φ.

Given any smooth function $\xi : \bar{\Omega} \to \mathbb{R}$ with support in $B_\rho(y)$ and which is $\geqslant 0$, there exists $\varepsilon_3 = \varepsilon_3(\xi) > 0$ such that

$$\varphi^\varepsilon = \varphi + \varepsilon \xi (n^\varphi \circ \varphi) \in \Phi \quad \text{for } -\varepsilon_3 \leqslant \varepsilon \leqslant 0 .$$

Thus

$$\varepsilon \left\{ \int_{\Gamma_2} \xi \{ \hat{T}(\nabla\varphi) n \cdot (n^\varphi \circ \varphi) \} \, da + O(\varepsilon) \right\} \geqslant 0 \quad \text{for } -\varepsilon_3 \leqslant \varepsilon \leqslant 0 ,$$

from which we deduce

$$\int_{\Gamma_2} \xi \{ \hat{T}(\nabla\varphi) n \cdot (n^\varphi \circ \varphi) \} \, da \leqslant 0 .$$

Since this relation must hold for all smooth functions $\xi : \bar{\Omega} \to \mathbb{R}$ with support in $B_\rho(y)$ that are $\geqslant 0$, we must have $\hat{T}(\nabla\varphi) n \cdot (n^\varphi \circ \varphi) \leqslant 0$. Since we have just shown that the vector $\hat{T}(\nabla\varphi) n$ is parallel to the vector n^φ, the last inequality forces the vector $\hat{T}(\nabla\varphi) n$ to be of the form λn^φ with $\lambda \leqslant 0$ in this case. ■

Remark. If $C = \mathbb{R}^3$, the above problem reduces to a displacement–

traction problem, with the boundary conditions $\hat{T}(\nabla\varphi)n = g$ on Γ_1, $\hat{T}(\nabla\varphi)n = 0$ on Γ_2. ■

In order to interpret the boundary conditions on Γ_2 found in the above theorem, we recall (Theorem 1.7-1) that the first Piola–Kirchhoff stress tensor $T = \hat{T}(\nabla\varphi)$, the normal vector n, and the area element $\mathrm{d}a$, at a point $x \in \Gamma$ and the corresponding Cauchy stress tensor T^φ, the normal vector n^φ, and the area element $\mathrm{d}a^\varphi$, at the point $\varphi(x)$ are related by $Tn\,\mathrm{d}a = T^\varphi n^\varphi\,\mathrm{d}a^\varphi$. In particular, the Piola–Kirchhoff stress vector Tn and the Cauchy stress vector $T^\varphi n^\varphi$ are parallel. Hence *the boundary condition on Γ_2 can be equivalently expressed as a boundary condition on the Cauchy stress vector $T^\varphi n^\varphi$ on $\varphi(\Gamma_2)$*, namely:

$$\boxed{\begin{aligned} &T^\varphi(x^\varphi)n^\varphi(x^\varphi) = o \quad \text{if} \quad x^\varphi \in \operatorname{int} C\,,\\ &T^\varphi(x^\varphi)n^\varphi(x^\varphi) = \lambda^\varphi(x^\varphi)n^\varphi(x^\varphi)\,, \quad \lambda^\varphi(x^\varphi) \leqslant 0\,, \quad \text{if} \quad x^\varphi \in \partial C\,. \end{aligned}}$$

Since the Cauchy stress vector $T^\varphi n^\varphi$, which measures the density of the applied surface force per unit area of the deformed surface $\varphi(\Gamma)$, is normal to the surface ∂C and directed inside the set C at those points of $\varphi(\Gamma_2)$ where contact with ∂C has occurred, *the unilateral boundary condition of place on Γ_2 constitutes a model of* **contact without friction with the obstacle** ∂C (Fig. 5.3-1). In this respect, the function $\lambda^\varphi : \varphi(\Gamma_2) \to \mathbb{R}$, which measures the intensity of the contact load, is nothing but the *Kuhn & Tucker multiplier associated with the constraint $\varphi(\Gamma_2) \subset C$* (for details about this well known notion in optimization theory, see e.g. Ciarlet [1982]).

We conclude this section by examining the special case where $\Gamma_0 = \emptyset$, which *corresponds to a pure traction problem with a unilateral boundary condition of place*. Such a situation is particularly worthy of interest because it corresponds to a vast majority of actual problems. Following Ciarlet & Nečas [1985], we shall in this case mathematically substantiate the following natural, but vague, idea: *If the applied forces make "on the average" an obtuse angle with the "directions in which the body can escape"* (these notations need of course to be given specific meaning), *then we expect solutions to exist, without imposing an additional condition such as $\int_\Omega \varphi\,\mathrm{d}x = e$.*

To make this idea more precise, assume that a body is subjected to "vertical" forces and that the set C is an "infinite goblet", as suggested in Fig. 5.3-2, (where, for simplicity, it is assumed that $g = o$). Clearly there

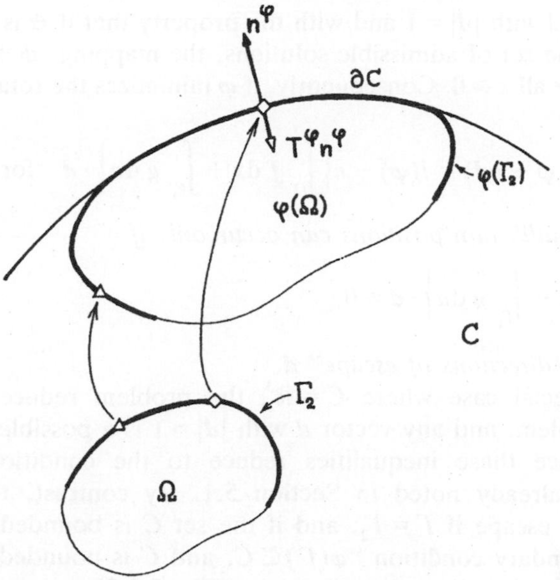

Fig. 5.3-1. A boundary condition of contact without friction: At those points where the deformed surface $\varphi(\Gamma_2)$ and the surface ∂C have a common tangent plane, the Cauchy stress vector $T^\varphi n^\varphi$ is of the form $\lambda^\varphi n^\varphi$, $\lambda^\varphi \leq 0$. The Cauchy stress vector vanishes at those points of the deformed surface $\varphi(\Gamma_2)$ which are in the interior of the set C.

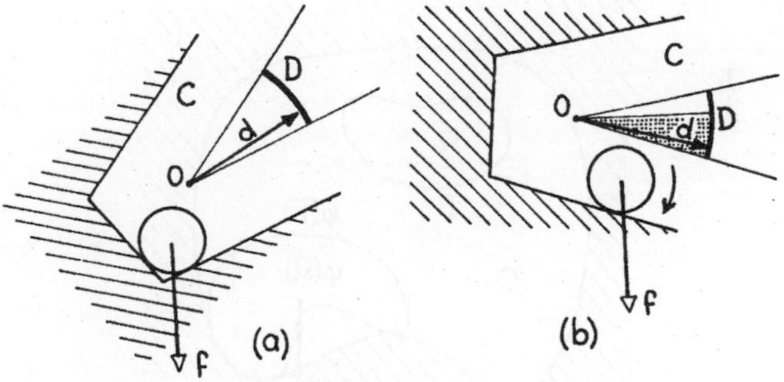

Fig. 5.3-2. (a) All possible "directions of escape" d make obtuse angles with the direction of the applied force f. (b) Some "directions of escape" make an acute angle with the direction of the applied force f: There is no equilibrium position (unless some additional condition is imposed).

exist vectors $\boldsymbol{d}$ with $|\boldsymbol{d}| = 1$ and with the property that if $\boldsymbol{\psi}$ is an arbitrary element of the set of admissible solutions, the mappings $\boldsymbol{\psi} + \varepsilon \boldsymbol{d}$ are also admissible for all $\varepsilon \geqslant 0$. Consequently, if $\boldsymbol{\varphi}$ minimizes the total energy, we must have

$$I(\boldsymbol{\varphi}) \leqslant I(\boldsymbol{\varphi} + \varepsilon \boldsymbol{d}) = I(\boldsymbol{\varphi}) - \varepsilon\left\{\int_\Omega \boldsymbol{f}\,\mathrm{d}x + \int_{\Gamma_1} \boldsymbol{g}\,\mathrm{d}a\right\} \cdot \boldsymbol{d} \quad \text{for all } \varepsilon \geqslant 0 .$$

Therefore *equilibrium positions can occur only if*

$$\left\{\int_\Omega \boldsymbol{f}\,\mathrm{d}x + \int_{\Gamma_1} \boldsymbol{g}\,\mathrm{d}a\right\} \cdot \boldsymbol{d} \leqslant 0$$

for all such "directions of escape" $\boldsymbol{d}$.

In the special case where $C = \mathbb{R}^3$, the problem reduces to a pure traction problem, and any vector $\boldsymbol{d}$ with $|\boldsymbol{d}| = 1$ is a possible direction of escape. Hence these inequalities reduce to the condition $\int_\Omega \boldsymbol{f}\,\mathrm{d}x + \int_\Gamma \boldsymbol{g}\,\mathrm{d}a = \boldsymbol{o}$ already noted in Section 5.1. By contrast, there are no directions of escape if $\Gamma = \Gamma_2$, and if the set C is bounded; the corresponding boundary condition "$\boldsymbol{\varphi}(\Gamma) \subset C$, and C is bounded" is called a **boundary condition of total confinement** (Fig. 5.3-3).

In Sect. 7.8, we shall further exploit these ideas, notably by attaching to any confinement problem a specific *set D of "directions of escape"* (which in the situation considered in Fig. 5.3-2 corresponds to the above "definition"), and by showing that existence holds under the slightly stronger assumption that the *strict* inequalities $\{\int_\Omega \boldsymbol{f}\,\mathrm{d}x + \int_{\Gamma_1} \boldsymbol{g}\,\mathrm{d}a\} \cdot \boldsymbol{d} < 0$ are valid for all vectors $\boldsymbol{d} \in D$ (Theorem 7.8-2).

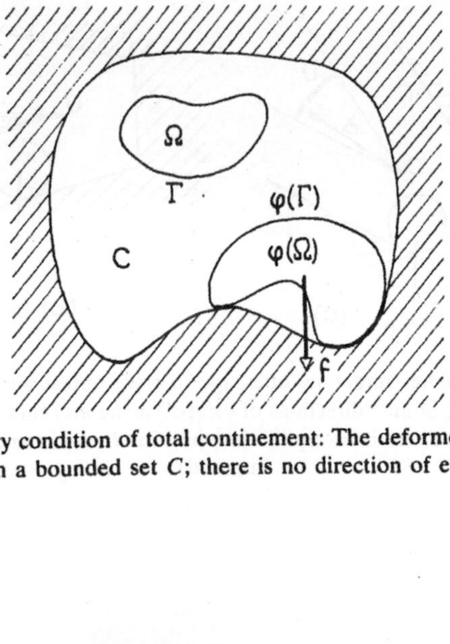

Fig. 5.3-3. A boundary condition of total continement: The deformed boundary $\boldsymbol{\varphi}(\Gamma)$ must lie in a bounded set C; there is no direction of escape.

*5.4. THE TOPOLOGICAL DEGREE IN $\mathbb{R}^n$

We continue this chapter by discussing the orientation-preserving character ($\det \nabla\varphi > 0$ in Ω) and interior injectivity property (the mapping $\varphi : \bar{\Omega} \to \mathbb{R}^3$ is injective, except possibly on Γ) that a deformation φ must evidently possess in order to be physically acceptable. Since (as shown in the next section) a powerful tool for proving the injectivity of a mapping is the *topological degree*, we first briefly present this fundamental notion (due to Brouwer [1912], then extended to infinite-dimensional spaces by Leray & Schauder [1934]), which is also very useful for proving other properties of mappings, such as surjectivity, fixed points, multiplicity of solutions, etc. For proofs and further results, see notably Rado & Reichelderfer [1955], Schwartz [1967, Chapter VI], Nirenberg [1974], Rabinowitz [1975], Berger [1977], Lloyd [1978], Doubrovine, Novikov & Fomenko [1982b, Ch. 3], Deimling [1985], Zeidler [1986].

The definition of the topological degree relies on the following interesting *per se* property of continuously differentiable functions, which has many uses in analysis.

Theorem 5.4-1 (Sard's theorem). *Let Ω be a bounded open subset in $\mathbb{R}^n$. Given a mapping $\varphi \in \mathscr{C}^1(\Omega; \mathbb{R}^n)$, let*

$$S_\varphi = \{x \in \Omega; \det \nabla\varphi(x) = 0\}, \quad \text{where } \nabla\varphi(x) = \left(\frac{\partial \varphi_i}{\partial x_j}(x)\right).$$

Then

$$dx - \text{meas}\, \varphi(S_\varphi) = 0. \quad \blacksquare$$

Let then a bounded open subset Ω of $\mathbb{R}^n$, a mapping $\varphi \in \mathscr{C}^0(\bar{\Omega}; \mathbb{R}^n) \cap \mathscr{C}^1(\Omega; \mathbb{R}^n)$, and a point $b \in \varphi(\bar{\Omega}) - \varphi(\partial\Omega \cup S_\varphi)$, be given, where the set S_φ is defined as in Sard's theorem. Thus the inverse image $\varphi^{-1}(b) = \{x \in \bar{\Omega}; \varphi(x) = b\}$ is a nonempty subset of the set Ω (the assumption that Ω is open implies $\Omega \cup \partial\Omega = \bar{\Omega}$, and $\Omega \cap \partial\Omega = \emptyset$) and all the points $x \in \varphi^{-1}(b)$ verify $\det \nabla\varphi(x) \neq 0$. In addition, the set $\varphi^{-1}(b)$ is finite: To see this, we note that by the local inversion theorem (Theorem 1.2-4), each point $x \in \varphi^{-1}(b) \in \Omega$ possesses a neighborhood $V_x \subset \Omega$ such that the restriction $\varphi|_{V_x} \to \mathbb{R}^n$ is a $\mathscr{C}^1$-diffeomorphism onto a neighborhood W_x of b. Since in particular $y \not\in \varphi^{-1}(b)$ for all $y \in V_x - \{x\}$, the set $\varphi^{-1}(b)$ is discrete: Each one of its points x possesses a neighborhood V_x such that $(V_x - \{x\}) \cap$

$\varphi^{-1}(b)=\emptyset$. On the other hand, the set $\varphi^{-1}(b)$ is compact (it is closed since the mapping φ is continuous on the set $\bar{\Omega}$, and bounded since the set Ω is bounded). Thus the set $\varphi^{-1}(b)$ is finite, and consequently the expression

$$\boxed{\begin{aligned}&\deg(\varphi, \Omega, b) := \sum_{x\in\varphi^{-1}(b)} \operatorname{sgn}\{\det \nabla\varphi(x)\}\\ &\quad \text{if } b \in \varphi(\bar{\Omega}) - \varphi(\partial\Omega \cup S_\varphi)\,,\end{aligned}}$$

where $\operatorname{sgn}\{\alpha\} = +1$ if $\alpha > 0$ and $\operatorname{sgn}\{\alpha\} = -1$ if $\alpha < 0$, unambiguously defines an integer $\det(\varphi, \Omega, b) \in \mathbb{Z}$. Let next

$$\boxed{\deg(\varphi, \Omega, b) := 0 \quad \text{if } b \in \mathbb{R}^n - \varphi(\bar{\Omega})\,.}$$

Then one can prove that *if b and b' are two points of the set $\varphi(\bar{\Omega}) - \varphi(\partial\Omega \cup S_\varphi)$ in the same connected component of the set $\mathbb{R}^n - \varphi(\Omega)$*, then $\deg(\varphi, \Omega, b) = \deg(\varphi, \Omega, b')$. With this result, the definition of the number $\deg(\varphi, \Omega, b)$ can be extended to all points $b \in \mathbb{R}^n - \varphi(\partial\Omega)$: Given a point $b \in \varphi(S_\varphi) - \varphi(\partial\Omega)$ (this is the only case that remains to be considered), let C_b denote the connected component of b in the open set $\mathbb{R}^n - \varphi(\partial\Omega)$. Since it is nonempty ($b \in C_b$) and open, we necessarily have dx-meas $C_b > 0$. Hence by Sard's theorem (Theorem 5.4-1), the set C_b necessarily contains points $b' \not\in \varphi(S_\varphi)$. We are thus justified in letting

$$\boxed{\begin{aligned}&\deg(\varphi, \Omega, b) := \deg(\varphi, \Omega, b') \quad \text{if } b \in \varphi(S_\varphi) - \varphi(\partial\Omega)\,,\\ &\text{where } b' \text{ is any point in the set } C_b - \varphi(S_\varphi) \neq \emptyset\,,\end{aligned}}$$

and therefore, we have associated with any mapping $\varphi \in \mathscr{C}^0(\bar{\Omega}; \mathbb{R}^n) \cap \mathscr{C}^1(\Omega; \mathbb{R}^n)$ and any point $b \in \mathbb{R}^n - \varphi(\partial\Omega)$ an *integer* $\deg(\varphi, \Omega, b) \in \mathbb{Z}$, by whichever of the above formulas is appropriate.

The definition can be extended to mappings $\varphi: \bar{\Omega} \to \mathbb{R}^n$ that are only continuous on the set $\bar{\Omega}$, by using the following observation: Given a mapping $\varphi \in \mathscr{C}^0(\bar{\Omega}; \mathbb{R}^n)$ and a point $b \not\in \varphi(\partial\Omega)$, let $d(b, \varphi(\partial\Omega))$ denote the distance from b to the set $\varphi(\partial\Omega)$ (this distance is >0 since the set $\varphi(\partial\Omega)$ is compact). Then there exists a number $\varepsilon = \varepsilon(d(b, \varphi(\partial\Omega)))$ such that

$$0 < \varepsilon < d(b, \varphi(\partial\Omega)) \text{ and } \deg(\psi^1, \Omega, b) = \deg(\psi^2, \Omega, b)$$

for all $\psi^1, \psi^2 \in \mathscr{C}^0(\bar{\Omega}; \mathbb{R}^n) \cap \mathscr{C}^1(\Omega; \mathbb{R}^n)$ satisfying $\|\psi^\alpha - \varphi\|_{\mathscr{C}^0(\bar{\Omega}; \mathbb{R}^n)} \leq \varepsilon$, $\alpha = 1, 2$ (each number $\deg(\psi^\alpha, \Omega, b)$ is well defined since $\varepsilon < d(b, \varphi(\partial\Omega))$ implies $b \not\in \psi^\alpha(\partial\Omega)$). It thus suffices to let

$$\boxed{\begin{aligned} &\deg(\varphi, \Omega, b) = \lim_{k \to \infty} \deg(\psi^k, \Omega, b) \quad \text{for any sequence } (\psi^k) \text{ with} \\ &\psi^k \in \mathscr{C}^0(\bar{\Omega}; \mathbb{R}^n) \cap \mathscr{C}^1(\Omega; \mathbb{R}^n), \ \lim_{k \to \infty} \|\psi^k - \varphi\|_{\mathscr{C}^0(\bar{\Omega}; \mathbb{R}^n)} = 0 . \end{aligned}}$$

The integer $\deg(\varphi, \Omega, b) \in \mathbb{Z}$ defined in this fashion for any mapping $\varphi \in \mathscr{C}^0(\bar{\Omega}; \mathbb{R}^n)$ and any point $b \in \varphi(\partial\Omega)$ is called the **topological degree** *of the mapping* φ *at the point* b *with respect to the set* Ω. As suggested by its name, it is a *topological* notion, depending only on the notion of continuous functions. In this respect, the first formula used for its definition may be simply understood as a convenient means of *computing* the degree in the special case of continuously differentiable mapping.

Remark. When the mapping φ is continuously differentiable, the topological degree can be also defined by an integral of the form

$$\deg(\varphi, \Omega, b) = \int_\Omega \rho_b^\varepsilon(\varphi(x)) \det \nabla\varphi(x) \, dx$$
$$\text{if } b \in \varphi(\bar{\Omega}) - \varphi(\partial\Omega \cup S_\varphi),$$

where the function $\rho_b^\varepsilon \in \mathscr{C}^0(\mathbb{R}^n)$ satisfies $\int_{\mathbb{R}^n} \rho_b^\varepsilon(y) \, dy = 1$, $\operatorname{supp} \rho_b^\varepsilon \subset B_\varepsilon(b)$, and ε is sufficiently small (Exercise 5.6). ■

Let us now state various important properties of the topological degree (some of them have already been mentioned), partly illustrated in Figs. 5.4-1 ($n = 1$) and 5.4-2 ($n = 2$). Other basic features, such as the dependence with respect to partitions of the set Ω, the excision property, etc., have been omitted because we shall not need them in the sequel.

Theorem 5.4-2 (properties of the topological degree). *Let* Ω *be a bounded open subset of* $\mathbb{R}^n$, *and let* $\deg(\varphi, \Omega, b)$ *be the topological degree of a mapping* $\varphi \in \mathscr{C}^0(\bar{\Omega}; \mathbb{R}^n)$ *at a point* $b \not\in \varphi(\partial\Omega)$ *with respect to the set* Ω. *Then,*

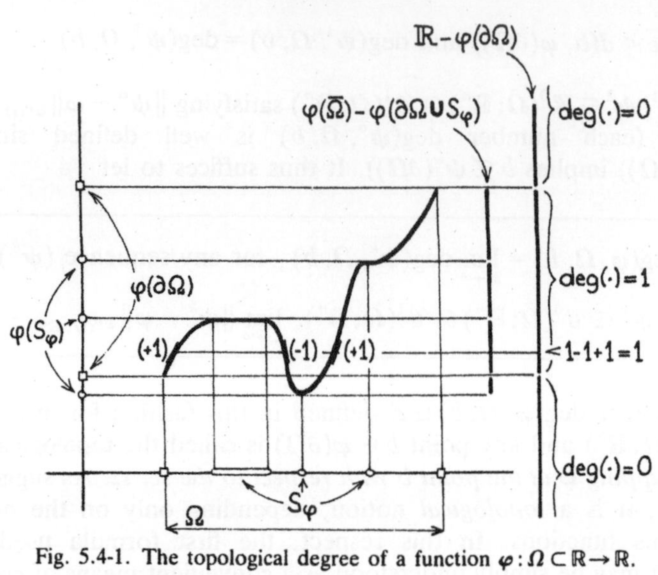

Fig. 5.4-1. The topological degree of a function $\varphi : \Omega \subset \mathbb{R} \rightarrow \mathbb{R}$.

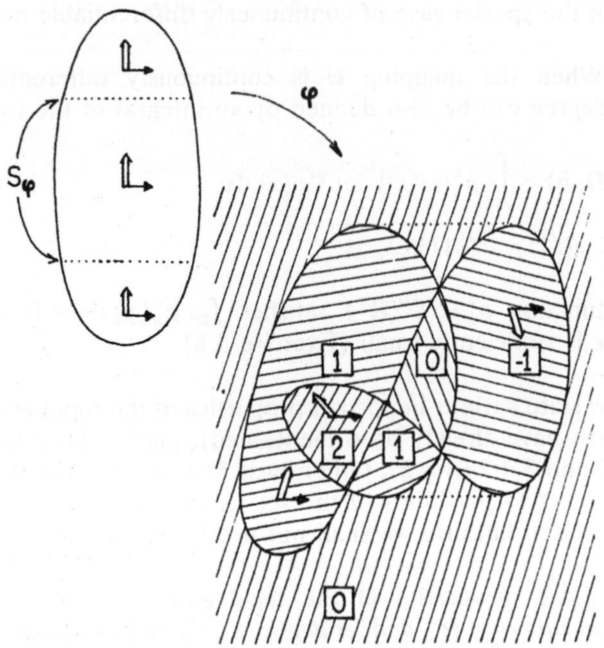

Fig. 5.4-2. The topological degree of a mapping $\varphi : \Omega \subset \mathbb{R}^2 \rightarrow \mathbb{R}^2$. Each hatched region is a connected component of $\mathbb{R}^2 - \varphi(\partial\Omega)$, in which the topological degree has a constant value, indicated in a box.

(a) $b \notin \varphi(\bar{\Omega}) \Rightarrow \deg(\varphi, \Omega, b) = 0$, *or equivalently*:

$$\deg(\varphi, \Omega, b) \neq 0 \Rightarrow b \in \varphi(\Omega).$$

(b) *If b and b' are in the same connected component of the set* $\mathbb{R}^n - \varphi(\partial\Omega)$, *then* $\deg(\varphi, \Omega, b) = \deg(\varphi, \Omega, b')$.

(c) *Continuity with respect to the mapping* φ: *Let* $\varphi \in \mathscr{C}^0(\bar{\Omega}; \mathbb{R}^n)$ *and* $b \notin \varphi(\partial\Omega)$ *be given. Then there exists* $\varepsilon > 0$ *such that*

$$\psi \in \mathscr{C}^0(\bar{\Omega}; \mathbb{R}^n) \text{ and } \|\psi - \varphi\|_{\mathscr{C}^0(\bar{\Omega}; \mathbb{R}^n)} \leq \varepsilon \Rightarrow b \notin \psi(\partial\Omega) \text{ and}$$
$$\deg(\psi, \Omega, b) = \deg(\varphi, \Omega, b)\,.$$

(d) *Homotopic invariance*: *Let*

$$t \in [0, 1] \rightarrow \varphi_t \in \mathscr{C}^0(\bar{\Omega}; \mathbb{R}^n)$$

be a continuous family of mappings such that $b \notin \varphi_t(\partial\Omega)$ *for all* $t \in [0, 1]$. *Then*

$$\deg(\varphi_t, \Omega, b) = \deg(\varphi_0, \Omega, b) \quad \text{for all } 0 \leq t \leq 1\,.$$

(e) *Assume* Ω *is connected and* $\varphi: \bar{\Omega} \rightarrow \mathbb{R}^n$ *is injective. Then either* $\deg(\varphi, \Omega, b) = +1$ *for all* $b \in \varphi(\Omega)$, *or* $\deg(\varphi, \Omega, b) = -1$ *for all* $b \in \varphi(\Omega)$.

(f) *If the mapping* φ *is continuously differentiable in the set* Ω, *then*

$$\deg(\varphi, \Omega, b) = \sum_{x \in \varphi^{-1}(b)} \operatorname{sgn}\{\det \nabla\varphi(x)\}$$
$$\text{for all } b \in \varphi(\bar{\Omega}) - \varphi(\partial\Omega \cup S_\varphi)\,,$$

where

$$S_\varphi = \{x \in \Omega; \det \nabla\varphi(x) = 0\}\,. \quad ■$$

A useful consequence of homotopic invariance is that "*the degree depends only on boundary values*". More specifically, let φ *and* ψ *be two mappings in the space* $\mathscr{C}^0(\bar{\Omega}; \mathbb{R}^n)$ *that satisfy*

$$\varphi = \psi \text{ on } \partial\Omega\,.$$

Then

$$\deg(\varphi, \Omega, b) = \deg(\psi, \Omega, b) \quad \text{for all } b \notin \varphi(\partial\Omega)\,.$$

To see this, it suffices to consider the continuous family of mappings

$$t \in [0,1] \rightarrow \varphi_t = (1-t)\varphi + t\psi, \quad 0 \leqslant t \leqslant 1 .$$

5.5. ORIENTATION-PRESERVING CHARACTER AND INJECTIVITY OF MAPPINGS

Our first observation is that *an orientation-preserving mapping* $\varphi \in \mathscr{C}^1(\bar{\Omega}; \mathbb{R}^3)$, *i.e., a mapping that satisfies* $\det \nabla\varphi(x) > 0$ *for all* $x \in \bar{\Omega}$, *is locally invertible*, i.e., each point in Ω possesses a neighborhood on which the restriction of the mapping φ is injective (this property holds for all $x \in \bar{\Omega}$ if φ is of class $\mathscr{C}^1$ on an open set containing $\bar{\Omega}$). This follows from the application of the implicit function theorem (Theorem 1.2-3) since the Fréchet derivative of the mapping φ is invertible at all points in $\bar{\Omega}$ (the derivative is represented by the matrix $\nabla\varphi$ in the canonical basis, and this matrix is invertible since its determinant is >0).

On the other hand, *local invertibility does not entail in general injectivity*: Consider for instance the mapping

$$\varphi : x \in \bar{\Omega} \subset \mathbb{R}^3 \rightarrow \varphi(x) = \left(x_1 \cos \frac{x_2}{l}, x_1 \sin \frac{x_2}{l}, x_3\right) \in \mathbb{R}^3 ,$$

where $\bar{\Omega}$ is a rectangular rod of length $2\theta l$ contained in the set $\{x \in \mathbb{R}^3; x_1 > 0\}$ and parallel to the vector e_2, as shown in Fig. 5.5-1. Then $\det \nabla\varphi(x) = x_1/l > 0$ for all $x \in \bar{\Omega}$, yet for $\theta \geqslant \pi$ the mapping is not injective since $\varphi(x_1, \pi l, x_3) = \varphi(x_1, -\pi l, x_3)$: For $\theta = \pi$, "the injectivity is lost on the boundary", while for $\theta > \pi$, "interpenetration" has occurred.

We now give two useful sufficient conditions that guarantee the injectivity of a mapping $\varphi : \bar{\Omega} \subset \mathbb{R}^n \rightarrow \mathbb{R}^n$. The first condition asserts that a mapping φ is orientation-preserving and injective if the norm of the gradient of the associated displacement $u = \varphi - id$ is sufficiently small in the set $\bar{\Omega}$. We recall that $|B| = \sup_{v \in \mathbb{R}^3} |Bv|/|v|$ denotes the matrix norm subordinate to the Euclidean vector norm.

Theorem 5.5-1 (sufficient conditions for preservation of orientation and injectivity). (a) *Let* $\varphi = id + u : \Omega \subset \mathbb{R}^n \rightarrow \mathbb{R}^n$ *be a mapping differentiable at a point* $x \in \Omega$. *Then*

$$|\nabla u(x)| < 1 \Rightarrow \det \nabla\varphi(x) > 0 .$$

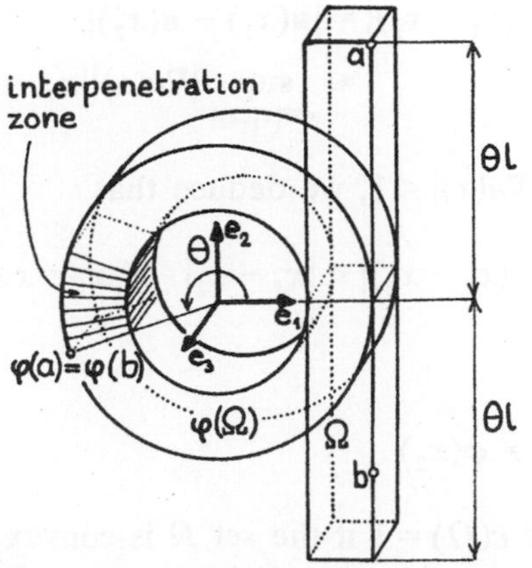

Fig. 5.5-1. An orientation-preserving mapping $\varphi : \bar{\Omega} \subset \mathbb{R}^3 \to \mathbb{R}^3$ that is not injective.

(b) *Let Ω be a domain in $\mathbb{R}^n$. There exists a constant $c(\Omega) > 0$ such that any mapping $\varphi = id + u \in \mathscr{C}^1(\bar{\Omega}; \mathbb{R}^n)$ satisfying*

$$\sup_{x \in \bar{\Omega}} |\nabla u(x)| < c(\Omega)$$

is injective.

Proof. Let x be a point where $|\nabla u(x)| < 1$. Then

$$\det(I + t\nabla u(x)) \neq 0 \quad \text{for } 0 \leq t \leq 1 ,$$

since all the matrices $(I + t\nabla u(x))$, $0 \leq t \leq 1$, are invertible. On the other hand, the function

$$\delta : t \in [0, 1] \to \delta(t) := \det(I + t\nabla u(x))$$

is continuous, so that $\delta([0, 1])$ is a closed interval of $\mathbb{R}$. Since $\delta([0, 1])$ contains $1 = \delta(0)$ but not 0, we conclude that $\det(I + \nabla u(x)) = \delta(1) > 0$, which proves (a).

To prove (b), let us first assume that the open set Ω is *convex* (so that the set $\bar{\Omega}$ is also convex; cf. Theorem 4.7-1), and let x_1 and x_2 be two arbitrary points in the set $\bar{\Omega}$. Applying the mean value theorem (Theorem 1.2-2) to the function $\varphi = id + u$, we obtain

$$|\varphi(x_1) - \varphi(x_2) - (x_1 - x_2)| = |u(x_1) - u(x_2)|$$
$$\leqslant \sup_{x \in]x_1, x_2[} |\nabla u(x)||x_1 - x_2| .$$

Therefore if $\sup_{x\in\bar{\Omega}}|\nabla u(x)| < 1$, we deduce that

$$|\varphi(x_1) - \varphi(x_2) - (x_1 - x_2)| < |x_1 - x_2| \quad \text{if } x_1 \neq x_2 ,$$

and hence

$$x_1 \neq x_2 \Rightarrow \varphi(x_1) \neq \varphi(x_2) .$$

We may thus choose $c(\Omega) = 1$ if the set Ω is convex.

If we drop the assumption that Ω is convex, it is not too hard (but somewhat tedious; cf. Exercise 1.9) to show that an open set that is a domain has the following geometrical property: there exists a number $c(\Omega) > 0$ such that, given arbitrary points x_1 and x_2 in the set $\bar{\Omega}$, there exists a finite number of points y_k, $1 \leqslant k \leqslant l+1$, such that

$$y_1 = x_1, \quad y_k \in \Omega \quad \text{for } 2 \leqslant k \leqslant l, \quad y_{l+1} = x_2 ,$$
$$]y_k, y_{k+1}[\subset \Omega \quad \text{for } 1 \leqslant k \leqslant l, \ \sum_{k=1}^{l} |y_k - y_{k+1}| \leqslant \frac{1}{c(\Omega)} |x_1 - x_2| .$$

Using this property and the assumed inequality $\sup_{x\in\Omega}|\nabla u(x)| < c(\Omega)$, we obtain

$$|\varphi(x_1) - \varphi(x_2) - (x_1 - x_2)| = |u(x_1) - u(x_2)| \leqslant \sum_{k=0}^{l} |y_k - y_{k+1}|$$
$$\leqslant \sup_{x\in\Omega}|\nabla u(x)| \sum_{k=0}^{l} |y_k - y_{k+1}|$$
$$< |x_1 - x_2| \quad \text{if } x_1 \neq x_2 .$$

Hence we again conclude in this case that the mapping φ is injective. ■

Remark. The proof of (b) when Ω is convex was communicated to the author by Franco Brezzi. ■

We next give a second set of sufficient conditions for the injectivity of a mapping $\varphi : \bar{\Omega} \subset \mathbb{R}^n \rightarrow \mathbb{R}^n$, asserting basically that *φ be orientation-preserving in Ω and that φ coincide with a continuous injective mapping $\varphi_0 : \bar{\Omega} \rightarrow \mathbb{R}^n$ on the boundary $\partial\Omega$.* While (as already noticed) *local* invertibility is a simple consequence of preservation of orientation, deriving *global* invertibility from this property is another matter. As we shall see, it requires significantly more subtle arguments, which rest notably on the properties of the *topological degree* (Sect. 5.4). We recall that the equality $\operatorname{int} \bar{\Omega} = \Omega$ (which is an assumption in the next theorem) is satisfied if Ω is a domain but that it may not hold for more general open sets (Exercise 1.7).

Theorem 5.5-2 (sufficient condition for injectivity). *Let Ω be a bounded open connected subset of $\mathbb{R}^n$ such that $\operatorname{int} \bar{\Omega} = \Omega$, let $\varphi_0 \in \mathscr{C}^0(\bar{\Omega}; \mathbb{R}^n)$ be an injective mapping, and let $\varphi \in \mathscr{C}^0(\bar{\Omega}; \mathbb{R}^n) \cap \mathscr{C}^1(\Omega; \mathbb{R}^n)$ be a mapping that satisfies*

$$\begin{cases} \det \nabla\varphi(x) > 0 & \text{for all } x \in \Omega\,, \\ \varphi(x) = \varphi_0(x) & \text{for all } x \in \partial\Omega\,. \end{cases}$$

Then the mapping $\varphi : \bar{\Omega} \rightarrow \varphi(\bar{\Omega})$ is a homeomorphism (in particular the mapping $\varphi : \bar{\Omega} \rightarrow \mathbb{R}^n$ is injective), the mapping $\varphi : \Omega \rightarrow \varphi(\Omega)$ is a $\mathscr{C}^1$-diffeomorphism, and finally,

$$\varphi(\Omega) = \varphi_0(\Omega),\ \varphi(\bar{\Omega}) = \varphi_0(\bar{\Omega})\,.$$

Proof. (i) By Theorem 5.4-2(e), the connectedness of the set Ω and the injectivity of the mapping φ_0 imply

$$\deg(\varphi_0, \Omega, b) = 1, \text{ or } -1, \quad \text{for all } b \in \varphi_0(\Omega)\,.$$

On the other hand (Theorem 5.4-2(a)):

$$\deg(\varphi_0, \Omega, b) = 0 \quad \text{for all } b \notin \varphi_0(\bar{\Omega})\,.$$

Since the mapping φ_0 is injective, and since $\varphi(\partial\Omega) = \varphi_0(\partial\Omega)$, any point b in the set $\varphi_0(\Omega)$ is not in the set $\varphi(\partial\Omega)$. Consequently, the property of homotopic invariance (Theorem 5.4-2(d)) coupled with the assumption $\varphi = \varphi_0$ on $\partial\Omega$, yields

$$\deg(\varphi, \Omega, b) = \deg(\varphi_0, \Omega, b) = \begin{cases} 1, \text{ or } -1, \text{ if } b \in \varphi_0(\Omega), \\ 0 \text{ if } b \notin \varphi_0(\bar{\Omega}). \end{cases}$$

(ii) Let $b \in \varphi_0(\Omega)$, so that $\deg(\varphi, \Omega, b)$ is either 1 or -1 by (i). We cannot have $b \notin \varphi(\bar{\Omega})$ for otherwise $\deg(\varphi, \Omega, b)$ would necessarily vanish (Theorem 5.4-2(a)), and we can neither have $b \in \varphi(\partial\Omega)$ as already observed. The only possibility is thus $b \in \varphi(\Omega)$, and hence

$$\varphi_0(\Omega) \subset \varphi(\Omega) .$$

Using the assumption $\det \nabla\varphi(x) > 0$ for all $x \in \Omega$ and Theorem 5.4-2(f), we know on the other hand that

$$\deg(\varphi, \Omega, b) = \sum_{x \in \varphi^{-1}(b)} \operatorname{sgn}\{\det \nabla\varphi(x)\} = \operatorname{card} \varphi^{-1}(b) \geq 1$$
$$\text{for all } b \in \varphi(\bar{\Omega}) - \varphi(\partial\Omega)$$

(the set S_φ is empty in this case). Since

$$b \in \varphi_0(\Omega) \Rightarrow b \in \varphi(\Omega) \subset \varphi(\bar{\Omega}) - \varphi(\partial\Omega) ,$$

we deduce that necessarily $\deg(\varphi, \Omega, b) = 1$ if $b \in \varphi_0(\Omega)$, i.e.,

$$\operatorname{card} \varphi^{-1}(b) = 1 \quad \text{for all } b \in \varphi_0(\Omega) .$$

(iii) Let $b \notin \varphi_0(\bar{\Omega})$, so that $\deg(\varphi, \Omega, b) = 0$ by (i). Since on the other hand $b \notin \varphi_0(\partial\Omega) = \varphi(\partial\Omega)$, we have either $\deg(\varphi, \Omega, b) \geq 1$ if $b \in \varphi(\bar{\Omega})$ or $\deg(\varphi, \Omega, b) = 0$ if $b \notin \varphi(\bar{\Omega})$. The only possibility is thus $b \notin \varphi(\bar{\Omega})$, and hence

$$\varphi(\bar{\Omega}) \subset \varphi_0(\bar{\Omega}) .$$

To sum up, we have established the following inclusions:

$$\varphi_0(\Omega) \subset \varphi(\Omega) \subset \varphi(\bar{\Omega}) \subset \varphi_0(\bar{\Omega}) .$$

(iv) By taking the closure in the above relation and noting that

$$\varphi(\bar{\Omega}) = \{\varphi(\Omega)\}^-$$

(Theorems 1.2-7 or 1.2-8), we first deduce that

$$\varphi(\bar{\Omega}) = \varphi_0(\bar{\Omega}) .$$

The invariance of domain theorem and the assumption int $\bar{\Omega} = \Omega$ next imply that (Theorem 1.2-8)

$$\varphi_0(\Omega) = \text{int}\{\varphi_0(\Omega)\}^- .$$

Since the interior of a set is the largest open set contained in it and since $\varphi(\Omega)$ is an open subset of $\{\varphi_0(\Omega)\}^-$ (by (iii) and the invariance of domain; cf. Theorem 1.2-6), we deduce that

$$\varphi(\Omega) = \varphi_0(\Omega) .$$

Hence the last property established in step (ii) shows that the mapping $\varphi : \Omega \to \varphi(\Omega)$ is injective. Since the set $\varphi(\partial\Omega) = \varphi_0(\partial\Omega)$ does not intersect the set $\varphi_0(\Omega)$ (the mapping φ_0 is injective on $\bar{\Omega}$ by assumption), we conclude that the mapping φ is injective on the set $\bar{\Omega}$. Hence it is a homeomorphism of $\bar{\Omega}$ onto the set $\varphi(\bar{\Omega})$ by a classical property of injective continuous mappings on compact sets. That its restriction to the set Ω is a $\mathscr{C}^1$-diffeomorphism onto the open set $\varphi(\Omega)$ is part of the invariance of domain theorem (Theorem 1.2-5). ■

Theorem 5.5-2 can be extended to Lipschitz-continuous mappings φ (Pourciau [1983]), and to mappings φ in the Sobolev spaces $W^{1,p}(\Omega)$, $p > n$ (Ball [1981b, Theorem 1]; see also Exercise 5.7).

Remarks. (1) While the condition $\varphi'(x) > 0$ for all points x of an interval Ω of $\mathbb{R}$ is a sufficient condition for the injectiveness of a function $\varphi \in \mathscr{C}^0(\bar{\Omega}) \cap \mathscr{C}^1(\Omega)$, it is clear that when $n \geqslant 2$ the hypothesis $\det \nabla\varphi(x) > 0$ in Ω has to be supplemented by another condition (the coincidence with an injective mapping on the boundary) in order to guarantee injectivity. In this respect, refer to the counter-example provided in Fig. 5.5-1.

(2) The special case where $\varphi_0 = \boldsymbol{id}$ is particularly worthy of interest. Observe that the seemingly "evident" conclusions $\varphi(\Omega) = \Omega$ and $\varphi(\bar{\Omega}) = \bar{\Omega}$ corresponding to this case are by no means easy to prove. ■

The applicability of Theorem 5.5-2 in elasticity is limited by the assumption that the mapping φ must be equal to an injective mapping on *all* the boundary $\partial\Omega$ (this was not the case of Theorem 5.5-1), while in actual situations the deformation is often only specified on a *strict subset* of the boundary (we shall see in the next section how the injectivity of the deformation can still be guaranteed in this case). Be that as it may, it is nevertheless a powerful result for studying the *pure displacement problem*, as we shall see in the next chapter.

Remarks. (1) In any situation where the assumption "$\det \nabla\varphi(x) > 0$ for all $x \in \Omega$" of Theorem 5.5-2 is obtained as a consequence of the stronger assumption "$|\nabla u(x)| < 1$ for all $x \in \Omega$" of Theorem 5.5-1, Theorem 5.5-1 has a wider applicability than Theorem 5.5-2 for proving the injectivity of a mapping φ, since it does not contain any assumption regarding the values of φ on the boundary $\partial\Omega$ (besides, it is much more simple to prove). An illustration of this situation will be provided by the proof of Theorem 6.9-1.

(2) There are, however, situations where Theorem 5.5-2 yields more powerful injectivity results than Theorem 5.5-1, for it does not require that $\sup_{x\in\Omega}|\nabla u(x)|$ be <1; consider for instance a pure displacement problem for an *incompressible material*, whose solution u satisfies $\det(I + \nabla u)(x) = 1$ for all $x \in \Omega$. ■

The assumptions that φ be equal on $\partial\Omega$ to a mapping that is injective on $\bar{\Omega}$ and that $\det \nabla\varphi(x) > 0$ for all $x \in \Omega$ can be weakened, as shown by the following result of Meisters & Olech [1963]:

Theorem 5.5-3 (sufficient condition for injectivity): *Let O be an open subset of $\mathbb{R}^n$, let K be a compact subset of O with a connected boundary ∂K, and let there be given a mapping $\varphi: O \subset \mathbb{R}^n \to \mathbb{R}^n$ that satisfies:*

$$\begin{cases} \varphi \in \mathscr{C}^1(O;\mathbb{R}^n); \\ \det \nabla\varphi > 0 \text{ in } \quad \operatorname{int} K \quad \text{except possibly on a finite subset;} \\ \det \nabla\varphi > 0 \quad \text{for at least one point of } \partial K; \\ \varphi_{|\partial K} \text{ is injective.} \end{cases}$$

Then the mapping $\varphi: K \to \mathbb{R}^n$ is injective. ■

This result is valid regardless of any assumed regularity of the boundary ∂K, but it relies on the assumption that the mapping φ be of class $\mathscr{C}^1$ on an open set containing K. Recall in this respect that the boundary of the open set Ω was assumed to be smooth enough in Theorem 5.5-1 (Ω was a domain) and that the assumption int $\bar{\Omega} = \Omega$ in Theorem 5.5-2 is in a sense a smoothness assumption on the boundary $\partial\Omega$. The weakened assumptions of Theorem 5.5-3 are similarly compensated by the hypothesis that the boundary ∂K be connected, an assumption that was not needed in Theorems 5.5-1 and 5.5-2. Weinstein [1985] has however shown that the assumption that ∂K be connected can be also removed.

5.6. INTERIOR INJECTIVITY, SELF-CONTACT, AND NONINTERPENETRATION IN HYPERELASTICITY

The injectivity of an orientation-preserving mapping $\varphi : \bar{\Omega} \to \mathbb{R}^3$ that solves a boundary value problem of elasticity can be deduced from Theorems 5.5-2 or 5.5-3 only if φ solves a *pure displacement problem*, since φ must then be equal to a known injective mapping on *all* of the boundary $\partial\Omega$. Except in this case, an additional condition is thus needed that guarantees the **interior injectivity** of φ, i.e., that φ is injective on the open set Ω (if self-contact is allowed, the mapping φ may loose its injectivity on $\partial\Omega$), as suggested by the displacement-pressure problem considered in Fig. 5.6-1.

We now show that for *hyperelastic materials*, such a condition consists

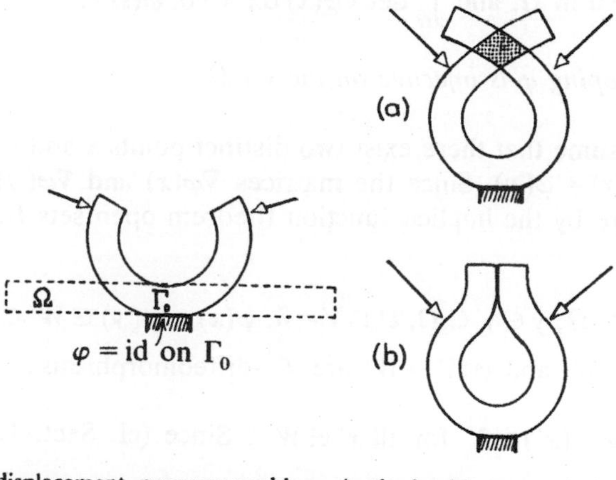

Fig. 5.6-1. A displacement–pressure problem: As the load increases, interpenetration may occur if the mathematical model only includes the orientation-preserving condition (a). In order to be physically realistic, the model must include an additional condition of interior injectivity, i.e., a condition of noninterpenetration that simultaneously allows self-contact (b).

in imposing, in addition to the orientation-preserving condition $\det \nabla\varphi > 0$ in $\bar{\Omega}$, the **injectivity condition:**

$$\boxed{\int_\Omega \det \nabla\varphi(x)\,\mathrm{d}x \leq \operatorname{vol} \varphi(\Omega)}$$

to the minimizers $\varphi : \bar{\Omega} \to \mathbb{R}^3$ of the energy, and more generally to all the admissible solutions. More specifically, we first show that the injectivity condition implies the interior injectivity of the mapping φ (Theorem 5.6-1); we next study the "geometry of self-contact" at those points of the boundary where the mapping φ loses its injectivity (Theorem 5.6-2); and finally, we establish the *interior injectivity of the minimizers of the energy in the case of mixed displacement–traction problems* (Theorem 5.6-3). Throughout this section, we follow Ciarlet & Nečas [1986].

We begin by establishing the injectivity on the open set Ω of an orientation-preserving mapping $\varphi : \bar{\Omega} \to \mathbb{R}^3$ that satisfies the injectivity condition.

Theorem 5.6-1 (sufficient condition for injectivity). *Let Ω be a domain in $\mathbb{R}^3$, and let $\varphi \in \mathscr{C}^1(\bar{\Omega}; \mathbb{R}^3)$ be such that*

$$\det \nabla\varphi > 0 \text{ in } \Omega, \text{ and } \int_\Omega \det \nabla\varphi(x)\,\mathrm{d}x \leq \operatorname{vol} \varphi(\Omega)\,.$$

Then the mapping φ is injective on the set Ω.

Proof. Assume that there exist two distinct points x and y in the set Ω such that $\varphi(x) = \varphi(y)$. Since the matrices $\nabla\varphi(x)$ and $\nabla\varphi(y)$ are invertible, there are by the implicit function theorem open sets U, V, W' such that

$$\begin{cases} x \in U \subset \Omega;\ y \in V \subset \Omega,\ U \cap V = \emptyset,\ \varphi(x) = \varphi(y) \in W' \subset \varphi(\Omega)\,, \\ \varphi : U \to W' \text{ and } \varphi : V \to W' \text{ are } \mathscr{C}^1\text{-diffeomorphisms}\,. \end{cases}$$

Hence $\operatorname{card} \varphi^{-1}(x') \geq 2$ for all $x' \in W'$. Since (cf. Sect. 1.5 for references)

$$\int_{\varphi(\Omega)} \operatorname{card} \varphi^{-1}(x')\,\mathrm{d}x' = \int_\Omega \det \nabla\varphi(x)\,\mathrm{d}x$$

whenever one of the two integrals exists (this is the case here since $\det \nabla\varphi \in \mathscr{C}^0(\bar{\Omega})$ by assumption), and since $\text{vol}\, W' > 0$, it follows that

$$\text{vol}\, \varphi(\Omega) = \int_{\varphi(\Omega)} \mathrm{d}x' < \int_{\varphi(\Omega)} \text{card}\, \varphi^{-1}(x')\, \mathrm{d}x' = \int_{\Omega} \det \nabla\varphi(x)\, \mathrm{d}x\,,$$

but this contradicts the inequality $\int_\Omega \det \nabla\varphi(x)\, \mathrm{d}x \leq \text{vol}\, \varphi(\Omega)$. Hence the mapping φ is injective in Ω. ■

In addition to the assumptions stated in Theorem 5.6-1, we shall also allow the admissible deformations φ to satisfy a **confinement condition** of the form consider by Noll [1978], viz.,

$$\boxed{\varphi(\bar{\Omega}) \subset B\,,}$$

where B is an arbitrary closed subset of $\mathbb{R}^3$. Since the boundary of the set B acts as an **obstacle**, a confinement condition resembles a unilateral boundary condition of place $\varphi(\Gamma_2) \subset C$ (cf. Sect. 5.2; see also Exercise 5.8). The definition of a boundary of class $\mathscr{C}^1$ used here is that of Nečas [1967] if the boundary is bounded (Sect. 1.6). If the boundary is unbounded (as may be the case for the set ∂B), the only modification in the definition is that the number of local coordinate systems is infinite.

Theorem 5.6-2. *Let Ω be a domain in $\mathbb{R}^3$ with a boundary $\partial\Omega$ of class $\mathscr{C}^1$, and let $\varphi \in \mathscr{C}^1(\bar{\Omega}; \mathbb{R}^3)$ be a mapping that satisfies*

$$\det \nabla\varphi > 0 \text{ in } \bar{\Omega}, \qquad \int_\Omega \det \nabla\varphi(x)\, \mathrm{d}x \leq \text{vol}\, \varphi(\Omega)\,,$$

and

$$\varphi(\bar{\Omega}) \subset B\,.$$

Then the mapping φ, which is injective on Ω by Theorem 5.6-1, further satisfies:

$$\varphi(\Omega) \subset \text{int}\, B, \quad \varphi(\Omega) \cap \varphi(\partial\Omega) = \emptyset\,,$$

$$x' \in \partial B \cap \varphi(\partial\Omega) \Rightarrow \text{card}\, \varphi^{-1}(x') = 1\,,$$

and the surface ∂B and $\varphi(\partial\Omega)$ have the same tangent plane at the point x',

$$x' \in \varphi(\partial\Omega) \cap \operatorname{int} B \Rightarrow \operatorname{card} \varphi^{-1}(x') = 1 \text{ or } 2 .$$

If $\varphi^{-1}(x') = \{x, y\}$, then $x \in \partial\Omega$, $y \in \partial\Omega$, and the outer normal vectors to the surface $\varphi(\partial\Omega)$ at the points $\varphi(x) = \varphi(y)$ have opposite directions (consequently, the tangent planes coincide).

Proof. Let $x \in \Omega$; there exists by the implicit function theorem an open set $U \subset \Omega$ containing x such that $\varphi: U \to \varphi(U)$ is a $\mathscr{C}^1$-diffeomorphism. The set $\varphi(U)$, which is open (by the invariance of domain theorem), contains $\varphi(x)$, and is contained in the set B, since $\varphi(\bar{\Omega}) \subset B$ by assumption. Hence $\varphi(x) \in \operatorname{int} B$, and this proves the inclusion $\varphi(\Omega) \subset \operatorname{int} B$.

Let $x \in \partial\Omega$; since the boundary $\partial\Omega$ is of class $\mathscr{C}^1$ and since the mapping φ is in the space $\mathscr{C}^1(\bar{\Omega}; \mathbb{R}^3)$, we can find an open set $A \subset \mathbb{R}^3$ containing x and an extension (still denoted) $\varphi \in \mathscr{C}^1(\bar{\Omega} \cup A; \mathbb{R}^3)$ such that $\det \nabla\varphi > 0$ in $\bar{\Omega} \cup A$. This extension can be constructed using the extension techniques for smooth boundaries described in Nečas [1967]. Then by the implicit function theorem, there exist open sets $U \subset A$ and U' containing x and $x' := \varphi(x)$ respectively, such that $\varphi: U \to U'$ is a $\mathscr{C}^1$-diffeomorphism.

Assume there exists $y \in \Omega$ such that $x' = \varphi(y)$. After reducing U' if necessary, we can find a neighborhood V of y such that U and V are disjoint and $\varphi: V \to U'$ is a $\mathscr{C}^1$-diffeomorphism. Since U is a neighborhood of $x \in \partial\Omega$, it contains a point $z \in \Omega$, and thus the set $\varphi^{-1}(\varphi(z))$ contains at least two elements; but this contradicts the injectivity of φ on the set Ω (Theorem 5.6-1).

Let next $x' \in \partial B \cap \varphi(\partial\Omega)$; if the surfaces ∂B and $\varphi(\partial\Omega)$ had different tangent planes at the point x', we would have $\varphi(\Omega) \cap \partial B \neq \emptyset$, in contradiction with the inclusion $\varphi(\Omega) \subset \operatorname{int} B$ (to see this, let $\boldsymbol{n}'$ denote the unit outer normal vector to the surface ∂B at the point x', and consider a plane containing the vector $\boldsymbol{n}'$ and intersecting the two tangent planes along two distinct lines). Hence the tangent planes coincide and since $\varphi(\bar{\Omega}) \subset B$, the unit outer normal vectors to the surfaces ∂B and $\varphi(\partial\Omega)$ necessarily have the same direction.

Assume that the set $\varphi^{-1}(x')$ contains another point $y \neq x$, which also belongs to $\partial\Omega$ since $\varphi(\Omega) \cap \varphi(\partial\Omega) = \emptyset$. Using the implicit function theorem, we can find open sets U_x, U_y, U' containing x, y, x' respectively such that $U_x \cap U_y = \emptyset$ and such that $\varphi: U_x \cap \Omega \to U' \cap \varphi(\Omega)$ and $\varphi: U_y \cap \Omega \to U' \cap \varphi(\Omega)$ are both $\mathscr{C}^1$-diffeomorphisms. But this implies that the mapping φ is not injective on Ω, in contradiction with Theorem 5.6-1. Hence the set $\varphi^{-1}(x')$ contains exactly one point.

Let $x' \in \varphi(\partial\Omega) \cap \operatorname{int} B$ be such that $\operatorname{card} \varphi^{-1}(x') \geq 2$. Then there exist $x \in \partial\Omega$ and $y \in \bar{\Omega}$ such that $x \neq y$ and $x' = \varphi(x) = \varphi(y)$. If $y \in \Omega$, we infer as before that the mapping φ is not injective on Ω, again in contradiction with Theorem 5.6-1; we thus conclude that $y \in \partial\Omega$.

Next, there exist open sets V and U' containing y and x' respectively such that $\varphi: V \cap \bar{\Omega} \to U' \cap \varphi(\bar{\Omega})$ is a $\mathscr{C}^1$-diffeomorphism. It then suffices to reproduce the above arguments, the set $U' \cap \varphi(\Omega)$ playing the rôle of the set $\operatorname{ext} B$: This proves that x is the only point other than y in the inverse image $\varphi^{-1}(x')$, and that the tangent planes to the surface $\varphi(\partial\Omega)$ at the points $\varphi(x)$ and $\varphi(y)$ coincide and have opposite outer normal vectors. ∎

Remarks. (1) While Theorem 5.6-1 holds under the assumption that $\det \nabla\varphi$ is >0 in Ω, the stronger assumption that $\det \nabla\varphi$ be >0 in $\bar{\Omega}$ is needed in Theorem 5.6-2 in order to insure that the extension of φ over the open set denoted A in the proof still satisfies $\det \nabla\varphi > 0$ on A.

(2) The regularity assumptions in Theorem 5.6-2 are essential. In particular, it is easy to construct counter-examples to the two implications stated in this theorem with less smooth mapping φ and boundaries $\partial\Omega$ and ∂B. ∎

We now give a formal interpretation of a minimizer of the total energy over a set of admissible solutions $\psi: \bar{\Omega} \to \mathbb{R}^3$ that, in addition to being orientation-preserving, also satisfy the injectivity condition $\int_\Omega \det \nabla\psi(x)\, dx \leq \operatorname{vol} \psi(\Omega)$. For simplicity, we assume that there are no applied surface forces.

Theorem 5.6-3 (displacement–traction problem with an injectivity condition). *Let Ω be a domain in $\mathbb{R}^3$ with a sufficiently smooth boundary Γ, let Γ_0 and Γ_1 be disjoint relatively open subsets of Γ such that* $\operatorname{area}\{\Gamma - (\Gamma_0 \cup \Gamma_1)\} = 0$, *and let B be a closed subset of $\mathbb{R}^3$ with a sufficiently smooth boundary ∂B. Let the set of admissible solutions be of the form*

$$\boxed{\begin{aligned}\Phi = \{\psi: \bar{\Omega} \to \mathbb{R}^3;\ \det \nabla\psi > 0 \text{ in } \bar{\Omega},\ \psi = \varphi_0 \text{ on } \Gamma_0,\\ \psi(\bar{\Omega}) \subset B,\ \int_\Omega \det \nabla\psi\, dx \leq \operatorname{vol} \psi(\Omega)\},\end{aligned}}$$

assume that $\varphi_0(\Gamma_0) \subset \partial B$, and let the total energy be defined by

$$I(\psi) = \int_\Omega \hat{W}(\nabla\psi)\,\mathrm{d}x - \int_\Omega f\cdot\psi\,\mathrm{d}x\,.$$

A smooth enough solution φ of the minimization problem:

$$\varphi \in \Phi \text{ and } I(\varphi) = \inf_{\psi\in\Phi} I(\psi)\,,$$

is injective on Ω, and it solves the following boundary value problem:

$$\begin{aligned}
&-\mathbf{div}\,\hat{T}(\nabla\varphi) = f \text{ in } \Omega, \quad \text{with } \hat{T}(F) = \frac{\partial \hat{W}}{\partial F}(F) \quad \text{for all } F \in \mathbb{M}^3_+\,,\\
&\varphi = \varphi_0 \text{ on } \Gamma_0\,,\\
&\varphi(\bar{\Omega}) \subset B\,,\\
&\hat{T}(\nabla\varphi(x))n(x) = \lambda(x)n^\varphi(x^\varphi) \quad \text{with } \lambda(x) \leq 0,\ x \in \Gamma_1\,,
\end{aligned}$$

the latter case corresponding to one of the following three mutually exclusive cases (Fig. 5.6-2):

$$\begin{aligned}
&\text{case } (a)\text{: } \varphi(x) \in \operatorname{int} B \text{ and } \varphi^{-1}(\varphi(x)) = \{x\}; \text{ then } \lambda(x) = 0\,;\\
&\text{case } (b)\text{: } \varphi(x) \in \partial B; \text{ then } \varphi^{-1}(\varphi(x)) = \{x\}\,;\\
&\text{case } (c)\text{: } \varphi(x) \in \operatorname{int} B \text{ and } \varphi^{-1}(\varphi(x)) = \{x, y\} \text{ with } y \in \Gamma_1;\\
&\qquad \text{then } n^\varphi(x) + n^\varphi(y) = \mathbf{0} \text{ and } \lambda(x)\,\mathrm{d}a(x) = \lambda(y)\,\mathrm{d}a(y)\,,
\end{aligned}$$

where $n(z)$ and $n^\varphi(z^\varphi)$ denote in general the unit outer normal vectors along the surfaces Γ and $\varphi(\Gamma)$ at the points z and $z^\varphi = \varphi(z)$ respectively, and $\mathrm{d}a(z)$ *denotes the area element along Γ at the point z.*

Proof. The proof consists in showing that *adequate "variations"* $\varphi^\varepsilon = \varphi + \varepsilon\theta$ *around a minimizer φ of the energy can be constructed in such a way that they satisfy not only the conditions* $\det \nabla\varphi^\varepsilon > 0$ *in* $\bar{\Omega}$, $\varphi^\varepsilon = \varphi_0$ *on* Γ_0, $\varphi^\varepsilon(\bar{\Omega}) \subset B$, *but also the injectivity condition* $\int_\Omega \det \nabla\varphi^\varepsilon\,\mathrm{d}x \leq \operatorname{vol}\varphi^\varepsilon(\Omega)$, *or equivalently, that they be injective over the set Ω* (by Theorem 5.6-1; the converse is clear).

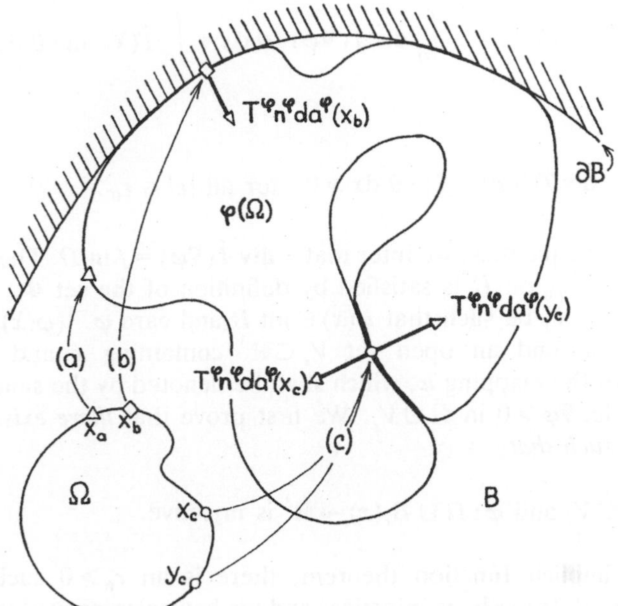

Fig. 5.6-2. The different types of boundary conditions along Γ_1 described in Theorem 5.6-3. The surface force $T^{\varphi}n^{\varphi}\,da^{\varphi}$ vanishes in case (a), and is normal to, and directed away from, the obstacle in case (b) (contact without friction with the obstacle); in case (c), the surface forces $T^{\varphi}(x_c)n^{\varphi}(x_c)\,da^{\varphi}(x_c)$ and $T^{\varphi}(y_c)n^{\varphi}(y_c)\,da^{\varphi}(y_c)$ are normal to the surface $\varphi(\Gamma_1)$ and directed inward, and take opposite values (self-contact without friction).

(i) Let φ be a smooth enough solution of the minimization problem, let x be a point in Ω, and let $B_r(x)$ be an open ball centered in x such that $\overline{B_r(x)} \subset \Omega$. Given any function $\theta \in \mathcal{D}(B_r(x))$ and any $\varepsilon \in \mathbb{R}$, let $\varphi^{\varepsilon} = \varphi + \varepsilon\theta$. Clearly, $\varphi^{\varepsilon} = \varphi^0$ on Γ_0, and there is an $\varepsilon_0 = \varepsilon_0(\theta) > 0$ such that $\det \nabla\varphi^{\varepsilon} > 0$ in $\bar{\Omega}$ and $\varphi^{\varepsilon}(\bar{\Omega}) \subset B$ if $|\varepsilon| \leq \varepsilon_0$.

Applying Theorem 5.5-2 on the open set U, we find that φ^{ε} is injective on $\overline{B_r(x)}$ and that $\varphi^{\varepsilon}(B_r(x)) = \varphi(B_r(x))$ for $|\varepsilon| \leq \varepsilon_0$ (by Theorem 5.6-1, the mapping φ is injective on the set Ω, which contains the set $\overline{B_r(x)}$ by assumption). The injectivity on Ω follows by observing that φ^{ε} is injective on $\Omega - B_r(x)$ (since $\varphi = \varphi^{\varepsilon}$ on $\Omega - B_r(x)$), and that

$$\varphi^{\varepsilon}(\Omega - B_r(x)) \cap \varphi^{\varepsilon}(B_r(x)) = \varphi(\Omega - B_r(x)) \cap \varphi(B_r(x)) = \emptyset .$$

Stating that $I(\varphi^{\varepsilon}) \geq I(\varphi)$, and using Green's formula

$$\int_\Omega \hat{\boldsymbol{T}}(\nabla\boldsymbol{\varphi}) : \nabla\boldsymbol{\theta}\, dx = -\int_\Omega \mathbf{div}\, \hat{\boldsymbol{T}}(\nabla\boldsymbol{\varphi}) \cdot \boldsymbol{\theta}\, dx + \int_\Gamma \hat{\boldsymbol{T}}(\nabla\boldsymbol{\varphi})\boldsymbol{n} \cdot \boldsymbol{\theta}\, da\,,$$

we find that

$$\varepsilon \int_\Omega \{-\mathbf{div}\, \hat{\boldsymbol{T}}(\nabla\boldsymbol{\varphi}) - \boldsymbol{f}\} \cdot \boldsymbol{\theta}\, dx \geq 0 \quad \text{for all } |\varepsilon| \leq \varepsilon_0\,.$$

From these inequalities, we infer that $-\mathbf{div}\, \hat{\boldsymbol{T}}(\nabla\boldsymbol{\varphi}) = \boldsymbol{f}$ in Ω. The boundary condition $\boldsymbol{\varphi} = \boldsymbol{\varphi}_0$ on Γ_0 is satisfied by definition of the set $\boldsymbol{\Phi}$.

(ii) Let $x \in \Gamma_1$ be such that $\boldsymbol{\varphi}(x) \in \operatorname{int} B$ and $\operatorname{card} \boldsymbol{\varphi}^{-1}(\boldsymbol{\varphi}(x)) = 1$ (case (a)). We can find an open set $V_x \subset \mathbb{R}^3$ containing x and a smooth extension of the mapping $\boldsymbol{\varphi}$, which shall be denoted by the same letter $\boldsymbol{\varphi}$, such that $\det \nabla\boldsymbol{\varphi} > 0$ in $\bar{\Omega} \cup V_x$. We first prove that *there exists an open ball $B_r(x)$ such that*

$$B_r(x) \subset V_x \text{ and } \boldsymbol{\varphi} : \Omega \cup B_r(x) \to \mathbb{R}^3 \text{ is injective.}$$

By the implicit function theorem, there is an $r_0 > 0$ such that the mapping $\boldsymbol{\varphi} : B_{r_0}(x) \to \mathbb{R}^3$ is injective, and we have also proved in Theorem 5.6-1 that the mapping $\boldsymbol{\varphi}$ is injective on Ω. If our assertion is false, there are sequences (x^k) and (y^k) such that:

$$x^k \in B_{1/k}(x),\ y^k \in \Omega - B_{1/k}(x),\ \boldsymbol{\varphi}(x^k) = \boldsymbol{\varphi}(y^k) \quad \text{for all } k \geq 1\,.$$

Since $\bar{\Omega}$ is bounded, there is a subsequence (y^l) that converges to a point $y \in \bar{\Omega}$. Since $\lim_{l\to\infty} x^l = x$, we have

$$\boldsymbol{\varphi}(x) = \lim_{l\to\infty} \boldsymbol{\varphi}(x^l) = \lim_{l\to\infty} \boldsymbol{\varphi}(y^l) = \boldsymbol{\varphi}(y)\,,$$

and thus the assumption that $\operatorname{card} \boldsymbol{\varphi}^{-1}(\boldsymbol{\varphi}(x)) = 1$ implies $x = y$. Hence the points x^l and y^l belong to the ball $B_{r_0}(x)$ for l large enough, which contradicts the injectivity of $\boldsymbol{\varphi}$ on $B_{r_0}(x)$, and our assertion is proved.

After further reducing the radius r if necessary, we can assume that $B_r(x) \cap \Gamma_0 = \emptyset$, that $\boldsymbol{\varphi}(\bar{\Omega} \cup \overline{B_r(x)}) \subset B$ (we assume here that $\boldsymbol{\varphi}(x) \in \operatorname{int} B$), and finally that $\boldsymbol{\varphi}$ is injective on $\Omega \cup \overline{B_r(x)}$ (this last property is needed in the ensuing argument).

Given any function $\boldsymbol{\theta} \in \mathfrak{D}(B_r(x))$ and any $\varepsilon \in \mathbb{R}$, let $\boldsymbol{\varphi}^\varepsilon = \boldsymbol{\varphi} + \varepsilon\boldsymbol{\theta}$. Clearly $\boldsymbol{\varphi}^\varepsilon = \boldsymbol{\varphi}^0$ on Γ_0, and there is an $\varepsilon_0 = \varepsilon_0(\boldsymbol{\theta}) > 0$ such that $\det \nabla\boldsymbol{\varphi}^\varepsilon > 0$ in $\bar{\Omega}$ and $\boldsymbol{\varphi}^\varepsilon(\bar{\Omega}) \subset B$ if $|\varepsilon| \leq \varepsilon_0$. Then the injectivity of the mapping $\boldsymbol{\varphi}^\varepsilon$

on the set $\Omega \cup B_r(x)$, hence *a fortiori* on the set Ω, follows from the injectivity of the mapping $\boldsymbol{\varphi}$ on $\Omega \cup B_r(x)$ by the same argument as in part (i).

Stating again that $I(\boldsymbol{\varphi}^\varepsilon) \geq I(\boldsymbol{\varphi})$ for such functions $\boldsymbol{\varphi}^\varepsilon$ and using Green's formula, we find that

$$\varepsilon\left\{\int_{\Gamma_1} \hat{\boldsymbol{T}}(\nabla\boldsymbol{\varphi})\boldsymbol{n} \cdot \boldsymbol{\theta}\, da + O(\varepsilon)\right\} \geq 0$$

for all $|\varepsilon| \leq \varepsilon_0$, from which we deduce that the boundary condition $\hat{\boldsymbol{T}}(\nabla\boldsymbol{\varphi}(x))\boldsymbol{n}(x) = \boldsymbol{o}$ holds in case (a).

(iii) Let $x \in \Gamma_1$ be such that $\boldsymbol{\varphi}(x) \in \partial B$, so that card $\boldsymbol{\varphi}^{-1}(\boldsymbol{\varphi}(x)) = 1$ by Theorem 5.6-2 (case (b)). Arguing as in part (ii), we can find a ball $B_r(x)$ and an extension $\boldsymbol{\varphi}$ such that $\boldsymbol{\varphi} : \Omega \cup B_r(x) \to \mathbb{R}^3$ is injective.

Let $(\boldsymbol{t}'_1, \boldsymbol{t}'_2, \boldsymbol{t}'_3)$ be a smooth field of linearly independent vectors defined in a neighborhood V' of the point $\boldsymbol{\varphi}(x)$ containing the open set $\boldsymbol{\varphi}(B_r(x))$ (this may require that r be decreased), with the following properties:

$$\begin{cases} |\boldsymbol{t}'_1(z)| = |\boldsymbol{t}'_2(z)| = |\boldsymbol{t}'_3(z)| = 1 \quad \text{for all } z \in V', \\ \boldsymbol{t}'_1(\boldsymbol{\varphi}(y)) \text{ and } \boldsymbol{t}'_2(\boldsymbol{\varphi}(y)) \text{ span the tangent plane at the surface } \partial\boldsymbol{\varphi}(\Omega) \\ \quad \text{at all points } \boldsymbol{\varphi}(y) \in \partial\boldsymbol{\varphi}(\Omega) \cap V', \\ \boldsymbol{t}'_3(\boldsymbol{\varphi}(y)) \text{ is the unit outer normal vector } \boldsymbol{n}'(\boldsymbol{\varphi}(y)) \text{ to the surface} \\ \quad \partial\boldsymbol{\varphi}(\Omega) \text{ at all points } \boldsymbol{\varphi}(y) \in \partial\boldsymbol{\varphi}(\Omega) \cap V'. \end{cases}$$

Given any functions $\theta_1, \theta_2 \in \mathcal{D}(B_r(x))$, there exists $\varepsilon_0 = \varepsilon_0(\theta_1, \theta_2) > 0$ and there are functions $\lambda_1^\varepsilon, \lambda_2^\varepsilon : \bar{\Omega} \to \mathbb{R}$ such that

$$\operatorname{supp} \lambda_\alpha^\varepsilon \subset B_r(x), \quad \sup_{y \in B_r(x)} |\lambda_\alpha^\varepsilon(y)| = O(\varepsilon) \quad \text{for all } |\varepsilon| \leq \varepsilon_0,\ \alpha = 1, 2,$$

and such that the functions

$$\boldsymbol{\varphi}_\alpha^\varepsilon = \boldsymbol{\varphi} + \varepsilon(\theta_\alpha \boldsymbol{t}'_\alpha \circ \boldsymbol{\varphi} + \lambda_\alpha^\varepsilon \boldsymbol{t}'_3 \circ \boldsymbol{\varphi}),\ \alpha = 1, 2,$$

verify

$$\boldsymbol{\varphi}_\alpha^\varepsilon(\bar{\Omega}) \subset B \text{ and } \det \nabla\boldsymbol{\varphi}_\alpha^\varepsilon > 0 \text{ in } \bar{\Omega} \quad \text{for all } |\varepsilon| \leq \varepsilon_0,\ \alpha = 1, 2.$$

Hence the same argument as in part (ii) shows that the functions $\boldsymbol{\varphi}_\alpha^\varepsilon$ are injective on the set Ω. Another application of the Green formula then shows that

$$\varepsilon\left\{\int_{\Gamma_1} \theta_\alpha \{\hat{T}(\nabla\varphi)\boldsymbol{n}\cdot(\boldsymbol{t}'_\alpha\circ\varphi)\}\,\mathrm{d}a + O(\varepsilon)\right\} \geq 0 \quad \text{for all } |\varepsilon| \leq \varepsilon_0\,,$$

and thus we must have $\hat{T}(\nabla\varphi)\boldsymbol{n}\cdot\boldsymbol{t}'_\alpha = 0$ for $\alpha = 1, 2$, i.e., *the vector* $\hat{T}(\nabla\varphi)\boldsymbol{n}$ *must be parallel to the unit outer normal vector* $\boldsymbol{n}^\varphi$ *along* Γ_1 *at the point* x.

We can similarly prove that, given any function $\theta \in \mathscr{D}(B_r(x))$ that is ≥ 0, there is an $\varepsilon_0 = \varepsilon_0(\theta) > 0$ such that

$$\varphi^\varepsilon = \varphi + \varepsilon\theta\boldsymbol{t}'_3\circ\varphi \in \Phi \quad \text{for } -\varepsilon_0 \leq \varepsilon \leq 0\,.$$

Observing that $\boldsymbol{t}'_3 = \boldsymbol{n}^\varphi$ along Γ_1 and using Green's formula, we obtain the inequality

$$\varepsilon\left\{\int_{\Gamma_1} \theta\{\hat{T}(\nabla\varphi)\boldsymbol{n}\cdot(\boldsymbol{n}^\varphi\circ\varphi)\}\,\mathrm{d}a + O(\varepsilon)\right\} \geq O \quad \text{for } -\varepsilon_0 \leq \varepsilon \leq 0\,,$$

which shows that *the vector* $\hat{T}(\nabla\varphi)\boldsymbol{n}$ *must be of the form* $\lambda\boldsymbol{n}^\varphi$ *with* $\lambda \leq 0$ (we already know that $\hat{T}(\nabla\varphi)\boldsymbol{n}$ is parallel to $\boldsymbol{n}^\varphi$). Hence we have proved that the boundary condition

$$\hat{T}(\nabla\varphi(x))\boldsymbol{n}(x) = \lambda(x)\boldsymbol{n}^\varphi(x^\varphi) \quad \text{with } \lambda(x) \leq 0,\ x \in \Gamma_1\,,$$

holds in case (b).

(iv) Let $x \in \Gamma_1$ be such that $\varphi^{-1}(\varphi(x)) = \{x, y\}$ with $y \in \Gamma_1$ and $y \neq x$ (case (c)). By Theorem 5.6-2, the tangent planes to the surface $\varphi(\partial\Omega)$ at the point $x' := \varphi(x) = \varphi(y)$ coincide, and their unit outer normal vectors have opposite directions. By the implicit function theorem, there are disjoint open balls $B_{2r}(x)$ and $B_r(y)$ such that the restrictions of φ to both sets $B_{2r}(x) \cap \bar{\Omega}$ and $B_r(y) \cap \bar{\Omega}$ are $\mathscr{C}^1$-diffeomorphisms. Moreover, we have

$$\varphi(B_{2r}(x) \cap \Omega) \cap \varphi(B_r(y) \cap \Omega) = \emptyset\,,$$

since φ is injective on Ω by Theorem 5.6-1 Then we argue as in part (iii): We first consider a smooth field of linearly independent vectors $(\boldsymbol{t}'_1, \boldsymbol{t}'_2, \boldsymbol{t}'_3)$ in a neighborhood V' of x' with the same properties as in (iii), the set $\varphi(\Gamma_1 \cap B_{2r}(x))$ playing the rôle of the set $\partial\varphi(\Omega) \cap V'$; then we consider variations of the form

$$\varphi^\varepsilon_\alpha = \varphi + \varepsilon(\theta_\alpha\boldsymbol{t}'_\alpha\circ\varphi + \lambda^\varepsilon_\alpha\boldsymbol{t}'_3\circ\varphi),\ \alpha = 1, 2\,,$$

with functions $\theta_1, \theta_2 \in \mathcal{D}(B_r(x))$. This means that the set $\varphi(B_r(y) \cap \bar{\Omega})$ is kept fixed, while such variations with supports in $\dot{B}_r(x) \cap \bar{\Omega}$ are allowed. These variations are injective on the set $\overline{B_{2r}(x) \cap \Omega}$ if $|\varepsilon|$ is small enough.

Moreover, we can find functions $\lambda_1^\varepsilon, \lambda_2^\varepsilon : \bar{\Omega} \to \mathbb{R}$ such that $\sup_{z \in B_r(x)} |\lambda_\alpha^\varepsilon(z)| = O(\varepsilon)$ and

$$\varphi_\alpha^\varepsilon(B_r(x) \cap \Omega) \cap \varphi_\alpha^\varepsilon(B_r(y) \cap \Omega) = \emptyset$$

(this is analogous to proving that $\varphi_\alpha^\varepsilon(\bar{\Omega}) \subset B$ in part (iii)). Hence these variations $\varphi_\alpha^\varepsilon$ are injective on Ω. Considering next variations of the form

$$\varphi^\varepsilon = \varphi + \varepsilon\theta t_3' \circ \varphi \quad \text{for } -\varepsilon_0 \leq \varepsilon \leq 0\,,$$

and arguing as in step (iii), we conclude that the boundary condition

$$\hat{T}(\nabla\varphi(x))n(x) = \lambda(x) n^\varphi(x^\varphi) \quad \text{with } \lambda(x) \leq 0\,,$$

also holds in case (c).

Finally, let $r' > 0$ be so chosen that

$$\varphi^{-1}(B_{r'}(x') \cap \varphi(\bar{\Omega})) \subset (B_r(x) \cap \bar{\Omega}) \cup (B_r(y) \cap \bar{\Omega})\,,$$

let a vector field (t_3') be defined as in part (iii) in a neighborhood of x' that contains the ball $B_{r'}(x')$, with $\varphi(\Gamma_1 \cap B_r(x))$ playing the rôle of the set $\partial\varphi(\Omega) \cap V'$, and let $\eta \in \mathcal{D}(B_{r'}(x'))$ be given. Then for $|\varepsilon|$ small enough, the variations φ^ε defined by

$$\varphi^\varepsilon(z) = \varphi(z) + \varepsilon\eta(\varphi(z)) t_3'(\varphi(z)) \quad \text{for } z \in \bar{\Omega}\,,$$

are injective (this is seen by invoking the same types of arguments as before). Stating that $I(\varphi^\varepsilon) \geq I(\varphi)$, we obtain

$$\varepsilon\Big(\int_{\Gamma_1 \cap B_r(x)} (\eta \circ \varphi)\{\hat{T}(\nabla\varphi)n \cdot (t_3' \circ \varphi)\}\,da +$$

$$\int_{\Gamma_1 \cap B_r(y)} (\eta \circ \varphi)\{\hat{T}(\nabla\varphi)n \cdot (t_3' \circ \varphi)\}\,da + O(\varepsilon)\Big) \geq 0$$

for $|\varepsilon|$ small enough. Using the relations

$$t_3'(\varphi(z)) = n^\varphi(z^\varphi) \quad \text{for } z \in \Gamma_1 \cap B_r(x)\,,$$

$$\boldsymbol{n}^{\varphi}(u^{\varphi}) + \boldsymbol{n}^{\varphi}(v^{\varphi}) = \boldsymbol{o} \text{ if } u \in \Gamma_1 \cap B_r(x),\ v \in \Gamma_1 \cap B_r(y)\,,$$
$$\text{and } \boldsymbol{\varphi}(u) = \boldsymbol{\varphi}(v)\,,$$
$$\hat{\boldsymbol{T}}(\nabla\boldsymbol{\varphi}(u)) = \lambda(u)\boldsymbol{n}^{\varphi}(u^{\varphi}) \quad \text{for } u \in \Gamma_1\,,$$
$$\lambda(u) = 0 \text{ if } u \in \Gamma_1 \cap B_r(x), \text{ and } \boldsymbol{\varphi}^{-1}(\boldsymbol{\varphi}(u)) = \{u\}\,,$$

we obtain

$$\int_{\Gamma_1 \cap B_r(x)} (\eta \circ \boldsymbol{\varphi})\lambda \, \mathrm{d}a - \int_{\Gamma_1 \cap B_r(y)} (\eta \circ \boldsymbol{\varphi})\lambda \, \mathrm{d}a = 0\,,$$

and since $(\eta \circ \boldsymbol{\varphi})(x) = (\eta \circ \boldsymbol{\varphi})(y)$, we conclude that $\lambda(x)\,\mathrm{d}a(x) = \lambda(y)\,\mathrm{d}a(y)$, and thus all relations arising in case (c) are proved. ■

Remark. If we had not assumed that $\boldsymbol{\varphi}_0(\Gamma_0)$ were part of the boundary ∂B, it would remain to consider the case where $x \in \Gamma_1$ is such that $\boldsymbol{\varphi}^{-1}(\boldsymbol{\varphi}(x)) = \{x, y\}$ with $y \in \Gamma_0$. This situation could then be treated as case (c). ■

To interpret the boundary condition found along Γ_1, we recall that the first Piola–Kirchhoff stress tensor $\hat{\boldsymbol{T}}(\nabla\boldsymbol{\varphi}(x)) = (\partial \hat{W}/\partial \boldsymbol{F})(\nabla\boldsymbol{\varphi}(x))$ at a point x of the boundary Γ of the reference configuration $\bar{\Omega}$ and the Cauchy stress tensor $\boldsymbol{T}^{\varphi}(x^{\varphi})$ at the corresponding point $x^{\varphi} = \boldsymbol{\varphi}(x)$ of the boundary of the deformed configuration $\boldsymbol{\varphi}(\bar{\Omega})$ are related by

$$\hat{\boldsymbol{T}}(\nabla\boldsymbol{\varphi}(x))\boldsymbol{n}(x)\,\mathrm{d}a(x) = \boldsymbol{T}^{\varphi}(x^{\varphi})\boldsymbol{n}^{\varphi}(x^{\varphi})\,\mathrm{d}a^{\varphi}(x^{\varphi})\,,$$

where $\boldsymbol{n}(x)$, $\boldsymbol{n}^{\varphi}(x^{\varphi})$, $\mathrm{d}a(x)$ are defined as in Theorem 5.6-3, and $\mathrm{d}a^{\varphi}(x^{\varphi})$ denotes the area element along $\partial\boldsymbol{\varphi}(\bar{\Omega})$ at the point $\boldsymbol{\varphi}(x)$ (Theorem 1.7-1). Hence the boundary condition along Γ_1 can be also written as

$$\boldsymbol{T}^{\varphi}(x^{\varphi})\boldsymbol{n}^{\varphi}(x^{\varphi}) = \lambda^{\varphi}(x^{\varphi})\boldsymbol{n}^{\varphi}(x^{\varphi}) \text{ with } \lambda^{\varphi}(x^{\varphi}) \leq 0 \quad \text{for } x \in \Gamma_1\,.$$

Since the Cauchy stress vector $\boldsymbol{T}^{\varphi}\boldsymbol{n}^{\varphi}$ measures the density of the surface force per unit area of the boundary of the deformed configuration, the various situations found in Theorem 5.6-3 can be interpreted as follows:

Case (a): $\boldsymbol{\varphi}(x) \in \operatorname{int} B$ and $\boldsymbol{\varphi}^{-1}(\boldsymbol{\varphi}(x)) = \{x\}$. There is neither contact with the obstacle nor self-contact, and the surface force $\boldsymbol{T}^{\varphi}(x^{\varphi})\boldsymbol{n}^{\varphi}\,\mathrm{d}a^{\varphi}$ vanishes, as in a zero boundary condition of traction.

Case (*b*): $\varphi(x) \in \partial B$. There is contact with the obstacle, and the surface force $T^\varphi(x^\varphi)n^\varphi\, da^\varphi$ is normal to, and directed away from, the obstacle: This is a model of **contact without friction with the obstacle** ∂B.

Case (*c*): *There are two distinct points* $x, y \in \Gamma_1$ *such that* $\varphi(x) = \varphi(y)$. Then

$$n^\varphi(x^\varphi) + n^\varphi(y^\varphi) = o, \; \lambda(x)\, da(x) = \lambda(y)\, da(y),$$

and thus

$$\boxed{T^\varphi(x^\varphi)n^\varphi(x^\varphi)\, da^\varphi(x^\varphi) + T^\varphi(y^\varphi)n^\varphi(y^\varphi)\, da^\varphi(y^\varphi) = o.}$$

Hence the surface forces at the points x^φ and y^φ are normal to the surface $\partial\varphi(\Gamma_1)$, they are directed inside the deformed configuration, and they take opposite values: This is a model of **self-contact without friction**.

Finally, notice that *the function* $\lambda : \Gamma_1 \to]-\infty, 0]$ *found in Theorem 5.6-3, or the associated function* $\lambda^\varphi : \varphi(\Gamma_1) \to]-\infty, 0]$, *can be thought of as a Kuhn & Tucker multiplier associated with the constraints* $\varphi(\bar{\Omega}) \subset B$ (cases (a) and (b)) and $\int_\Omega \det \nabla\varphi\, dx \leq \operatorname{vol} \varphi(\Omega)$ (case (c)). Observe also that the function λ^φ can be interpreted in mechanical terms as the *reaction of contact*.

Remark. The injectivity condition $\int_\Omega \det \nabla\varphi\, dx \leq \operatorname{vol} \varphi(\Omega)$ can be equivalently replaced by the condition $\int_\Omega \det \nabla\varphi\, dx = \operatorname{vol} \varphi(\Omega)$ since the opposite inequality always holds. ■

5.7. INTERNAL AND EXTERNAL GEOMETRICAL CONSTRAINTS ON THE ADMISSIBLE DEFORMATIONS

In addition to being orientation-preserving and injective on Ω, a deformation $\varphi : \bar{\Omega} \to \mathbb{R}^3$ in an actual problem may be subjected to further specific *geometrical conditions*, depending on the particular material constituting the body under consideration, or on the action of the "outside world".

In particular, φ may be subjected to an **internal constraint**, which takes into account a *geometrical property of the material, valid for all bodies made of the same material*. For instance, an *incompressible body* is made of an **incompressible material**, i.e., a material that cannot undergo any change of volume. Hence in order to be physically acceptable, the deformations of an incompressible body must satisfy the **incompressibility condition**

$$\boxed{\det \nabla\varphi = 1 \text{ in } \bar{\Omega}\,.}$$

Notice in passing that the deformations associated with an incompressible body are automatically orientation-preserving (the associated boundary value problem is given in Exercise 5.9). By contrast, a body, or a material, for which the only restriction on the numbers $\det \nabla\varphi(x)$, $x \in \Omega$, is that they be >0 (preservation of the orientation) is called **compressible**, although this qualification is usually omitted in elasticity. A good introduction to incompressible materials and their response functions is given in Le Dret [1985]. See also Cohen & Wang [1987].

Hence the *orientation-preserving condition* $\det \nabla\varphi > 0$ in $\bar{\Omega}$, the *incompressibility condition* $\det \nabla\varphi = 1$ in $\bar{\Omega}$, and the *injectivity condition* $\int_\Omega \det \nabla\varphi \, dx \leqslant \operatorname{vol} \varphi(\Omega)$ (Sect. 5.6), are examples of internal constraints.

Another instance of internal constraint, introduced by Prager [1957, 1958, 1964] and studied by Duvaut & Lions [1972, p. 269], Demengel [1985a], Demengel & Suquet [1986] in linearized elasticity, is given by a material that may become "locked" when some "measure of the strain" reaches a certain level. More specifically, such a material is subjected to a **locking constraint** of the form

$$\boxed{\hat{L}(\nabla\varphi) \leqslant 0 \text{ in } \bar{\Omega}\,,}$$

where the *locking function* $\hat{L} : \mathbb{M}^3_+ \to \mathbb{R}$ is given (notice that the inequality is no longer strict as in the orientation-preserving condition). In the nonlinearly elastic case, Ciarlet & Nečas [1985] have proposed the locking function:

$$\hat{L} : F \in \mathbb{M}^3_+ \to \hat{L}(F) = E^{\mathrm{d}} : E^{\mathrm{d}} - \alpha, \ \alpha > 0\,,$$

where $E^{\mathrm{d}} = E - \frac{1}{3}(\operatorname{tr} E)I$ denotes the *deviatoric part* of the Green–St Venant strain tensor $E = \frac{1}{2}(\nabla\varphi^{\mathrm{T}}\nabla\varphi - I)$. This example is discussed in Exercise 5.10.

An "extreme" example of internal constraint is given by

$$\nabla\varphi(x) = Q \in \mathbb{O}^3_+ \quad \text{for all } x \in \bar{\Omega}\,.$$

It corresponds to *rigid bodies*, i.e., bodies that can only undergo rigid deformations (Sect. 1.8). Understandably, the mechanics of rigid bodies (Exercise 5.11) is far simpler than that of elastic bodies!

By contrast with an internal constraint, an **external constraint** *is a geometric limitation on the positions that a body can occupy in a given problem, irrespectively of the material constituting the body*. For instance, a *boundary condition of place* (Sects. 2.6 and 4.1) consists in specifying the deformation on a subset Γ_0 of the boundary Γ of the reference configuration: A mapping $\varphi_0 : \Gamma_0 \to \mathbb{R}^3$ is given, and the deformation is required to satisfy the boundary condition

$$\varphi = \varphi_0 \text{ on } \Gamma_0 ,$$

i.e., the body is "glued" on the portion $\varphi_0(\Gamma_0)$ of its boundary. Such a condition may be weakened, in that the deformation may be only "partially" prescribed along a subset of the boundary Γ, as we have seen in Sect. 5.2 about a plate problem (Fig. 5.2-2).

An external constraint can also take the form of a *unilateral boundary condition of place* $\varphi(\Gamma_2) \subset C$, $\Gamma_2 \subset \Gamma$ (Sect. 5.3), or a *confinement condition* $\varphi(\bar{\Omega}) \subset B$ (Sect. 5.6).

We are now in a position to define the **set of admissible deformations** associated with a given problem: *It consists of all the smooth enough deformations that satisfy all the internal and external constraints found in the problem*. For instance, the set of admissible deformations corresponding to the plate problem illustrated in Fig. 5.2-2 is (we keep the smoothness unspecified for the time being):

$$\begin{aligned} \boldsymbol{\Phi} = \Big\{ & \boldsymbol{\psi} : \bar{\Omega} \to \mathbb{R}^3;\ \boldsymbol{\psi} = \boldsymbol{id} \text{ on } \Gamma_0,\ \boldsymbol{\psi} = \boldsymbol{id} + \alpha \boldsymbol{e}_2 \text{ on } \Gamma_0', \\ & \alpha \in \mathbb{R},\ \psi_3 \leqslant h \text{ on } \Gamma_2,\ \det \nabla\boldsymbol{\psi} > 0 \text{ in } \bar{\Omega}, \\ & \int_\Omega \det \nabla\boldsymbol{\psi}\, dx \leqslant \operatorname{vol} \boldsymbol{\psi}(\Omega) \Big\} . \end{aligned}$$

Remark. If the material is hyperelastic and polyconvex, we shall see in Chapter 7 that the smoothness of the admissible deformations is essentially governed by the coerciveness inequality satisfied by the stored energy function. More specifically, we shall see that, if

$$\begin{aligned} & \hat{W}(\boldsymbol{F}) \geqslant \alpha\{\|\boldsymbol{F}\|^p + \|\mathbf{Cof}\, \boldsymbol{F}\|^q + (\det \boldsymbol{F})^r\} + \beta, \\ & \qquad \alpha > 0,\ p \geqslant 2,\ q \geqslant \frac{p}{p-1},\ r > 1 , \end{aligned}$$

for all $\boldsymbol{F} \in \mathbb{M}^3_+$, the admissible deformations $\boldsymbol{\psi}$ should be such that

$$\boldsymbol{\psi} \in \boldsymbol{W}^{1,p}(\Omega), \ \mathbf{Cof}\, \boldsymbol{\nabla\psi} \in \boldsymbol{L}^q(\Omega), \ \mathbf{det}\, \boldsymbol{\nabla\psi} \in L^r(\Omega)$$

(the Sobolev spaces $W^{1,p}(\Omega)$ are defined in the next chapter). ■

5.8. PHYSICAL EXAMPLES OF NONUNIQUENESS

One of the most striking features of three-dimensional elasticity is the *lack of uniqueness observed in physical situations*. Consequently, if the boundary value problems of three-dimensional elasticity are to be qualified as "good" mathematical models, they must be compatible with the existence of several possible distinct solutions, even infinitely many in some cases. The purpose of this section is to illustrate this characteristic by means of several examples borrowed from the most immediate physical evidence. We successively consider pure traction problems, displacement–traction problems, and pure displacement problems, In each case, it is assumed that the reference configuration is a natural state.

As our first example, adapted from Noll [1978], we consider a *pure traction problem* corresponding to an experiment that everyone can easily perform: A tennis ball is cut above its equator as indicated in Fig. 5.8-1(a). When pushed hard enough from below around the point b, the larger part of the cut ball snaps into an *everted state* indicated in Fig. 5.8-1(b), which clearly corresponds to a different deformation. The state of stress is also different: If a small cut is done in a neighborhood of the point $a = \varphi_1(a)$, the lips of the cut remain attached which the lips of the same cut around the point $\varphi_2(a)$ separate.

Remarks. (1) By contrast, if the same kind of experiment is performed on the upper, and smaller; part, the cut ball snaps back to its original state.

(2) Noll [1978] describes a similar experiment, where even more distinct solutions can be experimentally produced. ■

Another feasible experiment, albeit considerably more difficult to perform, consists in everting a cylindrical tube of rubber (usually modeled as a nearly incompressible hyperelastic material) as shown in Fig. 5.8-2. This example, which is neatly, and humorously, discussed in Truesdell [1978], is another example of a *pure traction problem* (with zero applied

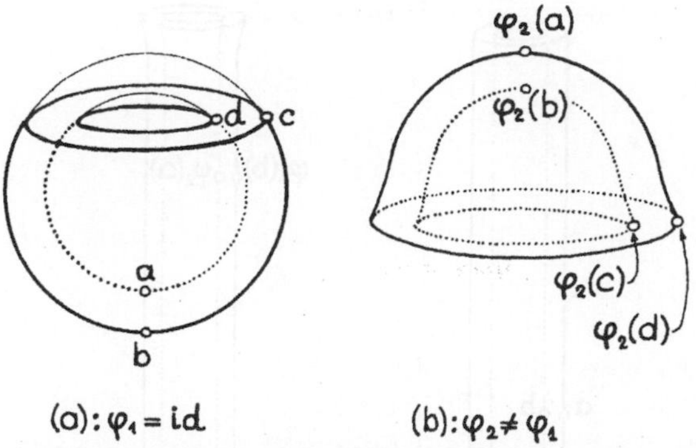

Fig. 5.8-1. Noll's experiment shows that there may exist two distinct solutions to a pure traction problem.

surface forces) that possesses at least two physically realistic distinct solutions. For more details on the eversion of cylindrical tubes and spherical shells, see Ericksen [1955b], Chadwick [1972], Chadwick & Haddon [1972], Wang & Truesdell [1973, p. 310 ff.], Antman [1979], Adeleke [1983].

While in these experiments, the evidence of nonuniqueness relies on physical experience, there are cases where it can be predicted that some combinations of boundary conditions must certainly entail nonuniqueness, without actually performing the corresponding experiment. For instance, J.L. Ericksen has observed that the *pure traction problem* modeling a rectangular block (the reference configuration; cf. Fig. 5.8-3(a)) under axial tension (Fig. 5.8-3(b)) must also account for the block under axial compression obtained by turning upside down the reference configuration (Fig. 5.8-3(c)), since both deformations φ_1 and φ_2 should be solutions of the same problem (we assume for definiteness that body forces can be neglected and that surface forces are dead loads; the notations are those of Fig. 5.8-3):

$$\begin{cases} -\mathbf{div}\, \hat{T}(\nabla\varphi) = \boldsymbol{o} \text{ in } \Omega\,, \\ \hat{T}(\nabla\varphi)\boldsymbol{e}_3 = \pi \boldsymbol{e}_3 \text{ on } \Gamma_1 \cup \Gamma_1'\,,\ \pi > 0\,, \\ \hat{T}(\nabla\varphi)\boldsymbol{e}_3 = \boldsymbol{o} \text{ on } \Gamma - (\Gamma_1 \cup \Gamma_1')\,. \end{cases}$$

As our next example, due to Gurtin [1978], consider a rubber strip with fixed ends, as indicated in Fig. 5.8-4. If we assume that the lateral surface is free from applied surface forces and that the weight can be neglected,

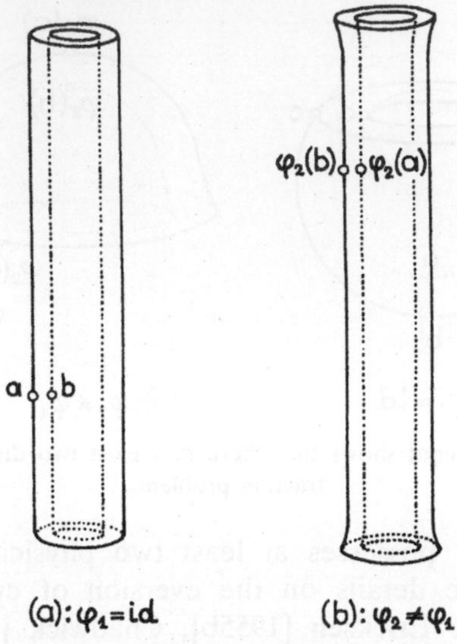

Fig. 5.8-2. Truesdell's experiment is a second instance of a pure traction problem with two distinct solutions. The everted tube is a little longer, its exterior diameter and thickness are slightly smaller, and it has flaring ends.

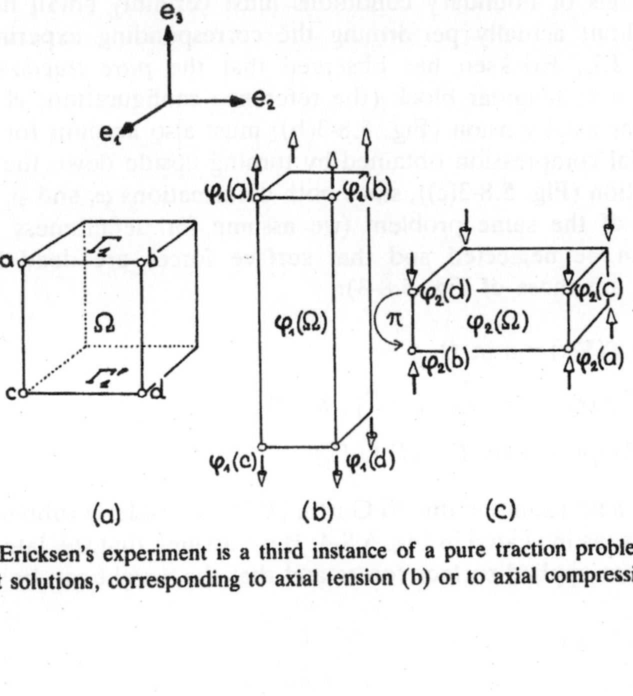

Fig. 5.8-3. Ericksen's experiment is a third instance of a pure traction problem with two distinct solutions, corresponding to axial tension (b) or to axial compression (c).

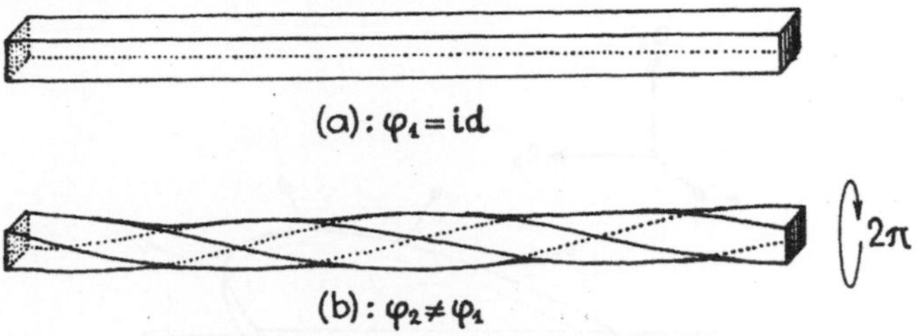

Fig. 5.8-4. Gurtin's experiment is an example of a displacement–traction problem with an infinite number of distinct solutions.

the particular deformation $\varphi_1 = \boldsymbol{id}$ should be a solution (Fig. 5.8-4(a)) of the associated *displacement–traction problem*. Imagine now that we twist the right end of the strip through an angle of 2π (or the left end through an angle of -2π) as indicated in Fig. 5.8-4(b). Then the corresponding deformation φ_2, which again satisfies the boundary condition $\varphi = \boldsymbol{id}$ at both ends, is clearly a different solution. In fact, since we might as well twist either end through an angle of $2k\pi$ for any $k \in \mathbb{Z}$, this is an example of a problem with an *infinite* number of solutions, with cardinality $\geqslant \aleph_0$ (that of the set $\mathbb{N}$).

It is intuitively clear that a way to dispose of this indeterminacy would be to consider the associated *dynamical* problem, where the deformation is known at an "initial" time, and where the possible twisting of the ends is taken care of by a boundary condition of place that varies continuously with time (but this approach seems to be out of reach at the present time). For the static problems we are interested in, this kind of ambiguity seems almost as hard to overcome. However, when this type of problem is approximated by a one-dimensional model, a successful approach has been proposed by Alexander & Antman [1982], who introduced an appropriate *measure of twist*, based on the *Gauss linking number*.

Another aspect of nonuniqueness in nonlinear elasticity is illustrated by out next example, which corresponds to an actual experiment (Fig. 5.8-5). Consider a circular steel plate of thickness 2ε whose weight is neglected, and which is subjected to the boundary conditions

$$\begin{cases} \dfrac{1}{2\varepsilon}\displaystyle\int_{-\varepsilon}^{\varepsilon} \boldsymbol{T}^{\varphi}\boldsymbol{n}^{\varphi}\,\mathrm{d}x_3 = -\lambda \boldsymbol{n}^{\varphi}\,, \\ u_\alpha \text{ independent of } x_3,\ u_3 = 0\,, \end{cases}$$

on its lateral surface Γ_0 (such boundary conditions are discussed at length

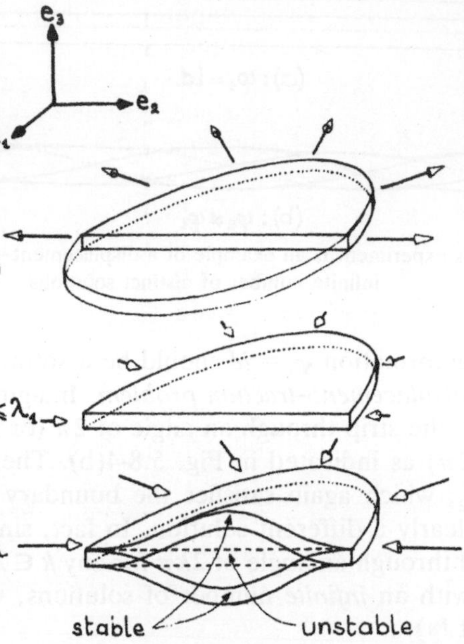

Fig. 5.8-5. The buckling of a plate (for convenience, only half of the plate is represented): For sufficiently large compressive applied forces, there exist three distinct solutions (two "stable", and one "unstable").

in Vol. II). Then one observes experimentally that $\varphi = id$ is the only solution when the lateral surface of the plate is either pulled ($\lambda < 0$) or pressed by small enough forces ($0 \leqslant \lambda \leqslant \lambda_1$). But for sufficiently large applied surface forces ($\lambda_1 < \lambda$), the solution $\varphi = id$ (which always exists) becomes unstable: Any perturbation, whatever small, will cause the plate to suddenly buckle, i.e., to occupy either one of two symmetric stable solutions (we do not attach to the adjectives "stable" or "unstable" any other meaning than their intuitive one). Thus, for λ large enough, we should expect at least *three* solutions, counting the unstable one (in fact, for larger values of λ we may even observe a larger number of distinct solutions). As we shall see in Vol. II, such nonuniqueness properties for plates, i.e., those connected with *buckling*, are satisfactorily modeled by the two-dimensional *von Kármán equations* (Berger [1967, 1977], Berger & Fife [1968], Ciarlet & Rabier [1980]).

The nonuniqueness properties observed in the two preceding problems are different in nature. While in the latter problem nonuniqueness due to buckling occurs without any "temporary" alteration of the boundary conditions, nonuniqueness due to the rotation of either end of the strip (Fig. 5.8-4) in the former problem requires some "intermediary unglueing and re-glueing" that must actually take place, even though this episode is not "seen" by the mathematical model.

Let us conclude by a counter-example to uniqueness for a *pure displacement problem*, which is due to John [1964]: Consider an elastic body that occupies in its reference configuration the volume enclosed between two concentric spheres (Fig. 5.8-6), and which satisfies the boundary condition of place $\varphi = id$ on the two bounding spheres. Hence if the applied body forces are neglected, $\varphi_1 = id$ is a solution. By rotating either bounding sphere by an angle of $2k\pi$, $k \in \mathbb{Z}$, about any axis passing through their center (and possibly by combining these operations in any sequence), we obtain an infinite number of distinct solutions, with cardinality $\geqslant \aleph_1$ (that of the set $\mathbb{R}$).

In addition to the references already quoted, there is an abundant of literature on the uniqueness or nonuniqueness of solutions of the boundary value problems of nonlinear elasticity, and on the related problems of finding *a priori* bounds on, and of continuously following, the solutions of the associated minimization problems. For uniqueness or nonuniqueness, see Ericksen & Toupin [1956], Hill [1957, 1961], Truesdell & Toupin [1963], Aron [1978], Gurtin [1978, 1982], Gurtin & Spector [1979], Spector [1980, 1982], Capriz & Podio-Guidugli [1981], Knops & Stuart [1984], Stuart [1986]; for various *a priori* bounds, see Villagio [1972], Aron & Roseman [1977], Breuer & Roseman [1978, 1979, 1980], Roseman [1981] and, especially, the cstimates of John [1961, 1972a,

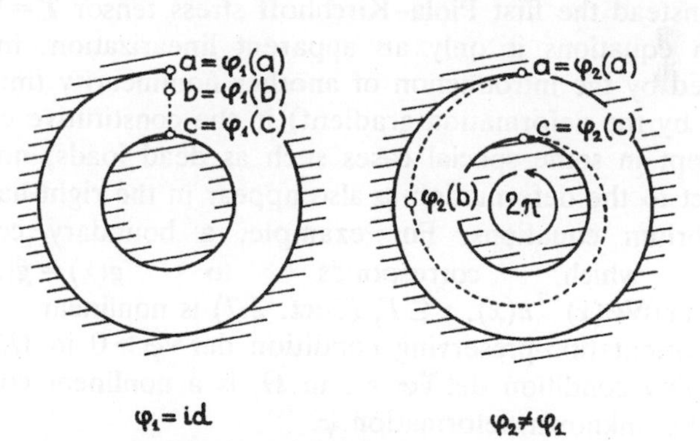

Fig. 5.8-6. John's ideal experiment is an example of a pure displacement problem with an infinite number of distinct solutions.

1972b, 1975] and Kohn [1982]; for the continuous dependence of the minimizers in terms of a parameter, see Aron [1979], Breuer & Aron [1982].

In conclusion, notice that, had we not already formulated the boundary value problems of three-dimensional elasticity, we could predict in view of the above examples that *these problems must certainly be nonlinear*, since uniqueness of the solution (possibly in quotient spaces) is a characteristic of *linear* problems.

5.9. THE NONLINEARITIES IN THREE-DIMENSIONAL ELASTICITY; THE ELASTICITY TENSOR

The assertion that the boundary value problem of three-dimensional elasticity is *nonlinear* is the archetype of an understatement, since nonlinearities occur in many different ways in its formulation:

(i) The right Cauchy–Green strain tensor $\boldsymbol{C} = \boldsymbol{\nabla}\boldsymbol{\varphi}^{\mathrm{T}}\boldsymbol{\nabla}\boldsymbol{\varphi}$ is a nonlinear (quadratic) function of the unknown deformation $\boldsymbol{\varphi}$; equivalently, the Green–St Venant strain tensor $\boldsymbol{E} = \frac{1}{2}(\boldsymbol{\nabla u}^{\mathrm{T}} + \boldsymbol{\nabla u} + \boldsymbol{\nabla u}^{\mathrm{T}}\boldsymbol{\nabla u})$ is a nonlinear function of the unknown displacement $\boldsymbol{u}$.

(ii) The constitutive equation $\boldsymbol{\Sigma}(x) = \tilde{\boldsymbol{\Sigma}}(x, \boldsymbol{C}) = \check{\boldsymbol{\Sigma}}(x, \boldsymbol{E})$ is nonlinear with respect to $\boldsymbol{C}$, or equivalently with respect to $\boldsymbol{E} = \frac{1}{2}(\boldsymbol{C} - \boldsymbol{I})$, except in the case of a St Venant–Kirchhoff material.

(iii) The left-hand sides of the equilibrium equations: $-\mathbf{div}\, \boldsymbol{\nabla\varphi}(x)\boldsymbol{\Sigma}(x)$, $x \in \Omega$, and $\boldsymbol{\nabla\varphi}(x)\boldsymbol{\Sigma}(x)\boldsymbol{n}$, $x \in \Gamma_1$, if there is a boundary condition of traction along a subset Γ_1 of Γ, are nonlinear (bilinear) functions of $\boldsymbol{\varphi}$ and $\boldsymbol{\Sigma}$. Using instead the first Piola–Kirchhoff stress tensor $\boldsymbol{T} = \boldsymbol{\nabla\varphi\Sigma}$ in the equilibrium equations is only an apparent linearization, in that it is compensated by the introduction of another nonlinearity (multiplication on the left by the deformation gradient) in the constitutive equation.

(iv) Except in some special cases such as dead loads, nonlinearities with respect to the deformation $\boldsymbol{\varphi}$ also appear in the right-hand sides of the equilibrium equations. For example, a boundary condition of pressure, which corresponds to $\boldsymbol{g}(x) = \hat{\boldsymbol{g}}(x, \boldsymbol{\nabla\varphi}(x)) = -\pi(\det \boldsymbol{\nabla\varphi}(x))\boldsymbol{\nabla\varphi}(x)^{-\mathrm{T}}\boldsymbol{n}(x)$, $x \in \Gamma_1$ (Sect. 2.7) is nonlinear.

(v) The orientation-preserving condition $\det \boldsymbol{\nabla\varphi} > 0$ in Ω, or the incompressibility condition $\det \boldsymbol{\nabla\varphi} = 1$ in Ω, is a nonlinear condition imposed on the unknown deformation $\boldsymbol{\varphi}$.

(vi) The injectivity condition $\int_\Omega \det \boldsymbol{\nabla\varphi}\, \mathrm{d}x \leq \operatorname{vol} \boldsymbol{\varphi}(\Omega)$ is nonlinear with respect to $\boldsymbol{\varphi}$.

(vii) Other internal constraints such as locking constraints, or external constraints such as a unilateral boundary condition of place (Sect. 5.7), are likewise expressed as nonlinear conditions on the unknown deformation.

In components, the equilibrium equation in Ω, viz.,

$$-\mathbf{div}\, \hat{T}(x, \nabla\varphi(x)) = \hat{f}(x, \varphi(x)), \quad x \in \Omega ,$$

takes the form of a *system of three nonlinear partial differential equations of the second order with respect to the components* φ_i *of the unknown deformation*:

$$-\frac{\partial}{\partial x_j} \hat{t}_{ij}(x, \nabla\varphi(x)) = \hat{f}_i(x, \varphi(x)) .$$

Assuming appropriate differentiability, we can also write these equations as

$$\boxed{-\frac{\partial \hat{t}_{ij}}{\partial F_{kl}}(x, \nabla\varphi(x)) \frac{\partial^2 \varphi_k}{\partial x_j \partial x_l}(x) - \frac{\partial \hat{t}_{ij}}{\partial x_j}(x, \nabla\varphi(x)) = \hat{f}_i(x, \varphi(x)) ,}$$

or, if the material is hyperelastic,

$$\boxed{\begin{aligned} -\frac{\partial^2 \hat{W}}{\partial F_{ij} \partial F_{kl}}(x, \nabla\varphi(x)) \frac{\partial^2 \varphi_k}{\partial x_j \partial x_l}(x) - \frac{\partial^2 \hat{W}}{\partial x_j \partial F_{ij}}(x, \nabla\varphi(x)) \\ = \hat{f}_i(x, \varphi(x)) .\end{aligned}}$$

Hence the equations fall short of being "completely nonlinear". It is their "divergence structure" that renders them linear with respect to the terms of highest order, i.e., the partial derivatives $\partial^2\varphi_k/\partial x_j \partial x_l$.

In the terminology of partial differential equations, such equations, whose leading terms are nonlinear but depend linearly on the derivatives of highest order, are labeled *quasilinear*, by contrast with the easier to analyze *semilinear equations*, where the higher-order terms are linear and the nonlinearities occur only in the lower-order terms. As we shall see in Vol. II, it is somewhat surprising that the equations that model the two-dimensional nonlinear plate models are semilinear, yet they are directly derived from the present quasilinear three-dimensional equations. As a consequence, the mathematical analysis of plate models is comparatively easier and also more complete. A wealth of examples of, and methods for solving, semilinear problems is given in Lions [1969].

The higher-order terms in the equations of equilibrium are governed by *a fourth-order tensor* $\mathsf{A}(x, \boldsymbol{F})$, defined for all points $x \in \Omega$ and all matrices $\boldsymbol{F} \in \mathbb{M}^3_+$ by

$$\mathsf{A}(x, \boldsymbol{F}) = (a_{ijkl}(x, \boldsymbol{F})) := \left(\frac{\partial \hat{t}_{ij}}{\partial F_{kl}}(x, \boldsymbol{F})\right) = \left(\frac{\partial^2 \hat{W}}{\partial F_{ij} \partial F_{kl}}(x, \boldsymbol{F})\right).$$

This tensor is called the **elasticity tensor** and its position as the leading term of the equations makes it amenable to various mathematical assumptions, borrowed in particular from the theory of elliptic systems, which we shall next briefly review. We shall also see in Sect. 6.2 that its particular value for $\boldsymbol{F} = \boldsymbol{I}$, which is given by

$$\mathsf{A}(\boldsymbol{I}) = (\lambda \delta_{ij} \delta_{kl} + \mu(\delta_{ik} \delta_{jl} + \delta_{il} \delta_{jk}))$$

for a homogeneous isotropic material whose reference configuration is a natural state, plays a crucial role in *linearized elasticity* (see also Exercise 5.17).

5.10. CONSTITUTIVE ASSUMPTIONS

Since the main purpose of the second part of the book is to mathematically overcome the nonlinearities inherent to our problem, especially those that appear in the left-hand side $-\mathbf{div}\, \hat{\boldsymbol{T}}(\boldsymbol{\nabla}\boldsymbol{\varphi})$ of the equilibrium equations, it is appropriate that we conclude this first part by gathering *all the information at our disposal, old and new, about the constitutive equation*, expressed indifferently in terms of the first or the second Piola–Kirchhoff stress tensor, or in terms of the stored energy function if the material is hyperelastic. Such assumptions bear the generic name of **constitutive assumptions**. They may be suggested by our perception of the "real world" (e.g., material frame-indifference, behavior of the stress for large strains, etc.), they may be thought of as purely mathematical conditions (e.g. polyconvexity, seen as a convenient way of obtaining existence results, etc.), or they may combine the two aspects (e.g., the behavior for large strains expressed as a coerciveness condition on the stored energy function, etc.).

Accordingly, we first briefly recapitulate four types of constitutive assumptions already encountered in the text, and we mention a new one. In order to keep the notation simple, we consider the case of homogeneous materials, but it should be kept in mind that each constitutive assumptions applies as well to inhomogeneous materials.

(i) *Isotropy*: When combined with material frame-indifference (which is an axiom, not an assumption), this assumption results in the following forms of the response function for the second Piola–Kirchhoff stress (Theorem 3.6-2):

$$\boxed{\hat{\Sigma}(F) = \tilde{\Sigma}(C) = \gamma_0(\iota_C)I + \gamma_1(\iota_C)C + \gamma_2(\iota_C)C^2, \quad C = F^{\mathrm{T}}F\,,}$$

and of the stored energy function (Theorem 4.4-1):

$$\boxed{\hat{W}(F) = \dot{W}(\iota_C)\,.}$$

(ii) *Behavior for small strains*: If we assume that the reference configuration is a natural state, and that the material is homogeneous and isotropic, we must have (Theorem 3.8-1)

$$\boxed{\hat{\Sigma}(F) = \check{\Sigma}(E) = \lambda(\operatorname{tr} E)I + 2\mu E + o(\|E\|), \quad F^{\mathrm{T}}F = I + 2E\,,}$$

and of the stored energy function (Theorem 4.4-1):

$$\boxed{\hat{W}(F) = \frac{\lambda}{2}(\operatorname{tr} E)^2 + \mu \operatorname{tr} E^2 + o(\|E\|^2)\,,}$$

and experimental evidence shows that the Lamé constants λ and μ are >0 (Sect. 3.8).

(iii) *Behavior for large strains*: As shown in Sect. 4.6, the property that "infinite stress must accompany extreme strains" (Antman [1983]) is expressed mathematically by the assumptions that the stored energy function satisfies

$$\boxed{\lim \hat{W}(F) = +\infty \text{ as } \det F \to 0^+, \quad F \in \mathbb{M}^3_+\,,}$$

and by a *coerciveness inequality* of the form

$$\hat{W}(F) \geq \alpha(\|F\|^p + \|\mathrm{Cof}\, F\|^q + (\det F)^r) + \beta \quad \text{for all } F \in \mathbb{M}^3_+ ,$$

for some constants $\alpha > 0, p > 0, q > 0, r > 0, \beta \in \mathbb{R}$.

(iv) *Polyconvexity of the stored energy function*: The "impossible convexity" of the stored energy function $\hat{W}$ (Theorem 4.8-1) can be advantageously replaced by the weaker requirement that it be *polyconvex*, in the sense that there exists a convex function $\mathbb{W}: \mathbb{M}^3 \times \mathbb{M}^3 \times]0, +\infty[$ such that (Sect. 4.9)

$$\hat{W}(F) = \mathbb{W}(F, \mathrm{Cof}\, F, \det F) \quad \text{for all } F \in \mathbb{M}^3_+ .$$

Polyconvexity can be in turn replaced by the even weaker requirement of *quasiconvexity* of the stored energy function. This fundamental notion in the calculus of variations is due to Morrey [1952], who showed that it is a necessary and sufficient condition for the sequential weak lower semicontinuity of the associated integral (see Exercise 7.1). The definition of quasiconvexity and its relation to polyconvexity are given in Exercise 5.14.

Polyconvexity can also be viewed as a special case of *rank-one convexity*, which is even more general than quasiconvexity; the definition of rank-one convex functions and some of their properties are given in Exercise 5.15.

(v) *Ellipticity conditions on the elasticity tensor*. A constitutive assumption directly borrowed from the theory of *elliptic systems* consists in requiring that the elasticity tensor (Sect. 5.9)

$$\mathsf{A}(F) = (a_{ijkl}(F)) = \left(\frac{\partial \hat{T}_{ij}}{\partial F_{kl}}(F)\right) = \left(\frac{\partial^2 \hat{W}}{\partial F_{ij} \partial F_{kl}}(F)\right)$$

satisfy at a particular matrix $F \in \mathbb{M}^3_+$, or for all matrices $F \in \mathbb{M}^3_+$, the following **ellipticity condition**, also called the **Legendre–Hadamard condition:**

$$a_{ijkl}(F) b_i b_k c_j c_l \geq 0 \quad \text{for all } b = (b_i) \in \mathbb{R}^3, \; c = (c_j) \in \mathbb{R}^3 .$$

This condition becomes the *strong ellipticity condition* or the *strong Legendre–Hadamard condition* if the same inequality holds *strictly* for all nonzero vectors b and c. Here are various possible justifications of such conditions:

The strong Legendre–Hadamard condition at $\boldsymbol{F}=\boldsymbol{I}$ implies that a satisfactory existence and uniqueness theory can be developed for the equations of *linearized elasticity* (Sect. 6.3), including the dynamical case (Duvaut & Lions [1972, Ch. 3, Sect. 4], Marsden & Hughes [1983, Sect. 6.3]). In linearized elasticity again, it is a necessary and sufficient condition for the existence of traveling waves with real wave speeds (cf. for instance Marsden & Hughes [1983, p. 240]) and historically, it is this property that motivated the early works of Legendre and Hadamard. For this reason, the elasticity tensor is also called the *acoustic tensor*.

In nonlinear elasticity, the strong Legendre–Hadamard condition is a sufficient condition for the existence "in the small", i.e., only for "short" times, of solutions of the pure displacement problem in the *dynamical case* (Hughes, Kato & Marsden [1977], Kato [1977], Wheeler [1977]), and in the *visco-elastic* case (Potier-Ferry [1982]). It also plays a crucial rôle in the calculus of variations in general for studying the *regularity* of solutions. See Hildebrandt [1977], Giusti [1983], Giaquinta [1983a, 1983b], and especially Nečas [1981, 1983a, 1983b] for the applications to elasticity.

Another interesting justification of the strong ellipticity condition has been given by Antman [1978], who has shown that it mathematically substantiates the experimental evidence according to which "*a body is elongated in the direction of the applied forces*". In the same vein, Owen [1986, 1987] has shown that the strong ellipticity condition, appropriately reduced to a one-dimensional condition as in Antman [1976b], plays a crucial rôle in the modeling of *necking deformations* in rods. Further discussions on this condition and its relations to elasticity are given in Antman [1978a, 1983, 1984], Ericksen [1983], Knowles & Sternberg [1975, 1977, 1978], Simpson & Spector [1983], Zee & Sternberg [1983]. See also Fosdick & MacSithigh [1986] for the *Legendre–Hadamard condition in incompressible elasticity*.

Other justifications of ellipticity conditions, notably through their relations with other constitutive assumptions such as polyconvexity, are indicated in Exercises 5.14 to 5.17.

EXERCISES

5.1. Consider the pure traction problem: $-\mathbf{div}\,\hat{\boldsymbol{T}}(\boldsymbol{\nabla}\boldsymbol{\varphi})=\boldsymbol{f}$ in Ω, $\hat{\boldsymbol{T}}(\boldsymbol{\nabla}\boldsymbol{\varphi})\boldsymbol{n}=\boldsymbol{g}$ on Γ. Show that if $\boldsymbol{\varphi}$ is a solution, then necessarily

$$\int_\Omega \boldsymbol{\varphi} \wedge \boldsymbol{f}\, \mathrm{d}x + \int_\Gamma \boldsymbol{\varphi} \wedge \boldsymbol{g}\, \mathrm{d}a = \boldsymbol{o}\,.$$

Remark. By constrast with the necessary condition $\int_\Omega \boldsymbol{f}\, \mathrm{d}x + \int_\Gamma \boldsymbol{g}\, \mathrm{d}a = \boldsymbol{o}$ that must be satisfied *a priori* by the applied forces in a pure traction problem, this condition is automatically satisfied by any solution of the problem (in this respect, see the discussion given in Gurtin [1981, p. 183]).

5.2. Let the total energy be of the form

$$I(\boldsymbol{\psi}) = \int_\Omega \hat{W}(\nabla\boldsymbol{\psi})\, \mathrm{d}x - \left\{\int_\Omega \boldsymbol{f}\cdot\boldsymbol{\psi}\, \mathrm{d}x + \int_\Gamma \boldsymbol{g}\cdot\boldsymbol{\psi}\, \mathrm{d}a\right\},$$

let

$$\boldsymbol{\Phi} = \left\{\boldsymbol{\psi} : \bar{\Omega} \to \mathbb{R}^3;\ \det \nabla\boldsymbol{\psi} > 0 \text{ in } \Omega,\ \int_\Omega \boldsymbol{\psi}\, \mathrm{d}x = \boldsymbol{e}\right\},$$

where $\boldsymbol{e}$ is a fixed vector in $\mathbb{R}^3$, and let

$$\boldsymbol{b} := \int_\Omega \boldsymbol{f}\, \mathrm{d}x + \int_\Gamma \boldsymbol{g}\, \mathrm{d}a\,.$$

Show that the functional I is stationary at a point $\boldsymbol{\varphi} \in \boldsymbol{\Phi}$ if and only if $\boldsymbol{\varphi}$ satisfies:

$$\begin{cases} -\operatorname{div} \dfrac{\partial \hat{W}}{\partial \boldsymbol{F}}(\nabla\boldsymbol{\varphi}) = \boldsymbol{f} - \boldsymbol{b} \text{ in } \Omega\,, \\ \dfrac{\partial \hat{W}}{\partial \boldsymbol{F}}(\nabla\boldsymbol{\varphi})\boldsymbol{n} = \boldsymbol{g} \text{ on } \Gamma\,. \end{cases}$$

Hints. First prove that the functional I is stationary with respect to "variations" $\boldsymbol{\theta} : \bar{\Omega} \to \mathbb{R}^3$ that satisfy $\int_\Omega \boldsymbol{\theta}\, \mathrm{d}x = \boldsymbol{o}$ (i.e., $I'(\boldsymbol{\varphi})\boldsymbol{\theta} = 0$ for all such variations $\boldsymbol{\theta}$). The converse property is an application of the *Lagrange multiplier theorem* (cf. e.g. Ciarlet [1982, Theorem 7.2-2]), which shows that *the vector* $\boldsymbol{b}$ *is the Lagrange multiplier associated with the constraint* $\int_\Omega \boldsymbol{\psi}\, \mathrm{d}x = \boldsymbol{e}$.

5.3. Consider the *balloon problem* described in Fig. 5.2-1. Let $\boldsymbol{\varphi} \in \mathscr{C}^1(\bar{\Omega}; \mathbb{R}^3)$ be a mapping that is injective on $\bar{\Omega}$. How can one compute the enclosed volume $v_i(\boldsymbol{\varphi})$?

5.4. Let ω be a domain, with boundary γ, of the plane spanned by the unit vectors e_1 and e_2, and let

$$\Omega = \omega \times]-\varepsilon, \varepsilon[, \ \Gamma_0 = \gamma \times [-\varepsilon, \varepsilon],$$

$$\Gamma_1 = (\omega \times \{-\varepsilon\}) \cup (\omega \times \{\varepsilon\}), \ \varepsilon > 0,$$

so that the reference configuration $\bar{\Omega}$ is a plate.

(1) State and prove the analog of Theorem 5.2-1 when the deformed plate is subjected to two kinds of dead loads: body forces in Ω and horizontal forces on the lateral surface Γ_0, of which only the resultant after integration across the thickness of the plate is known. In other words, one has

$$\frac{1}{2\varepsilon} \int_{-\varepsilon}^{\varepsilon} \sum_{\alpha,\beta=1}^{2} \hat{T}_{\alpha\beta}(\nabla\varphi) n_\beta \, dx_3 = h_\alpha \text{ on } \gamma, \ \alpha = 1, 2,$$

where the two functions $h_\alpha : \gamma \to \mathbb{R}$, are given.

(2) Show that the following boundary conditions on the lateral surface Γ_0 are also amenable to the framework of Theorem 5.2-1:

$$\begin{cases} \varphi = id + \sum_{\alpha=1}^{2} u_\alpha e_\alpha, \ u_\alpha \text{ independent of } x_3, \\ \dfrac{1}{2\varepsilon} \displaystyle\int_{-\varepsilon}^{\varepsilon} T^\varphi n^\varphi \, dx_3 = -\pi n^\varphi \text{ on } \gamma, \ \pi \in \mathbb{R} \text{ given}. \end{cases}$$

Remarks. Case (2) is to case (1) what a pressure load is to an applied surface force that is a dead load. Such boundary conditions naturally arise in the derivation of the von Kármán equations from three-dimensional models (cf. Ciarlet [1980], Ciarlet & Paumier [1986], for case (1), Blanchard & Ciarlet [1983] for case (2), and Volume II).

5.5. This exercise is a complement to Theorem 5.3-1. The notation is identical, and the boundary of the set C is assumed to be as smooth as necessary.

(1) Show that the set

$$\Phi := \{\psi \in \mathscr{C}^1(\bar{\Omega}; \mathbb{R}^3); \det \nabla\psi > 0 \text{ in } \Omega, \ \psi = \varphi_0 \text{ on } \Gamma_0, \ \psi \in C \text{ on } \Gamma_2\}$$

can be given a structure of a manifold, and infer from the proof of Theorem 5.3-1 the definition of the tangent space at a point $\varphi \in \Phi$.

(2) Assume in addition that the set C is convex, and show that any

point in the set Φ contains a convex neighborhood. Use this observation and the characterization of minima over convex sets (Theorem 4.7-5) to give an alternate proof to Theorem 5.3-1.

5.6. Let there be given a bounded open subset Ω of $\mathbb{R}^n$, a mapping $\varphi \in \mathscr{C}^0(\bar{\Omega}; \mathbb{R}^n) \cap \mathscr{C}^1(\Omega; \mathbb{R}^n)$, and a point $b \in \varphi(\bar{\Omega}) - \varphi(\partial\Omega \cup S_\varphi)$, where the set S_φ is defined as in Theorem 5.4-1. For each $\varepsilon > 0$, let ρ_b^ε be a function satisfying

$$\rho_b^\varepsilon \in \mathscr{C}^0(\mathbb{R}^n), \quad \int_{\mathbb{R}^n} \rho_b^\varepsilon(y)\, dy = 1, \quad \operatorname{supp} \rho_b^\varepsilon \subset B_\varepsilon(b)\,.$$

Show that there exists $\varepsilon(b) > 0$ such that

$$\deg(\varphi, \Omega, b) = \int_\Omega \rho_b^\varepsilon(\varphi(x)) \det \nabla\varphi(x)\, dx \text{ for all } 0 < \varepsilon \leq \varepsilon(b)\,,$$

where the topological degree $\deg(\varphi, \Omega, b)$ is defined as in Section 5.4.

Remark. A more general integral representation is given in Nirenberg [1974, Theorem 1.5.5].

5.7. The following delicate extension of Theorem 5.5-2 is due to Ball [1981b, Theorems 1 and 2]. Let Ω be a domain in $\mathbb{R}^n$, let $\varphi_0 \in \mathscr{C}^0(\bar{\Omega}; \mathbb{R}^n)$ be injective on Ω, and let $\varphi : \bar{\Omega} \to \mathbb{R}^n$ be a mapping that satisfies:

$$\begin{cases} \varphi \in W^{1,p}(\Omega) & \text{for some } p > n\,, \\ \det \nabla\varphi > 0 \text{ a.e. in } \Omega\,, \\ \varphi(x) = \varphi_0(x) & \text{for all } x \in \partial\Omega\,. \end{cases}$$

(1) Show that $\varphi(\bar{\Omega}) = \varphi_0(\bar{\Omega})$ and that $\varphi : \bar{\Omega} \to \mathbb{R}^3$ is *injective almost everywhere*, in the sense that

$$\operatorname{card} \varphi^{-1}(x') = 1 \quad \text{for almost all } x' \in \varphi(\bar{\Omega})\,.$$

(2) Assume in addition that the open set $\varphi_0(\Omega)$ is a domain, and that

$$\int_\Omega \|(\nabla\varphi(x))^{-1}\|^q \det \nabla\varphi(x)\, dx < +\infty \quad \text{for some } q > n\,.$$

Show that $\varphi : \bar{\Omega} \to \varphi(\bar{\Omega})$ is a homeomorphism.

5.8. Let Ω be a bounded open subset of $\mathbb{R}^3$, let B be a closed subset of $\mathbb{R}^3$, and let $\varphi \in \mathscr{C}^0(\bar{\Omega}; \mathbb{R}^3)$ be a mapping that is injective on Ω and that satisfies the unilateral boundary condition of place $\varphi(\Gamma) \subset B$. Does it satisfy the confinement condition $\varphi(\bar{\Omega}) \subset B$?

5.9. Let the total energy be of the form

$$I(\psi) = \int_\Omega \hat{W}(\nabla\psi)\, \mathrm{d}x - \int_\Omega \boldsymbol{f} \cdot \psi \, \mathrm{d}x \,,$$

and let

$$\Phi = \{\psi : \bar{\Omega} \to \mathbb{R}^3;\ \det \nabla\psi = 1 \cdot \text{in } \Omega,\ \psi = \varphi_0 \text{ on } \Gamma\} \,,$$

so that the associated elastic material is *incompressible* (Sect. 5.7). Show that the functional I is stationary at a point $\varphi \in \Phi$ if and only if φ is a solution of the *pure displacement problem of incompressible elasticity*:

$$\begin{cases} -\mathbf{div}\, \hat{T}(\nabla\varphi) + \mathbf{div}(p\, \mathbf{Cof}\, \nabla\varphi) = \boldsymbol{f} \text{ in } \Omega \,, \\ \det \nabla\varphi = 1 \text{ in } \Omega \,, \\ \varphi = \boldsymbol{id} \text{ on } \Gamma \,, \end{cases}$$

where $p : \Omega \to \mathbb{R}$ is the *Lagrange multiplier* associated with the constraint $\det \nabla\psi = 1$ (a similar situation was encountered in Exercise 5.2). The function p is called the *hydrostatic pressure*.

5.10. Ciarlet & Nečas [1985] have proposed the following example of locking function (which generalizes those introduced by Prager [1957] in linearized elasticity):

$$\hat{\Lambda}(\nabla\varphi) = \boldsymbol{E}^d : \boldsymbol{E}^d - \alpha, \quad \alpha > 0 \,,$$

where

$$\boldsymbol{E}^d = \boldsymbol{E} - \tfrac{1}{3}(\operatorname{tr} \boldsymbol{E})\boldsymbol{I}, \quad \boldsymbol{E} = \tfrac{1}{2}(\nabla\varphi^{\mathrm{T}}\nabla\varphi - \boldsymbol{I}) \,,$$

denotes the *deviatoric part* of the Green–St Venant strain tensor $\boldsymbol{E}$. Observe that such a locking function is frame-indifferent (Sect. 3.3) since it satisfies

$$\hat{\Lambda}(\boldsymbol{QF}) = \hat{\Lambda}(\boldsymbol{Q}) \quad \text{for all } \boldsymbol{F} \in \mathbb{M}^3_+,\ \boldsymbol{Q} \in \mathbb{O}^3_+ .$$

(1) Show that the number $\boldsymbol{E}^d : \boldsymbol{E}^d$ can also be written as

$$\boldsymbol{E}^d : \boldsymbol{E}^d = \tfrac{1}{3}\{(\lambda_2 - \lambda_3)^2 + (\lambda_3 - \lambda_1)^2 + (\lambda_1 - \lambda_2)^2\},\ \lambda_i = \lambda_i(\boldsymbol{E}) .$$

Remarks. This relation shows that the number $\boldsymbol{E}^d : \boldsymbol{E}^d$ is an *invariant of the tensor* $\boldsymbol{E}$, and that it can be viewed as *a measure of the maximum "shear strains"* (cf. the discussion given in Nečas & Hlaváček [1981, pp. 27–28] for the stress tensor). This last observation thus provides an interpretation of the associated locking constraint $\hat{\Lambda}(\nabla\boldsymbol{\varphi}) \leqslant 0$ in $\bar{\Omega}$.

(2) In view of establishing an existence result (Theorem 7.8-1), we need locking functions $\hat{L} : \mathbb{M}^3_+ \to \mathbb{R}$ that are *polyconvex*, in the sense that there exists a *convex* function $\mathbb{L} : \mathbb{M}^3 \times \mathbb{M}^3 \times]0, +\infty[\to \mathbb{R}$ such that (Sect. 4.9):

$$\hat{L}(\boldsymbol{F}) = \mathbb{L}(\boldsymbol{F}, \mathbf{Cof}\, \boldsymbol{F}, \det \boldsymbol{F}) \quad \text{for all } \boldsymbol{F} \in \mathbb{M}^3_+ .$$

Show that there exist a constant ν and a function $\hat{L} : \mathbb{M}^3_+ \to \mathbb{R}$ of the form

$$\hat{L}(\boldsymbol{F}) = \alpha + a\|\boldsymbol{F}\|^2 + b\|\mathbf{Cof}\, \boldsymbol{F}\|^2 + \Gamma(\det \boldsymbol{F}),\ \boldsymbol{F} \in \mathbb{M}^3_+ ,$$

where $\alpha \in \mathbb{R}$, the constants a and b are >0, and the function $\Gamma :]0, +\infty[\to \mathbb{R}$ is convex, such that

$$\hat{L}(\boldsymbol{F}) = -\nu + \boldsymbol{E}^d : \boldsymbol{E}^d + O(\|\boldsymbol{E}\|^3) .$$

In other words, we have constructed a frame-indifferent locking function $\hat{L}$ that agrees with the locking function $\hat{\Lambda}$ considered in (1) to within the second order with respect to $\boldsymbol{E}$, and which is polyconvex.

Hints. A simple computation shows that

$$\boldsymbol{E}^d : \boldsymbol{E}^d = -\tfrac{1}{3}(\operatorname{tr} \boldsymbol{E})^2 + \operatorname{tr} \boldsymbol{E}^2 ,$$

which is an expression reminiscent of that found for the strain energy of a St Venant–Kirchhoff material. Then argue as in Theorem 4.10-2 (a simplification arises in that the condition $\lim_{\delta \to 0^+} \Gamma(\delta) = +\infty$ is no longer needed).

5.11. Consider a *rigid body*, for which all the admissible deformations are necessarily rigid, i.e., of the form $\varphi(x) = a + Qox$ for all $x \in \bar{\Omega}$, for some $a \in \mathbb{R}^3$ and $Q \in \mathbb{O}^3_+$. Express the axioms of force and moment balance, the equations of equilibrium, the constitutive equation, etc., for such a body (observe that a rigid deformation depends only on six independent parameters).

5.12. (1) Show that the set

$$\Phi := \{\psi \in \mathscr{C}^1(\bar{\Omega}; \mathbb{R}^3);\ \det \nabla\psi > 0 \text{ in } \Omega\}$$

is not convex.

(2) Show that the set Φ may even consist of a countable union of mutually disjoint convex sets if the set Ω is not simply connected (Antman [1976a, p. 313]; see also Pierce & Whitman [1980]).

5.13. Given a fourth-order tensor $\mathsf{A} = (a_{ijkl})$ and a matrix $B = (b_{ij})$, let

$$B : \mathsf{A} : B = b_{ij} a_{ijkl} b_{kl}\,,$$

so that the strong Legendre–Hadamard condition at a matrix $F \in \mathbb{M}^3_+$ can be written as

$$bc^{\mathrm{T}} : \mathsf{A}(F) : bc^{\mathrm{T}} > 0 \quad \text{for all } b \neq o,\ c \neq o\,.$$

The stronger constitutive assumption

$$B : \mathsf{A}(F) : B > 0 \quad \text{for all matrices } B \neq O\,,$$

has been proposed by Coleman & Noll [1959], and for this reason is known as the *Coleman–Noll inequality*.

(1) Show that the Coleman–Noll inequality implies a monotonicity-type condition on the first Piola–Kirchhoff stress tensor, in the sense that

$$(\hat{T}(F_2) - \hat{T}(F_1)) : (F_2 - F_1) > 0 \quad \text{for all } F_2 = WF_1,\ W \in \mathbb{S}^3_>,\ W \neq I\,.$$

(2) Show that for a homogeneous isotropic hyperelastic material, the Coleman–Noll inequality implies that the stored energy function $\hat{W}(F)$ is strictly convex. Thus it is not a realistic constitutive assumption (Theorem 4.8-1), as observed by its authors themselves (Coleman & Noll [1963]).

(3) Show that a St Venant–Kirchhoff material does not satisfy the Coleman–Noll inequality.

Remark. The implication of (2) was noted by Ogden [1970, 1972].

5.14. A concept of major importance introduced by Morrey [1952] in the calculus of variations is the basis for defining a general *constitutive assumption*, which contains polyconvexity as a special case: The stored energy function $\hat{W}$ of a homogeneous material is *quasiconvex* at a particular matrix $\boldsymbol{F} \in \mathbb{M}^3_+$ if, for all bounded open subsets $V \subset \mathbb{R}^3$ and for all functions $\boldsymbol{\theta} \in \mathcal{D}(V)$ such that $\{\boldsymbol{F} + \boldsymbol{\nabla\theta}(x)\} \in \mathbb{M}^3_+$ for all $x \in V$,

$$\frac{1}{\text{vol}\, V} \int_V \hat{W}(\boldsymbol{F} + \boldsymbol{\nabla\theta}(x))\, \mathrm{d}x \geqslant \hat{W}(\boldsymbol{F}),$$

where $\mathcal{D}(V)$ denotes the space of functions in $\mathscr{C}^\infty(V)$ with compact supports in V.

(1) Let the set of admissible deformations be of the form

$$\boldsymbol{\Phi} = \{\boldsymbol{\psi} \in \mathscr{C}^1(\bar{\Omega});\ \det \boldsymbol{\nabla\psi} > 0 \text{ in } \Omega,\ \boldsymbol{\psi} = \boldsymbol{\varphi}_0 \text{ on } \Gamma_0\},$$

and let $\boldsymbol{\varphi} \in \boldsymbol{\Phi}$ be such that, for some $\varepsilon > 0$,

$$I(\boldsymbol{\varphi}) \leqslant I(\boldsymbol{\psi}) \quad \text{for all } \boldsymbol{\psi} \in \boldsymbol{\Phi} \text{ that satisfy } (\boldsymbol{\varphi} - \boldsymbol{\psi}) \in \mathcal{D}(\Omega) \text{ and } \|\boldsymbol{\varphi} - \boldsymbol{\psi}\|_{\mathscr{C}^0(\bar{\Omega})} \leqslant \varepsilon,$$

where

$$I(\boldsymbol{\psi}) = \int_\Omega \hat{W}(\boldsymbol{\nabla\psi})\, \mathrm{d}x.$$

Show that the stored energy function is quasiconvex at each matrix $\boldsymbol{\nabla\varphi}(x)$, $x \in \Omega$ (Ball [1977, Theorem 3.1]).

(2) Show that if $\hat{W} : \boldsymbol{F} \in \mathbb{M}^3_+ \to \mathbb{R}$ is a convex function, then it is quasiconvex at all $\boldsymbol{F} \in \mathbb{M}^3_+$.

(3) More generally, show that if the stored energy function $\hat{W}$ is polyconvex, then it is quasiconvex (Ball [1977, Theorem 4.5]).

(4) Show that if $\hat{W}$ is quasiconvex and of class $\mathscr{C}^2$, the elasticity tensor

$$\left(\frac{\partial^2 \hat{W}}{\partial F_{ij} \partial F_{kl}}(\boldsymbol{F})\right)$$

satisfies the Legendre–Hadamard condition at all $\boldsymbol{F} \in \mathbb{M}^3_+$.

(5) The definition of quasiconvexity can be extended in a natural way to functions defined on open subsets of the space of matrices with m rows and n columns. Show that for $m = 1$, or for $n = 1$, quasiconvexity is equivalent to convexity.

Remarks. In general, quasiconvexity does not imply polyconvexity, and thus quasiconvexity is a more general assumption. However, "the quasiconvexity condition is not a pointwise condition on the [stored energy function], and is therefore difficult to verify in particular cases" (Ball [1977, p. 356]). The relations between the two notions are discussed at length in Ball [1977], Marcellini [1984], Dacorogna [1982a, 1982b, 1986, 1987]. See also Serre [1981a, 1981b, 1983]. The connection between quasiconvexity and the regularity of minimizers is discussed in Evans [1986], Acerbi & Fusco [1987].

5.15. Let $\mathbb{U}$ be an open subset of $\mathbb{M}^n$. A function $\hat{W} : \mathbb{U} \to \mathbb{R}$ is *rank-one convex* if it is "convex along all directions spanned by matrices of rank one", i.e., if

$$\hat{W}(\lambda \boldsymbol{F} + (1 - \lambda)\boldsymbol{G}) \leq \lambda \hat{W}(\boldsymbol{F}) + (1 - \lambda)\hat{W}(\boldsymbol{G}), \ 0 \leq \lambda \leq 1 ,$$

for all $\boldsymbol{F}, \boldsymbol{G} \in \mathbb{M}^n$ such that $[\boldsymbol{F}, \boldsymbol{G}] \subset \mathbb{U}$ and rank $(\boldsymbol{F} - \boldsymbol{G}) \leq 1$ (this definition may be extended to functions of rectangular matrices).

(1) Show that if a function is quasiconvex, according to the definition of Exercise 5.14 (with $\mathbb{M}^3_+$ replaced by $\mathbb{U}$), it is rank-one convex.

(2) Show that a function $\hat{W} \in \mathscr{C}^2(\mathbb{U}; \mathbb{R})$ is rank-one convex if and only if it satisfies the Legendre–Hadamard condition

$$\frac{\partial^2 \hat{W}}{\partial F_{ij} \partial F_{kl}} (\boldsymbol{F}) b_i b_k c_j c_l \geq 0 \quad \text{for all } \boldsymbol{b} = (b_i),\ \boldsymbol{c} = (c_j) \in \mathbb{R}^n ,$$

at all $\boldsymbol{F} \in \mathbb{U}$.

Remark. Proofs and further properties are given in Ball [1977], Dacorogna [1982a, 1986, 1987], Aubert [1987].

5.16. The purpose of this exercise is to relate the Legendre–Hadamard condition to the minimizers of the energy. Let the set of admissible deformations be of the form

$$\boldsymbol{\Phi} = \{\boldsymbol{\psi} \in \mathscr{C}^1(\bar{\Omega}); \det \boldsymbol{\nabla}\boldsymbol{\psi} > 0 \text{ in } \Omega,\ \boldsymbol{\psi} = \boldsymbol{\varphi}_0 \text{ on } \Gamma_0\} ,$$

and let $\boldsymbol{\varphi} \in \boldsymbol{\Phi}$ be a "weak local minimum" of the energy

$$I(\boldsymbol{\psi}) = \int_{\Omega} \hat{W}(\nabla\boldsymbol{\psi})\,\mathrm{d}x - \left\{\int_{\Omega} \boldsymbol{f} \cdot \boldsymbol{\psi}\,\mathrm{d}x + \int_{\Gamma_1} \boldsymbol{g} \cdot \boldsymbol{\psi}\,\mathrm{d}a\right\},$$

in the sense that, for each $\boldsymbol{\theta} \in \mathscr{C}^1(\bar{\Omega})$ satisfying $\boldsymbol{\theta} = \boldsymbol{o}$ on Γ_0, there exists $\varepsilon_0(\boldsymbol{\theta}) > 0$ such that

$$I(\boldsymbol{\varphi}) \leq I(\boldsymbol{\varphi} + \varepsilon\boldsymbol{\theta}) \quad \text{for all } |\varepsilon| \leq \varepsilon_0(\boldsymbol{\theta}).$$

This being the case, show that the elasticity tensor

$$\left(\frac{\partial^2 \hat{W}}{\partial F_{ij}\,\partial F_{kl}}(\nabla\boldsymbol{\varphi}(x))\right)$$

satisfies the Legendre–Hadamard condition at all points $x \in \Omega$.

5.17. Let $\mathsf{A}(\boldsymbol{I}) = ((\partial\hat{t}_{ij}/\partial F_{kl})(\boldsymbol{I}))$ denote the elasticity tensor of a homogeneous elastic material whose reference configuration is a natural state. If $\boldsymbol{H} = (h_{kl})$ denotes an arbitrary matrix of order three, we let $\mathsf{A}(\boldsymbol{I})\boldsymbol{H}$ denote the matrix $((\partial\hat{t}_{ij}/\partial F_{kl})(\boldsymbol{I})h_{kl})$.

(1) Show that one also has $\mathsf{A}(\boldsymbol{I}) = ((\partial\hat{t}^{\mathrm{D}}_{ij}/\partial F_{kl})(\boldsymbol{I}))$, where $\hat{\boldsymbol{T}}^{\mathrm{D}} = (\hat{t}^{\mathrm{D}}_{ij})$ denotes the response function for the Cauchy stress.

(2) Show that the matrix $\mathsf{A}(\boldsymbol{I})\boldsymbol{H}$ is symmetric for all $\boldsymbol{H} \in \mathbb{M}^3$.

(3) Show that $\mathsf{A}(\boldsymbol{I})\boldsymbol{H} = \boldsymbol{0}$ if the matrix $\boldsymbol{H}$ is skew-symmetric, i.e., if $\boldsymbol{H} = -\boldsymbol{H}^{\mathrm{T}}$.

(4) Show that

$$\mathsf{A}(\boldsymbol{I})\boldsymbol{H} : \boldsymbol{H} = \mathsf{A}(\boldsymbol{I})\{\tfrac{1}{2}(\boldsymbol{H} + \boldsymbol{H}^{\mathrm{T}})\} : \{\tfrac{1}{2}(\boldsymbol{H} + \boldsymbol{H}^{\mathrm{T}})\} \quad \text{for all } \boldsymbol{H} \in \mathbb{M}^3.$$

(5) Assume that the material is isotropic. Show that the tensor $\mathsf{A}(\boldsymbol{I})$ satisfies the strong ellipticity condition if and only if $\mu > 0$ and $\lambda > -2\mu$.

(6) Assume that the material is isotropic. Show that

$$\mathsf{A}(\boldsymbol{I})\boldsymbol{B} : \boldsymbol{B} > 0 \text{ for all } \boldsymbol{B} \in \mathbb{S}^3 - \{\boldsymbol{0}\} \Leftrightarrow \mu > 0 \text{ and } \lambda > -\tfrac{2}{3}\mu.$$

Remark. These properties are proved in Gurtin [1981a, Ch. 9] and Gurtin [1981b, Sect. 29].

5.18. A *Hadamard–Green material* has a stored energy function of the form

$$\hat{W}(F) = \frac{\alpha}{2} \|F\|^2 + \frac{\beta}{4} \{\|F\|^4 - \|FF^{\mathrm{T}}\|^2\} + \Gamma(\det F),$$

where the function $\Gamma :]0, +\infty[\to \mathbb{R}$ is smooth enough, and α, β are real constants.

(1) Show that the reference configuration is a natural state and the elasticity tensor $\mathsf{A}(I) = (a_{ijkl}(I))$ satisfies

$$b_{ij} a_{ijkl}(I) b_{kl} > 0 \quad \text{for all } B = (b_{ij}) \in \mathbb{S}^3 - \{0\}$$

if and only if:

$$\Gamma'(1) + \alpha + 2\beta = 0,\ \alpha + \beta > 0,\ 3\Gamma''(1) - \alpha + 2\beta > 0.$$

(2) Assume that

$$\alpha > 0,\ \beta \geq 0,\ \Gamma'(1) + \alpha + 2\beta = 0,\ (\delta\Gamma'(\delta))' \geq 0 \quad \text{for all } \delta \in]0, 1].$$

Show that the elasticity tensor $\mathsf{A}(F)$ satisfies the strong ellipticity condition for all $F \in \mathbb{M}^3_+$ such that $\det F \leq 1$.

(3) Assume that the function $\Gamma :]0, +\infty[\to \mathbb{R}$ is convex. Are there values of α, β for which the function $\hat{W}$ is polyconvex?

Remark. The results of (1) and (2) are due to Simpson & Spector [1984a].

PART B

MATHEMATICAL METHODS IN THREE-DIMENSIONAL ELASTICITY

CHAPTER 6

EXISTENCE THEORY BASED ON THE IMPLICIT FUNCTION THEOREM

INTRODUCTION

The question of *existence* of solutions to the nonlinear boundary value problem of three-dimensional elasticity can be approached in two ways:

In one approach, it is assumed that the material is hyperelastic, so that particular solutions are obtained as minimizers of the energy over a set of admissible deformations with appropriate smoothness; the existence results obtained in this fashion are discussed in the next chapter.

Another approach consists in applying the *implicit function theorem* directly to the boundary value problem of three-dimensional elasticity. More specifically, consider the *pure displacement problem*:

$$-\mathbf{div}\{(\boldsymbol{I} + \nabla \boldsymbol{u})\check{\boldsymbol{\Sigma}}(\boldsymbol{E}(\boldsymbol{u}))\} = \boldsymbol{f} \text{ in } \Omega\,,$$
$$\boldsymbol{u} = \boldsymbol{o} \text{ on } \Gamma\,,$$

which possesses the particular solution $\boldsymbol{u} = \boldsymbol{o}$ corresponding to $\boldsymbol{f} = \boldsymbol{o}$ (we assume that the reference configuration is a natural state). Under mild smoothness assumptions on the response function $\check{\boldsymbol{\Sigma}}$, it can be shown that the nonlinear operator

$$\boldsymbol{A} : \boldsymbol{u} \to \boldsymbol{A}(\boldsymbol{u}) := -\mathbf{div}\{(\boldsymbol{I} + \nabla \boldsymbol{u})\check{\boldsymbol{\Sigma}}(\boldsymbol{E}(\boldsymbol{u}))\}$$

maps the Sobolev space $\boldsymbol{W}^{2,p}(\Omega)$ into the space $\boldsymbol{L}^p(\Omega)$ for each $p > 3$ (cf. Sects. 6.5 and 6.6, where the differentiability properties of the mapping $\boldsymbol{A}$ are also studied; the key result used there is that the Sobolev space $W^{1,p}(\Omega)$ is an algebra for $p > 3$). Hence solving the pure displacement problem consists in finding a displacement

$$\boldsymbol{u} \in \boldsymbol{V}^p(\Omega) := \{\boldsymbol{v} \in \boldsymbol{W}^{2,p}(\Omega);\, \boldsymbol{v} = \boldsymbol{o} \text{ on } \Gamma\}$$

such that

$$\boldsymbol{A}(\boldsymbol{u}) = \boldsymbol{f}\,.$$

In order to use the implicit function theorem in a neighborhood of the origin in both spaces $\boldsymbol{V}^p(\Omega)$ and $\boldsymbol{L}^p(\Omega)$, we must verify that the derivative $\boldsymbol{A}'(\boldsymbol{o})$ is an isomorphism between the spaces $\boldsymbol{V}^p(\Omega)$ and $\boldsymbol{L}^p(\Omega)$.

But the equation $\boldsymbol{A}'(\boldsymbol{o})\boldsymbol{u} = \boldsymbol{f}$ is precisely a boundary value problem of *linearized elasticity* (cf. Sect. 6.2; we assume that the material is homogeneous and isotropic):

$$\begin{cases} -\mathbf{div}\{\lambda(\operatorname{tr} \boldsymbol{e}(\boldsymbol{u}))\boldsymbol{I} + 2\mu\boldsymbol{e}(\boldsymbol{u})\} = \boldsymbol{f} \text{ in } \Omega\,, \quad \text{with } \boldsymbol{e}(\boldsymbol{u}) = \tfrac{1}{2}(\boldsymbol{\nabla u} + \boldsymbol{\nabla u}^{\mathrm{T}})\,, \\ \boldsymbol{u} = \boldsymbol{o} \text{ on } \Gamma\,. \end{cases}$$

Accordingly, we study this problem in Sect. 6.3: We first establish the existence and uniqueness of a *weak solution* in the space $\boldsymbol{H}_0^1(\Omega)$ (Theorem 6.3-5); we then show that, if the boundary Γ is smooth enough, the operator $\boldsymbol{A}'(\boldsymbol{o}): \boldsymbol{V}^p(\Omega) \to \boldsymbol{L}^p(\Omega)$ is *onto*, i.e., that we have a *regularity result* of the form

$$\boldsymbol{A}'(\boldsymbol{o})\boldsymbol{u} \in \boldsymbol{L}^p(\Omega) \Rightarrow \boldsymbol{u} \in \boldsymbol{V}^p(\Omega)\,.$$

This regularity result holds essentially because the boundary condition does not change along Γ (Theorem 6.3-6): *It is the lack of such a regularity result when a boundary condition of place* $\boldsymbol{u} = \boldsymbol{o}$ *and a linearized boundary condition of traction are imposed along connecting portions of* Γ *that prevents the applicability of the implicit function theorem to genuine displacement–traction problems*, except in some very special cases.

Combining this regularity result with the implicit function theorem, we obtain an *existence theorem in the space* $\boldsymbol{W}^{2,p}(\Omega)$: For each number $p > 3$, there exists a neighborhood $\boldsymbol{F}^p$ of $\boldsymbol{o}$ in $\boldsymbol{L}^p(\Omega)$ and a neighborhood $\boldsymbol{U}^p$ of $\boldsymbol{o}$ in $\boldsymbol{V}^p(\Omega)$ such that, for each $\boldsymbol{f} \in \boldsymbol{F}^p$, the displacement boundary value problem has exactly one solution $\boldsymbol{u}(\boldsymbol{f}) \in \boldsymbol{U}^p$ (Theorem 6.7-1; cf. also Theorem 6.4-1 in the simpler case of St Venant–Kirchhoff materials). We also establish that the solution $\boldsymbol{u}_{\text{lin}}(\boldsymbol{f})$ of the corresponding linearized displacement *problem* provides an approximation to the exact solution $\boldsymbol{u}(\boldsymbol{f})$ for small enough forces, as shown by the error estimate (Theorem 6.8-1):

$$\|\boldsymbol{u}(\boldsymbol{f}) - \boldsymbol{u}_{\text{lin}}(\boldsymbol{f})\|_{2,p,\Omega} = O(|\boldsymbol{f}|^2_{0,p,\Omega})\,.$$

Finally, we prove that the associated mapping $\boldsymbol{\varphi} = \boldsymbol{id} + \boldsymbol{u}$ is orientation-preserving and injective on $\bar{\Omega}$, provided the forces are small enough (Theorem 6.9-1).

For a given $f \in F^p$ (we assume here that F^p is convex), each problem

$$A(\tilde{u}(\lambda)) = \lambda f, 0 \leq \lambda \leq 1,$$

has a solution $\tilde{u}(\lambda) := u(\lambda f) \in U^p$; hence by differentiating, we conclude that the mapping $\tilde{u} : [0,1] \to F^p \subset V^p(\Omega)$ defined in this fashion also satisfies

$$\begin{cases} \tilde{u}'(\lambda) = \{A'(\tilde{u}(\lambda))\}^{-1} f, 0 \leq \lambda \leq 1, \\ \tilde{u}(0) = o. \end{cases}$$

This *ordinary differential equation in the Banach space* $V^p(\Omega)$ can be studied on its own, and this provides in particular another proof of the existence of a solution of the pure displacement problem (Theorem 6.12-1). But another interest of this observation is that *Euler's method* (or for that matter any other discrete method applicable to an ordinary differential equation) may be applied to approximate the exact solutions $\tilde{u}(\lambda)$, $0 \leq \lambda \leq 1$, hence in particular $\tilde{u}(1) = u(f)$: Given a partition $0 = \lambda^0 < \lambda^1 < \cdots < \lambda^N = 1$, we define a sequence $u^0, u^1, \ldots, u^N$ by the recursion formula:

$$\begin{cases} \dfrac{u^{n+1} - u^n}{\lambda^{n+1} - \lambda^n} = \{A'(u^n)\}^{-1} f, \quad 0 \leq n \leq N-1, \\ u^0 = o. \end{cases}$$

We then show (Theorem 6.13-1) that

$$\max_{0 \leq n \leq N} \|u^n - \tilde{u}(\lambda^n)\|_{2,p,\Omega} = O(\max_{0 \leq n \leq N-1} |\lambda^{n+1} - \lambda^n|),$$

and thus Euler's method is convergent.

Since the approximate displacements $u^{n+1} = u^n + \delta u^n$ are computed recursively by solving a sequence of *linear* problems, viz.,

$$A'(u^n)\delta u^n = \delta f^n := (\lambda^{n+1} - \lambda^n) f, \quad 0 \leq n \leq N-1,$$

Euler's method may be also viewed as a *Lagrangian incremental method* for approximating the displacement boundary value problem $A(u) = f$ (Sects. 6.10 and 6.11). Consequently, the convergence of Euler's method implies the *convergence of the incremental method*.

*6.1. SOBOLEV SPACES

As a preparation to our study of questions of existence, uniqueness or nonuniqueness, regularity, etc., for the boundary value problems of three-dimensional elasticity, we briefly review some fundamental properties of the *Sobolev spaces*, which have become, since the pioneering work of Sobolev [1938, 1950], an indispensable tool for the study of linear and nonlinear partial differential equations and of problems of the calculus of variations. Detailed studies of these spaces may be found in Lions & Magenes [1968, Ch. 1] and Dautray & Lions [1984, Ch. 4] in the Hilbertian case, Lions [1965, Ch. 1–3], Nečas [1967, Ch. 2], Adams [1975]. An excellent introduction is given in Brezis [1983, Ch. 9].

Let Ω be an open subset of $\mathbb{R}^n$, and let p be an extended real number satisfying $1 \leqslant p \leqslant \infty$. We denote by $L^p(\Omega)$ the space of (equivalence classes of dx-almost everywhere equal, dx-measurable) functions $v: \Omega \to \mathbb{R}$ that satisfy

$$|v|_{0,p,\Omega} := \left\{ \int_\Omega |v(x)|^p \, \mathrm{d}x \right\}^{1/p} < +\infty \text{ if } 1 \leqslant p < \infty ,$$

$$|v|_{0,\infty,\Omega} := \inf\{\alpha \geqslant 0; \mathrm{d}x\text{-meas}\{x \in \Omega; |v(x)| > \alpha\} = 0\} < +\infty \text{ if } p = \infty .$$

If $p = 2$, we use the shorter notation

$$|v|_{0,\Omega} := |v|_{0,2,\Omega} .$$

We shall not recall here the basic properties of the spaces $L^p(\Omega)$, for which the reader is referred to, e.g., Brezis [1983, Ch. 4].

Let Ω be an open subset of $\mathbb{R}^n$. We let $\mathcal{D}(\Omega)$ denote the space of real-valued functions $\theta \in \mathscr{C}^\infty(\Omega)$ whose support is a compact subset of Ω, which may vary from one function in $\mathcal{D}(\Omega)$ to another; we recall that the support of a function $\theta: \Omega \to \mathbb{R}$ is the set

$$\operatorname{supp} \theta = cl\{x \in \Omega; \theta(x) \neq 0\} .$$

The space $\mathcal{D}(\Omega)$ does not reduce to the null function: Let $\omega: \mathbb{R}^n \to \mathbb{R}$ be defined by

$$\omega(x) = \exp\left(\frac{1}{|x|^2 - 1}\right) \text{ if } |x| < 1, \quad \omega(x) = 0 \text{ if } |x| \geqslant 1 ,$$

and let $B_r(a)$ be an open ball contained in Ω; then the function

$$x \in \Omega \rightarrow \omega\left(\frac{x-a}{r}\right)$$

belongs to the space $\mathcal{D}(\Omega)$.

A **distribution** *on* Ω is a linear form $T : \mathcal{D}(\Omega) \rightarrow \mathbb{R}$ with the following property: Given any compact subset K of Ω, there exist a constant $C(K)$ and an integer $m(K) \geq 0$ such that (we use here the multi-index notation for partial derivatives; cf. Sect. 1.3):

$$|T(\theta)| \leq C(K) \sup_{\left\{\substack{|\alpha| \leq m(K) \\ x \in K}\right.} |\partial^\alpha \theta(x)| \quad \text{for all } \theta \in \mathcal{D}(\Omega) \text{ with supp } \theta \subset K\,.$$

The space formed by all distributions on Ω *is denoted* $\mathcal{D}'(\Omega)$.

The space $\mathcal{D}(\Omega)$ is equipped in a natural way with an "inductive limit" topology, which makes it a locally convex topological vector space. In this topology, a sequence (θ^k) converges to θ if and only if there exists a compact subset K of Ω such that:

$$\begin{cases} \text{supp } \theta^k \subset K \quad \text{for all } k\,, \\ (\partial^\alpha \theta^k) \text{ converges uniformly on } K \text{ to } \partial^\alpha \theta \text{ as } k \rightarrow \infty\,, \\ \quad \text{for all multi-index } \alpha\,. \end{cases}$$

Note, however, that the topology of the space $\mathcal{D}(\Omega)$ is *not* metrizable, so that it cannot be defined by convergent sequences alone.

It can then be shown that *the space* $\mathcal{D}'(\Omega)$ *is the dual space of* $\mathcal{D}(\Omega)$, i.e., $\mathcal{D}'(\Omega)$ consists of all the linear forms on $\mathcal{D}(\Omega)$ that are continuous with respect to the topology of $\mathcal{D}(\Omega)$. As a dual space, $\mathcal{D}'(\Omega)$ is equipped in a natural fashion with the "weak-star" topology, defined by the family of semi-norms $T \in \mathcal{D}'(\Omega) \rightarrow T(\theta)$, $\theta \in \mathcal{D}(\Omega)$. In this topology, which is again *not* metrizable, a sequence (T^k) converges to T if and only if

$$T^k(\theta) \xrightarrow[k \rightarrow \infty]{} T(\theta) \quad \text{for all } \theta \in \mathcal{D}(\Omega)\,,$$

and if this is the case, we say that *the sequence* (T^k) *converges to* T *in the sense of distributions*. For more details about the topologies of the spaces $\mathcal{D}(\Omega)$ and $\mathcal{D}'(\Omega)$, see Choquet-Bruhat [1973], Hörmander [1983a, Ch. 1–7], Vo-Khac [1972], Yosida [1966], and of course, the celebrated treatise of Schwartz [1966], who created and formalized the theory of distributions.

Let f be a *locally integrable function on* Ω, that is, a function $f:\Omega\to\mathbb{R}$ with the property that for each compact subset K of Ω, $f\in L^1(K)$; in particular any function $f\in L^p(\Omega)$, $1\leq p\leq\infty$, is locally integrable on Ω. Then the linear form

$$T_f:\theta\in\mathcal{D}(\Omega)\to T_f(\theta):=\int_\Omega f\theta\,\mathrm{d}x$$

defines a distribution on Ω, since for any compact subset K of Ω, and for any function $\theta\in\mathcal{D}(\Omega)$ with $\operatorname{supp}\theta\subset K$,

$$|T_f(\theta)|\leq|f|_{0,1,K}\sup_{x\in K}|\theta(x)|\,.$$

The distribution T_f is called the *distribution associated with the locally integrable function f*.

Another source of distributions is provided by the *differentiation in the sense of distributions*: Let T be a distribution, and let $\alpha=(\alpha_1,\alpha_2,\dots,\alpha_n)$ be a multi-index, with $|\alpha|=\Sigma_{i=1}^n\alpha_i$. Then the linear form

$$\partial^\alpha T:\theta\in\mathcal{D}(\Omega)\to\partial^\alpha T(\theta):=(-1)^{|\alpha|}T(\partial^\alpha\theta)$$

is again a distribution (this is a simple consequence of the definition), called the *partial derivative of order* α *of* T. In particular, the ith *partial derivative of* T is defined by

$$\partial_i T:\theta\in\mathcal{D}(\Omega)\to\partial_i T(\theta):=-T(\partial_i\theta)\,.$$

Notice that, *if f is continuously differentiable on* Ω, an integration by parts shows that

$$\partial_i T_f(\theta)=-T_f(\partial_i\theta)=-\int_\Omega f\partial_i\theta\,\mathrm{d}x=\int_\Omega(\partial_i f)\theta\,\mathrm{d}x\quad\text{for all }\theta\in\mathcal{D}(\Omega)\,,$$

i.e., $\partial_i T_f$ is precisely the distribution associated with the locally integrable function $\partial_i f$. In this sense, differentiation in the sense of distributions is a generalization of the "classical" differentiation of smooth functions.

For any integer $m\geq 0$ and any extended real number p satisfying $1\leq p\leq\infty$, the **Sobolev space** $W^{m,p}(\Omega)$ consists of those (equivalence classes of) functions $v\in L^p(\Omega)$ for which all partial derivatives, in the sense of distributions, of order $\leq m$ also belong to the space $L^p(\Omega)$. In other words, for each multi-index α with $|\alpha| \leq m$, there exists a function $v^\alpha \in L^p(\Omega)$ such that

$$\int_\Omega v^\alpha \theta \, dx = (-1)^{|\alpha|} \int_\Omega v \partial^\alpha \theta \, dx \quad \text{for all } \theta \in \mathcal{D}(\Omega) .$$

Notice that, in particular, $W^{0,p}(\Omega) = L^p(\Omega)$.

Although *the functions* v^α *are not in general partial derivatives in the usual sense* (except if the function v is smooth enough, as was pointed out earlier), it is customary to write $\partial^\alpha v$ instead of v^α, and we shall comply with this usage; special care should be taken, however, not to blithely attribute to these "*generalized derivatives*", as they are sometimes called, all properties of "usual" derivatives.

We now list various fundamental properties of Sobolev spaces that we shall need later on; observe that the assumptions on the open set Ω become more and more severe as we proceed. We recall that a normed vector space is *reflexive* if it can be identified with the dual space of its dual space.

Theorem 6.1-1. *Let Ω be an open subset of $\mathbb{R}^n$. Equipped with the norm*

$$v \to \|v\|_{m,p,\Omega} := \left\{ \int_\Omega \sum_{|\alpha| \leq m} |\partial^\alpha v|^p \, dx \right\}^{1/p} \quad \text{if } 1 \leq p < \infty ,$$

$$v \to \|v\|_{m,\infty,\Omega} := \max_{|\alpha| \leq m} |\partial^\alpha v|_{0,\infty,\Omega} \quad \text{if } p = \infty ,$$

the Sobolev space $W^{m,p}(\Omega)$ is a Banach space, which is reflexive if $1 < p < \infty$. The Sobolev space $H^m(\Omega) := W^{m,2}(\Omega)$ is a Hilbert space. ∎

We shall also use the following *semi-norms*:

$$v \to |v|_{m,p,\Omega} = \left\{ \int_\Omega \sum_{|\alpha| = m} |\partial^\alpha v|^p \, dx \right\}^{1/p} \quad \text{if } 1 \leq p < \infty ,$$

$$v \to |v|_{m,\infty,\Omega} = \max_{|\alpha| = m} |\partial^\alpha v|_{0,\infty,\Omega} \quad \text{if } p = \infty ,$$

over the space $W^{m,p}(\Omega)$; notice that $\|\cdot\|_{0,p,\Omega} = |\cdot|_{0,p,\Omega}$. When $p = 2$, the index p is usually dropped in the notation used for the norms and semi-norms, which then become

$$\|\cdot\|_{m,\Omega} := \|\cdot\|_{m,2,\Omega}, \quad |\cdot|_{m,\Omega} := |\cdot|_{m,2,\Omega}.$$

Notice that, if $v \in \mathscr{C}^m(\bar{\Omega})$ (Sect. 1.3), one also has

$$\|v\|_{m,\infty,\Omega} = \max_{|\alpha| \leq m} \sup_{x \in \bar{\Omega}} |\partial^\alpha v(x)|, \quad |v|_{m,\infty,\Omega} = \max_{|\alpha| \leq m} \sup_{x \in \bar{\Omega}} |\partial^\alpha v(x)|.$$

While the space $\mathscr{D}(\Omega)$ is dense in the space $L^p(\Omega)$, $1 \leq p < \infty$, it is no longer dense in the space $W^{m,p}(\Omega)$, $m \geq 1, 1 \leq p < \infty$, except if the set $\mathbb{R}^n - \Omega$ is "very small" (in particular, $\mathscr{D}(\mathbb{R}^n)$ is dense in $W^{m,p}(\mathbb{R}^n)$; if $\Omega \neq \mathbb{R}^n$, a necessary, but not sufficient, condition that $\mathscr{D}(\Omega)$ be dense in $W^{m,p}(\Omega)$ is that dx-meas$\{\mathbb{R}^n - \Omega\} = 0$; cf. Lions [1965, p. 20 ff.]). This observation leads to the following definition: Let Ω be an open subset of $\mathbb{R}^n$; for any integer $m \geq 0$ and any real number $p \geq 1$ ($p = \infty$ is excluded), we define the **Sobolev spaces**

$$\boxed{W_0^{m,p}(\Omega) := \{\mathscr{D}(\Omega)\}^-, \quad H_0^m(\Omega) := W_0^{m,2}(\Omega),}$$

the closure being taken with respect to the norm $\|\cdot\|_{m,p,\Omega}$.

A subset Ω of $\mathbb{R}^n$ is of *finite width* if it lies between two parallel hyperplanes (in particular, a bounded set is of finite width). As shown by the next theorem, the semi-norm $|\cdot|_{m,p,\Omega}$ becomes in this case a norm on the space $W_0^{m,p}(\Omega)$, which is in addition equivalent to the norm $\|\cdot\|_{m,p,\Omega}$.

Theorem 6.1-2. (a) **(Poincaré inequality).** *Let Ω be an open subset of $\mathbb{R}^n$ of finite width. Then for each p with $1 \leq p < \infty$, there exists a constant b such that*

$$|v|_{0,p,\Omega} \leq b|v|_{1,p,\Omega} \quad \text{for all } v \in W_0^{1,p}(\Omega).$$

(b) *As a consequence, there exists a constant c such that*

$$\boxed{|v|_{m,p,\Omega} \leq \|v\|_{m,p,\Omega} \leq c|v|_{m,p,\Omega} \quad \text{for all } v \in W_0^{m,p}(\Omega).}$$

■

It is easy to prove that, for $m \geq 1$, the space $W^{m,p}(\Omega)$ is strictly contained in the space $L^p(\Omega)$. This means that some extra "smoothness" is acquired by any function in $L^p(\Omega)$ that possesses generalized derivatives in $L^p(\Omega)$. For instance, let Ω be a domain in $\mathbb{R}^2$; then a function in the space $W^{1,p}(\Omega)$ is necessarily continuous if $p>2$, is in any space $L^q(\Omega)$, $1\leq q<\infty$ if $p=2$, and is in the space $L^{2p/(2-p)}(\Omega)$ if $1\leq p<2$; that a function in the space $H^1(\Omega)$, $\Omega\subset\mathbb{R}^2$, is not necessarily continuous is illustrated in Exercise 6.1 by an example of a function that is even discontinuous on a dense subset of Ω. This is one instance of the general *imbeddings* stated in the next theorem; the notation $X\hookrightarrow Y$ indicates that the normed vector space X is *continuously imbedded* in the normed vector space Y, in the sense that $X\subset Y$ and that there exists a constant c such that $\|v\|_Y\leq c\|v\|_X$ for all $v\in X$. We recall that the spaces $\mathscr{C}^{m,\lambda}(\bar{\Omega})$ have been defined in Sect. 1.3 and that a domain in $\mathbb{R}^n$ is a bounded open connected subset of $\mathbb{R}^n$ with a Lipschitz-continuous boundary (Sect. 1.6).

Theorem 6.1-3 (Sobolev imbedding theorem). *Let Ω be a domain in $\mathbb{R}^n$, let $m\geq 0$ be an integer, and let $1\leq p<\infty$. Then the following continuous imbeddings hold*:

$$W^{m,p}(\Omega)\hookrightarrow L^{p^*}(\Omega)\quad\text{with } \frac{1}{p^*}=\frac{1}{p}-\frac{m}{n}\ \text{if } m<\frac{n}{p},$$

$$W^{m,p}(\Omega)\hookrightarrow L^{q}(\Omega)\quad\text{for all } q \text{ with } 1\leq q<\infty \text{ if } m=\frac{n}{p},$$

$$W^{m,p}(\Omega)\hookrightarrow \mathscr{C}^{0,m-n/p}(\bar{\Omega})\quad\text{if } \frac{n}{p}<m<\frac{n}{p}+1,$$

$$W^{m,p}(\Omega)\hookrightarrow \mathscr{C}^{0,\lambda}(\bar{\Omega})\quad\text{for all } \lambda\in\,]0,1[\ \text{if } m=\frac{n}{p}+1,$$

$$W^{m,p}(\Omega)\hookrightarrow \mathscr{C}^{0,1}(\bar{\Omega})\quad\text{if } \frac{n}{p}+1<m.$$

■

Remark. Some care should be taken in interpreting these continuous imbeddings (or, for that matter, the compact imbeddings of Theorem 6.1-5) since an element of a Sobolev space is an *equivalence class* of dx-almost everywhere equal functions. For instance, the imbedding $W^{m,p}(\Omega)\hookrightarrow\mathscr{C}^{0,\lambda}(\bar{\Omega})$ means that there is a constant c such that, in each equivalence class v of the space $W^{m,p}(\Omega)$, there is a (unique) representative $\bar{v}$ in the space $\mathscr{C}^{0,\lambda}(\bar{\Omega})$ that satisfies $\|\bar{v}\|_{\mathscr{C}^{0,\lambda}(\bar{\Omega})}\leq c\|v\|_{m,p,\Omega}$. ■

An important consequence of the Sobolev imbedding theorem is that the same inequality that guarantees that the imbedding $W^{m,p}(\Omega)\hookrightarrow\mathscr{C}^{0}(\bar{\Omega})$ holds, viz., $mp>n$, also guarantees that the Sobolev space $W^{m,p}(\Omega)$ is a *Banach algebra*, i.e., the product of two functions in $W^{m,p}(\Omega)$ also belongs to $W^{m,p}(\Omega)$ if $mp > n$, and the product mapping defined in this fashion is continuous with respect to the norm $\|\cdot\|_{m,p,\Omega}$; more specifically:

Theorem 6.1-4 (**$W^{m,p}(\Omega)$ is a Banach algebra if $mp > n$**). *Let Ω be a domain in $\mathbb{R}^n$, let $m \geq 0$ be an integer, and let $1 \leq p < \infty$. If $mp > n$,*

$$u, v \in W^{m,p}(\Omega) \Rightarrow uv \in W^{m,p}(\Omega),$$

and there exists a constant c such that

$$\|uv\|_{m,p,\Omega} \leq c\|u\|_{m,p,\Omega}\|v\|_{m,p,\Omega} \quad \text{for all } u, v \in W^{m,p}(\Omega). \qquad \blacksquare$$

A normed vector space X is *compactly imbedded* in a normed vector space Y if $X \hookrightarrow Y$ and if the continuous injection $\iota : x \in X \to \iota(x) = x \in Y$ is a *compact* linear operator, i.e., if ι maps each bounded sequence (x^k) into a sequence $(\iota(x^k))$ that contains a subsequence converging to some limit in Y; such a compact imbedding is denoted $X \Subset Y$. The next result specifies whether the continuous imbeddings of Theorem 6.1-3 are compact, a property that is often useful; the number p^* is defined as in Theorem 6.1-3.

Theorem 6.1-5 (Rellich–Kondrašov imbedding theorem). *Let Ω be a domain in $\mathbb{R}^n$, let $m \geq 0$ be an integer, and let $1 \leq p < \infty$. Then the following compact imbeddings hold*:

$$W^{m,p}(\Omega) \Subset L^q(\Omega) \quad \text{for all } q \text{ with } 1 \leq q < p^* \text{ if } m < \frac{n}{p},$$

$$W^{m,p}(\Omega) \Subset L^q(\Omega) \quad \text{for all } q \text{ with } 1 \leq q < \infty \text{ if } m = \frac{n}{p},$$

$$W^{m,p}(\Omega) \Subset \mathscr{C}^0(\bar{\Omega}) \quad \text{if } \frac{n}{p} < m. \qquad \blacksquare$$

As a special case of the Rellich–Kondrašov theorem, note that *the compact imbedding*

$$H^1(\Omega) \Subset L^2(\Omega),$$

always holds, independently of the dimension n.

Another important property of functions in the Sobolev spaces

$W^{m,p}(\Omega)$ is that, under mild assumptions on the set Ω, they can be approximated by smooth functions. The space $\mathscr{C}^\infty(\bar{\Omega})$ has been defined in Sect. 1.3.

Theorem 6.1-6 (approximation by smooth functions). *Let Ω be a domain in $\mathbb{R}^n$, let $m \geq 0$ be an integer, and let $1 \leq p < \infty$. Then*

$$\{\mathscr{C}^\infty(\bar{\Omega})\}^- = W^{m,p}(\Omega),$$

the closure being taken with respect to the norm $\|\cdot\|_{m,p,\Omega}$. ∎

From now on, we shall consider various properties which, for simplicity, are only presented in the special case $m = 1$, i.e., for functions in the Sobolev space $W^{1,p}(\Omega)$.

Consider a continuous function $v : \bar{\Omega} \to \mathbb{R}$. Then its *trace* $\operatorname{tr} v$ on the boundary Γ of the set Ω is classically defined by $\operatorname{tr} v(x) = v(x)$ for all $x \in \Gamma$. A remarkable property of functions in Sobolev spaces is that generalized "traces" can still be defined, even when the functions are not continuous. The basis for this extesion is the following observation: Let Ω be a domain in $\mathbb{R}^n$; then one can show that the mapping

$$\operatorname{tr} : \mathscr{C}^\infty(\bar{\Omega}) \to L^q(\Gamma)$$

is well defined and continuous if the space $\mathscr{C}^\infty(\bar{\Omega})$ is endowed with the norm $\|\cdot\|_{1,p,\Omega}$, with

$$1 \leq p < n, \text{ and } \frac{1}{q} = \frac{1}{p} - \frac{1}{(n-1)}\left(\frac{p-1}{p}\right)$$

(the spaces $L^q(\Gamma)$ have been defined in Sect. 1.6). Since the space $\mathscr{C}^\infty(\bar{\Omega})$ is dense in the space $W^{1,p}(\Omega)$ (Theorem 6.1-6) and since the space $L^p(\Gamma)$ is complete, it follows from a classical result of analysis that there exists a unique continuous linear extension from the space $W^{1,p}(\Omega)$ into the space $L^q(\Gamma)$ that coincides with the "classical" function tr on the subspace $\mathscr{C}^\infty(\bar{\Omega})$. This extension, which shall still be denoted by the symbol tr, is called the **trace operator**. We now state various continuity and compactness properties of the trace operator; we also indicate its relation to the space $W_0^{1,p}(\Omega)$.

Theorem 6.1-7 (properties of the trace operator). (a) *Let Ω be a domain in $\mathbb{R}^n$.*

(a) *Let* $1 \leqslant p < \infty$. *Then*

$$\operatorname{tr} \in \mathscr{L}(W^{1,p}(\Omega), L^{p^\#}(\Gamma)) \quad \text{with } \frac{1}{p^\#} = \frac{1}{p} - \frac{1}{(n-1)}\left(\frac{p-1}{p}\right) \quad \text{if } 1 \leqslant p < n\,,$$

$$\operatorname{tr} \in \mathscr{L}(W^{1,p}(\Omega), L^{q}(\Gamma)) \quad \text{for all } q \text{ with } 1 \leqslant q < \infty \text{ if } p = n\,,$$

$$\operatorname{tr} \in \mathscr{L}(W^{1,p}(\Omega), \mathscr{C}^0(\Gamma)) \text{ if } n < p\,.$$

(b) *If* $1 < p < n$, *the trace operator* $\operatorname{tr}: W^{1,p}(\Omega) \to L^q(\Gamma)$ *is compact for all* q *such that* $1 \leqslant q < p^\#$.

(c) *Let* $1 \leqslant p < n$. *Then*

$$\operatorname{tr}(W^{1,p}(\Omega)) \subsetneqq L^{p^\#}(\Gamma)\,,$$

$$\{\operatorname{tr}(W^{1,p}(\Omega))\}^- = L^p(\Gamma)\,,$$

the closure being taken with respect to the norm of the space $L^p(\Gamma)$.

(d) *Let* $1 \leqslant p < \infty$. *Then*

$$W_0^{1,p}(\Omega) = \{v \in W^{1,p}(\Omega);\ \operatorname{tr} v = 0\}\,. \quad \blacksquare$$

The relation $\operatorname{tr}(W^{1,p}(\Omega)) \subsetneqq L^{p^\#}(\Gamma)$ is the basis for defining the *trace space*

$$W^{1-1/p,p}(\Gamma) := \operatorname{tr}(W^{1,p}(\Omega)) \quad \text{for } 1 < p < n\,,$$

which thus consists of the traces of all the functions in $W^{1,p}(\Omega)$.

If $1 \leqslant p < n$, the relation $\operatorname{tr} v = 0$ in (d) means that the function $\operatorname{tr} v$ is the zero function of the space $L^{p^\#}(\Gamma)$ if $1 \leqslant p < n$. Since it is customary to omit the symbol "tr" whenever no confusion should arise, and since it is useful to recall that da-almost everywhere equal functions are identified, we shall rather rewrite relation (d) as

$$W_0^{1,p}(\Omega) = \{v \in W^{1,p}(\Omega);\ v = 0\ da\text{-a.e. on } \Gamma\}\,.$$

We shall also encounter spaces such as

$$V = \{v \in W^{1,p}(\Omega);\ \operatorname{tr} v = 0 \text{ on } \Gamma_0\}$$

where Γ_0 is a da-measurable subset of Γ; then the relation "tr $v = 0$ on Γ_0" similarly means that, as a function in the space $L^{p^\#}(\Gamma)$, the function tr v vanishes on the subset Γ_0; then for the same reasons, we shall rather rewrite such a space V as

$$V = \{v \in W^{1,p}(\Omega);\ v = 0 \text{ d}a\text{-a.e. on } \Gamma_0\}.$$

The Poincaré inequality of Theorem 6.1-2 admits two useful generalizations, one of which involves the trace operator:

Theorem 6.1-8 (generalized Poincaré inequalities). *Let Ω be a domain in $\mathbb{R}^n$, and let $1 \leq p < \infty$.*

(a) *There exists a constant c_0 such that*

$$\int_\Omega |v|^p \,dx \leq c_0 \left\{ \int_\Omega |\mathbf{grad}\, v|^p \,dx + \left| \int_\Omega v \,dx \right|^p \right\}$$

for all $v \in W^{1,p}(\Omega)$.

(b) *Let Γ_0 be a* da-*measurable subset of Γ with* da-meas $\Gamma_0 > 0$. *Then there exists a constant c_1 such that*

$$\int_\Omega |v|^p \,dx \leq c_1 \left\{ \int_\Omega |\mathbf{grad}\, v|^p \,dx + \left| \int_{\Gamma_0} v \,da \right|^p \right\}$$

for all $v \in W^{1,p}(\Omega)$. ■

We conclude this review by the extension to functions with values in a Sobolev space of the fundamental Green's formula for smooth functions (cf. Sect. 1.6; we recall that a unit outer normal vector exists da-almost everywhere along the boundary of a domain).

Theorem 6.1-9 (Green's formula in Sobolev spaces). *Let Ω be a domain in $\mathbb{R}^n$, and let $\boldsymbol{n} = (n_i)$ denote the unit outer normal vector along $\partial\Omega$. Let $u \in W^{1,p}(\Omega)$ and $v \in W^{1,q}(\Omega)$* with

$$\frac{1}{p} + \frac{1}{q} \leq 1 + \frac{1}{n} \text{ if } 1 \leq p < n \text{ and } 1 \leq q < n,$$

$$1 < q \text{ if } n \leq p,$$

$$1 < p \text{ if } n \leq q.$$

Then, for $i = 1, 2, \ldots, n$, the function uvn_i belongs to the space $L^1(\partial\Omega)$, and

$$\int_\Omega (\partial_i u) v \, dx = -\int_\Omega u \, \partial_i v \, dx + \int_{\partial\Omega} uvn_i \, da . \qquad \blacksquare$$

Let $\Omega \subset \mathbb{R}^3$. If $X(\Omega)$ denotes any one of the vector spaces encountered in this section, we shall denote by $\boldsymbol{X}(\Omega)$ the corresponding space of vector-valued, or tensor-valued, functions with components in $X(\Omega)$, and if $\|\cdot\|$, or $|\cdot|$, denotes a norm, or a semi-norm, in the space $X(\Omega)$, we shall use the same symbols to denote the corresponding product norm, or semi-norm, in the space $\boldsymbol{X}(\Omega)$. For instance,

$$\boldsymbol{\mathcal{D}}(\Omega) = \{\boldsymbol{\theta} = (\theta_i);\ \theta_i \in \mathcal{D}(\Omega),\ 1 \le i \le 3\} ,$$

$$\boldsymbol{L}^2(\Omega) = \{\boldsymbol{e} = (e_{ij});\ e_{ij} \in L^2(\Omega),\ 1 \le i, j \le 3\} ,$$

$$|\boldsymbol{e}|_{0,\Omega} = \Big\{ \sum_{i,j=1}^{3} |e_{ij}|^2_{0,\Omega} \Big\}^{1/2} ,$$

$$\boldsymbol{W}^{1,p}(\Omega) = \{\boldsymbol{\Sigma} = (\sigma_{ij});\ \sigma_{ij} \in W^{1,p}(\Omega),\ 1 \le i, j \le 3\} ,$$

$$\|\boldsymbol{\Sigma}\|_{1,p,\Omega} = \Big\{ \sum_{i,j=1}^{3} \|\sigma_{ij}\|^p_{1,p,\Omega} \Big\}^{1/p} , \text{ etc} .$$

6.2. THE BOUNDARY VALUE PROBLEMS OF LINEARIZED ELASTICITY

Consider a *displacement–traction problem for a homogeneous isotropic elastic material whose reference configuration is a natural state*. We assume that the applied forces are *dead loads* and that *the boundary condition of place is of the form $\boldsymbol{\varphi} = \boldsymbol{id}$ on Γ_0*. Expressed in terms of the displacement vector $\boldsymbol{u}$, the associated boundary value problem then takes the following form (Sect. 5.1):

$$-\mathbf{div}\{(\boldsymbol{I} + \nabla\boldsymbol{u})\check{\boldsymbol{\Sigma}}(\boldsymbol{E}(\boldsymbol{u}))\} = \boldsymbol{f} \text{ in } \Omega ,$$

$$\boldsymbol{u} = \boldsymbol{o} \text{ on } \Gamma_0 ,$$

$$(\boldsymbol{I} + \nabla\boldsymbol{u})\check{\boldsymbol{\Sigma}}(\boldsymbol{E}(\boldsymbol{u}))\boldsymbol{n} = \boldsymbol{g} \text{ on } \Gamma_1 .$$

The response function $\check{\boldsymbol{\Sigma}}$ for the second Piola–Kirchhoff stress is defined in a neighborhood $\mathbb{V}(\boldsymbol{0})$ of the origin in $\mathbb{S}^3$(Theorem 1.8-3) and, under mild regularity assumptions, it satisfies (Theorem 3.8-1):

$$\boxed{\check{\boldsymbol{\Sigma}}(\boldsymbol{E}) = \lambda(\operatorname{tr}\boldsymbol{E})\boldsymbol{I} + 2\mu\boldsymbol{E} + o(\|\boldsymbol{E}\|)\,,}$$

where λ and μ are the *Lamé constants* of the material. The Green–St Venant strain tensor $\boldsymbol{E}(\boldsymbol{u})$ is given in terms of the displacement gradient $\boldsymbol{\nabla u}$ by

$$\boxed{\boldsymbol{E}(\boldsymbol{u}) = \tfrac{1}{2}(\boldsymbol{\nabla u}^{\mathrm{T}} + \boldsymbol{\nabla u} + \boldsymbol{\nabla u}^{\mathrm{T}}\boldsymbol{\nabla u})\,.}$$

Note that the analysis in this section applies to *pure displacement problems* ($\Gamma = \Gamma_0$), but *not* to *pure traction problems* ($\Gamma = \Gamma_1$), whose linearization poses specific difficulties, which are briefly discussed in Sect. 6.7.

Our objective is to formally compute the derivative at $\boldsymbol{u} = \boldsymbol{o}$ *of the* **operator of nonlinear elasticity** $\mathcal{A}$ *represented by the left-hand sides of the displacement-traction problem*. "Formally" means that the assumptions made in the next theorem require in particular that both the functions in the space denoted $\boldsymbol{W}(\Omega)$ and the response function $\check{\boldsymbol{\Sigma}}$ be smooth enough in order that the operator $\mathcal{A}$ be well defined; besides, these assumptions alone do not guarantee the differentiability of $\mathcal{A}$, whose proof requires substantial effort, as we shall see in Sect. 6.6; there, we shall establish that the mapping $\mathcal{A}$ is differentiable if the following spaces are chosen, for some $p > 3$:

$$\boldsymbol{W}(\Omega) = \{\boldsymbol{v} \in \boldsymbol{W}^{2,p}(\Omega);\ \boldsymbol{v} = \boldsymbol{o} \text{ on } \Gamma_0\},$$

$$\boldsymbol{F}(\Omega) = \boldsymbol{L}^p(\Omega),\ \boldsymbol{G}(\Gamma_1) = \mathscr{C}^0(\Gamma_1)\,,$$

Theorem 6.2-1 (linearization of a displacement–traction problem). *Let* $\check{\boldsymbol{\Sigma}} : \mathbb{V}(\boldsymbol{0}) \subset \mathbb{S}^3 \to \mathbb{S}^3$ *denote the response function of a homogeneous isotropic elastic material, whose reference configuration is a natural state, and assume that we can find:*

(a) *a neighborhood* $\boldsymbol{V}(\boldsymbol{o})$ *of the origin in a normed vector space* $\boldsymbol{W}(\Omega)$ *of vector-valued function* $\boldsymbol{v} : \bar{\Omega} \to \mathbb{R}^3$ *that satisfy the boundary condition* $\boldsymbol{v} = \boldsymbol{o}$ *on* Γ_0, *in such a way that* $\boldsymbol{E}(\boldsymbol{v}) \in \mathbb{V}(\boldsymbol{0})$ if $\boldsymbol{v} \in \boldsymbol{V}(\boldsymbol{o})$,

(b) *a normed vector space* $\boldsymbol{F}(\Omega)$ *of vector-valued functions defined on* Ω *such that*

$$\boldsymbol{v} \in \boldsymbol{V}(\boldsymbol{o}) \subset \boldsymbol{W}(\Omega) \Rightarrow \boldsymbol{A}(\boldsymbol{v}) := -\mathbf{div}\{(\boldsymbol{I} + \nabla\boldsymbol{v})\check{\boldsymbol{\Sigma}}(\boldsymbol{E}(\boldsymbol{v}))\} \in \boldsymbol{F}(\Omega),$$

(c) *a normed vector space* $\boldsymbol{G}(\Gamma_1)$ *of vector-valued functions defined on* Γ_1 *such that*

$$\boldsymbol{v} \in \boldsymbol{V}(\boldsymbol{o}) \subset \boldsymbol{W}(\Omega) \Rightarrow \boldsymbol{B}(\boldsymbol{v}) := (\boldsymbol{I} + \nabla\boldsymbol{v})\check{\boldsymbol{\Sigma}}(\boldsymbol{E}(\boldsymbol{v}))\boldsymbol{n}|_{\Gamma_1} \in \boldsymbol{G}(\Gamma_1),$$

in such a way that the operator of nonlinear elasticity

$$\mathcal{A} : \boldsymbol{v} \in \boldsymbol{V}(\boldsymbol{o}) \subset \boldsymbol{W}(\Omega) \to \mathcal{A}(\boldsymbol{v}) := (\boldsymbol{A}(\boldsymbol{v}), \boldsymbol{B}(\boldsymbol{v})) \in \boldsymbol{F}(\Omega) \times \boldsymbol{G}(\Gamma_1)$$

is differentiable at $\boldsymbol{v} = \boldsymbol{o}$. *Then the action of the Fréchet derivative* $\mathcal{A}'(\boldsymbol{o}) = (\boldsymbol{A}'(\boldsymbol{o}), \boldsymbol{B}'(\boldsymbol{o}))$ *on an arbitrary element* $\boldsymbol{v} \in \boldsymbol{W}(\Omega)$ *is given by*

$$\mathcal{A}'(\boldsymbol{o})\boldsymbol{v} = (-\mathbf{div}\{\lambda(\operatorname{tr}\boldsymbol{e}(\boldsymbol{v}))\boldsymbol{I} + 2\mu\boldsymbol{e}(\boldsymbol{v})\}, \{\lambda(\operatorname{tr}\boldsymbol{e}(\boldsymbol{v}))\boldsymbol{I} + 2\mu\boldsymbol{e}(\boldsymbol{v})\}\boldsymbol{n}) \in \boldsymbol{F}(\Omega) \times \boldsymbol{G}(\Gamma_1).$$

where

$$\boldsymbol{e}(\boldsymbol{v}) := \tfrac{1}{2}(\nabla\boldsymbol{v}^{\mathrm{T}} + \nabla\boldsymbol{v}).$$

Proof. In order to compute $\mathcal{A}'(\boldsymbol{o})\boldsymbol{v}$, it suffices, once the derivative $\mathcal{A}'(\boldsymbol{o})$ is assumed to exist, to compute the terms that are linear with respect to $\boldsymbol{v}$ in the difference $\{\mathcal{A}(\boldsymbol{v}) - \mathcal{A}(\boldsymbol{o})\}$ when $\boldsymbol{E}(\boldsymbol{v})$ and $\check{\boldsymbol{\Sigma}}(\boldsymbol{E})$ are replaced by their first-order expansions, viz.,

$$\boldsymbol{E}(\boldsymbol{v}) = \boldsymbol{e}(\boldsymbol{v}) + o(\|\boldsymbol{v}\|_{\boldsymbol{W}(\Omega)}),$$
$$\check{\boldsymbol{\Sigma}}(\boldsymbol{E}) = \lambda(\operatorname{tr}\boldsymbol{E})\boldsymbol{I} + 2\mu\boldsymbol{E} + o(\|\boldsymbol{E}\|).$$

In this fashion, we find that

$$\boldsymbol{A}(\boldsymbol{v}) - \boldsymbol{A}(\boldsymbol{o}) = -\mathbf{div}\{\lambda(\operatorname{tr}\boldsymbol{e}(\boldsymbol{v}))\boldsymbol{I} + 2\mu\boldsymbol{e}(\boldsymbol{v})\} + o(\|\boldsymbol{v}\|_{\boldsymbol{W}(\Omega)}) \text{ in } \boldsymbol{F}(\Omega),$$
$$\boldsymbol{B}(\boldsymbol{v}) - \boldsymbol{B}(\boldsymbol{o}) = \{\lambda(\operatorname{tr}\boldsymbol{e}(\boldsymbol{v}))\boldsymbol{I} + 2\mu\boldsymbol{e}(\boldsymbol{v})\}\boldsymbol{n} + o(\|\boldsymbol{v}\|_{\boldsymbol{W}(\Omega)}) \text{ in } \boldsymbol{G}(\Gamma_1),$$

and the assertion follows. ∎

Assume that we are in a situation where we can apply Theorem 6.2-1. Then, by definition, the **linearized displacement–traction problem** associated with the displacement-traction problem that we are considering is the following *linear boundary value problem*: Given $(\boldsymbol{f}, \boldsymbol{g}) \in \boldsymbol{F}(\Omega) \times \boldsymbol{G}(\Gamma_1)$, find $\boldsymbol{u} \in \boldsymbol{W}(\Omega)$ such that

$$\boldsymbol{\mathcal{A}}'(\boldsymbol{o})\boldsymbol{u} = (\boldsymbol{f}, \boldsymbol{g}) \Leftrightarrow \begin{cases} \boldsymbol{A}'(\boldsymbol{o})\boldsymbol{u} = \boldsymbol{f}\,, \\ \boldsymbol{B}'(\boldsymbol{o})\boldsymbol{u} = \boldsymbol{g}\,, \end{cases}$$

or, more explicitly, such that

$$\begin{aligned} &-\mathbf{div}\{\lambda(\operatorname{tr} \boldsymbol{e}(\boldsymbol{u}))\boldsymbol{I} + 2\mu \boldsymbol{e}(\boldsymbol{u})\} = \boldsymbol{f} \text{ in } \Omega\,, \\ &\boldsymbol{u} = \boldsymbol{o} \text{ on } \Gamma_0\,, \\ &\{\lambda(\operatorname{tr} \boldsymbol{e}(\boldsymbol{u}))\boldsymbol{I} + 2\mu \boldsymbol{e}(\boldsymbol{u})\}\boldsymbol{n} = \boldsymbol{g} \text{ on } \Gamma_1\,, \end{aligned}$$

where

$$\boldsymbol{e}(\boldsymbol{u}) := \tfrac{1}{2}(\boldsymbol{\nabla u}^{\mathrm{T}} + \boldsymbol{\nabla u})$$

denotes the **linearized strain tensor**. The operator

$$\boldsymbol{A}'(\boldsymbol{o}): \boldsymbol{v} \rightarrow -\mathbf{div}\{\lambda(\operatorname{tr} \boldsymbol{e}(\boldsymbol{v}))\boldsymbol{I} + 2\mu \boldsymbol{e}(\boldsymbol{v})\}$$

is called the **operator of linearized elasticity**. Observe that its expression is independent of the topologies of the spaces $\boldsymbol{W}(\Omega)$, $\boldsymbol{F}(\Omega)$, $\boldsymbol{G}(\Gamma_1)$.

Componentwise, the linearized displacement–traction problem reads:

$$\begin{aligned} &-\partial_j(a_{ijpq} e_{pq}(u)) = f_i \text{ in } \Omega\,, \\ &u_i = 0 \text{ on } \Gamma_0\,, \\ &a_{ijpq} e_{pq}(u) n_j = g_i \text{ on } \Gamma_1\,, \end{aligned}$$

with

$$\begin{aligned} &a_{ijpq} = \lambda \delta_{ij}\delta_{pq} + \mu(\delta_{ip}\delta_{jq} + \delta_{iq}\delta_{jp})\,, \\ &e_{pq}(u) = \tfrac{1}{2}(\partial_p u_q + \partial_q u_p)\,. \end{aligned}$$

Notice that *the constants* a_{ijpq} *are the components of the elasticity tensor* $\mathsf{A}(\boldsymbol{I})$ (Sect. 5.9).

We strongly emphasize that, in our view, *this linear problem represents the "derivative at the origin" of the displacement–traction problem of three-dimensional elasticity*. As such, it is an extremely useful mathematical intermediary for establishing existence results in nonlinear elasticity (as we shall see in this chapter); it is also an incomparable means of numerically approximating the solutions of problems in engineering (see e.g., Ciarlet [1978], Zienkiewicz [1977]); it is, finally, an invaluable tool for the qualitative appraisal of elastic structures (see e.g., von Kármán & Biot [1940], Landau & Lifchitz [1967], Fraeijs de Veubeke [1979], Valid [1977], Bamberger [1981]). But otherwise *it cannot be considered as a model*; among other things, *it does not satisfy the axiom of material frame-indifference*, as shown by Fosdick & Serrin [1979]; cf. Exercise 3.7). For these reasons, we shall avoid the misleading, but common, terminology: boundary value problem of "linear" elasticity. *Elasticity cannot be linear*!

Likewise, we should not think of the unknown $\boldsymbol{u}$ *appearing in the linearized displacement–traction problem as a displacement, nor of the linearized strain tensor* $\boldsymbol{e}(\boldsymbol{u})$ *as a strain tensor, nor of the expression* $\{\lambda(\operatorname{tr}\boldsymbol{e}(\boldsymbol{u}))\boldsymbol{I} + 2\mu\boldsymbol{e}(\boldsymbol{u})\}$ *as representing any one of the three stress tensors*. What *can* be expected is that, in carefully circumscribed situations, the unknowns $\boldsymbol{u}$ and $\{\lambda(\operatorname{tr}\boldsymbol{e}(\boldsymbol{u}))\boldsymbol{I} + 2\mu\boldsymbol{e}(\boldsymbol{u})\}$ can be shown to be *approximations* of the "true" displacement vector and second Piola–Kirchhoff stress tensor fields, and indeed, the description of such situations is one of our objectives in this chapter (Theorem 6.8-1).

Remark. The linearized constitutive equation $\boldsymbol{u} \to \lambda(\operatorname{tr}\boldsymbol{e}(\boldsymbol{u}))\boldsymbol{I} + 2\mu\boldsymbol{e}(\boldsymbol{u})$ is often called *Hooke's law*. ∎

Notice how crucial in our derivation of this linear problem is the assumption that the reference configuration is a *natural state*: This assumption is reflected by the absence of zero-order term in the expansion of the response function $\check{\boldsymbol{\Sigma}}$ around $\boldsymbol{E} = \boldsymbol{0}$; besides, it implies that $\boldsymbol{u} = \boldsymbol{o}$ *is a particular solution corresponding* to $\boldsymbol{f} = \boldsymbol{o}$ *and* $\boldsymbol{g} = \boldsymbol{o}$, a crucial fact for the application of the implicit function theorem, as we shall see in Sect. 6.7.

The linearized displacement–traction problem that we have derived, and the **linearized pure displacement problem** or **linearized pure traction problem** that can be similarly derived (with an already noted specific difficulty in the latter case) constitute the **boundary value problems of linearized elasticity**.

*6.3. A BRIEF MATHEMATICAL REVIEW OF LINEARIZED ELASTICITY

There are numerous mathematical treatments of linearized elasticity, among which the treatise of Gurtin [1972] stands as a classic. Other modern, mathematically-oriented, treatments are given in Knops & Payne [1971], Fichera [1972a, 1972b], Duvaut & Lions [1972], Villagio [1977], Nečas & Hlaváček [1981]. Here we shall simply review some basic properties of the linearized displacement–traction problem, as regards questions of existence, uniqueness, and regularity of its solutions. To this end, we shall proceed along lines that are familiar in the mathematical analysis of partial differential equations: By a simple integration by parts, this linear problem is first formally written in a *weak form* (Theorem 6.3-1), which then makes it amenable to an *existence result in the space* $\boldsymbol{H}^1(\Omega)$ (Theorem 6.3-5); this existence result is easy in its principle (Theorem 6.3-2) but not so easy in its application, which requires in particular the crucial *Korn inequality* (Theorem 6.3-3). Finally, a *regularity result* is established (Theorem 6.3-6), which shows that in some special situations, the *weak solution found in the space* $\boldsymbol{H}^1(\Omega)$ possesses enough additional regularity to be considered as a "classical" solution (that is, differentiable in the usual sense).

Theorem 6.3-1 (weak form of the linearized displacement–traction problem). *Finding a solution* $\boldsymbol{u}$ *of the linear boundary value problem*:

$$
\boxed{\begin{aligned}
&-\mathbf{div}\{\lambda(\operatorname{tr} \boldsymbol{e}(\boldsymbol{u}))\boldsymbol{I} + 2\mu \boldsymbol{e}(\boldsymbol{u})\} = \boldsymbol{f} \text{ in } \Omega\,,\\
&\boldsymbol{u} = \boldsymbol{o} \text{ on } \Gamma_0\,,\\
&\{\lambda(\operatorname{tr} \boldsymbol{e}(\boldsymbol{u}))\boldsymbol{I} + 2\mu \boldsymbol{e}(\boldsymbol{u})\}\boldsymbol{n} = \boldsymbol{g} \text{ on } \Gamma_1\,,
\end{aligned}}
$$

is formally equivalent to finding a solution $\boldsymbol{u}$ *of the equations*

$$
\boxed{B(\boldsymbol{u}, \boldsymbol{v}) = L(\boldsymbol{v}) \quad \text{for all } \boldsymbol{v} \in \boldsymbol{V}\,,}
$$

where

$$
\boxed{\begin{aligned}
&B(\boldsymbol{u}, \boldsymbol{v}) := \int_\Omega \{\lambda \operatorname{tr} \boldsymbol{e}(\boldsymbol{u}) \operatorname{tr} \boldsymbol{e}(\boldsymbol{v}) + 2\mu \boldsymbol{e}(\boldsymbol{u}) : \boldsymbol{e}(\boldsymbol{v})\}\, \mathrm{d}x\,,\\
&L(\boldsymbol{v}) := \int_\Omega \boldsymbol{f} \cdot \boldsymbol{v}\, \mathrm{d}x + \int_{\Gamma_1} \boldsymbol{g} \cdot \boldsymbol{v}\, \mathrm{d}a\,,
\end{aligned}}
$$

and V denotes a space of smooth enough vector-valued functions $\boldsymbol{v}: \bar{\Omega} \to \mathbb{R}^3$ that vanish on Γ_0.

Proof. Using the fundamental Green's formula, we deduce that, for any smooth enough symmetric tensor field $\boldsymbol{S}: \bar{\Omega} \to \mathbb{S}^3$ and vector field $\boldsymbol{v}: \bar{\Omega} \to \mathbb{R}^3$, the *Green formula*

$$\int_\Omega \mathbf{div}\, \boldsymbol{S} \cdot \boldsymbol{v}\, dx = -\int_\Omega \boldsymbol{S} : \nabla \boldsymbol{v}\, dx + \int_\Gamma \boldsymbol{S}\boldsymbol{n} \cdot \boldsymbol{v}\, da$$
$$= -\int_\Omega \boldsymbol{S} : \boldsymbol{e}(\boldsymbol{v})\, dx + \int_\Gamma \boldsymbol{S}\boldsymbol{n} \cdot \boldsymbol{v}\, da$$

holds. The proof then follows along the same lines as those of Theorems 2.4-1 or 2.6-1. ∎

Remark. The equations "$B(\boldsymbol{u}, \boldsymbol{v}) = L(\boldsymbol{v})$ for all $\boldsymbol{v} \in V$" are sometimes referred to as another "principle of virtual work". This terminology is erroneous: The true principles of virtual work are those of Theorems 2.4-1 or 2.6-1. ∎

We next prove an existence result that applies to a large class of *abstract problems* of the form found in Theorem 6.3-1, viz., find $\boldsymbol{u} \in V$ such that $B(\boldsymbol{u}, \boldsymbol{v}) = L(\boldsymbol{v})$ for all $\boldsymbol{v} \in V$, where B is a *continuous symmetric bilinear form* and L a *continuous linear form* defined on a space V. Under the crucial assumptions that the space V is *complete* and that the form B is *V-elliptic* (the definition is given in the next theorem), we show that the existence of a solution is a direct consequence of the Riesz representation theorem in a Hilbert space, and that the solution is also the unique minimizer of the *quadratic functional* $J(\boldsymbol{v}) = \frac{1}{2}B(\boldsymbol{v}, \boldsymbol{v}) - L(\boldsymbol{v})$.

Theorem 6.3-2. *Let V be a Banach space with norm $\|\cdot\|$, let $L: V \to \mathbb{R}$ be a continuous linear form, and let $B: V \times V \to \mathbb{R}$ be a symmetric and continuous bilinear form that is* **V-elliptic**, *in the sense that there exists a constant β such that*

$$\boxed{\beta > 0 \text{ and } B(\boldsymbol{v}, \boldsymbol{v}) \geq \beta \|\boldsymbol{v}\|^2 \quad \text{for all } \boldsymbol{v} \in V\,,}$$

Then the problem: Find $\boldsymbol{u} \in V$ such that

$$\boxed{B(\boldsymbol{u}, \boldsymbol{v}) = L(\boldsymbol{v}) \quad \text{for all } \boldsymbol{v} \in V\,,}$$

has one and only one solution, which is also the unique solution of the problem: Find $\boldsymbol{u} \in V$ ***such that***

$$\boxed{J(\boldsymbol{u}) = \inf_{\boldsymbol{v} \in V} J(\boldsymbol{v}), \quad \text{where } J : \boldsymbol{v} \in V \to J(\boldsymbol{v}) := \tfrac{1}{2} B(\boldsymbol{v}, \boldsymbol{v}) - L(\boldsymbol{v}) .}$$

Proof. (i) By the V-ellipticity and the continuity of the bilinear form,

$$\beta \|\boldsymbol{v}\|^2 \leq B(\boldsymbol{v}, \boldsymbol{v}) \leq \|B\| \, \|\boldsymbol{v}\|^2 \quad \text{for all } \boldsymbol{v} \in V ;$$

hence the symmetric bilinear form B is an inner product over the space V, and the associated norm $\boldsymbol{v} \in V \to \{B(\boldsymbol{v}, \boldsymbol{v})\}^{1/2}$ is equivalent to the given norm. Thus the space V is a Hilbert space when it is equipped with this inner product. By the *Riesz representation theorem* (for a proof, cf. e.g. Brezis [1983, p. 81]), there exists one and only one element $\boldsymbol{l} \in V$ such that

$$L(\boldsymbol{v}) = B(\boldsymbol{l}, \boldsymbol{v}) \quad \text{for all } \boldsymbol{v} \in V ,$$

and thus $\boldsymbol{u} = \boldsymbol{l}$ is the unique solution to our problem.

(ii) Since $J'(\boldsymbol{u})\boldsymbol{v} = B(\boldsymbol{u}, \boldsymbol{v}) - L(\boldsymbol{v})$ and $J''(\boldsymbol{u})(\boldsymbol{v}, \boldsymbol{v}) = B(\boldsymbol{v}, \boldsymbol{v})$, any one of the Taylor expansions of the functional J reduces to (Theorem 1.3-3):

$$J(\boldsymbol{u} + \boldsymbol{v}) = J(\boldsymbol{u}) + \{B(\boldsymbol{u}, \boldsymbol{v}) - L(\boldsymbol{v})\} + \tfrac{1}{2} B(\boldsymbol{v}, \boldsymbol{v})$$

(this identity can be also directly verified). Therefore,

$$B(\boldsymbol{u}, \boldsymbol{v}) = L(\boldsymbol{v}) \quad \text{for all } \boldsymbol{v} \in V \Rightarrow$$

$$J(\boldsymbol{u} + \boldsymbol{v}) - J(\boldsymbol{u}) = \tfrac{1}{2} B(\boldsymbol{v}, \boldsymbol{v}) \geq \frac{\alpha}{2} \|\boldsymbol{v}\|^2 \quad \text{for all } \boldsymbol{v} \in V ,$$

i.e., $\boldsymbol{u}$ is a minimizer of the functional J in this case. Conversely, let $\boldsymbol{u}$ be a minimizer of J and let $\boldsymbol{v}$ be an arbitrary element of V. The inequalities

$$0 \leq J(\boldsymbol{u} + \theta \boldsymbol{v}) - J(\boldsymbol{u}) = \theta(\{B(\boldsymbol{u}, \boldsymbol{v}) - L(\boldsymbol{v})\} + \frac{\theta}{2} B(\boldsymbol{v}, \boldsymbol{v}))$$
$$\text{for all } \theta \in \mathbb{R} ,$$

then imply that $B(\boldsymbol{u}, \boldsymbol{v}) = L(\boldsymbol{v})$ (if $B(\boldsymbol{u}, \boldsymbol{v}) - L(\boldsymbol{v})$ were $\neq 0$, the difference $J(\boldsymbol{u} + \theta \boldsymbol{v}) - J(\boldsymbol{u})$ would be <0 for sufficiently small values of $|\theta|$). Note that the same conclusions can be also drawn from Theorems 1.3-1 and 1.3-2. ∎

A solution $\boldsymbol{u}$ to the equations "$B(\boldsymbol{u}, \boldsymbol{v}) = L(\boldsymbol{v})$ for all $\boldsymbol{v} \in V$" still exists if the bilinear form is not symmetric (these equations are no longer associated with a minimization problem, however). This generalization is the celebrated *Lax–Milgram lemma* (Lax & Milgram [1954]).

Another generalization consists in minimizing the functional J over a *nonempty closed convex subset* U of the space V. The minimization problem is then equivalent (Theorem 4.7-8) to finding $\boldsymbol{u} \in U$ that solves the *variational inequalities*:

$$B(\boldsymbol{u}, \boldsymbol{v} - \boldsymbol{u}) \geqslant L(\boldsymbol{v} - \boldsymbol{u}) \quad \text{for all } \boldsymbol{v} \in U .$$

For a proof of the existence of a solution, see e.g. Ciarlet [1978, Theorem 1.1.1]; the original existence proof of Lions & Stampacchia [1967] is also applicable to variational inequalities where the bilinear form is no longer symmetric.

The theory of variational inequalities is now a well established field, with numerous applications to "constrained" problems in *linearized elasticity*; see in particular Duvaut & Lions [1972], Fichera [1972b], Glowinski, Lions & Trémolières [1976], Baiocchi & Capelo [1978], Kinderlehrer & Stampacchia [1980], Glowinski [1984], Panagiotopoulos [1985], Troianiello [1987], Rodrigues [1987].

The consideration of the same constraints in genuine three-dimensional elasticity is a field of considerable research interest, as exemplified by the question of contact without friction with an obstacle considered in Sect. 5.3 (see also Sect. 7.8).

Remark. The functional J is *convex* (Theorem 4.7-7) and it satisfies a *coerciveness inequality*:

$$J(\boldsymbol{v}) = \tfrac{1}{2}B(\boldsymbol{v}, \boldsymbol{v}) - L(\boldsymbol{v}) \geqslant \frac{\beta}{2}\|\boldsymbol{v}\|^2 - \|L\|\,\|\boldsymbol{v}\| \quad \text{for all } \boldsymbol{v} \in V . \qquad \blacksquare$$

In order to decide in which particular space V we should seek a solution of the weak form of the linearized displacement–traction problem, we first notice that *the symmetric bilinear form B of Theorem* 6.3.1 *is continuous with respect to the norm* $\|\cdot\|_{1,\Omega}$ (this follows from repeated applications of Cauchy–Schwarz inequality and from the definition of the norm $\|\cdot\|_{1,\Omega}$; cf. Sect. 6.1). Hence we are naturally led to let

$$V := \{\boldsymbol{v} \in \boldsymbol{H}^1(\Omega);\ \boldsymbol{v} = \boldsymbol{o} \text{ da-a.e. on } \Gamma_0\} .$$

We then observe that

$$B(v, v) = \int_\Omega \{\lambda \operatorname{tr} e(v) \operatorname{tr} e(v) + 2\mu e(v) : e(v)\}\, dx$$
$$\geqslant 2\mu \int_\Omega e(v) : e(v)\, dx$$

since the Lamé constants λ, μ of actual materials are >0 (Sect. 3.8); hence the V-ellipticity of B will follow if we can show that, *on the space* V, *the semi-norm*

$$v \in H^1(\Omega) \to |e(v)|_{0,\Omega} := \left\{\int_\Omega e(v) : e(v)\, dx\right\}^{1/2}$$

is a norm, equivalent to the norm $\|\cdot\|_{1,\Omega}$. This result will in turn be a consequence of the following fundamental inequality, which is due to Korn [1907, 1908, 1914].

Theorem 6.3-3 (Korn's inequality). *Let Ω be a domain in $\mathbb{R}^3$. For each $v = (v_i) \in H^1(\Omega)$, let*

$$e(v) := (\tfrac{1}{2}(\partial_j v_i + \partial_i v_j)) \in L^2(\Omega)\,.$$

Then there exists a constant $c > 0$ such that

$$\boxed{\|v\|_{1,\Omega} \leqslant c\{|v|^2_{0,\Omega} + |e(v)|^2_{0,\Omega}\}^{1/2} \quad \text{for all } v \in H^1(\Omega)\,,}$$

and thus, on the space $H^1(\Omega)$, the mapping

$$v \to \{|v|^2_{0,\Omega} + |e(v)|^2_{0,\Omega}\}^{1/2}$$

is a norm, equivalent to the norm $\|\cdot\|_{1,\Omega}$. ■

Various proofs have been given of this delicate inequality; see in particular Friedrichs [1947], Gobert [1962], Hlaváček & Nečas [1970a, 1970b], Duvaut & Lions [1972, p. 110], Fichera [1972a, Sect. 12], Nečas & Hlaváček [1981, Sect. 6.3]; in Temam [1983], Korn's inequality is proved in the space $W^{1,p}(\Omega)$, $1 < p < \infty$; an elementary proof is given in Nitsche [1981] (see also Miyoshi [1985, Appendix (A)]).

The essence of this remarkable inequality is that *the two Hilbert spaces*

$$H^1(\Omega) = \{v = (v_i) \in L^2(\Omega);\ \partial_j v_i \in L^2(\Omega),\ 1 \leq i, j \leq 3\}$$

and

$$K(\Omega) := \{v = (v_i) \in L^2(\Omega);\ e_{ij}(v) = \tfrac{1}{2}(\partial_i v_j + \partial_j v_i) \in L^2(\Omega),\ 1 \leq i \leq j \leq 3\}\,.$$

are identical. This being the case, Korn's inequality is a consequence of the closed graph theorem (for a proof, see e.g. Brezis [1983, p. 19]) applied to the identity mapping from $H^1(\Omega)$ into $K(\Omega)$, which is bijective since $H^1(\Omega) = K(\Omega)$, and continuous, since there clearly exists a constant d such that

$$\|v\|_{K(\Omega)} := \{|v|_{0,\Omega}^2 + |e(v)|_{0,\Omega}^2\}^{1/2} \leq d\|v\|_{1,\Omega} \text{ for all } v \in H^1(\Omega).$$

Notice the definitely unexpected character of the inclusion $K(\Omega) \subset H^1(\Omega)$ (the inclusion $H^1(\Omega) \subset K(\Omega)$ clearly holds), since a function in $K(\Omega)$ has only 6 independent linear combinations of partial derivatives in the space $L^2(\Omega)$, while a function $v \in H^1(\Omega)$ must have all its 9 partial derivatives $\partial_j v_i$ in $L^2(\Omega)$.

With Korn's inequality, we now prove the V-ellipticity of the bilinear form $v \to \int_\Omega e(v) : e(v)\, dx$, whence that of the bilinear form B.

Theorem 6.3-4. *Let Ω be a domain in $\mathbb{R}^3$, let Γ_0 be a da-measurable subset of $\Gamma = \partial\Omega$. Then the space*

$$V := \{v \in H^1(\Omega);\ v = o \text{ da-a.e. on } \Gamma_0\}$$

is a closed subspace of $H^1(\Omega)$. If area $\Gamma_0 > 0$, there exists a constant $c > 0$ such that

$$c^{-1}\|v\|_{1,\Omega} \leq |e(v)|_{0,\Omega} \leq c\|v\|_{1,\Omega} \quad \text{for all } v \in V\,,$$

i.e., on the space V, the semi-norm $v \to |e(v)|_{0,\Omega}$ is a norm, equivalent to the norm $\|\cdot\|_{1,\Omega}$.

Proof. (i) Let (v^k) be a sequence of elements $v^k \in V$ such that $v^k \to v$ in $H^1(\Omega)$. By Theorem 6.1-7, $\operatorname{tr} v^k \to \operatorname{tr} v$ in $L^4(\Gamma)$, and so there exists a subsequence (v^l) such that $\operatorname{tr} v(x) = \lim_{l\to\infty} \operatorname{tr} v^l(x)$ for da-almost every

$x \in \Gamma$. Hence, $\mathrm{tr}\, \boldsymbol{v} = \boldsymbol{o}$ $\mathrm{d}a$-a.e. on Γ_0, and thus the space $\boldsymbol{V}$ is closed in $\boldsymbol{H}^1(\Omega)$.

(ii) We next establish that the *semi-norm* $\boldsymbol{v} \to |\boldsymbol{e}(\boldsymbol{v})|_{0,\Omega}$ *is a norm on the space* $\boldsymbol{V}$, by showing that $|\boldsymbol{e}(\boldsymbol{v})|_{0,\Omega} = 0$ implies $\boldsymbol{v} = \boldsymbol{0}$ if $\boldsymbol{v} \in \boldsymbol{V}$. To this end, we observe that the following identities hold:

$$\partial_{jk} v_i = \partial_j e_{ik}(\boldsymbol{v}) + \partial_k e_{ij}(\boldsymbol{v}) - \partial_i e_{jk}(\boldsymbol{v}) \text{ in } \mathcal{D}'(\Omega) \quad \text{for } 1 \leq i, j, k \leq 3 .$$

Hence we deduce that

$$|\boldsymbol{e}(\boldsymbol{v})|_{0,\Omega} = 0 \Rightarrow \boldsymbol{e}(\boldsymbol{v}) = \boldsymbol{0} \Rightarrow \partial_{jk} v_i = 0 \text{ in } \mathcal{D}'(\Omega) \quad \text{for } 1 \leq i, j, k \leq 3 .$$

By a classical result from distribution theory (cf. Schwartz [1966, p. 60]; this requires that the set Ω be connected, as is the case for a domain), we deduce that each function v_i is a polynomial of degree ≤ 1 in the variables x_i, i.e., $v_i(x) = a_i + b_{ij} x_j$. Since $e_{ij}(\boldsymbol{v}) = 0$ further implies that $b_{ij} = -b_{ji}$, we reach the conclusion that there are constants a_i and b_i such that

$$v_1(x) = a_1 - b_3 x_2 + b_2 x_3 ,$$
$$v_2(x) = a_2 + b_3 x_1 - b_1 x_3 ,$$
$$v_3(x) = a_3 - b_2 x_1 + b_1 x_2 ;$$

equivalently, there exist two vectors $\boldsymbol{a}, \boldsymbol{b} \in \mathbb{R}^3$ such that

$$\boldsymbol{v}(x) = \boldsymbol{a} + \boldsymbol{b} \wedge \boldsymbol{ox} \quad \text{for all } x \in \bar{\Omega} .$$

Hence the set

$$\{x \in \mathbb{R}^3;\, \boldsymbol{v}(x) = \boldsymbol{o}\} = \begin{cases} \left\{ \boldsymbol{ox} = \dfrac{\boldsymbol{b} \wedge \boldsymbol{a}}{|\boldsymbol{b}|^2} + t\boldsymbol{b};\, t \in \mathbb{R} \right\} \text{ if } \boldsymbol{b} \neq \boldsymbol{o} \text{ and } \boldsymbol{a} \cdot \boldsymbol{b} = 0 , \\ \emptyset \text{ if } \boldsymbol{b} \neq \boldsymbol{o} \text{ and } \boldsymbol{a} \cdot \boldsymbol{b} \neq 0 , \\ \emptyset \text{ if } \boldsymbol{b} = \boldsymbol{o} \text{ and } \boldsymbol{a} \neq \boldsymbol{o} , \end{cases}$$

is always of zero area, and thus

$$\text{area } \Gamma_0 > 0 \Rightarrow \{\boldsymbol{v} \in \boldsymbol{H}^1(\Omega);\, \boldsymbol{e}(\boldsymbol{v}) = \boldsymbol{0},\, \boldsymbol{v} = \boldsymbol{o} \text{ d}a\text{-a.e. on } \Gamma_0\} = \{\boldsymbol{o}\} ,$$

as was to be proven.

(iii) The inequality $|\boldsymbol{e}(\boldsymbol{v})|_{0,\Omega} \leq c\|\boldsymbol{v}\|_{1,\Omega}$ clearly holds for all $\boldsymbol{v} \in \boldsymbol{V}$ (in fact, for all $\boldsymbol{v} \in \boldsymbol{H}^1(\Omega)$). If the other equality is false, there exists a sequence $(\boldsymbol{v}^k)$ of functions $\boldsymbol{v}^k \in \boldsymbol{V}$ such that

$$\|\boldsymbol{v}^k\|_{1,\Omega} = 1 \quad \text{for all } k, \quad \text{and } \lim_{k\to\infty} |\boldsymbol{e}(\boldsymbol{v}^k)|_{0,\Omega} = 0 .$$

Since the sequence $(\boldsymbol{v}^k)$ is then bounded in the space $\boldsymbol{H}^1(\Omega)$, there exists a subsequence $(\boldsymbol{v}^l)$ that converges in the space $\boldsymbol{L}^2(\Omega)$, as a consequence of the *Rellich–Kondrašov theorem* (Theorem 6.1-5). Since the subsequence $(\boldsymbol{e}(\boldsymbol{v}^l))$ also converges in the spaces $L^2(\Omega)$ (to $\boldsymbol{0}$, but this fact is not needed in this part of the argument), we conclude that the sequence $(\boldsymbol{v}^l)$ is a *Cauchy sequence* with respect to the norm

$$\boldsymbol{v} \to \{|\boldsymbol{v}|_{0,\Omega}^2 + |\boldsymbol{e}(\boldsymbol{v})|_{0,\Omega}^2\}^{1/2} .$$

By *Korn's inequality* (Theorem 6.3-3), this norm is equivalent to the norm $\|\cdot\|_{1,\Omega}$ on the space $\boldsymbol{H}^1(\Omega)$. Hence this Cauchy sequence converges to some element $\boldsymbol{v} \in \boldsymbol{V}$, since the space $\boldsymbol{V}$ is *complete* (as a closed subspace of $\boldsymbol{H}^1(\Omega)$; cf. part (i)). Therefore, the limit $\boldsymbol{v}$ satisfies

$$|\boldsymbol{e}(\boldsymbol{v})|_{0,\Omega} = \lim_{l\to\infty} |\boldsymbol{e}(\boldsymbol{v}^l)|_{0,\Omega} = 0 ,$$

and hence $\boldsymbol{v} = \boldsymbol{o}$ (part (ii)). But this contradicts the equalities $\|\boldsymbol{v}^l\|_{1,\Omega} = 1$ for all l.

(iv) *If $\Gamma = \Gamma_0$, in which case $\boldsymbol{V} = \boldsymbol{H}_0^1(\Omega)$, the equivalence of norms can be obtained by a simple proof that by-passes Korn's inequality*; for in this case, we shall show by a direct computation that

$$|\boldsymbol{v}|_{1,\Omega} \leq \sqrt{2}|\boldsymbol{e}(\boldsymbol{v})|_{0,\Omega} \quad \text{for all } \boldsymbol{v} \in \mathcal{D}(\Omega) ,$$

and the conclusion will follow since, by definition, the space $\mathcal{D}(\Omega)$ is dense in $\boldsymbol{H}_0^1(\Omega)$ (both semi-norms appearing in the inequality are continuous with respect to the norm $\|\cdot\|_{1,\Omega}$) and since, on the space $\boldsymbol{H}_0^1(\Omega)$, the semi-norm $|\cdot|_{1,\Omega}$ is a norm equivalent to $\|\cdot\|_{1,\Omega}$ by Poincaré's inequality (Theorem 6.1-2).

Let $\boldsymbol{v} : \bar{\Omega} \to \mathbb{R}^3$ be a smooth vector field; a simple computation shows that

$$2\boldsymbol{e}(\boldsymbol{v}) : \boldsymbol{e}(\boldsymbol{v}) - \boldsymbol{\nabla v} : \boldsymbol{\nabla v} = \operatorname{div}\{(\boldsymbol{\nabla v})\boldsymbol{v} - (\operatorname{div} \boldsymbol{v})\boldsymbol{v}\} + (\operatorname{div} \boldsymbol{v})^2 .$$

Since

$$\boldsymbol{v} = \boldsymbol{o} \text{ on } \Gamma \Rightarrow \int_\Omega \operatorname{div}\{(\boldsymbol{\nabla v})\boldsymbol{v} - (\operatorname{div} \boldsymbol{v})\boldsymbol{v}\}\, dx = 0$$

by the divergence theorem for vector fields, we conclude that

$$2|\boldsymbol{e}(\boldsymbol{v})|_{0,\Omega}^2 - |\boldsymbol{v}|_{1,\Omega}^2 = 2\int_\Omega \boldsymbol{e}(\boldsymbol{v}) : \boldsymbol{e}(\boldsymbol{v})\, dx - \int_\Omega \boldsymbol{\nabla v} : \boldsymbol{\nabla v}\, dx$$

$$= \int_\Omega (\operatorname{div} \boldsymbol{v})^2\, dx \geqslant 0 \quad \text{for all } \boldsymbol{v} \in \mathcal{D}(\Omega),$$

as was to be proven. ∎

By step (ii), a vector field $\boldsymbol{v} \in \mathcal{D}(\Omega)$ satisfies $\boldsymbol{e}(\boldsymbol{v}) = \boldsymbol{0}$ if and only if there exist two vectors $\boldsymbol{a}, \boldsymbol{b} \in \mathbb{R}^3$ such that $\boldsymbol{v}(x) = \boldsymbol{a} + \boldsymbol{b} \wedge \boldsymbol{ox}$ for all $x \in \Omega$. Such a vector field is called an *infinitesimal rigid displacement*; its relation to a *rigid deformation*, i.e., a deformation $\boldsymbol{\psi} = \boldsymbol{id} + \boldsymbol{v}$ that satisfies $\boldsymbol{E}(\boldsymbol{v}) = \boldsymbol{0}$ (Theorem 1.8-1), is the object of Exercise 6.2.

Assembling the previous results, we next establish the existence of a solution $\boldsymbol{u} \in \boldsymbol{H}^1(\Omega)$ of the weak form of the linearized displacement–traction problem; such a solution is called a **weak solution** (the linearized pure traction problem is considered in Exercise 6.3).

Theorem 6.3-5 (existence of a weak solution). *Let Ω be a domain in $\mathbb{R}^3$, let Γ_0 be a* d*a-measurable subset of $\Gamma = \partial\Omega$ with* area *$\Gamma_0 > 0$, and let the constants $\lambda > 0$, $\mu > 0$ and the functions $\boldsymbol{f} \in \boldsymbol{L}^{6/5}(\Omega)$, $\boldsymbol{g} \in \boldsymbol{L}^{4/3}(\Gamma_1)$, with $\Gamma_1 = \Gamma - \Gamma_0$, be given. Then there is one and only one function $\boldsymbol{u}$ in the space*

$$\boxed{V = \{\boldsymbol{v} \in \boldsymbol{H}^1(\Omega);\ \boldsymbol{v} = \boldsymbol{o} \text{ da-a.e. on } \Gamma_0\}}$$

that satisfies

$$\boxed{B(\boldsymbol{u}, \boldsymbol{v}) = L(\boldsymbol{v}) \quad \text{for all } \boldsymbol{v} \in V,}$$

where

$$B(u, v) = \int_\Omega \{\lambda \operatorname{tr} e(u) \operatorname{tr} e(v) + 2\mu e(u) : e(v)\}\, dx ,$$

$$L(v) = \int_\Omega f \cdot v\, dx + \int_{\Gamma_1} g \cdot v\, da .$$

In addition

$$J(u) = \inf_{v \in V} J(v) , \quad \text{where } J(v) = \tfrac{1}{2} B(v, v) - L(v) .$$

Proof. The Sobolev imbedding theorem (Theorem 6.1-3) and the continuity of the trace operator (Theorem 6.1-7) imply that the linear form L is continuous on the space $H^1(\Omega)$ if $f \in L^{6/5}(\Omega)$ and $g \in L^{4/3}(\Gamma_1)$. Since $\lambda > 0$, $\mu > 0$, and area $\Gamma_0 > 0$, the symmetric and continuous bilinear form B is V-elliptic by Theorem 6.3-4. Hence the conclusion follows by Theorem 6.3-2. ∎

Remark. The result of Theorem 6.3-5 remains valid if the assumptions $\lambda > 0$, $\mu > 0$, are weakened as $\lambda > -\frac{2}{3}\mu$, $\mu > 0$ (Exercise 6.4). ∎

We finally show that, *if* $\Gamma = \Gamma_0$, the weak solution found in Theorem 6.3-5 possesses *additional regularity* if the data (the boundary of Ω and the right-hand side f) also possess additional regularity. Since in this case the boundary condition is $u = o$ on Γ, we are solving a *linearized pure displacement problem*, and the space V is the space $H_0^1(\Omega)$.

Theorem 6.3-6 (regularity of the weak solution for the linearized pure displacement problem). *Let Ω be a domain in $\mathbb{R}^3$ with a boundary Γ of class $\mathscr{C}^2$, let $f \in L^p(\Omega)$, $p \geq \frac{6}{5}$, and let $\Gamma = \Gamma_0$. Then the weak solution $u \in H_0^1(\Omega)$ of the linearized pure displacement problem is in the space $W^{2,p}(\Omega)$, and it satisfies*

$$-\mathbf{div}\{\lambda(\operatorname{tr} e(u)) I + 2\mu e(u)\} = f \text{ in } L^p(\Omega) .$$

Let $m \geq 1$ be an integer. If the boundary Γ is of class $\mathscr{C}^{m+2}$ and if $f \in W^{m,p}(\Omega)$, the weak solution $u \in H_0^1(\Omega)$ is in the space $W^{m+2,p}(\Omega)$.

Proof. We only sketch very briefly the main steps of the proof, which is otherwise long and delicate.

(i) Because the operator of linearized elasticity is *strongly elliptic*, the implication

$$f \in L^2(\Omega) \Rightarrow u \in H^2(\Omega) \cap H_0^1(\Omega)$$

holds if the boundary Γ is of class $\mathscr{C}^2$ (Nečas [1967, p. 260]). Hence the announced regularity holds for $m = 0$, $p = 2$.

(ii) Because the linearized pure displacement problem is *uniformly elliptic* and satisfies *the supplementary and complementing conditions*, according to the definitions of Agmon, Douglis & Nirenberg [1964], it follows from Geymonat [1965, Theorem 3.5] that *the mapping*

$$A'(o): v \in V^p(\Omega) := \{v \in W^{2,p}(\Omega);\ v = o \text{ on } \Gamma\}$$
$$\to -\mathbf{div}\{\lambda(\operatorname{tr} e(v))I + 2\mu e(v)\} \in L^p(\Omega)$$

has an index $\operatorname{ind} A'(o)$ *that is independent of* $p \in]1, \infty[$. We recall that

$$\operatorname{ind} A'(o) = \dim \operatorname{Ker} A'(o) - \dim \operatorname{Coker} A'(o)\,,$$

where $\operatorname{Coker} A'(o)$ is the quotient space of the space $L^p(\Omega)$ by the space $\operatorname{Im} A'(o)$ (the index is well defined only if both space $\operatorname{Ker} A'(o)$ and $\operatorname{Coker} A'(o)$ are finite-dimensional). In our case, we know by (i) that $\operatorname{ind} A'(o) = 0$ for $p = 2$ since $A'(o)$ is a bijection in this case ($\operatorname{Ker} A'(o) = \{o\}$ if and only if $A'(o)$ is injective, and $\operatorname{Coker} A'(o) = \{o\}$ if and only if $A'(o)$ is surjective).

Since $V^p(\Omega) \hookrightarrow H_0^1(\Omega)$ for $p \geq \frac{6}{5}$ (Theorem 6.1-3), the mapping $A'(o): V^p(\Omega) \to L^p(\Omega)$ is injective for these values of p (if $f \in L^{6/5}(\Omega)$, the weak solution is unique in the space $H_0^1(\Omega)$; cf. Theorem 6.3-5); hence $\dim \operatorname{Ker} A'(o) = 0$. Since $\operatorname{ind} A'(o) = 0$ on the other hand, the mapping $A'(o)$ is also surjective in this case. Hence the regularity result holds for $m = 0$, $p \geq \frac{6}{5}$ (for the case $1 < p < \frac{6}{5}$, see Exercise 6.5).

(iii) The weak solution $u \in W^{2,p}(\Omega) \cap H_0^1(\Omega)$ satisfies

$$\int_\Omega \{\lambda(\operatorname{tr} e(u))I + 2\mu e(u)\} : e(v)\, dx = \int_\Omega f \cdot v\, dx \quad \text{for all } v \in \mathscr{D}(\Omega)\,.$$

Hence we can apply the Green formula in Sobolev spaces (Theorem

6.1-9) to the left-hand sides; this gives

$$\int_{\Omega} \{\lambda(\operatorname{tr} \boldsymbol{e}(\boldsymbol{u}))\boldsymbol{I} + 2\mu \boldsymbol{e}(\boldsymbol{u})\} : \boldsymbol{e}(\boldsymbol{v})\, \mathrm{d}x$$

$$= -\int_{\Omega} \operatorname{\mathbf{div}}\{\lambda(\operatorname{tr} \boldsymbol{e}(\boldsymbol{u}))\boldsymbol{I} + 2\mu \boldsymbol{e}(\boldsymbol{u})\} \cdot \boldsymbol{v}\, \mathrm{d}x$$

(there is no boundary term since $\boldsymbol{v} = \boldsymbol{o}$ on Γ), and the conclusion follows since $\{\mathcal{D}(\Omega)\}^{-} = \boldsymbol{L}^p(\Omega)$.

(iv) Once the regularity result

$$\boldsymbol{f} \in \boldsymbol{W}^{m,p}(\Omega) \Rightarrow \boldsymbol{u} \in \boldsymbol{W}^{m+2,p}(\Omega)$$

has been established for $m = 0$, it follows from Agmon, Douglis & Nirenberg [1964] and Geymonat [1965] that it also holds for higher values of the integer m if the boundary Γ is of class $\mathscr{C}^{m+2}$. ■

The regularity results of Theorem 6.3-6 can be extended to *linearized displacement–traction problems*, but *only if the closures of the sets Γ_0 and Γ_1 do not intersect*; for example, they apply if

$$\Omega = \{x \in \mathbb{R}^3 : 0 < r_0 < |ox| < r_1\},\ \Gamma_0 = S_{r_0},\ \Gamma_1 = S_{r_1}.$$

They also apply to *pure traction problems* (for $m = 2$, the right-hand side $\boldsymbol{g}$ is then assumed to belong to the trace space $W^{1-(1/p),p}(\Gamma)$ introduced in Sect. 6.1; cf. Exercise 6.3).

Otherwise the weak solution still possesses "interior regularity" in open subsets Ω' such that $\bar{\Omega}' \subset \Omega$, but lacks regularity at the intersecting points of $\bar{\Gamma}_0$ and $\bar{\Gamma}_1$, i.e., *where the boundary condition changes its type*. Regularity is also lost at *corners* along the boundary even if the boundary condition does not change its type there. See Grisvard [1985] for scalar equations, and Grisvard [1987] for the linearized pure traction problem.

There is a vast literature on the regularity, or lack thereof, of elliptic systems that include the boundary value problems of linearized elasticity as special cases. In addition to the references to Agmon, Douglis & Nirenberg [1964], Geymonat [1965], Nečas [1967] already quoted, we mention in particular Koselev [1958], Lions & Magenes [1968], Fichera [1972a, 1972c], Višik [1976], Giaquinta & Modica]1982].

6.4. OUTLINE OF THE EXISTENCE THEORY BASED ON THE IMPLICIT FUNCTION THEOREM

We begin by considering the *pure displacement problem* for a *St Venant–Kirchhoff material.* The restriction to a pure displacement problem is essential (as we shall see, cf. Sect. 6.7), but the consideration of this particular elastic material (which, as a model, otherwise suffers from serious drawbacks; cf. Sect. 3.9) is not restrictive since it displays the main *mathematical* features of the method, while being much easier to handle. Our motivation in this section is thus essentially pedagogical.

Assuming that the applied body forces are dead loads and that the boundary condition of place corresponds to a vanishing displacement, we have to solve the following boundary value problem:

$$\begin{cases} -\mathbf{div}\{(\boldsymbol{I} + \boldsymbol{\nabla u})\boldsymbol{\Sigma}\} = \boldsymbol{f} \text{ in } \Omega\,, \\ \boldsymbol{\Sigma} = \check{\boldsymbol{\Sigma}}(\boldsymbol{E}(\boldsymbol{u})) = \lambda(\operatorname{tr} \boldsymbol{E}(\boldsymbol{u}))\boldsymbol{I} + 2\mu \boldsymbol{E}(\boldsymbol{u}) \text{ in } \Omega\,, \\ \boldsymbol{u} = \boldsymbol{o} \text{ on } \Gamma = \partial\Omega\,, \end{cases}$$

or, componentwise,

$$\begin{cases} -\partial_j(\sigma_{ij} + \sigma_{kj}\partial_k u_i) = f_i \text{ in } \Omega\,, \\ \sigma_{ij} = a_{ijpq}E_{pq}(\boldsymbol{u}) \text{ in } \Omega\,, \\ u_i = 0 \text{ on } \Gamma\,, \end{cases}$$

where Ω is a domain in $\mathbb{R}^3$ with a sufficiently smooth boundary. We recall that $\boldsymbol{u} = (u_i)$ is the unknown displacement vector, $\boldsymbol{\Sigma} = (\sigma_{ij})$ is the unknown second Piola–Kirchhoff stress tensor, $\boldsymbol{f} = (f_i)$ is the given density per unit volume of the applied body forces, the constants a_{ijpq} are given by

$$a_{ijpq} = \lambda\delta_{ij}\delta_{pq} + \mu(\delta_{ip}\delta_{jq} + \delta_{iq}\delta_{jp})\,,$$

the Lamé constants λ and μ satisfy the inequalities

$$\lambda > 0\,, \quad \mu > 0$$

for actual materials (cf. Sect. 3.8), and the Green–St Venant strain tensor $\boldsymbol{E}(\boldsymbol{u}) = (E_{ij}(\boldsymbol{u}))$ is given by

$$E_{ij}(\boldsymbol{u}) = \tfrac{1}{2}(\partial_i u_j + \partial_j u_i + \partial_i u_m \partial_j u_m)\,.$$

Remark. Only the weaker inequalities $\mu > 0$, $\lambda > -\frac{2}{3}\mu$ are in fact needed in the proof of our existence results (Exercise 6.4). ■

The above problem can be converted into a problem where the displacement vector is the only unknown: Find a vector field $\boldsymbol{u}:\bar{\Omega} \to \mathbb{R}^3$ that satisfies

$$\begin{cases} \boldsymbol{A}(\boldsymbol{u}) = \boldsymbol{f} \text{ in } \Omega, \\ \boldsymbol{u} = \boldsymbol{o} \text{ on } \Gamma, \end{cases}$$

where the nonlinear operator $\boldsymbol{A} = (A_i)$ is defined by

$$A_i(\boldsymbol{u}) := -\partial_j(a_{ijpq}e_{pq}(\boldsymbol{u}) + \tfrac{1}{2}a_{ijpq}\partial_p u_m \partial_q u_m + a_{kjpq}\partial_p u_q \partial_k u_i + \tfrac{1}{2}a_{kjpq}\partial_p u_m \partial_q u_m \partial_k u_i),$$

and

$$e_{pq}(\boldsymbol{u}) = \tfrac{1}{2}(\partial_p u_q + \partial_q u_p)$$

are the components of the linearized strain tensor $\boldsymbol{e}(\boldsymbol{u})$. Observe that we have used the relations

$$a_{kjpq}e_{pq}(\boldsymbol{u}) = a_{kjpq}\partial_p u_q,$$

which follow from the symmetry property $a_{ijpq} = a_{ijqp}$.

We now prove our first existence result.

Theorem 6.4-1. *Let Ω be a domain in $\mathbb{R}^3$ with a boundary Γ of class $\mathscr{C}^2$, and assume that the constitutive equation is of the form*

$$\check{\boldsymbol{\Sigma}}(\boldsymbol{E}) = \lambda(\operatorname{tr} \boldsymbol{E})\boldsymbol{I} + 2\mu\boldsymbol{E}, \quad \text{with } \lambda > 0,\ \mu > 0.$$

Then for each number $p > 3$, there exist a neighborhood $\boldsymbol{F}^p$ of the origin in the space $\boldsymbol{L}^p(\Omega)$ and a neighborhood $\boldsymbol{U}^p$ of the origin in the space

$$\boldsymbol{V}^p(\Omega) := \{\boldsymbol{v} \in \boldsymbol{W}^{2,p}(\Omega);\ \boldsymbol{v} = \boldsymbol{o} \text{ on } \Gamma\},$$

such that, for each $\boldsymbol{f} \in \boldsymbol{F}^p$, the boundary value problem

$$\boldsymbol{A}(\boldsymbol{u}) := -\operatorname{\mathbf{div}}\{(\boldsymbol{I} + \nabla\boldsymbol{u})\check{\boldsymbol{\Sigma}}(\boldsymbol{E}(\boldsymbol{u}))\} = \boldsymbol{f},$$

has exactly one solution $\boldsymbol{u}$ in $\boldsymbol{U}^p$.

Proof. The key observation is that the *Sobolev space* $W^{1,p}(\Omega)$ *is a Banach algebra for* $p>3$ *if* Ω *is a domain* (Theorem 6.1-4). As a consequence, *the nonlinear operator* A *maps the space* $W^{2,p}(\Omega)$ *into the space* $L^p(\Omega)$ *and it is infinitely differentiable between these two spaces*, since it is a sum of continuous linear, bilinear, and trilinear mappings (hence all its derivatives of order ≥ 4 vanish).

Since $u = o$ is clearly a solution corresponding to $f = o$, a natural idea consists in showing that *the mapping* A *is locally invertible*, i.e., in a neighborhood of this particular solution. In other words, the objective is to apply the *implicit function theorem* (Theorem 1.2-3), or its corollary, the *local inversion theorem* (Theorem 1.2-4), in a neighborhood of the origin in the space $V^p(\Omega) \times L^p(\Omega)$. The only assumption that needs to be checked is that *the derivative* $A'(o)$ *at the origin is an isomorphism between the spaces* $V^p(\Omega)$ *and* $L^p(\Omega)$. But the problem: Find u such that

$$A'(o)u = f$$

is precisely the *linearized pure displacement problem*:

$$\begin{cases} -\partial_j(a_{ijpq}e_{pq}(u)) = f_i \text{ in } \Omega\,, \\ u_i = 0 \text{ on } \Gamma\,, \end{cases}$$

i.e., $A'(o)$ *is the operator of linearized elasticity* (Sect. 6.2).

Since Γ is of class $\mathscr{C}^2$, the *regularity theory* developed earlier applies: For each $f \in L^p(\Omega)$, $p>3$, the problem $A'(o)u = f$ has one and only one solution u in the space $H_0^1(\Omega) \cap W^{2,p}(\Omega)$ (Theorems 6.3-5 and 6.3-6).

Hence the continuous linear operator $A'(o): V^p(\Omega) \to L^p(\Omega)$ is *bijective*. Therefore its inverse is also continuous, since, by the *closed graph theorem* (for a proof, see Brezis [1983, p. 19] for instance), a continuous, bijective, linear operator between two Banach spaces is an isomorphism, i.e., its inverse is also continuous. ■

Our program now essentially consists in extending the *differentiability properties* of the operator A used in Theorem 6.4-1 to nonlinear operators

$$A: u \to A(u) = -\mathbf{div}\{(I + \nabla u)\check{\Sigma}(E(u))\}$$

corresponding to more general (and more realistic) response functions $\check{\Sigma}$.

6.5. THE MAPPING $\boldsymbol{E} \in \boldsymbol{V}(\boldsymbol{0}) \subset \boldsymbol{W}^{1,p}(\Omega) \rightarrow \check{\boldsymbol{\Sigma}}(\boldsymbol{E}) \in \boldsymbol{W}^{1,p}(\Omega)$, $p>3$

In view of extending the analysis of Sect. 6.4 to a general constitutive equation, we must specify under which conditions $\check{\boldsymbol{\Sigma}}(\boldsymbol{E})$ is a well defined member of the space $\boldsymbol{W}^{1,p}(\Omega)$ when $\boldsymbol{E}$ belongs to the subset

$$\boldsymbol{V}(\boldsymbol{0}) := \{\boldsymbol{E} \in \boldsymbol{W}^{1,p}(\Omega);\ \boldsymbol{E}(x) \in \mathbb{V}(\boldsymbol{0}) \text{ for all } x \in \Omega\},$$

which, by the Sobolev imbedding $\boldsymbol{W}^{1,p}(\Omega) \hookrightarrow \boldsymbol{\mathscr{C}}^0(\bar{\Omega})$ for $p>3$, is a neighborhood of the origin in the space $\boldsymbol{W}^{1,p}(\Omega)$ (the tensor $\check{\boldsymbol{\Sigma}}(\boldsymbol{E})$ is only defined for those tensors $\boldsymbol{E}$ that belong to the neighborhood $\mathbb{V}(\boldsymbol{0})$ of the origin in $\mathbb{S}^3$ found in Theorem 1.8-3), and we must next specify under which additional hypotheses the mapping

$$\check{\boldsymbol{\Sigma}} : \boldsymbol{E} \in \boldsymbol{V}(\boldsymbol{0}) \subset \boldsymbol{W}^{1,p}(\Omega) \rightarrow \check{\boldsymbol{\Sigma}}(\boldsymbol{E}) \in \boldsymbol{W}^{1,p}(\Omega)$$

determined in this fashion is of class $\mathscr{C}^1$. Notice that, for notational simplicity, we use the *same symbol* $\check{\boldsymbol{\Sigma}}$ for denoting the response function that maps the subset $\mathbb{V}(\boldsymbol{0})$ of the space $\mathbb{S}^3$ of symmetric tensors into $\mathbb{S}^3$ and the function that maps the subset $\boldsymbol{V}(\boldsymbol{0})$ of the space $\boldsymbol{W}^{1,p}(\Omega)$ of symmetric tensor-valued functions into $\boldsymbol{W}^{1,p}(\Omega)$.

The following *result of differentiability* is due to Valent [1978c]. For ease of exposition, it is stated for mappings $\check{\boldsymbol{\Sigma}}$ that are defined for *all* matrices (symmetric or not); it should be clear that, except for notational technicalities, its "local version" (i.e., for mappings $\check{\boldsymbol{\Sigma}}$ that are only defined in a neighborhood of the origin in $\mathbb{S}^3$) that will be needed later is proved in exactly the same way.

Theorem 6.5-1. *Let $p>3$ and let Ω be a domain in $\mathbb{R}^3$. Given a tensor field $\boldsymbol{E} \in \boldsymbol{W}^{1,p}(\Omega)$ and a mapping $\check{\boldsymbol{\Sigma}} \in \mathscr{C}^1(\mathbb{M}^3; \mathbb{M}^3)$, the matrix-valued function $\check{\boldsymbol{\Sigma}}(\boldsymbol{E}) : x \in \bar{\Omega} \rightarrow \check{\boldsymbol{\Sigma}}(\boldsymbol{E}(x))$ belongs to the space $\boldsymbol{W}^{1,p}(\Omega)$, and*

$$\partial_q(\check{\sigma}_{ij}(\boldsymbol{E}))(x) = \frac{\partial \check{\sigma}_{ij}}{\partial E_{kl}}(\boldsymbol{E}(x))\,\partial_q E_{kl}(x) \quad \text{for almost all } x \in \Omega .$$

If the mapping $\check{\boldsymbol{\Sigma}} : \mathbb{M}^3 \rightarrow \mathbb{M}^3$ is of class $\mathscr{C}^{m+1}$, $m \geq 0$, the mapping

$$\check{\boldsymbol{\Sigma}} : \boldsymbol{E} \in \boldsymbol{W}^{1,p}(\Omega) \rightarrow \check{\boldsymbol{\Sigma}}(\boldsymbol{E}) \in \boldsymbol{W}^{1,p}(\Omega)$$

defined in this fashion is of class $\mathscr{C}^m$ and its mth-derivative mapping $\check{\boldsymbol{\Sigma}}^{(m)}$ is continuous, and bounded in the sense that

$$\sup_{\|E\|_{1,p,\Omega} \leq r} \|\check{\Sigma}^{(m)}(E)\| < +\infty \quad \text{for each } r > 0 .$$

For $m = 1$, *one has*

$$\check{\sigma}'_{ij}(E)G = \frac{\partial \check{\sigma}_{ij}}{\partial E_{kl}}(E)G_{kl} \quad \text{for all } E, G \in W^{1,p}(\Omega) .$$

Proof. The proof essentially relies on the fact that, if Ω is a domain in $\mathbb{R}^3$ and if $p > 3$, the Sobolev space $W^{1,p}(\Omega)$ is continuously imbedded in the space $\mathscr{C}^0(\bar{\Omega})$ and is a Banach algebra. For convenience, the proof is divided in several steps.

(i) *Assuming* $\check{\Sigma} \in \mathscr{C}^1(\mathbb{M}^3; \mathbb{M}^3)$, *we show that*

$$E \in W^{1,p}(\Omega) \Rightarrow \check{\Sigma}(E) \in W^{1,p}(\Omega) ,$$

with

$$\partial_q(\check{\sigma}_{ij}(E))(x) = \frac{\partial \check{\sigma}_{ij}}{\partial E_{kl}}(E(x))\partial_q E_{kl}(x) .$$

For any $E \in W^{1,p}(\Omega)$, the function

$$\check{\Sigma}(E) : x \in \bar{\Omega} \to (\check{\Sigma}(E))(x) := \check{\Sigma}(E(x)) \in \mathbb{M}^3$$

is continuous since the inclusion $W^{1,p}(\Omega) \hookrightarrow \mathscr{C}^0(\bar{\Omega})$ holds. Therefore it is also in the space $L^p(\Omega)$. In order to show that it is in the space $W^{1,p}(\Omega)$, we must first find natural "candidates" for the derivatives $\partial_q(\check{\sigma}_{ij}(E))$ in the space $L^p(\Omega)$, and then show that these are indeed the derivatives of the functions $\check{\sigma}_{ij}(E)$ in the sense of distributions. Clearly, we are led to "try"

$$\partial_q(\check{\sigma}_{ij}(E))(x) := \frac{\partial \check{\sigma}_{ij}}{\partial E_{kl}}(E(x))\partial_q E_{kl}(x) .$$

For any $E \in W^{1,p}(\Omega) \hookrightarrow \mathscr{C}^0(\bar{\Omega})$, each function

$$x \in \bar{\Omega} \to \frac{\partial \check{\sigma}_{ij}}{\partial E_{kl}}(E(x))$$

is continuous since $\check{\Sigma} \in \mathscr{C}^1(\mathbb{M}^3; \mathbb{M}^3)$, and each function E_{kl} is in $L^p(\Omega)$. Therefore the functions $\partial_q(\check{\sigma}_{ij}(E))$ defined above belong to the space

$L^p(\Omega)$. We must next check that

$$\int_\Omega \check{\sigma}_{ij}(\boldsymbol{E}(x))\partial_q\theta(x)\,dx = -\int_\Omega \frac{\partial\check{\sigma}_{ij}}{\partial E_{kl}}(\boldsymbol{E}(x))\partial_q E_{kl}(x)\theta(x)\,dx$$

for all $\boldsymbol{E} \in \boldsymbol{W}^{1,p}(\Omega)$ and all $\theta \in \mathcal{D}(\Omega)$. We first notice that the above formula holds for functions $\boldsymbol{E} \in \mathscr{C}^1(\bar{\Omega}; \mathbb{M}^3)$ since it reduces to a simple Green formula for smooth functions. We shall therefore resort to a density argument based on the relation $\{\mathscr{C}^1(\bar{\Omega})\}^- = W^{1,p}(\Omega)$: Given $\boldsymbol{E} \in \boldsymbol{W}^{1,p}(\Omega)$, let $(\boldsymbol{E}^n)$ be a sequence that satisfies

$$\boldsymbol{E}^n \in \mathscr{C}^1(\bar{\Omega}; \mathbb{M}^3),\ \boldsymbol{E}^n \to \boldsymbol{E} \text{ in } \boldsymbol{W}^{1,p}(\Omega)\,,$$

so that we have

$$\int_\Omega \check{\sigma}_{ij}(\boldsymbol{E}^n(x))\partial_q\theta(x)\,dx = -\int_\Omega \frac{\partial\check{\sigma}_{ij}}{\partial E_{kl}}(\boldsymbol{E}^n(x))\partial_q E^n_{kl}(x)\theta(x)\,dx \quad \text{for all } n\,.$$

Let θ be a fixed function in the space $\mathcal{D}(\Omega)$. By the inclusion $W^{1,p}(\Omega) \hookrightarrow \mathscr{C}^0(\bar{\Omega})$, the sequence $(\boldsymbol{E}^n)$ converges uniformly (i.e., in the space $\mathscr{C}^0(\bar{\Omega})$) and so does the sequence $(\check{\sigma}_{ij}(\boldsymbol{E}^n))$ (to $\check{\sigma}_{ij}(\boldsymbol{E})$) since the functions $\check{\sigma}_{ij}: \mathbb{M}^3 \to \mathbb{M}^3$ are continuous. The uniform convergence of the integrands therefore implies that

$$\lim_{n\to\infty}\int_\Omega \check{\sigma}_{ij}(\boldsymbol{E}^n(x))\partial_q\theta(x)\,dx = \int_\Omega \check{\sigma}_{ij}(\boldsymbol{E}(x))\partial_q\theta(x)\,dx\,.$$

Since the functions $\partial\check{\sigma}_{ij}/\partial E_{kl}: \mathbb{M}^3 \to \mathbb{M}^3$ are also continuous by assumption,. each sequence $((\partial\check{\sigma}_{ij}/\partial E_{kl})(\boldsymbol{E}^n))$ likewise converges uniformly, to $(\partial\check{\sigma}_{ij}/\partial E_{kl})(\boldsymbol{E})$. Since the sequence $(\partial_q E^n_{kl})$ converges to $\partial_q E_{kl}$ in $L^p(\Omega)$, the sequence

$$\left(\frac{\partial\check{\sigma}_{ij}}{\partial E_{kl}}(\boldsymbol{E}^n)\partial_q E^n_{kl}\right)$$

converges to $(\partial\check{\sigma}_{ij}/\partial E_{kl})(\boldsymbol{E})\partial_q E_{kl}$ in $L^p(\Omega)$, and whence in $L^1(\Omega)$ (Ω is bounded). Thus

$$\lim_{n\to\infty} \int_\Omega \frac{\partial \check{\sigma}_{ij}}{\partial E_{kl}} (E^n(x)) \partial_q E^n_{kl}(x) \theta(x)\, dx$$
$$= \int_\Omega \frac{\partial \check{\sigma}_{ij}}{\partial E_{kl}} (E(x)) \partial_q E_{kl}(x) \theta(x)\, dx .$$

(ii) *Assuming again that* $\check{\Sigma} \in \mathscr{C}^1(\mathbb{M}^3; \mathbb{M}^3)$, *we show that the mapping* $\check{\Sigma}: W^{1,p}(\Omega) \to W^{1,p}(\Omega)$ *defined in step* (i) *is continuous, and bounded in the sense that*

$$\sup_{\|E\|_{1,p,\Omega} \leq r} \|\check{\Sigma}(E)\|_{1,p,\Omega} < +\infty \quad \text{for all } r \geq 0 ,$$

so that the theorem will be proved in the case $m = 0$. To begin with, we show that:

$$E^n \to E \text{ in } W^{1,p}(\Omega) \Rightarrow \check{\Sigma}(E^n) \to \check{\Sigma}(E) \text{ in } W^{1,p}(\Omega) .$$

Since the sequence $(\check{\sigma}_{ij}(E^n))$ then converges uniformly to $\check{\sigma}_{ij}(E)$ (step (i)), it *a fortiori* converges in $L^p(\Omega)$. By step (i), we can write

$$\partial_q \check{\sigma}_{ij}(E^n) = \frac{\partial \check{\sigma}_{ij}}{\partial E_{kl}} (E^n) \partial_q E^n_{kl} \quad \text{for each } n .$$

The sequence $((\partial \check{\sigma}_{ij} / \partial E_{kl})(E^n))$ converges uniformly to $(\partial \check{\sigma}_{ij}/\partial E_{kl})(E)$ and the sequence $(\partial_q E^n_{kl})$ converges in $L^p(\Omega)$ to $\partial_q E_{kl}$. Therefore the sequence $(\partial_q \check{\sigma}_{ij}(E^n))$ converges in $L^p(\Omega)$ to $(\partial \check{\sigma}_{ij}/\partial E_{kl})(E) \partial_q E_{kl} = \partial_q \check{\sigma}_{ij}(E)$.

To show that the mapping $\check{\Sigma}$ is bounded in the sense indicated, let

$$C(r) = \sup_{\|E\|_{1,p,\Omega} \leq r} |E|_{0,\infty,\Omega} \text{ for all } r \geq 0 .$$

Then

$$\sup_{\|E\|_{1,p,\Omega} \leq r} |\check{\Sigma}(E)|_{0,\infty,\Omega} \leq \sup_{\substack{G \in \mathbb{M}^3 \\ |G| \leq C(r)}} |\check{\Sigma}(G)| < +\infty .$$

Since $\partial_q \check{\sigma}_{ij}(E) = (\partial \check{\sigma}_{ij} / \partial E_{kl})(E) \partial_q E_{kl}$, we infer that

$$\sup_{\|E\|_{1,p,\Omega}\leq r} |\partial_q \check{\sigma}_{ij}(E)|_{0,p,\Omega} \leq \sup_{\left\{\substack{G\in\mathbb{M}^3 \\ |G|\leq C(r)}\right.} \left|\frac{\partial\check{\sigma}_{ij}}{\partial E_{kl}}(G)\right| |\partial_q E_{kl}|_{0,p,\Omega} ,$$

and step (ii) is proved.

(iii) *Assuming that* $\check{\boldsymbol{\Sigma}} \in \mathscr{C}^2(\mathbb{M}^3; \mathbb{M}^3)$, *we show that the mapping* $\check{\boldsymbol{\Sigma}}: \boldsymbol{W}^{1,p}(\Omega) \rightarrow \boldsymbol{W}^{1,p}(\Omega)$ *is of class* $\mathscr{C}^1$, *that its derivative* $\check{\boldsymbol{\Sigma}}' = (\check{\sigma}'_{ij})$ *is given by*

$$\check{\sigma}'_{ij}(E)G = \frac{\partial\check{\sigma}_{ij}}{\partial E_{kl}}(E)G_{kl} \quad \text{for all } E, G \in \boldsymbol{W}^{1,p}(\Omega) ,$$

and finally that its first derivative mapping

$$\check{\boldsymbol{\Sigma}}' : \boldsymbol{W}^{1,p}(\Omega) \rightarrow \mathscr{L}(\boldsymbol{W}^{1,p}(\Omega))$$

is bounded in the sense that

$$\sup_{\|E\|_{1,p,\Omega}\leq r} \|\check{\boldsymbol{\Sigma}}'(E)\| < +\infty \quad \text{for each } r > 0 .$$

By step (i), we infer that each function $(\partial\check{\sigma}_{ij}/\partial E_{kl})(E)$ is well defined in the space $W^{1,p}(\Omega)$ if $E \in \boldsymbol{W}^{1,p}(\Omega)$, since $\partial\check{\sigma}_{ij}/\partial E_{kl} \in \mathscr{C}^1(\mathbb{M}^3; \mathbb{M}^3)$. Thus the function $(\partial\check{\sigma}_{ij}/\partial E_{kl})(E)G_{kl}$ belongs to the space $W^{1,p}(\Omega)$ for $E, G \in \boldsymbol{W}^{1,p}(\Omega)$ since this space is an algebra, and for the same reason, the mapping

$$G \in \boldsymbol{W}^{1,p}(\Omega) \rightarrow \frac{\partial\check{\sigma}_{ij}}{\partial E_{kl}}(E)G_{kl} \in W^{1,p}(\Omega)$$

defined in this fashion is continuous for each $E \in \boldsymbol{W}^{1,p}(\Omega)$. In order to conclude that

$$\check{\sigma}'_{ij}(E)G = \frac{\partial\check{\sigma}_{ij}}{\partial E_k}(E)G_{kl} ,$$

it remains to show that, for arbitrary $E, G \in \boldsymbol{W}^{1,p}(\Omega)$,

$$\check{\sigma}_{ij}(E + G) - \check{\sigma}_{ij}(E) - \frac{\partial\check{\sigma}_{ij}}{\partial E_{kl}}(E)G_{kl} = o(\|G\|_{1,p,\Omega}) .$$

We can write, for $x \in \Omega$,

$$\check{\sigma}_{ij}(E+G)(x) - \check{\sigma}_{ij}(E)(x) - \frac{\partial \check{\sigma}_{ij}}{\partial E_{kl}}(E)(x)G_{kl}(x)$$

$$= G_{kl}(x)\int_0^1 \left\{ \frac{\partial \check{\sigma}_{ij}}{\partial E_{kl}}(E+tG)(x) - \frac{\partial \check{\sigma}_{ij}}{\partial E_{kl}}(E)(x) \right\} \mathrm{d}t\,.$$

For fixed indices i, j, k, l, the function

$$\varepsilon_{kl}^{ij} : (E, G) \in \mathbb{M}^3 \times \mathbb{M}^3 \rightarrow$$

$$\varepsilon_{kl}^{ij}(E, G) := \int_0^1 \left\{ \frac{\partial \check{\sigma}_{ij}}{\partial E_{kl}}(E+tG) - \frac{\partial \check{\sigma}_{ij}}{\partial E_{kl}}(E) \right\} \mathrm{d}t \in \mathbb{R}$$

is of class $\mathscr{C}^1$ since $\partial \check{\sigma}_{ij}/\partial E_{kl} \in \mathscr{C}^1(\mathbb{M}^3; \mathbb{M}^3)$ and thus *we may apply the result of step* (i), *with* $\mathbb{M}^3$ *replaced by* $\mathbb{M}^3 \times \mathbb{M}^3$: The corresponding mapping (again denoted by the same letter ε_{kl}^{ij}):

$$\varepsilon_{kl}^{ij} : (E, G) \in W^{1,p}(\Omega) \times W^{1,p}(\Omega) \rightarrow \varepsilon_{kl}^{ij}(E, G) \in W^{1,p}(\Omega)$$

is well defined and continuous, so that in particular

$$\lim_{\|G\|_{1,p,\Omega} \rightarrow 0} \|\varepsilon_{kl}^{ij}(E, G)\|_{1,p,\Omega} = \|\varepsilon_{kl}^{ij}(E, \boldsymbol{0})\|_{1,p,\Omega} = 0$$

for a fixed $E \in W^{1,p}(\Omega)$.

Since $W^{1,p}(\Omega)$ is an algebra, there exists a constant c such that

$$\|\check{\sigma}_{ij}(E+G) - \check{\sigma}_{ij}(E) - \frac{\partial \check{\sigma}_{ij}}{\partial E_{kl}}(E)G_{kl}\|_{1,p,\Omega}$$

$$\leq c\|G_{kl}\|_{1,p,\Omega}\|\varepsilon_{kl}^{ij}(E, G)\|_{1,p,\Omega}\,,$$

and the differentiability is established. The continuity of the first derivative mapping $\check{\Sigma}' : W^{1,p}(\Omega) \rightarrow \mathscr{L}(W^{1,p}(\Omega))$ follows from the continuity of the mappings

$$E \in W^{1,p}(\Omega) \rightarrow \frac{\partial \check{\sigma}_{ij}}{\partial E_{kl}}(E) \in W^{1,p}(\Omega)\,,$$

which itself follows from step (ii); the boundedness of the mapping $\check{\Sigma}'$ also follows from step (ii) and from the inequality

$$\|\check{\sigma}'_{ij}(E)\|_{\mathscr{L}(W^{1,p}(\Omega);W^{1,p}(\Omega))} = \sup_{\|G\|_{1,p,\Omega}\leq 1} \left\| \frac{\partial \check{\sigma}_{ij}}{\partial E_{kl}}(E)G_{kl} \right\|_{1,p,\Omega}$$

$$\leq 9c \sup_{\left\{\begin{smallmatrix}\|G\|_{1,p,\Omega}\leq 1 \\ 1\leq k,l\leq 3\end{smallmatrix}\right.} \left\| \frac{\partial \check{\sigma}_{ij}}{\partial E_{kl}}(E) \right\|_{1,p,\Omega},$$

which is again valid because the space $W^{1,p}(\Omega)$ is an algebra for $p>3$.

(iv) The proof for $m\geq 2$ proceeds by induction along the same lines and for this reason is omitted. ■

Observe that the boundedness of the derivatives of the continuous mappings $\check{\Sigma}^{(m)}$ *cannot* be derived from a compactness argument since the unit ball in an infinite-dimensional space (here, the space $W^{1,p}(\Omega)$) is never compact.

6.6. THE MAPPING $A: u\in V(o)\subset W^{2,p}(\Omega)\to -\mathrm{div}\{(I+\nabla u)\check{\Sigma}(E(u))\}\in L^p(\Omega)$, $p>3$

As already noted, the tensor $\check{\Sigma}(E)$ is defined only for the tensors E that belong to the set

$$\boxed{\mathbb{V}(0) := \{\tfrac{1}{2}(C-I)\in\mathbb{S}^3;\ C\in\mathbb{S}^3_>\},}$$

which is a *neighborhood of the origin in the space* $\mathbb{S}^3$ (Theorem 1.8-3). Since for $p>3$, the Sobolev imbedding $W^{1,p}(\Omega)\hookrightarrow \mathscr{C}^0(\bar{\Omega})$ holds (Theorem 6.1-3) and the mapping $v\in W^{2,p}(\Omega)\to E(v)\in W^{1,p}(\Omega)$ is continuous (the space $W^{1,p}(\Omega)$ is a Banach algebra for $p>3$; cf. Theorem 6.1-4), we conclude that *the set*

$$\boxed{V(0) := \{E\in W^{1,p}(\Omega);\ E(x)\in\mathbb{V}(0)\quad \text{for all } x\in\bar{\Omega}\}}$$

is a neighborhood of the origin 0 *in the space* $W^{1,p}(\Omega)$, *and that the set*

$$\boxed{V(o) := \{v\in W^{2,p}(\Omega);\ E(v)\in V(0)\}}$$

is a neighborhood of the origin o *in the space* $W^{2,p}(\Omega)$.

We now establish various *differentiability properties of the mapping*

$$A: u \in V(o) \subset W^{2,p}(\Omega) \to A(u) := -\mathrm{div}\{(I + \nabla u)\check{\Sigma}(E(u))\} \in L^p(\Omega),$$

when it is well defined. Those established under the assumption that the mapping $\check{\Sigma}: \mathbb{V}(0) \subset \mathbb{S}^3 \to \mathbb{S}^3$ is of class $\mathscr{C}^2$ will be used in proving an *existence result* for the boundary value problem: Find u such that $A(u) = f$ in Ω and $u = o$ on Γ (Theorem 6.7-1), while the properties established under the assumption that the same mapping is of class $\mathscr{C}^3$ will be needed in Sect. 6.13 for establishing the *convergence of the incremental methods*. The assumption that $\check{\Sigma}(E) = \lambda(\mathrm{tr}\, E)I + 2\mu E + o(E)$ with $\lambda > 0$, $\mu > 0$ holds for all homogeneous, isotropic, elastic materials whose reference configuration is a natural state (Sect. 3.8). Notice that for mappings $\check{\Sigma}: \mathbb{V}(0) \subset \mathbb{S}^3 \to \mathbb{S}^3$ that are of class $\mathscr{C}^2$, it implies (Theorem 1.3-3) the stronger relation $\check{\Sigma}(E) = \lambda(\mathrm{tr}\, E)I + 2\mu E + O(\|E\|^2)$ assumed in Theorems 6.6-1 and 6.7-1.

Theorem 6.6-1. *Let $p > 3$, let Ω be a domain in $\mathbb{R}^3$ with a boundary Γ of class $\mathscr{C}^2$, and let a mapping $\check{\Sigma} \in \mathscr{C}^2(\mathbb{V}(0); \mathbb{S}^3)$ be given. Then the associated operator of nonlinear elasticity*

$$A: u \in V(o) \subset W^{2,p}(\Omega) \to A(u) = -\mathrm{div}\{(I + \nabla u)\check{\Sigma}(E(u))\} \in L^p(\Omega)$$

is well defined and of class $\mathscr{C}^1$, and if in addition the mapping $\check{\Sigma}$ satisfies

$$\check{\Sigma}(E) = \lambda(\mathrm{tr}\, E)I + 2\mu E + O(\|E\|^2),$$

with

$$\lambda > 0, \quad \mu > 0,$$

then $A'(o)$ is the operator of linearized elasticity, and

$$A'(o) \in \mathscr{I}som(V^p(\Omega); L^p(\Omega)),$$

where

$$V^p(\Omega) := \{v \in W^{2,p}(\Omega);\ v = o \text{ on } \Gamma\}.$$

If the mapping $\check{\Sigma}: \mathbb{V}(0) \subset \mathbb{S}^3 \to \mathbb{S}^3$ is of class $\mathscr{C}^3$ and if $A'(o) \in \mathscr{I}som(V^p(\Omega); L^p(\Omega))$, there exists a number $\rho_0^p > 0$ such that, for any $\rho < \rho_0^p$,

$$v \in B_\rho^p \Rightarrow A'(v) \in \mathcal{I}som(V^p(\Omega); L^p(\Omega)),$$

$$\gamma_\rho^p := \sup_{v \in B_\rho^p} \|\{A'(v)\}^{-1}\| < +\infty,$$

$$L_\rho^p := \sup_{\substack{v,w \in B_\rho^p \\ v \neq w}} \frac{\|\{A'(v)\}^{-1} - \{A'(w)\}^{-1}\|}{\|v - w\|_{2,p,\Omega}} < +\infty,$$

where

$$B_\rho^p := \{v \in V^p(\Omega); \|v\|_{2,p,\Omega} \leq \rho\}.$$

Proof. The mapping $A : V(o) \subset W^{2,p}(\Omega) \to L^p(\Omega)$ is obtained by composition of several mappings:

$$v \in V(o) \subset W^{2,p}(\Omega) \to \nabla v = (\partial_j v_i) \in W^{1,p}(\Omega),$$
$$\nabla v \in W^{1,p}(\Omega) \to E(v) \in W^{1,p}(\Omega),$$
$$E \in V(0) \subset W^{1,p}(\Omega) \to \check{\Sigma}(E) \in W^{1,p}(\Omega),$$
$$(v, w) \in W^{1,p}(\Omega) \times W^{1,p}(\Omega) \to vw \in W^{1,p}(\Omega),$$
$$T \in W^{1,p}(\Omega) \to \mathbf{div}\, T \in L^p(\Omega),$$

which are of class $\mathscr{C}^\infty$, except possibly the mapping $\check{\Sigma}$; whence the smoothness of A is that of the mapping $\check{\Sigma} : W^{1,p}(\Omega) \to W^{1,p}(\Omega)$. In particular if $\check{\Sigma} \in \mathscr{C}^2(\mathbb{V}(0); \mathbb{S}^3)$, the mapping A is of class $\mathscr{C}^1$ (Theorem 6.5-1); in order to compute its derivative, which is then known to exist, it suffices to compute the terms that are linear with respect to v in the difference $\{A(u + v) - A(u)\}$. In this fashion, we find that

$$A_i'(u)v = -\partial_j\{(\check{\sigma}'_{ij}(E(u)) + \partial_k u_i \check{\sigma}'_{kj}(E(u)))(\nabla v^{\mathrm{T}} + \nabla v + \nabla v^{\mathrm{T}} \nabla u + \nabla u^{\mathrm{T}} \nabla v) + \check{\sigma}_{kj}(E(u))\partial_k v_i\}$$

for all $u, v \in W^{2,p}(\Omega)$. Combining this relation with the assumption that $\check{\Sigma}(E) = \lambda(\operatorname{tr} E)I + 2\mu E + o(E)$, we obtain

$$A_i'(0)v = -\partial_j\left\{\check{\sigma}'_{ij}(0)\left(\frac{\nabla v^{\mathrm{T}} + \nabla v}{2}\right)\right\} = -\partial_j\{\lambda \operatorname{tr} e(v)\delta_{ij} + 2\mu e_{ij}(v)\},$$

where $e(v) = \frac{1}{2}(\nabla v^T + \nabla v)$ is the linearized strain tensor. In other words, *the derivative* $A'(o)$ *is simply the operator of linearized elasticity* (Sect. 6.2), just as in the case of St Venant–Kirchhoff materials. Thus the same argument as in the proof of Theorem 6.4-1 again shows that the derivative $A'(o)$ is an isomorphism between the spaces $V^p(\Omega)$ and $L^p(\Omega)$.

If now $\dot{\Sigma} \in \mathscr{C}^3(\mathbb{V}(\mathbf{0}); \mathbb{S}^3)$, we infer from Theorem 6.5-1 that the associated mapping $\check{\Sigma} : V(\mathbf{0}) \subset W^{1,p}(\Omega) \to W^{1,p}(\Omega)$ is of class $\mathscr{C}^2$ and that its second derivative mapping is bounded in a ball of radius $r_0 > 0$ contained in $V(\mathbf{0})$, i.e.,

$$\sup_{\|E\|_{1,p,\Omega} \le r} \|\check{\Sigma}''(E)\| < +\infty \quad \text{for all } 0 \le r \le r_0 .$$

It is easily seen that the other mappings that are used in the definition of the mapping A also have all their derivatives bounded, so that the mapping A, which is also of class $\mathscr{C}^2$, has a bounded second derivative in a ball of radius $\rho_0 > 0$ contained in $V(o)$. Let

$$M^p(\rho) := \sup_{\|v\|_{2,p,\Omega} \le \rho} \|A''(v)\| \quad \text{for all } 0 \le \rho \le \rho_0 .$$

Then

$$\sup_{\|v\|_{2,p,\Omega} \le \rho} \|A'(v) - A'(o)\| \le \rho M^p(\rho), \; 0 \le \rho \le \rho_0 ,$$

by the mean value theorem (Theorem 1.2-2). If $A'(o) \in \mathscr{I}som(V^p(\Omega); L^p(\Omega))$, we can write

$$A'(v) = A'(o)(I + \{A'(o)\}^{-1}(A'(v) - A'(o)))$$

for all $v \in V^p(\Omega)$. Hence it follows that

$$A'(v) \in \mathscr{I}som(V^p(\Omega); L^p(\Omega))$$

for those $v \in V^p(\Omega)$ that satisfy

$$\|v\|_{2,p,\Omega} \le \rho \quad \text{with } \rho \le \rho_0, \quad \text{and } \rho M^p(\rho) < (\gamma_0^p)^{-1} ,$$

$$\text{where } \gamma_0^p := \|\{A'(o)\}^{-1}\| ,$$

since for such functions $\boldsymbol{v}$,

$$\|\{A'(\boldsymbol{o})\}^{-1}(A'(\boldsymbol{v})-A'(\boldsymbol{o}))\|\leq\gamma_0^p\rho M^p(\rho)<1 .$$

Consequently, the number ρ_0^p mentioned in the statement of the theorem can be any number that satisfies $0<\rho_0^p\leq\rho_0$ and $\rho_0^pM^p(\rho_0^p)\leq(\gamma_0^p)^{-1}$ (these inequalities can always be solved since the function $\rho\rightarrow\rho M^p(\rho)$ is continuous for $0\leq\rho\leq\rho_0$ and vanishes for $\rho=0$). We also have

$$\gamma_0^p := \sup_{\boldsymbol{v}\in B_\rho^p}\|\{A'(\boldsymbol{v})\}^{-1}\|\leq\frac{\gamma_0^p}{1-\gamma_0^p\rho M^p(\rho)}<+\infty \quad \text{for } \rho<\rho_0^p ,$$

where $B_\rho^p=\{\boldsymbol{v}\in V^p(\Omega);\ \|\boldsymbol{v}\|_{2,p,\Omega}\leq\rho\}$. Given two elements $\boldsymbol{v},\boldsymbol{w}\in B_\rho^p$, $\rho<\rho_0^p$, the relation

$$\{A'(\boldsymbol{v})\}^{-1}-\{A'(\boldsymbol{w})\}^{-1}=\{A'(\boldsymbol{v})\}^{-1}(A'(\boldsymbol{w})-A'(\boldsymbol{v}))\{A'(\boldsymbol{w})\}^{-1}$$

combined with another application of the mean value theorem shows that

$$\|\{A'(\boldsymbol{v})\}^{-1}-\{A'(\boldsymbol{w})\}^{-1}\|\leq(\gamma_0^p)^2M^p(\rho)\|\boldsymbol{v}-\boldsymbol{w}\|_{2,p,\Omega} ,$$

and thus

$$L_\rho^p := \sup_{\substack{\boldsymbol{v},\boldsymbol{w}\in B_\rho^p\\ \boldsymbol{v}\neq\boldsymbol{w}}}\frac{\|\{A'(\boldsymbol{v})\}^{-1}-\{A'(\boldsymbol{w})\}^{-1}\|}{\|\boldsymbol{v}-\boldsymbol{w}\|_{2,p,\Omega}}<+\infty \quad \text{for } \rho<\rho_0^p . \qquad \blacksquare$$

Note that, if $\check{\boldsymbol{\Sigma}}\in\mathscr{C}^2(\mathbb{V}(\boldsymbol{0});\mathbb{S}^3)$, it follows from Theorem 6.6-1 that *the operator of nonlinear elasticity*

$$\mathcal{A}:\boldsymbol{v}\in V(\boldsymbol{o})\subset\boldsymbol{W}^{2,p}(\Omega)\rightarrow(\boldsymbol{A}(\boldsymbol{v}),\boldsymbol{B}(\boldsymbol{v}))\in\boldsymbol{L}^p(\Omega)\times\mathscr{C}^0(\Gamma_1),\ p>3 ,$$

where

$$(\boldsymbol{A}(\boldsymbol{v}),\boldsymbol{B}(\boldsymbol{v}))=(-\mathbf{div}\{(\boldsymbol{I}+\nabla\boldsymbol{v})\check{\boldsymbol{\Sigma}}(\boldsymbol{E}(\boldsymbol{v}))\},\{(\boldsymbol{I}+\nabla\boldsymbol{v})\check{\boldsymbol{\Sigma}}(\boldsymbol{E}(\boldsymbol{v}))\}\boldsymbol{n}) ,$$

is of class $\mathscr{C}^1$ (for $p>3$, the trace operator $\mathrm{tr}: W^{1,p}(\Omega)\rightarrow\mathscr{C}^0(\Gamma)$ is continuous; cf. Theorem 6.1-7), and therefore it is differentiable at $\boldsymbol{v}=\boldsymbol{o}$. This observation thus provides examples of spaces $\boldsymbol{W}(\Omega)$, $\boldsymbol{F}(\Omega)$, $\boldsymbol{G}(\Gamma_1)$ that satisfy the assumptions of Theorem 6.2-1.

6.7. EXISTENCE RESULTS IN THE SPACES $W^{2,p}(\Omega)$, $p>3$

We are now in a position to extend the existence result established for St Venant–Kirchhoff materials (Theorem 6.4-1) to general homogeneous isotropic materials whose reference configuration is a natural state.

Theorem 6.7-1 (existence for pure displacement problems). *Let Ω be a domain in $\mathbb{R}^3$ with a boundary Γ of class $\mathscr{C}^2$, and let there be given a mapping $\check{\boldsymbol{\Sigma}} \in \mathscr{C}^2(\mathbb{V}(\boldsymbol{0}); \mathbb{S}^3)$ that satisfies*

$$\check{\boldsymbol{\Sigma}}(\boldsymbol{E}) = \lambda(\operatorname{tr} \boldsymbol{E})\boldsymbol{I} + 2\mu\boldsymbol{E} + O(\|\boldsymbol{E}\|^2), \quad \text{with } \lambda>0, \mu>0.$$

Then for each number $p>3$, there exist a neighborhood $\boldsymbol{F}^p$ of the origin in the space $\boldsymbol{L}^p(\Omega)$ and a neighborhood $\boldsymbol{U}^p$ of the origin in the space

$$\boldsymbol{V}^p(\Omega) = \{\boldsymbol{v} \in \boldsymbol{W}^{2,p}(\Omega); \boldsymbol{v} = \boldsymbol{o} \text{ on } \Gamma\}$$

such that for each $\boldsymbol{f} \in \boldsymbol{F}^p$, the boundary value problem

$$\boldsymbol{A}(\boldsymbol{u}) = -\operatorname{\mathbf{div}}\{(\boldsymbol{I} + \nabla\boldsymbol{u})\check{\boldsymbol{\Sigma}}(\boldsymbol{E}(\boldsymbol{u}))\} = \boldsymbol{f}$$

has exactly one solution $\boldsymbol{u}$ in $\boldsymbol{U}^p$.

Proof. Since the assumptions on the set Ω and on the mapping $\check{\boldsymbol{\Sigma}}$ imply that $\boldsymbol{A}'(\boldsymbol{o}) \in \mathscr{I}som(\boldsymbol{V}^p(\Omega); \boldsymbol{L}^p(\Omega))$ (Theorem 6.6-1), the proof consists in applying the implicit function theorem in the space $\boldsymbol{V}^p(\Omega) \times \boldsymbol{L}^p(\Omega)$ ($\boldsymbol{u} = \boldsymbol{o}$ is a solution corresponding to $\boldsymbol{f} = \boldsymbol{o}$) as in Theorem 6.4-1. ∎

Remark. A more explicit delineation of the neighborhood $\boldsymbol{F}^p$ will be provided in Theorem 6.12-1. ∎

The idea of using the implicit function theorem for obtaining local existence results in nonlinear elasticity goes back to Stoppelli [1954, 1955] and van Buren [1968]; more details are provided later about their results and their subsequent extensions. Existence results analogous to that of Theorem 6.7-1 have been obtained by Valent [1978c, 1985], Marsden & Hughes [1978, pp. 204 ff.], and Ciarlet & Destuynder [1979], for the *pure displacement problem*: Similar existence results in the spaces $\mathscr{C}^{2,\lambda}(\bar{\Omega})$ are also proved in Valent [1978c]; applied body force densities of the form $f(x) = \hat{f}(x, \varphi(x))$, $x \in \Omega$, are considered in Valent [1982] (see also Exercise 6.8). These results apply as well to nonhomogeneous or anisotropic elastic bodies; it then suffices to assume that the response function has adequate smoothness with respect to the variable $x \in \bar{\Omega}$, and that the bilinear form of the associated linearized problem is again $\boldsymbol{H}_0^1(\Omega)$-elliptic (Sect. 6.3), in which case such generalizations are straightforward. In these references, assumptions are usually made on the response function for the first Piola–Kirchhoff stress rather than on the response function for the second Piola–Kirchhoff stress as here.

Remarks. (1) For open subsets in $\mathbb{R}^n$ with sufficiently smooth boundaries, the $\boldsymbol{W}^{2,p}(\Omega)$-regularity, $p > n$, of the solution of the Dirichlet problem for linear second-order systems that are elliptic in the sense of Petrovskii [1950] and with right-hand sides in $\boldsymbol{L}^p(\Omega)$ has been proved by Košelev [1958]. This author has then applied this regularity result to obtain existence results for nonlinear systems whose associated operator is differentiable between the spaces $\boldsymbol{W}^{2,p}(\Omega)$ and $\boldsymbol{L}^p(\Omega)$.

(2) The implicit function theorem has also been used by Nečas [1976] for obtaining existence results for plane deformation problems for hyperelastic materials. ■

A landmark in the mathematical analysis of three-dimensional elasticity is the work of Signorini [1930, 1949], who considered the *pure traction problem*:

$$\begin{cases} -\operatorname{\mathbf{div}} \hat{\boldsymbol{T}}(\boldsymbol{I} + \nabla \boldsymbol{u}) = \boldsymbol{f} \text{ in } \Omega\,, \\ \hat{\boldsymbol{T}}(\boldsymbol{I} + \nabla \boldsymbol{u})\boldsymbol{n} = \boldsymbol{g} \text{ on } \Gamma\,. \end{cases}$$

Signorini assumed that the densities $\boldsymbol{f}$ and $\boldsymbol{g}$ are analytic functions of a parameter ε that vanish at $\varepsilon = 0$:

$$\boldsymbol{f} = \sum_{n=1}^{\infty} \varepsilon^n \boldsymbol{f}^n\,, \quad \boldsymbol{g} = \sum_{n=1}^{\infty} \varepsilon^n \boldsymbol{g}^n\,,$$

and that satisfy $\int_\Omega \boldsymbol{f}\,\mathrm{d}x + \int \boldsymbol{g}\,\mathrm{d}a = \boldsymbol{o}$. *Signorini's perturbation method* then consists in seeking a solution in the form

$$\boldsymbol{u} = \sum_{n=1}^{\infty} \varepsilon^n \boldsymbol{u}^n\,,$$

where the nth term u^n is calculated by recursively solving n linear problems, obtained by equating to zero the coefficients of ε^n when f, g and u are replaced in the pure traction problem by their formal expansions (it is further assumed that the response function $\hat{T}$ is analytic with respect to F). In this fashion, Signorini obtained a uniqueness theorem and he described compatibility conditions that the applied forces must satisfy in order that the first order term u^1 in the expansion of u be a solution of the associated linearized pure traction problem; he did not, however, consider the question of existence. Modern expositions of Signorini's method are given by Grioli [1962, Ch. 4] and Truesdell & Noll [1965, Sects. 63 and 64]. For more recent studies on this method, see Bharatha & Levinson [1977], Capriz & Podio-Guidugli [1974, 1979, 1982], Grioli [1983].

The first successful attempt to prove existence results in three-dimensional elasticity is due to Stoppelli [1954, 1955], who considered *pure traction problems* of the form

$$\begin{cases} -\mathbf{div}\,\hat{T}(I+\nabla u) = \varepsilon f \text{ in } \Omega\,, \\ \hat{T}(I+\nabla u)n = \varepsilon g \text{ on } \Gamma\,. \end{cases}$$

Stoppelli's idea was to apply the implicit function theorem in the spaces $\mathscr{C}^{2,\lambda}(\bar{\Omega})$, so as to obtain an existence theorem for $|\varepsilon|$ small enough. In his work, the $\mathscr{C}^{2,\lambda}(\bar{\Omega})$-regularity of the solution of the corresponding linearized boundary value problem is left as an assumption. But, considering the time where Stoppelli's results were published, one can only sympathize with this approach for, proving that the solution is in the space $\mathscr{C}^{2,\lambda}(\bar{\Omega})$ requires the same delicate analysis as proving that it is in the space $W^{2,p}(\Omega)$ (cf. Theorem 6.3-6, where we briefly outlined the proof; in Valent [1978c], the application of the implicit function theorem in the spaces $\mathscr{C}^{2,\lambda}(\bar{\Omega})$ is described for the pure displacement problem). It is assumed that the applied force densities satisfy

$$\int_\Omega f\,\mathrm{d}x + \int_\Gamma g\,\mathrm{d}a = o\,.$$

In addition, any solution u must satisfy *Signorini's compatibility condition*

$$\int_\Omega \varphi(x)\wedge f(x)\,\mathrm{d}x + \int_\Gamma \varphi(x)\wedge g(x)\,\mathrm{d}a = o\,, \quad \text{where } \varphi = id + u\,,$$

which directly follows from the axiom of moment balance (Exercise 2.2).

After linearizing these conditions, one can unambiguously define the linearized problem, whose solution is unique to within an infinitesimal rigid displacement (Exercise 6.3), while by contrast, the set of solutions of the nonlinear pure traction problem is either the set of all rigid deformations if all applied forces vanish, or a discrete set (save in some exceptional cases); this is but one indication of the specific difficulties that arise in the linearization of a pure traction problem. In order to overcome these difficulties, Stoppelli assumed that the applied forces do not possess an *axis of equilibrium*, around which an arbitrary rotation does not alter the static equilibrium of the body when the applied force densities are held fixed (Stoppelli partially removed this assumption in later works). More detailed expositions of Stoppelli's method are given in Grioli [1962, Ch. 5] and Truesdell & Noll [1965, Sect. 46].

The existence theory for pure traction problems around a natural state has since then received considerable interest. We mention in particular the results of Lanza de Cristoforis & Valent [1982], who establish the existence of a solution in either spaces $\boldsymbol{W}^{2,p}(\Omega)$ or $\mathscr{C}^{2,\lambda}(\bar{\Omega})$, and its uniqueness when the supplementary condition $\int_\Omega \boldsymbol{u}\, dx = \boldsymbol{o}$ is imposed (a similar condition is imposed for solving the pure traction problem in hyperelasticity; cf. Theorem 7.7-2).

Of particular interest is the *bifurcation of solutions of pure traction problems* pioneered by the study of the *Rivlin cube* (Rivlin [1948, 1974]; see also Sawyers [1976], Gurtin [1981a, Ch. 14]). This problem has been studied in depth by Chillingworth, Marsden & Wan [1982, 1983] and Marsden & Wan [1983], Wan & Marsden [1984]; and Wan [1986] for incompressible materials (an introduction to these works is given in Marsden & Hughes [1983, Sect. 73]). These authors thoroughly analyzed the bifurcation of solutions for small loads, in particular when the applied forces have an axis of equilibrium. As an illustration of the complexity of their results, we mention that, for some particular instances of applied forces and constitutive equations, they found up to 40 distinct solutions near a natural state! More recently, Le Dret [1987] has taken a novel approach, by considering Signorini's compatibility condition as defining a manifold in the product space $\boldsymbol{W}^{2,p}(\Omega) \times \boldsymbol{L}^p(\Omega) \times \boldsymbol{W}^{1-1/p,p}(\Gamma)$ (where the triple $(\boldsymbol{\varphi}, \boldsymbol{f}, \boldsymbol{g})$ lies). Using various tools from differential geometry, Le Dret then gives a simple proof of existence for the pure traction problem with dead loads, and for the *balloon problem* (Sect. 5.2); in addition, his analysis applies to *incompressible materials* as well. Finally, Le Dret also clarifies the relationship between Signorini's compatibility condition and the nonexistence results observed in linearized elasticity when the residual stress tensor does not vanish (Ericksen [1963, 1965]).

The implicit function theorem in the space $\mathscr{C}^{2,\lambda}(\bar{\Omega})$ was first used by van Buren [1968] for obtaining an existence theorem (for $|\varepsilon|$ small enough) for the *pure displacement problem*

$$\begin{cases} -\mathbf{div}\, \hat{\boldsymbol{T}}(\boldsymbol{I}+\nabla \boldsymbol{u}) = \varepsilon \boldsymbol{f} \text{ in } \Omega\,, \\ \boldsymbol{u} = \varepsilon \boldsymbol{u}_0 \text{ on } \Gamma\,, \end{cases}$$

but, as in Stoppelli's work, the existence of a solution of the linearized problem that is in the space $\mathscr{C}^{2,\lambda}(\bar{\Omega})$ was left as an assumption.

The successful application of the implicit function theorem to existence results for pure displacement problems depends on two keystones, *the differentiability of the operator of nonlinear elasticity*

$$\begin{aligned} \boldsymbol{A}: \boldsymbol{u} \in \boldsymbol{V}^p(\Omega) = \{\boldsymbol{v} \in \boldsymbol{W}^{2,p}(\Omega);\, \boldsymbol{v} = \boldsymbol{o} \text{ on } \Gamma\} \rightarrow \\ \boldsymbol{A}(\boldsymbol{u}) = -\mathbf{div}\{(\boldsymbol{I}+\nabla \boldsymbol{u})\check{\boldsymbol{\Sigma}}(\boldsymbol{E}(\boldsymbol{u}))\} \in \boldsymbol{L}^p(\Omega) \end{aligned}$$

for $p>3$, and the fact that *the derivative* $\boldsymbol{A}'(\boldsymbol{o})$ *is an isomorphism between the spaces* $\boldsymbol{V}^p(\Omega)$ *and* $\boldsymbol{L}^p(\Omega)$; or equivalently, that the weak solution of the linearized pure displacement problem lies in the space $\boldsymbol{W}^{2,p}(\Omega)$ if the right-hand side is in $\boldsymbol{L}^p(\Omega)$. Hence the extensions of this approach are limited to situations where both properties still hold.

Solutions of *linearized pure traction problems* possess a similar $\boldsymbol{W}^{2,p}(\Omega)$-regularity, so that analogous existence results can be obtained in this case too (once the specific difficulties inherent to this case are overcome). Note, however, that the solution of a linearized pure displacement problem or of a linearized pure traction problem may loose this extra regularity *if the boundary* Γ *is not smooth*, in which case the implicit function cannot be applied. Likewise, *it is the lack of* $\boldsymbol{W}^{2,p}(\Omega)$*-regularity results for linearized displacement–traction problems* (except in the very special case where $\bar{\Gamma}_0 \cap \bar{\Gamma}_1 = \emptyset$; cf. Sect. 6.3) *that prevents the use of the implicit function theorem for genuine displacement–traction problems*. A noticeable exception to these rules has however been found by Paumier [1985], who has shown that the solutions of linearized plate problems with appropriate boundary conditions of "sliding edges" possess the needed regularity.

To overcome the lack of regularity of the solution of the linearized problem, one could conceivably try to apply the implicit function theorem in Sobolev spaces "of lower order", where the solution of the linearized problem is known to lie anyway. For example, even if the boundary Γ is only Lipschitz-continuous (as is the case if Ω is a domain), the weak solution of the linearized pure displacement problem (for definiteness) always lies in the space $H_0^1(\Omega)$ as long as the right-hand side f lies in the space $H^{-1}(\Omega) := (H_0^1(\Omega))'$ (the dual space of $H_0^1(\Omega)$); it is easy to verify that the existence results of Theorems 6.3-2 and 6.3-5 apply as well to this situation). Hence the questions naturally arise: Under what conditions is the nonlinear operator

$$A : u \in W_0^{1,p}(\Omega) \to Au := -\mathbf{div}\, \hat{S}(\nabla u) \in W^{-1,p}(\Omega) := (W_0^{1,p}(\Omega))'$$

well defined? When it is, under what conditions is it differentiable at $u = o$? If this were the case, the derivative $A'(o)$ would again be the operator of linearized elasticity, so that the existence of a solution to the nonlinear problem in the space $H_0^1(\Omega)$ could be established.

The following result gives a positive answer to the first question: Let $\hat{S} : \mathbb{M}^3 \to \mathbb{M}^3$ be a differentiable mapping that satisfies a *linear growth condition* of the form $|\hat{S}(F)| \leq a + b|F|$ for all $F \in \mathbb{M}^3$. Then (see e.g. Krasnoselskii, Zabreyko, Pustylnik & Sbolevskii [1976]) for each $p > 1$, *the operator* $u \in W_0^{1,p}(\Omega) \to \hat{S}(\nabla u) \in L^p(\Omega)$ *is continuous*; therefore *the operator*

$$A : u \in W_0^{1,p}(\Omega) \to Au := -\mathbf{div}\, \hat{S}(\nabla u) \in W^{-1,p'}(\Omega) := (W_0^{1,p}(\Omega))'$$

is also continuous. Unfortunately, the above program cannot be carried out any further, in view of the following striking *result of nondifferentiability* (Valent & Zampieri [1977]): *Assume that the above operator A is a homeomorphism between a neighborhood of* o *in* $W_0^{1,p}(\Omega)$ *and a neighborhood of* o *in* $W^{-1,p'}(\Omega)$, *and that A is differentiable at* o. *Then the mapping* $\hat{S}$ *is necessarily affine*. In other words, any nonlinear operator is ruled out by this approach!

Operators such as

$$A : u \in X(\Omega) \to \{Au : x \to Au(x) = a(x, u(x))\} \in Y(\Omega)\,,$$

where the mapping $a : \Omega \times \mathbb{R}^m \to \mathbb{R}$ is given, are called *Nemytsky operators*, or *substitution operators*; their nondifferentiability properties have been known since Vainberg [1952] showed that if such an operator is Fréchet-differentiable from $X(\Omega) = L^2(\Omega)$ into $Y(\Omega) = L^2(\Omega)$, then the function a is *affine* in u, i.e., it is of the form

$$a(x, \zeta) = \alpha(x) + \beta(x)\zeta\,.$$

This is not to say, of course, that a Nemytsky operator is never differentiable. For instance, we showed in Sect. 6.5 that the Nemytsky operator

$\check{\boldsymbol{\Sigma}}: \boldsymbol{E} \in \boldsymbol{W}^{1,p}(\Omega) \rightarrow \check{\boldsymbol{\Sigma}}(\boldsymbol{E}) \in \boldsymbol{W}^{1,p}(\Omega)$ is differentiable if $\check{\boldsymbol{\Sigma}}$ is of class $\mathscr{C}^2$ (the key assumptions there are the inclusion $W^{1,p}(\Omega) \hookrightarrow \mathscr{C}^0(\bar{\Omega})$ and the fact that $W^{1,p}(\Omega)$ is a Banach algebra, for $n = 3$, $p > 3$; cf. also Exercise 6.9). For more information about the nondifferentiability of Nemytsky operators and the related question of deciding whether the linearization of a nonlinear operator is admissible, see Vainberg [1956], Valent & Zampieri [1977], Valent [1978a, 1978b], Dahlberg [1979], Ambrosio, Buttazzo & Leaci [1987], Keeling [1987]. The differentiability, or nondifferentiability, of *functionals* is discussed in Martini [1976].

In this framework, we also mention the approach of Oden [1979], who finds a weak solution in the space $\boldsymbol{W}_0^{1,p}(\Omega)$ for nonlinear problems whose right-hand sides are in $\boldsymbol{W}^{-1,p'}(\Omega)$ when the associated nonlinear operator satisfies a *generalized Gårding's inequality*.

To conclude, we mention the extension by Le Dret [1985] of the present approach to the *pure displacement problem of incompressible elasticity*, which consists in finding a field $(\boldsymbol{\varphi}, p): \bar{\Omega} \rightarrow \mathbb{R}^3 \times \mathbb{R}$ that solves the following boundary value problem:

$$\begin{cases} -\mathbf{div}\, \hat{\boldsymbol{T}}(\boldsymbol{\nabla}\boldsymbol{\varphi}) + \mathbf{div}(p\, \mathbf{Cof}\, \boldsymbol{\nabla}\boldsymbol{\varphi}) = \boldsymbol{f} \text{ in } \Omega\,, \\ \det \boldsymbol{\nabla}\boldsymbol{\varphi} = 1 \text{ in } \Omega\,, \\ \boldsymbol{\varphi} = \boldsymbol{id} \text{ on } \Gamma\,. \end{cases}$$

In these equations, $\boldsymbol{\varphi}$ is as usual the deformation, and p is the *hydrostatic pressure* (these equations are fully justified; see also Exercise 5.9). Le Dret's idea is to use the *implicit function theorem on a manifold*, such as

$$\{\boldsymbol{\varphi} \in \boldsymbol{W}^{2,p}(\Omega);\ \det \boldsymbol{\nabla}\boldsymbol{\varphi} = 1 \text{ in } \Omega,\ \boldsymbol{\varphi} = \boldsymbol{\varphi}_0 \text{ on } \Gamma\}\,.$$

In this fashion, a local existence result is obtained for small enough applied body forces.

6.8. COMPARISON WITH LINEARIZED ELASTICITY

The use of the implicit function theorem for proving the existence of a particular solution to the *nonlinear* boundary value problem $\boldsymbol{A}(\boldsymbol{u}) = \boldsymbol{f}$ for "small" enough right-hand sides $\boldsymbol{f}$ renders its comparison with the unique solution of the associated *linearized* problem $\boldsymbol{A}'(\boldsymbol{o})\boldsymbol{u} = \boldsymbol{f}$ particularly easy. In the following result, both the displacement vector field and the second Piola–Kirchhoff stress tensor field are compared.

Theorem 6.8-1. *Let Ω be a domain in $\mathbb{R}^3$ with a boundary of class $\mathscr{C}^2$, and let there be given a mapping $\check{\boldsymbol{\Sigma}} \in \mathscr{C}^3(\mathbb{V}(\boldsymbol{0}); \mathbb{S}^3)$ that satisfies*

$$\check{\boldsymbol{\Sigma}}(\boldsymbol{E}) = \lambda(\operatorname{tr} \boldsymbol{E})\boldsymbol{I} + 2\mu \boldsymbol{E} + O(\|\boldsymbol{E}\|^2) \quad \text{with } \lambda > 0, \mu > 0 .$$

With the same notation as in Theorem 6.7-1, let $\boldsymbol{u}(\boldsymbol{f})$ denote for each $\boldsymbol{f} \in \boldsymbol{F}^p$, $p > 3$, the unique solution in $\boldsymbol{U}^p \subset \boldsymbol{V}^p(\Omega)$ of the boundary value problem

$$\boldsymbol{A}(\boldsymbol{u}) = -\operatorname{\mathbf{div}}\{(\boldsymbol{I} + \boldsymbol{\nabla u})\check{\boldsymbol{\Sigma}}(\boldsymbol{E}(\boldsymbol{u}))\} = \boldsymbol{f} ,$$

and let $\boldsymbol{u}_{\text{lin}}(\boldsymbol{f})$ denote the unique solution in the space $\boldsymbol{V}^p(\Omega)$ of the associated linearized pure displacement problem:

$$\boldsymbol{A}'(\boldsymbol{o})\boldsymbol{u} = -\operatorname{\mathbf{div}}\{\lambda \operatorname{tr} \boldsymbol{e}(\boldsymbol{u})\boldsymbol{I} + 2\mu \boldsymbol{e}(\boldsymbol{u})\} = \boldsymbol{f} .$$

Then

$$\boxed{\begin{aligned} &\|\boldsymbol{u}(\boldsymbol{f}) - \boldsymbol{u}_{\text{lin}}(\boldsymbol{f})\|_{2,p,\Omega} = O(|\boldsymbol{f}|^2_{0,p,\Omega}) , \\ &\|\boldsymbol{\Sigma}(\boldsymbol{f}) - \boldsymbol{\Sigma}_{\text{lin}}(\boldsymbol{f})\|_{1,p,\Omega} = O(|\boldsymbol{f}|^2_{0,p,\Omega}) , \end{aligned}}$$

where

$$\boldsymbol{\Sigma}(\boldsymbol{f}) = \check{\boldsymbol{\Sigma}}(\boldsymbol{E}(\boldsymbol{u}(\boldsymbol{f}))) \text{ and } \boldsymbol{\Sigma}_{\text{lin}}(\boldsymbol{f}) := \lambda \operatorname{tr} \boldsymbol{e}(\boldsymbol{u}_{\text{lin}}(\boldsymbol{f}))\boldsymbol{I} + 2\mu \boldsymbol{e}(\boldsymbol{u}_{\text{lin}}(\boldsymbol{f})) .$$

Proof. If the mapping $\check{\boldsymbol{\Sigma}} : \mathbb{V}(\boldsymbol{0}) \subset \mathbb{S}^3 \to \mathbb{S}^3$ is of class $\mathscr{C}^3$, the associated mapping $\boldsymbol{A} : \boldsymbol{V}(\boldsymbol{o}) \subset \boldsymbol{V}^p(\Omega) \to \boldsymbol{L}^p(\Omega)$ is of class $\mathscr{C}^2$ by Theorem 6.5-1, and thus the implicit function $\boldsymbol{f} \in \boldsymbol{F}^p \to \boldsymbol{u}(\boldsymbol{f}) \in \boldsymbol{U}^p \subset \boldsymbol{V}^p(\Omega)$ is also of class $\mathscr{C}^2$ (Theorem 1.2-3). To compute its first derivative, we differentiate the relation

$$\boldsymbol{A}(\boldsymbol{u}(\boldsymbol{f})) = \boldsymbol{f} \quad \text{for all } \boldsymbol{f} \in \boldsymbol{F}^p .$$

This gives

$$\boldsymbol{A}'(\boldsymbol{u}(\boldsymbol{f}))\boldsymbol{u}'(\boldsymbol{f}) = \boldsymbol{id}_{\boldsymbol{L}^p(\Omega)} ,$$

so that in particular

$$\boldsymbol{u}'(\boldsymbol{o}) = \{\boldsymbol{A}'(\boldsymbol{o})\}^{-1} ,$$

and therefore, by Taylor–Young's formula for twice differentiable functions (Theorem 1.3-3),

$$u(f) = u(o) + u'(o)f + O(|f|^2_{0,p,\Omega}) = u_{\text{lin}}(f) + O(|f|^2_{0,p,\Omega})$$

since $u_{\text{lin}}(f) = \{A'(o)\}^{-1}f$. Since the mappings $f \in F^p \to u(f) \in W^{2,p}(\Omega)$ and $E \in W^{1,p}(\Omega) \to \check{\Sigma}(E) \in W^{1,p}(\Omega)$ are of class $\mathscr{C}^2$, the mapping

$$f \in F^p \to \Sigma(f) := \check{\Sigma}(E(u(f))) \in W^{1,p}(\Omega)$$

is also of class $\mathscr{C}^2$, and thus

$$\Sigma(f) = \Sigma(o) + \Sigma'(o)f + O(|f|^2_{0,p,\Omega}) .$$

The chain rule then yields

$$\Sigma(o)f = \check{\Sigma}'(o)E'(o)u'(o)f = \lambda \operatorname{tr} e(u_{\text{lin}}(f))I + 2\mu e(u_{\text{lin}}(f)) ,$$

and the proof is complete. ∎

Remark. If the mapping $\check{\Sigma}$ is only of class $\mathscr{C}^2$, the mapping A is of class $\mathscr{C}^1$. In this case the estimates of Theorem 6.8-1 take the weaker form

$$\|u(f) - u_{\text{lin}}(f)\|_{2,p,\Omega} = O(|f|_{0,p,\Omega}) ,$$

$$\|\Sigma(f) - \Sigma_{\text{lin}}(f)\|_{1,p,\Omega} = O(|f|_{0,p,\Omega}) .$$ ∎

The results of Theorem 6.8-1 seem to be the only known mathematically rigorous comparison between the nonlinear elasticity problem and its linearization about the origin. It is indeed a challenging open problem to extend this kind of comparison to more realistic boundary value problems, in particular to *displacement–traction problems*, including the case where the *boundary is not smooth*.

6.9. ORIENTATION-PRESERVING CHARACTER AND INJECTIVITY OF THE ASSOCIATED DEFORMATION

We now prove that the mapping $\varphi = id + u : \bar{\Omega} \to \mathbb{R}^3$ associated with the displacement $u : \bar{\Omega} \to \mathbb{R}^3$ solution of the boundary value problem

$A(u) = f$ (Theorem 6.7-1) is a deformation, i.e., it is *orientation-preserving* and *injective*, if the applied body forces are small enough in the space $L^p(\Omega)$. The following result is due to Valent [1978c, 1982] and Ciarlet & Destuynder [1979, Th. 2.2].

Theorem 6.9-1. *The notation and assumptions are as in Theorem 6.7-1. For each $p > 3$, there exists a number $r_p > 0$ such that, if*

$$f \in F^p \text{ and } |f|_{0,p,\Omega} \leq r_p ,$$

the associated deformation $\varphi = id + u$, $u \in U^p$, satisfies

$$\boxed{\begin{aligned} &\det \nabla\varphi > 0 \text{ in } \bar{\Omega} , \\ &\varphi : \bar{\Omega} \to \mathbb{R}^3 \text{ is injective, and } \varphi(\Omega) = \Omega,\ \varphi(\bar{\Omega}) = \bar{\Omega} . \end{aligned}}$$

Proof. Since the implicit function

$$f \in F^p \subset L^p(\Omega) \to u \in U^p \subset W^{2,p}(\Omega)$$

is continuous (it is even continuously differentiable, but this is not needed here), and since the continuous injection

$$W^{2,p}(\Omega) \hookrightarrow \mathscr{C}^1(\bar{\Omega})$$

holds for $p > 3$, the norm $\|u\|_{\mathscr{C}^1(\bar{\Omega};\mathbb{R}^3)}$ is as small as we please provided the norm $|f|_{0,p,\Omega}$ is small enough. In particular, there exists a number $r_p > 0$ such that

$$|f|_{0,p,\Omega} < r_p \Rightarrow \sup_{x \in \bar{\Omega}} |\nabla u(x)| < 1 .$$

Hence we infer from Theorem 5.5-1 that $\det \nabla\varphi(x) > 0$ for all $x \in \bar{\Omega}$.

To prove the second assertion, we can again use Theorem 5.5-1: There exists a constant $c(\Omega)$ such that, if

$$\sup_{x \in \bar{\Omega}} |\nabla u(x)| < c(\Omega) ,$$

the mapping $\varphi = id + u : \bar{\Omega} \to \mathbb{R}^3$ is injective. It therefore suffices to choose the number r_p in such a way that

$$|f|_{0,p,\Omega} < r_p \Rightarrow \sup_{x \in \bar{\Omega}} |\nabla u(x)| < c(\Omega) .$$

The injectivity of the mapping φ can also be derived from Theorem 5.5-2 since $\det \nabla\varphi(x) > 0$ for all $x \in \Omega$ if r_p is small enough, and since the mapping φ is by assumption equal to an injective mapping (the identity mapping) on the boundary of Ω. The relations $\varphi(\Omega) = \Omega$ and $\varphi(\bar{\Omega}) = \bar{\Omega}$ follow from the same theorem. ■

Remark. Once it is known that $\varphi \in W^{2,p}(\Omega)$ and that $\det \nabla\varphi > 0$ in Ω, one can prove the injectivity of the mapping φ by using the result of Meisters & Olech [1963] stated in Theorem 5.5-3 (this requires, however, that the boundary Γ be connected): By the *Calderón extension theorem for Sobolev spaces* (see Calderón [1961], Nečas [1967, pp. 75–80], Adams [1975, p. 91]), there exists an extension $\varphi^* \in W^{2,p}(\mathbb{R}^3) \subset \mathscr{C}^1(\mathbb{R}^3; \mathbb{R}^3)$ of the function φ. It then suffices to apply Theorem 5.5-3 with $0 = \mathbb{R}^n$, $K = \bar{\Omega}$, since $\varphi|_{\partial K} = \boldsymbol{id}_{\partial K}$ by assumption. ■

6.10. DESCRIPTION OF AN INCREMENTAL METHOD

A major difficulty in the numerical computation of large deformations of elastic structures is the proper handling of the various *nonlinearities* that occur in the boundary value problems of elasticity. To obviate this difficulty, a widely used method consists in letting the forces vary by small "increments" from zero to the given ones and to compute corresponding *approximate solutions* by *successive linearizations*. This is the principle of the *incremental methods*, for which the interested reader may consult the lucid expositions of Oden [1972, §§16.5 and 17.5], Washizu [1975, p. 384], Pian [1976], Argyris & Kleiber [1977], Cescotto, Frey & Fonder [1979], Mason [1980]; note that, in actual computations, incremental methods are most often used in conjunction with the *finite element method* (see e.g. Ciarlet [1978]). Our exposition follows Bernadou, Ciarlet & Hu [1984].

Let us first describe the application of an incremental method, as it is commonly presented in the engineering literature. Following in particular Washizu [1975, Appendix I, Section 9], we describe here the *Lagrangian method*, in the case of a *pure displacement problem* for a *St Venant–Kirchhoff material* (another method is suggested in Exercise 6.10). In this case, the boundary value problem takes the form:

$$\begin{cases} -\partial_j(\sigma_{ij} + \sigma_{kj}\partial_k u_i) = f_i \text{ in } \Omega, \\ \sigma_{ij} = a_{ijpq}E_{pq}(u) \text{ in } \Omega, \\ u = o \text{ on } \Gamma, \end{cases}$$

with

$$a_{ijpq} = \lambda\delta_{ij}\delta_{pq} + \mu(\delta_{ip}\delta_{jq} + \delta_{iq}\delta_{jp}),\ \lambda > 0,\ \mu > 0,$$
$$E_{pq}(u) = \tfrac{1}{2}(\partial_p u_q + \partial_q u_p + \partial_p u_m \partial_q u_m).$$

Let there be given any *partition*

$$0 = \lambda^0 < \lambda^1 < \cdots < \lambda^n < \lambda^{n+1} < \cdots < \lambda^{N-1} < \lambda^N = 1$$

of the interval [0, 1]. With such a partition, we associate an **incremental method**, the basic idea being to let the body forces vary by "small" *force increments*

$$\delta f^n = (\delta f_i^n) := (\lambda^{n+1} - \lambda^n)f,\quad 0 \leq n \leq N-1,$$

from o to the given force f, and to compute successive *approximations* u^n to the *exact* solutions $u(\lambda^n f)$ corresponding to the successive forces $\lambda^n f$, each approximation being computed by an appropriate *linearization* around the previous approximation.

Remark. In order that the solution $u(\lambda^n f)$ exist and be unique, we may assume that there exists a number $p > 3$ such that the force density f belongs to the neighborhood F^p of Theorem 6.7-1 and that this neighborhood contains the whole segment $[o, f]$. Then by Theorem 6.7-1, the solutions $u(\lambda^n f)$ exist and are uniquely defined in the neighborhood U^p. ∎

Let, for $n = 0, 1, \ldots, N-1$,

$$\Delta u_i^n := u_i(\lambda^{n+1}f) - u_i(\lambda^n f),\quad \Delta E_{ij}^n := E_{ij}(u(\lambda^{n+1}f)) - E_{ij}(u(\lambda^n f))$$

denote the *displacement increments* and the corresponding *strain increments*, respectively, so that the corresponding *stress increments* take the form

$$\sigma_{ij}(\lambda^{n+1}f) - \sigma_{ij}(\lambda^n f) = a_{ijpq}\Delta E_{pq}^n,\quad \text{with}$$
$$\sigma_{ij}(\lambda^n f) := a_{ijpq}E_{pq}(u(\lambda^n f)).$$

Since, by definition,

$$-\partial_j(\sigma_{ij}(\lambda^{n+1}f) + \sigma_{kj}(\lambda^{n+1}f)\partial_k u_i(\lambda^{n+1}f)) = f_i^{n+1} = f_i^n + \delta f_i^n \text{ in } \Omega ,$$

$$-\partial_j(\sigma_{ij}(\lambda^n f) + \sigma_{kj}(\lambda^n f)\partial_k u_i(\lambda^n f)) = f_i^n \text{ in } \Omega ,$$

combining the above equations shows that *the nth displacement increment* $\boldsymbol{\Delta u}^n := (\Delta u_i^n)$ *satisfies the following boundary value problem* (notice that no approximation has been made so far):

$$\begin{cases} -\partial_j(a_{ijpq}\Delta E^n_{pq} + \sigma_{kj}(\lambda^n f)\partial_k \Delta u_i^n \\ \quad + a_{kjpq}\Delta E^n_{pq}\partial_k u_i(\lambda^n f) a_{kjpq}\Delta E^n_{pq}\partial_k \Delta u_i^n) = \delta f_i^n \text{ in } \Omega , \\ 2\Delta E^n_{pq} = \partial_p \Delta u_q^n + \partial_q \Delta u_p^n + \partial_p \Delta u_m^n \partial_q u_m(\lambda^n f) \\ \qquad + \partial_p u_m(\lambda^n f)\partial_q \Delta u_m^n + \partial_p \Delta u_m^n \partial_q \Delta u_m^n \text{ in } \Omega , \\ \boldsymbol{\Delta u}^n = \boldsymbol{o} \text{ on } \Gamma . \end{cases}$$

We are now in a position to define an *approximate problem*: Considering that the nth displacement $\boldsymbol{u}(\lambda^n f)$ is known, and replacing ΔE^n_{pq} and $\sigma_{kj}(\lambda^n f)$ by their above expressions, we obtain a nonlinear boundary value problem with respect to the unknown vector $\boldsymbol{\Delta u}^n$. Then *the approximation simply consists in deleting all the terms that are nonlinear with respect to the unknown* $\boldsymbol{\Delta u}^n$ *in the resulting problem*. In this fashion, we obtain that the *nth approximate displacement vector increment* $\boldsymbol{\delta u}^n = (\delta u_i^n)$ should be a solution of the following *linear* boundary value problem, where $\boldsymbol{u}^n = (u_j^n)$ denotes the *nth approximate displacement vector*:

$$\begin{cases} -\partial_j(a_{ijpq}\partial_p \delta u_q^n + a_{ijpq}\partial_p u_m^n \partial_q \delta u_m^n + a_{kjpq}\partial_k u_i^n \partial_p \delta u_q^n \\ \quad + a_{kjpq}\partial_k u_i^n \partial_p u_m^n \partial_q \delta u_m^n + a_{kjqp}E_{pq}(\boldsymbol{u}^n)\partial_k \delta u_i^n) = \delta f_i^n \text{ in } \Omega , \\ \boldsymbol{\delta u}^n = \boldsymbol{o} \text{ on } \Gamma . \end{cases}$$

Notice the *change of notation*, from the *exact* solutions $\boldsymbol{u}(\lambda^n f)$ and $\boldsymbol{\Delta u}^n$ to their respective *approximations* $\boldsymbol{u}^n$ and $\boldsymbol{\delta u}^n$. The above equations may be rewritten in a form more reminiscent of the equations of linearized elasticity, although they do *not* coincide (unless $\boldsymbol{u}^n = \boldsymbol{o}$):

$$-\partial_j(\hat{a}_{ijpq}(\boldsymbol{\nabla u}^n)\partial_p \delta u_q^n) = \delta f_i^n \text{ in } \Omega ,$$

with

$$\hat{a}_{ijpq}(\nabla u^n) = a_{ijpq} + a_{kjpq}\partial_k u_i^n + a_{ijrp}\partial_r u_q^n + a_{kjpr}\partial_r u_q^n \partial_k u_i^n + a_{pjsr}E_{sr}(u^n)\delta_{iq}\,,$$

or in *weak form*, with "trial" functions $v = (v_i)$ vanishing on the boundary Γ:

$$\int_\Omega \hat{a}_{ijpq}(\nabla u^n)\partial_p \delta u_q^n \partial_j v_i \,dx = \int_\Omega \delta f_i^n v_i \,dx\,.$$

This is exactly the problem obtained by Washizu [1975, eq. (I-9.42), p. 393].

Assuming that the above boundary value problem has a solution u^n, we define the $(n+1)$*st approximate displacement*

$$u^{n+1} = u^n + \delta u^n\,,$$

which in turn allows us to similarly compute the $(n+1)$st approximate displacement increment, etc. . . . In this fashion the incremental method is completely defined; it ends with the computation of the *Nth approximate displacement* u^N and of the *Nth approximate stress*

$$\Sigma^N := (a_{ijpq}E_{pq}(u^N))\,,$$

which are expected to approach the exact solutions $u = u(f)$ and $\Sigma = (a_{ijpq}E_{pq}(u))$ as $\max_{0\le n\le N-1}(\lambda^{n+1} - \lambda^n)$ approaches zero. The existence of the successive approximations u^n and their convergence are established in Theorem 6.13-1.

6.11. THE INCREMENTAL METHOD AS THE ITERATIVE METHOD $\delta u^n = \{A'(u^n)\}^{-1}\delta f^n$

The incremental method can be given a very simple expression in terms of the derivative of the operator of nonlinear elasticity

$$A : u \in V^p(\Omega) \to A(u) = -\mathbf{div}\{(I + \nabla u)\check{\Sigma}(E(u))\} \in L^p(\Omega)\,,$$

with

$$V^p(\Omega) = \{v \in W^{2,p}(\Omega);\ v = o \text{ on } \Gamma\}\,,$$

and, in the case of a St Venant–Kirchhoff material,

$$\check{\Sigma}(E) = \lambda(\operatorname{tr} E)I + 2\mu E .$$

In addition to being simpler and more condensed, this new expression will allow us to extend it to *more general constitutive equations* (the Gâteaux derivative $A'(u)v$ has been computed for a general response function $\check{\Sigma}$ in the proof of Theorem 6.6-1). In the case of a St Venant–Kirchhoff material, this derivative reduces to

$$A_i'(u)v = -\partial_j(a_{ijpq}\partial_p v_q + a_{ijpq}\partial_p u_m \partial_q v_m + a_{kjpq}\partial_k u_i \partial_p v_p$$
$$+ a_{kjpq}\partial_k u_i \partial_p u_m \partial_q v_m + a_{kjpq}E_{pq}(u)\partial_k v_i) .$$

Therefore, we realize by inspection that *one iteration of the incremental method described in Sect.* 6.10 *can be simply written as: Given* $u^n \in V^p(\Omega)$, *find* $\delta u^n := (u^{n+1} - u^n) \in V^p(\Omega)$ *such that*

$$A'(u^n)\delta u^n = \delta f^n .$$

Motivated by this special case, we are naturally led to define an *incremental method* for approximating the solution of the pure displacement problem corresponding to a *general* response function $\check{\Sigma}$: Given any partition of the interval [0, 1], the method consists in recursively solving the equations (we assume that the existence of the successive displacement increments can be rigorously established):

$$\boxed{A'(u^n)(u^{n+1} - u^n) = \delta f^n,\ 0 \le n \le N-1, \quad u^0 = o ,}$$

where $A : V(o) \subset V^p(\Omega) \to L^p(\Omega)$ is now the operator associated with any given response function. If the linear operator $A'(u^n)$ is invertible, one iteration can be written equivalently as:

$$\boxed{\delta u^n = \{A'(u^n)\}^{-1}\delta f^n .}$$

Remark. Such an approximate scheme is natural: Once the problem to be solved is written in the operator form $A(u) = f$, it appears as the *simplest approximation to the equations*

$$\delta f^n = A(u(\lambda^{n+1} f)) - A(u(\lambda^n f)) =$$
$$A'(u(\lambda^n f))(u(\lambda^{n+1} f) - u(\lambda^n f)) + o(u(\lambda^{n+1} f) - u(\lambda^n f)) .$$

This observation in turn suggests trying *any* well established scheme for approximating a nonlinear equation written in the form $A(u) = f$. For example, an application of *Newton's method* results in the iterative procedure

$$A'(u^n)(u^{n+1} - u^n) = f - A(u^n),\ n \geq 0,$$

which can be shown to converge provided the initial value u^0 is adequately chosen (Exercise 6.11). ■

6.12. THE ORDINARY DIFFERENTIAL EQUATION $\tilde{u}'(\lambda) = \{A'(\tilde{u}(\lambda))\}^{-1}f$

Since one iteration of the incremental method can be also written as

$$\frac{u^{n+1} - u^n}{\lambda^{n+1} - \lambda^n} = \{A'(u^n)\}^{-1}f,$$

an obvious, yet *crucial*, observation is that the above method is nothing but *Euler's method for approximating the ordinary differential equation*: *Find* $\tilde{u} \in \mathscr{C}^1([0,1]; V^p(\Omega))$ *such that*:

$$\tilde{u}'(\lambda) = \{A'(\tilde{u}(\lambda))\}^{-1}f,\ 0 \leq \lambda \leq 1,$$
$$\tilde{u}(0) = o.$$

Notice that we purposely employ a new notation, $\tilde{u}$, to designate the unknown function.

Remark. The adjective "ordinary" is classically used to distinguish an equation of the above type from a "partial" differential equation. It should be remembered, however, that the unknown function $\tilde{u}$ takes its values in an *infinite-dimensional* Banach space (the space $V^p(\Omega)$). ■

Our first objective consists in establishing an existence and uniqueness result for this equation and in showing its relation with the equation $A(u) = f$. The following classical proof of existence and uniqueness can be found in many texts; see e.g. Arnold [1974], Coddington & Levinson [1955], Crouzeix & Mignot [1984], Hartman [1964], Roseau [1976]. It is one of those rather uncommon occurrences in analysis where the proof in the infinite-dimensional case (assuming the underlying space is complete) is only marginally more involved than in the finite-dimensional, or even scalar, case.

Theorem 6.12-1. *Let* $p > 3$. *Let there be given: a mapping* $\check{\Sigma} \in \mathscr{C}^3(\mathbb{V}(\boldsymbol{0}); \mathbb{S}^3)$ *that satisfies*

$$\check{\Sigma}(E) = \lambda(\operatorname{tr} E)I + 2\mu E + O(\|E\|^2),\ \lambda > 0,\ \mu > 0,$$

its associated operator of nonlinear elasticity

$$A: u \in V(o) \subset W^{2,p}(\Omega) \to A(u) := -\operatorname{div}\{(I + \nabla u)\check{\Sigma}(E(u))\} \in L^p(\Omega),$$

and an element $f \in L^p(\Omega)$ *that satisfies*

$$|f|_{0,p,\Omega} \leq \rho(\gamma_\rho^p)^{-1} \quad \text{for some number } \rho < \rho_0^p,$$

where the numbers $\rho_0^p > 0$ and $\gamma_\rho^p > 0$ *are defined as in Theorem* 6.6-1.

Then the ordinary differential equation: Find a function

$$\tilde{u}: \lambda \in [0,1] \to \tilde{u}(\lambda) \in V^p(\Omega) = \{v \in W^{2,p}(\Omega);\ v = o \text{ on } \Gamma\}$$

that satisfies

$$\tilde{u}'(\lambda) = \{A'(\tilde{u}(\lambda))\}^{-1}f,\quad 0 \leq \lambda \leq 1,\quad \text{and}\quad \tilde{u}(0) = o,$$

has one and only one solution such that

$$\tilde{u}(\lambda) \in B_\rho^p = \{v \in V^p(\Omega);\ \|v\|_{2,p,\Omega} \leq \rho\},\quad 0 \leq \lambda \leq 1.$$

This solution satisfies

$$A(\tilde{u}(\lambda)) = \lambda f,\quad 0 \leq \lambda \leq 1,$$

i.e., $\tilde{u}(\lambda)$ *is the unique solution, heretofore denoted* $u(\lambda f)$, *to the equation* $A(u) = f$ *in the ball* B_ρ^p.

Proof. Using as basic assumptions the *completeness* of the space and the "*local*" *Lipschitz-continuity* of the right-hand side, we shall establish the existence of a continuous solution of an associated *integral equation* by the *contraction mapping theorem*, thereby avoiding the difficulty of looking for a differentiable function. For clarity, we divide the proof in several steps.

(i) *The mapping*

$$\boldsymbol{\Phi} : \tilde{\boldsymbol{v}} \in \mathscr{C}^0([0,1]; \boldsymbol{B}_\rho^p) \rightarrow \boldsymbol{\Phi}(\tilde{\boldsymbol{v}}) \in \mathscr{C}^0([0,1]; \boldsymbol{V}^p(\Omega)),$$

with

$$\boldsymbol{\Phi}(\tilde{\boldsymbol{v}})(\lambda) := \int_0^\lambda \{\boldsymbol{A}'(\tilde{\boldsymbol{v}}(\mu))\}^{-1} \boldsymbol{f} \, \mathrm{d}\mu, \quad 0 \leq \lambda \leq 1,$$

is well defined if $\rho < \rho_0^p$, *and it maps the space* $\mathscr{C}^0([0,1]; \boldsymbol{B}_\rho^p)$ *into itself if* $|\boldsymbol{f}|_{0,p,\Omega} \leq \rho(\gamma_\rho^p)^{-1}$. Given an element $\tilde{\boldsymbol{v}} \in \mathscr{C}^0([0,1]; \boldsymbol{B}_\rho^p)$, with $\rho < \rho_0^p$, we infer from Theorem 6.6-1 that

$$\boldsymbol{A}'(\tilde{\boldsymbol{v}}(\mu)) \in \mathscr{I}som(\boldsymbol{V}^p(\Omega); \boldsymbol{L}^p(\Omega)) \quad \text{for } 0 \leq \mu \leq 1.$$

Besides, the mapping

$$\mu \in [0,1] \rightarrow \{\boldsymbol{A}'(\tilde{\boldsymbol{v}}(\mu))\}^{-1} \boldsymbol{f} \in \boldsymbol{V}^p(\Omega),$$

is continuous, by composition of several continuous mappings:

$$\mu \in [0,1] \rightarrow \tilde{\boldsymbol{v}}(\mu) \in \boldsymbol{B}_\rho^p,$$
$$\tilde{\boldsymbol{v}} \in \boldsymbol{B}_\rho^p \rightarrow \boldsymbol{A}'(\tilde{\boldsymbol{v}}) \in \mathscr{L}(\boldsymbol{V}^p(\Omega); \boldsymbol{L}^p(\Omega)),$$
$$\boldsymbol{A} \in \mathscr{I}som(\boldsymbol{V}^p(\Omega); \boldsymbol{L}^p(\Omega)) \rightarrow \boldsymbol{A}^{-1} \in \mathscr{I}som(\boldsymbol{L}^p(\Omega); \boldsymbol{V}^p(\Omega)),$$
$$\boldsymbol{B} \in \mathscr{L}(\boldsymbol{L}^p(\Omega); \boldsymbol{V}^p(\Omega)) \rightarrow \boldsymbol{B}\boldsymbol{f} \in \boldsymbol{V}^p(\Omega).$$

Since $\boldsymbol{V}^p(\Omega)$ is a Banach space, the integral of a continuous function is well defined, and it is a continuous function of its upper bound. Thus the function $\boldsymbol{\Phi}(\tilde{\boldsymbol{v}})$ is also in the space $\mathscr{C}^0([0,1]; \boldsymbol{V}^p(\Omega))$.

If we equip the vector space $\mathscr{C}^0([0,1]; \boldsymbol{V}^p(\Omega))$ with the norm

$$|||\boldsymbol{\Psi}||| := \sup_{0 \leq \lambda \leq 1} \|\boldsymbol{\Psi}(\lambda)\|_{2,p,\Omega},$$

it becomes a Banach space, so that its subset $\mathscr{C}^0([0,1]; \boldsymbol{B}_\rho^p)$ is a *complete metric space*. We also have

$$\tilde{v} \in \mathscr{C}^0([0,1], B_\rho^p) \Rightarrow \|\Phi(\tilde{v})(\lambda)\|_{2,p,\Omega} \leq \lambda \gamma_\rho^p |f|_{0,p,\Omega}, 0 \leq \lambda \leq 1,$$

and thus

$$\tilde{v} \in \mathscr{C}^0([0,1]; B_\rho^p) \Rightarrow |||\Phi(\tilde{v})||| \leq \gamma_\rho^p |f|_{0,p,\Omega}.$$

Consequently, the mapping Φ maps the complete metric space $\mathscr{C}^0([0,1]; B_\rho^p)$ into itself if $\gamma_\rho^p |f|_{0,p,\Omega} \leq \rho$.

(ii) *Either the mapping* Φ *or an iterate thereof*: $\Phi^k = \Phi(\Phi(\cdots))$ *for some* $k \geq 2$, *is a contraction of the space* $\mathscr{C}^0([0,1]; B_\rho^p)$. Given two elements $\tilde{v}, \tilde{w} \in \mathscr{C}^0([0,1]; B_\rho^p)$,

$$(\Phi(\tilde{v}) - \Phi(\tilde{w}))(\lambda) = \int_0^\lambda (\{A'(\tilde{v}(\mu))\}^{-1} - \{A'(\tilde{w}(\mu))\}^{-1}) f \,\mathrm{d}\mu,$$

so that, using the Lipschitz-continuity of the mapping

$$\tilde{v} \in B_\rho^p \to \{A'(\tilde{v})\}^{-1} \in \mathscr{I}som(V^p(\Omega); L^p(\Omega))$$

established in Theorem 6.6-1, we obtain

$$\begin{aligned}\|(\Phi(\tilde{v}) - \Phi(\tilde{w}))(\lambda)\|_{2,p,\Omega} &\leq L_\rho^p \int_0^\lambda \|(\tilde{v} - \tilde{w})(\mu)\|_{2,p,\Omega} |f|_{0,p,\Omega} \,\mathrm{d}\mu \\ &\leq \lambda L_\rho^p |f|_{0,p,\Omega} |||\tilde{v} - \tilde{w}|||, 0 \leq \lambda \leq 1,\end{aligned}$$

and hence

$$|||\Phi(\tilde{v}) - \Phi(\tilde{w})||| \leq (L_\rho^p |f|_{0,p,\Omega}) |||\tilde{v} - \tilde{w}|||.$$

Then either $(L_\rho^p |f|_{0,p,\Omega}) < 1$ and the mapping Φ is a contraction, or we iterate the argument: Assume that we have established the inequality

$$\begin{aligned}&\|(\Phi^{k-1}(\tilde{v}) - \Phi^{k-1}(\tilde{w}))(\lambda)\|_{2,p,\Omega} \\ &\leq \frac{1}{(k-1)!} (\lambda L^p |f|_{0,p,\Omega})^{k-1} |||\tilde{v} - \tilde{w}|||\end{aligned}$$

for some integer $k \geq 2$ (it holds for $k = 2$ with $\Phi^1 := \Phi$). From the relation

$$(\boldsymbol{\Phi}^k(\tilde{\boldsymbol{v}}) - \boldsymbol{\Phi}^k(\tilde{\boldsymbol{w}}))(\lambda) =$$
$$\int_0^\lambda (\{\boldsymbol{A}'(\boldsymbol{\Phi}^{k-1}(\tilde{\boldsymbol{v}}))(\mu)\}^{-1} - \{\boldsymbol{A}'(\boldsymbol{\Phi}^{k-1}(\tilde{\boldsymbol{w}})(\mu)\}^{-1})\,\mathrm{d}\mu\,,$$

we deduce

$$\begin{aligned}\|(\boldsymbol{\Phi}^k(\tilde{\boldsymbol{v}}) - \boldsymbol{\Phi}^k(\tilde{\boldsymbol{w}}))(\lambda)\|_{2,p,\Omega} &\\ \leq \frac{1}{(k-1)!}(L_\rho^p|\boldsymbol{f}|_{0,p,\Omega})^k|||\tilde{\boldsymbol{v}} - \tilde{\boldsymbol{w}}|||\int_0^\lambda \mu^{k-1}\,\mathrm{d}\mu &\\ = \frac{1}{k!}(\lambda L_\rho^p|\boldsymbol{f}|_{0,p,\Omega})^k|||\tilde{\boldsymbol{v}} - \tilde{\boldsymbol{w}}|||\,,\end{aligned}$$

and thus the inequality

$$|||\boldsymbol{\Phi}^k(\tilde{\boldsymbol{v}}) - \boldsymbol{\Phi}^k(\tilde{\boldsymbol{w}})||| \leq \frac{1}{k!}(L_\rho^p|\boldsymbol{f}|_{0,p,\Omega})^k|||\tilde{\boldsymbol{v}} - \tilde{\boldsymbol{w}}|||$$

holds for all integers $k \geq 1$. Since the expression

$$\frac{1}{k!}(L_\rho^p|\boldsymbol{f}|_{0,p,\Omega})^k$$

is the general term of a convergent series, there exist an integer k and a number β such that

$$\beta < 1 \text{ and } |||\boldsymbol{\Phi}^k(\tilde{\boldsymbol{v}}) - \boldsymbol{\Phi}^k(\tilde{\boldsymbol{w}})||| \leq \beta|||\tilde{\boldsymbol{v}} - \tilde{\boldsymbol{w}}|||$$
$$\text{for all } \tilde{\boldsymbol{v}}, \tilde{\boldsymbol{w}} \in \mathscr{C}^0([0,1], \boldsymbol{B}_\rho^p)\,.$$

(iii) *The fixed point of the contraction*

$$\boldsymbol{\Phi}^k : \mathscr{C}^0([0,1]; \boldsymbol{B}_\rho^p) \to \mathscr{C}^0([0,1]; \boldsymbol{B}_\rho^p)$$

is the unique solution to the given differential equation in the space $\mathscr{C}^0([0,1], \boldsymbol{V}_\rho^p)$. By the completeness of the space $\mathscr{C}^0([0,1]; \boldsymbol{B}_\rho^p)$ and the contraction mapping theorem, there exists a unique element $\tilde{\boldsymbol{u}}$ such that

$$\tilde{\boldsymbol{u}} \in \mathscr{C}^0([0,1]; \boldsymbol{B}^p) \text{ and } \boldsymbol{\Phi}^k(\tilde{\boldsymbol{u}}) = \tilde{\boldsymbol{u}}\,.$$

Since this element $\tilde{u}$ satisfies

$$\Phi(\Phi^k(\tilde{u})) = \Phi^k(\Phi(\tilde{u})) = \Phi(\tilde{u}) ,$$

the element $\Phi(u) \in \mathscr{C}^0([0,1]; B_\rho^p)$ is also a fixed point of the mapping Φ^k. Such a fixed point being unique, *the element* $\tilde{u}$ *is also a fixed point of the mapping* Φ, and it is its only fixed point in the space $\mathscr{C}^0([0,1]; B_\rho^p)$ (each fixed point of Φ is a fixed point of Φ^k). Thus we have, by definition,

$$\tilde{u}(\lambda) = \int_0^\lambda \{A'(\tilde{u}(\mu))\}^{-1} f \, \mathrm{d}\mu \quad \text{for } 0 \leq \lambda \leq 1 ,$$

and consequently the function $\tilde{u}$ is differentiable, with

$$\tilde{u}'(\lambda) = \{A'(\tilde{u}(\lambda))\}^{-1} f, \quad 0 \leq \lambda \leq 1 .$$

Since it is clear that $\tilde{u}(0) = o$ on the other hand, the function u is indeed a solution of the differential equation.

Conversely, any solution $\tilde{u}$ of the above differential equation with $\tilde{u}(\lambda) \in B_\rho^p$ for $0 \leq \lambda \leq 1$ is in the space $\mathscr{C}^0([0,1], B_\rho^p)$ and it satisfies

$$\tilde{u}(\lambda) = \tilde{u}(0) + \int_0^\lambda u'(\mu) \, \mathrm{d}\mu = \int_0^\lambda \{A'(u(\mu))\}^{-1} f \, \mathrm{d}\mu, \quad 0 \leq \lambda \leq 1 ,$$

if $\tilde{u}(0) = o$. Thus it is also a fixed point of the mapping Φ in the space $\mathscr{C}^0([0,1]; B_\rho^p)$. In particular it is unique.

(iv) *The solution* $\tilde{u}$ *of the differential equation satisfies*

$$\tilde{u}(\lambda) \in B_\rho^p \text{ and } A(\tilde{u}(\lambda)) = \lambda f, \quad 0 \leq \lambda \leq 1 .$$

The relation

$$o = A'(\tilde{u}(\lambda))\tilde{u}'(\lambda) - f = \frac{\mathrm{d}}{\mathrm{d}\lambda}\{A(\tilde{u}(\lambda)) - \lambda f\}, \quad 0 \leq \lambda \leq 1 ,$$

implies that the mapping

$$\lambda \in [0,1] \rightarrow \{A(\tilde{u}(\lambda)) - \lambda f\} \in L^p(\Omega)$$

is a constant mapping, which vanishes since $\tilde{u}(0) = o$. ■

Notice that the above proof provides *another proof of the existence result of Theorem* 6.7-1, at the expense however of a slightly stronger regularity assumption on the mapping $\check{\Sigma}$; it also provides a more explicit description of the neighborhood F^p, which can be chosen to be any ball

$$\{f \in L^p(\Omega); |f|_{0,p,\Omega} \leq \rho(\gamma_\rho^p)^{-1}\} \quad \text{with } \rho < \rho_0^p .$$

The idea of imbedding the solution of a nonlinear problem (here, $A(u) = f$) in a one-parameter family of solutions of a differential equation (here, $\tilde{u}(\lambda) = \{A'(\tilde{u}(\lambda))\}^{-1} f$) is the basis of the *continuation by differentiation methods*, as described for instance in Rheinboldt [1974, 1986]. As illustrated by the next convergence analysis, such methods often provide efficient tools for approximating the solution of the original problem by means of the application of a convergent approximation scheme to the differential equation.

6.13. CONVERGENCE OF THE INCREMENTAL METHOD

The convergence result given here is due to Bernadou, Ciarlet & Hu [1984] (see also Bernadou, Ciarlet & Hu [1982]). Once it has been observed that the incremental method can be viewed as Euler's method for approximating a differential equation in a Sobolev space, establishing the convergence essentially consists in adapting to the present situation the classical proof of convergence of Euler's method (see e.g. Crouzeix & Mignot [1984], Gear [1971], Henrici [1962], Stetter [1973]).

Theorem 6.13-1. *The assumptions and notation are as in Theorem* 6.12-1. *Given any partition*

$$0 = \lambda^0 < \lambda^1 \cdots < \lambda^n < \lambda^{n+1} < \cdots < \lambda^{N-1} < \lambda^N = 1$$

of the interval $[0, 1]$, *there exists one and only one sequence* $(u^n)_{n=0}^N$ *that satisfies*

$$\boxed{\begin{aligned} &u^0 = o, \quad u^n \in B_\rho^p, \quad 0 \leq n \leq N, \\ &A'(u^n)(u^{n+1} - u^n) = (\lambda^{n-1} - \lambda^n) f, \quad 0 \leq n \leq N-1 . \end{aligned}}$$

In addition, there exists a constant $c = c(|f|_{0,p,\Omega}, \gamma_0^p, L_\rho^p)$ *such that*

$$\max_{0 \le n \le N} \|u^n - u(\lambda^n f)\|_{2,p,\Omega} \le c\Delta\lambda ,$$

$$\max_{0 \le n \le N} \|\Sigma^n - \Sigma(\lambda^n f)\|_{1,p,\Omega} \le c\Delta\lambda ,$$

$$\text{with } \Delta\lambda := \max_{0 \le n \le N-1} (\lambda^{n-1} - \lambda^n) ,$$

where $u(\lambda^n f)$ *denotes the unique solution in the ball* B_ρ^p *of the equation* $A(u) = \lambda^n f$, *and*

$$\Sigma^n := \check{\Sigma}(E(u^n)), \quad \Sigma(\lambda^n f) := \check{\Sigma}(E(u(\lambda^n f)), 0 \le n \le N .$$

Proof. For conciseness, let

$$\Delta\lambda^n = \lambda^{n+1} - \lambda^n, \quad 0 \le n \le N-1 .$$

(i) *The algorithm is well defined*: Assume that

$$u^l \in B_\rho^p, 0 \le l \le k ,$$

for some integer k with $0 \le k \le N-1$ (clearly $u^0 \in B_\rho^p$). Then

$$u^{k+1} = u^k + \Delta\lambda^k \{A'(u^k)\}^{-1} f$$

is a well defined member of the space $V^p(\Omega)$. We must show that $u^{k+1} \in B_\rho^p$ in order to pursue the algorithm. Adding the relations

$$u^0 = o \text{ and } u^{l+1} = u^l + \Delta\lambda^l \{A'(u^l)\}^{-1} f, 0 \le l \le k ,$$

we obtain

$$u^{k+1} = \sum_{l=0}^{k} \Delta\lambda^l \{(A'(u^l))\}^{-1} f ,$$

and thus $(\Sigma_{l=0}^k \Delta\lambda^l = \lambda^{k+1} \le 1)$

$$\|u^{k+1}\|_{2,p,\Omega} \le \max_{0 \le l \le k} \|\{A'(u^l)\}^{-1}\| \, |f|_{0,p,\Omega} \le \gamma_\rho^p |f|_{0,p,\Omega} \le \rho ,$$

since $\gamma_\rho^p = \sup_{v\in B_\rho^p} \|\{A'(v)\}^{-1}\|$ (Theorem 6.6-1) and $|f|_{0,p,\Omega} \leqslant \rho(\gamma_\rho^p)^{-1}$ by assumption.

(ii) Let us establish an *inequality relating two successive errors* ε^n *and* ε^{n+1}, *where* (recall that $\tilde{u}(\lambda^n) = u(\lambda^n f)$)

$$\varepsilon^n := \|u^n - \tilde{u}(\lambda^n)\|_{2,p,\Omega}, \quad 0 \leqslant n \leqslant N .$$

For this purpose, we write the difference between the relations

$$u^{n+1} = u^n + \Delta\lambda^n \{A'(u^n)\}^{-1} f ,$$

$$\tilde{u}(\lambda^{n+1}) = \tilde{u}(\lambda^n) + \int_{\lambda^n}^{\lambda^{n+1}} \tilde{u}'(\mu)\, d\mu$$

$$= \tilde{u}(\lambda^n) + \int_{\lambda^n}^{\lambda^{n+1}} \{A'(\tilde{u}(\mu))\}^{-1} f\, d\mu ,$$

as

$$u^{n+1} - \tilde{u}(\lambda^{n+1}) = u^n - \tilde{u}(\lambda^n) + \Delta\lambda^n(\{A'(u^n)\}^{-1} - \{A'(\tilde{u}(\lambda^n))\}^{-1})f$$
$$+ \int_{\lambda^n}^{\lambda^{n+1}} (\{A'(\tilde{u}(\lambda^n))\}^{-1} - \{A'(\tilde{u}(\mu))\}^{-1})f\, d\mu .$$

The Lipschitz-continuity of the mapping $v \in B_\rho^p \to \{A'(v)\}^{-1}$ (Theorem 6.6-1) then implies that

$$\|(\{A'(u^n)\}^{-1} - \{A'(\tilde{u}(\lambda^n))\}^{-1})f\|_{2,p,\Omega} \leqslant L_\rho^p \|u^n - \tilde{u}(\lambda^n)\|_{2,p,\Omega} |f|_{0,p,\Omega}$$

and

$$\left\| \int_{\lambda^n}^{\lambda^{n+1}} (\{A'(\tilde{u}(\lambda^n))\}^{-1} - \{A'(\tilde{u}(\mu))\}^{-1})f\, d\mu \right\|_{2,p,\Omega}$$
$$\leqslant \Delta\lambda^n L_\rho^p \sup_{\lambda^n \leqslant \mu \leqslant \lambda^{n+1}} \|\tilde{u}(\lambda^n) - \tilde{u}(\mu)\|_{2,p,\Omega} |f|_{0,p,\Omega}$$
$$\leqslant (\Delta\lambda^n)^2 L_\rho^p \sup_{\lambda^n \leqslant \lambda \leqslant \lambda^{n+1}} \|\tilde{u}'(\lambda)\|\, |f|_{0,p,\Omega} ,$$

the last inequality being obtained by an application of the mean value theorem. Since

$$\sup_{\lambda^n \leqslant \lambda \leqslant \lambda^{n+1}} \|\tilde{u}'(\lambda)\| \leqslant \sup_{0 \leqslant \lambda \leqslant 1} \|\tilde{u}'(\lambda)\| = \sup_{0 \leqslant \lambda \leqslant 1} \|\{A'(\tilde{u}(\lambda))\}^{-1} f\|_{2,p,\Omega}$$

$$\leqslant \gamma_\rho^p |f|_{0,p,\Omega} ,$$

we finally obtain an inequality of the form

$$\varepsilon^{n+1} \leqslant (1 + L_\rho^p |f|_{0,p,\Omega} \Delta\lambda^n)\varepsilon^n + L_\rho^p \gamma_0^p |f|_{0,p,\Omega}^2 (\Delta\lambda^n)^2 ,$$

or in short,

$$\varepsilon^{n+1} \leqslant (1 + a\Delta\lambda^n)\varepsilon^n + b^n , \quad 0 \leqslant n \leqslant N-1 ,$$

with

$$a = L_\rho^p |f|_{0,p,\Omega}, \quad b^n = L_\rho^p \gamma_0^p |f|_{0,p,\Omega}^2 (\Delta\lambda^n)^2 .$$

(iii) We next use the following result: *Let $(\varepsilon^n)_{n \geqslant 0}$ be a sequence of real numbers satisfying*

$$\varepsilon^{n+1} \leqslant (1 + a\Delta\lambda^n)\varepsilon^n + b^n, \; n \geqslant 0 ,$$

with

$$a \geqslant 0, \; \lambda^0 = 0, \; \Delta\lambda^n = \lambda^{n+1} - \lambda^n > 0, \; b^n \geqslant 0, \; n \geqslant 0 .$$

Then,

$$\varepsilon^n \leqslant e^{a\lambda^n} \varepsilon^0 + \sum_{l=0}^{n-1} e^{a(\lambda^n - \lambda^{l+1})} b^l, \; n \geqslant 1 .$$

The proof is simple: By combining the first n inequalities, we find that

$$\varepsilon^n \leqslant \prod_{l=0}^{n-1} (1 + a\Delta\lambda^l)\varepsilon^0 + \sum_{l=0}^{n-2} \left\{ \prod_{k=l+1}^{n-1} (1 + a\Delta\lambda^k) \right\} b^l + b^{n-1} ,$$

and it then suffices to apply the inequality $(1 + x) \leqslant e^x$ valid for all $x \in \mathbb{R}$. In our case

$$\varepsilon^0 = 0 \text{ and } b^l = \beta(\Delta\lambda^l)^2 \quad \text{with } \beta = L_\rho^p \gamma_0^p |f|_{0,p,\Omega}^2 ,$$

and thus

$$\varepsilon^n \leq \beta \sum_{l=0}^{n-1} \mathrm{e}^{a(\lambda^n - \lambda^{l+1})} (\Delta\lambda^l)^2 .$$

From the inequality

$$\mathrm{e}^{a(\lambda^n - \lambda^{l+1})} \Delta\lambda^l \leq \int_{\lambda_l}^{\lambda_{l+1}} \mathrm{e}^{a(\lambda^n - s)} \, \mathrm{d}s ,$$

we deduce, with $\Delta\lambda = \max_{0 \leq n \leq N-1} \Delta\lambda^n$,

$$\varepsilon^n = \|u^n - \tilde{u}(\lambda^n)\|_{2,p,\Omega} \leq \beta\Delta\lambda \int_0^{\lambda^n} \mathrm{e}^{a(\lambda^n - s)} \, \mathrm{d}s = \frac{\beta}{a} (\mathrm{e}^{a\lambda^n} - 1)\Delta\lambda, \quad 0 \leq n \leq N-1 ,$$

and the first error bound follows.

(iv) In order to *estimate the error concerning the stresses*, we combine the boundedness of the derivatives of the mappings

$$\boldsymbol{v} \in \boldsymbol{W}^{2,p}(\Omega) \rightarrow \boldsymbol{E}(\boldsymbol{v}) \in \boldsymbol{W}^{1,p}(\Omega) ,$$

$$\boldsymbol{E} \in \boldsymbol{V}(\boldsymbol{0}) \subset \boldsymbol{W}^{1,p}(\Omega) \rightarrow \check{\boldsymbol{\Sigma}}(\boldsymbol{E}) \in \boldsymbol{W}^{1,p}(\Omega)$$

(this property is clear for the first mapping since the space $W^{1,p}(\Omega)$ is an algebra; for the second mapping, it has been established in Theorem 6.5-1) with the mean value theorem and the chain rule. In this fashion we obtain

$$\begin{aligned} \|\boldsymbol{\Sigma}^n - \boldsymbol{\Sigma}(\lambda^n \boldsymbol{f})\|_{1,p,\Omega} &= \|\check{\boldsymbol{\Sigma}}(\boldsymbol{E}(\boldsymbol{u}^n)) - \check{\boldsymbol{\Sigma}}(\boldsymbol{E}(\boldsymbol{u}(\lambda^n \boldsymbol{f})))\|_{1,p,\Omega} \\ &\leq \sup_{\|\boldsymbol{E}\|_{1,p,\Omega} \leq \delta_\rho^p} \|\check{\boldsymbol{\Sigma}}'(\boldsymbol{E})\| \sup_{\boldsymbol{v} \in \boldsymbol{B}_\rho^p} \|\boldsymbol{E}'(\boldsymbol{v})\| \|\boldsymbol{u}^n - \boldsymbol{u}(\lambda^n \boldsymbol{f})\|_{2,p,\Omega} , \end{aligned}$$

where

$$\delta_\rho^p := \sup_{\boldsymbol{v} \in \boldsymbol{B}_\rho^p} \|\boldsymbol{E}(\boldsymbol{v})\|_{1,p,\Omega} ,$$

and the proof is complete. ■

Remarks. (1) The upper bound for the quantity ε^n established at the end of step (iii) in the proof provides the *sharper estimates*

$$\max_{0 \leqslant l \leqslant n} \|u^n - u(\lambda^n f)\|_{2,p,\Omega} \leqslant \frac{\beta}{a}(e^{a\lambda^{n-1}})\Delta\lambda, \quad 0 \leqslant n \leqslant N-1,$$

which substantiate the natural idea that, for a fixed N, the error is an increasing function of n.

(2) Euler's method is the simplest example of a *one-step method*, i.e., that involves only two successive approximations, for approximating the solution of an ordinary differential equation. There exist more sophisticated one-step methods, called *Runge–Kutta's methods*, and multi-step methods, such as *Adams' methods*, whose error is of the order $(\Delta\lambda)^p$ for some $p \geqslant 2$; for details, see for instance Crouzeix & Mignot [1983, Chapters V, VI]. Hence such methods may be applied to the present problem.

(3) The fact that the nth approximate equation

$$A'(u^n)(u^{n+1} - u^n) = (\lambda^{n+1} - \lambda^n)f$$

can be effectively solved in the ball $B_\rho^p \subset V^p(\Omega)$ is in essence an *existence and regularity result* for the *linear* problem:

$$A'(u^n)u = g \Leftrightarrow -\partial_j\{\hat{a}_{ijpq}(\nabla u^n)\partial_p u_q\} = g \text{ in } \Omega,$$

which has thus a unique solution in the space $V^p(\Omega)$ if the right-hand side g is in $L^p(\Omega)$.

(4) A situation where the applied body force is not a dead load is considered in Exercise 6.8.

(5) The convergence of incremental methods in nonlinear elasticity problems had been previously analyzed in some special cases, such as the one-dimensional model of a thin shallow spherical shell, by Anselone & Moore [1966], and some finite-dimensional structural problems, by Rheinboldt [1981]. ■

Incremental methods can be also applied, with appropriate modifications, to the *pure traction problem*, and to the *pure displacement problem of incompressible elasticity* described at the end of Sect. 6.7. Their convergence, which then requires a substantially more delicate analysis, has been established by Nzengwa [1987].

EXERCISES

6.1. Let $\Omega = \{(x_1, x_2) \in \mathbb{R}^2;\ x_1^2 + x_2^2 < \rho\}$ with $\rho \in]0, 1[$, and let $\alpha \in]0, \frac{1}{2}[$.

(1) Show that the function

$$v : x \in \Omega \to v(x) := |\mathrm{Log}(x_1^2 + x_2^2)|^\alpha$$

belongs to the Sobolev space $H^1(\Omega)$.

(2) Let $\bigcup_{k=0}^\infty \{b_k\}$ be a dense subset of Ω, and let $\beta_k > 0$ be such that $\Sigma_{k=0}^\infty \beta_k < \infty$. Show that the function

$$w : x \in \Omega \to w(x) := \sum_{k=0}^\infty \beta_k v(x - b_k)$$

belongs to the Sobolev space $H^1(\Omega)$.

Remark. This exercise provides examples of (equivalence classes of) functions in the Sobolev space $H^1(\Omega)$ that have no continuous representatives, the function w being even discontinuous on a dense subset of Ω.

6.2. (1) Let $\boldsymbol{v} : x \in \Omega \to \boldsymbol{v}(x) = \boldsymbol{a} + \boldsymbol{b} \wedge \boldsymbol{ox}$ be an infinitesimal rigid displacement. Show that there exists a skew tensor $\boldsymbol{W}$ ($\boldsymbol{W}^{\mathrm{T}} = -\boldsymbol{W}$) such that $\boldsymbol{v}(x) = \boldsymbol{a} + \boldsymbol{Wox}$ for all $x \in \Omega$.

(2) Show that a skew tensor $\boldsymbol{W}$ generates a one-parameter subgroup $t \in \mathbb{R} \to \exp(t\boldsymbol{W})$ of the group $\mathbb{O}^3_+$.

(3) Show that the tangent space at $\boldsymbol{I}$ of the manifold $\mathbb{O}^3$ can be identified with the space of skew tensors.

Remark. All relevant information may be found in Wang & Truesdell [1973, Ch. 1]. See also Lang [1962], Doubrovine, Novikov & Fomenko [1982a, §24].

6.3. The object of this problem is to extend to *linearized pure traction problems* the results of Theorems 6.3-5 and 6.3-6. The notation is as in Theorem 6.3-5.

(1) Let Ω be a domain in $\mathbb{R}^3$, let the constants $\lambda > 0$, $\mu > 0$ and the functions $\boldsymbol{f} \in \boldsymbol{L}^{6/5}(\Omega)$, $\boldsymbol{g} \in \boldsymbol{L}^{4/3}(\Gamma)$ be given. Show that the equations

$$B(\boldsymbol{u}, \boldsymbol{v}) = \int_\Omega \boldsymbol{f} \cdot \boldsymbol{v}\, \mathrm{d}x + \int_\Gamma \boldsymbol{g} \cdot \boldsymbol{v}\, \mathrm{d}a \quad \text{for all } \boldsymbol{v} \in \boldsymbol{H}^1(\Omega)$$

have a solution $u \in H^1(\Omega)$ if and only if $\int_\Omega f \cdot v \, dx + \int_\Gamma g \cdot v \, da = 0$ for all infinitesimal rigid displacements v.

Hints. Let W denote the space formed by all infinitesimal rigid displacements. Show that $v \to |e(v)|_{0,\Omega}$ is a norm over the quotient space $V := H^1(\Omega)/W$, equivalent to the quotient norm; then show that Theorem 6.3-2 can be applied with this choice of space V (see Duvaut & Lions [1972, p. 119 ff.]).

(2) Assume that the boundary is of class $\mathscr{C}^2$ and that $f \in L^p(\Omega)$, $g \in W^{1-1/p,p}(\Gamma)$, for some $p \geq \frac{6}{5}$. Show that the solution u found in (1) is in the space $W^{2,p}(\Omega)$, by verifying in particular that the complementing condition of Agmon, Douglis & Nirenberg [1964] is satisfied by the linearized pure traction problem (in this respect, see Thompson [1969], Lanza de Cristoforis & Valent [1982], Podio-Guidugli & Vergara-Caffarelli [1984], Le Dret [1986]).

6.4. Show that the results of Theorems 6.3-5, 6.3-6, 6.4-1, 6.7-1, still hold under the weaker assumptions $\mu > 0$, $\lambda > -\frac{2}{3}\mu$.

Hint. Use the result of Exercise 5.17 (another approach is proposed in Marsden & Hughes [1983, Sect. 6.1]).

6.5. The assumption $p \geq \frac{6}{5}$ in Theorem 6.3-6 guarantees that the mapping $A'(o): V^p(\Omega) \to L^p(\Omega)$ is injective for these values of p. Is it still injective for $1 < p < \frac{6}{5}$?

6.6. Assume that the weak solution $u \in V$ found in Theorem 6.3-5 belongs to the Sobolev space $W^{m,p}(\Omega)$ for some integer $m \geq 2$ and some $p \geq 1$. Using the Green formula in Sobolev spaces (Theorem 6.1-9), show that, in a specific sense, the function u is a solution of the linearized displacement–traction problem (the functions f and g may be assumed to be smoother than in Theorem 6.3-5).

6.7. (1) Show that any function $u \in V(o) \subset W^{2,p}(\Omega)$, $p > 3$, satisfies $\det(I + \nabla u)(x) > 0$ for all $x \in \bar{\Omega}$ (the neighborhood $V(o)$ is defined in Sect. 6.6).

(2) Is the following assertion true or false? In any neighborhood of the origin in $W^{2,p}(\Omega)$, $1 \leq p \leq 3$, one can find a function u such that $\mathrm{vol}\{x \in \Omega; \det(I + \nabla u)(x) \leq 0\} > 0$.

6.8. Consider the pure displacement problem where the applied body force is the *centrifugal force* (see Sect. 2.7 for the definition and the notation), viz.,

$$\begin{cases} -\mathbf{div}\{\nabla\varphi(x)\hat{\Sigma}(\nabla\varphi(x))\} = \omega^2\rho(x)(\varphi_2(x)e_2 + \varphi_3(x)e_3),\ x \in \Omega\,, \\ \varphi = id \text{ on } \Gamma\,, \end{cases}$$

where $\check{\Sigma}(F) = \check{\Sigma}(\frac{1}{2}(F^TF - I))$ for all $F \in \mathbb{M}^3_+$, and the assumptions on the response function $\check{\Sigma}$ are as in Theorem 6.7-1. The mass density $\rho : \bar{\Omega} \to \mathbb{R}$ may be assumed to be as smooth as necessary.

(1) Show that there exists $\omega_0 > 0$ such that this problem has a solution $u = \varphi - id \in V^p(\Omega)$, $p > 0$, for all $|\omega| < \omega_0$.

(2) This problem can be written in operator form as

$$A_{\#}(u) := A(u) + Lu = f\,,$$

where the nonlinear operator A is defined as in Sect. 6.6, and the linear operator L and the right-hand side f are respectively given by:

$$Lu = -\omega^2\rho(u_2e_2 + u_3e_3),\quad f(x) = \omega^2\rho(x)(x_2e_2 + x_3e_3)\,.$$

As in Sect. 6.11, an incremental method can be defined that consists in recursively solving the linear problems

$$A'_{\#}(u^n)(u^{n+1} - u^n) = \delta f^n,\ 0 \leq n \leq N-1,\quad u^0 = o\,,$$

where $\delta f^n := (\lambda^{n+1} - \lambda^n)f$. Show that, if $|\omega|$ is small enough, the incremental method is well defined and convergent.

6.9. Consider the Nemytsky operator $A : u \in H^m(\Omega) \to Au \in L^2(\Omega)$, where $Au(x) = a(x, u(x))$, the function $a : \Omega \times \mathbb{R} \to \mathbb{R}$ is as smooth as necessary, and Ω is a bounded open subset of $\mathbb{R}^n$.

(1) Let $m = 0$. Show that if A is Fréchet-differentiable, the function a is necessarily affine with respect to its second argument, i.e., $a(x, \zeta) = \alpha(x) + \beta(x)\zeta$.

(2) Show that A is Fréchet-differentiable if the integer m is such that $H^m(\Omega) \hookrightarrow \mathscr{C}^0(\bar{\Omega})$.

6.10. In Washizu [1975, p. 395], another incremental method is defined, which "combines the Lagrangian and Eulerian approaches". Describe and analyze the convergence of this incremental method along the lines followed here for the Lagrangian method.

6.11. *Newton's method* for solving a nonlinear equation of the form $A(u) = f$ consists in choosing an initial guess u^0 and in successively solving the linear problems

$$A'(u^n)(u^{n+1} - u^n) = f - A(u^n),\ n \geq 0 .$$

Hence Newton's method replaces the solution of a nonlinear problem by the solution of an infinite number of linear problems; this is a characteristic feature of approximate methods for nonlinear problems, shared in particular by the incremental method described in Sect. 6.10. The notation is as in Theorem 6.6-1.

(1) Assume that

$$u^0 = o,\ \rho\gamma_0^p M^p(\rho) < \tfrac{1}{3},\ |f|_{0,p,\Omega} \leq \rho(\gamma_0^p)^{-1}(1 - 3\rho\gamma_0^p M^p(\rho)) ,$$

for some number $p > 3$, and show that all the successive iterates u^n lie in the ball B_ρ^p, so that Newton's method is well defined.

(2) With the same assumptions as in (1), show that Newton's method converges and that the convergence is geometric, in the sense that the error $\|u^n - u\|_{2,p,\Omega}$ is bounded by the general term of a convergent geometric series. More specifically, show that there exists a number $c(\rho)$ independent of n such that

$$\|u^n - u\|_{2,p,\Omega} \leq c(\rho)(3\rho\gamma_0^p M^p(\rho))^n,\ n \geq 0 ,$$

where u is the unique solution to the equation $A(u) = f$ in the ball B_ρ^p.

(3) Since it is numerically costly to invert a different operator $A'(u^n)$ at each step, a natural variant consists in only inverting the operator $A'(u^0)$ corresponding to the initial value. The corresponding *generalized Newton's method* then takes the form:

$$A'(u^0)(u^{n+1} - u^n) = f - A(u^n),\ n \geq 0 .$$

Assuming that $u^0 = o$ and that

$$\rho\gamma_0^p M^p(\rho) < 1 \text{ and } |f|_{0,p,\Omega} \leq \rho(\gamma_0^p)^{-1}(1 - \rho\gamma_0^p M^p(\rho)) ,$$

for some number $p > 3$, show that this method is well defined and that it also converges geometrically.

Remark. The convergence of Newton's method and its variants is a classical result (see for instance Ciarlet [1983, Sect. 7.5]).

6.12. The notation and hypotheses are as in Theorem 6.12-2; it is assumed in addition that the mapping $\check{\boldsymbol{\Sigma}} : \mathbb{V}(\boldsymbol{O}) \to \mathbb{S}^3$ is analytic. Show that the mapping $\lambda \in I \to \tilde{\boldsymbol{u}}(\lambda) := \boldsymbol{u}(\lambda \boldsymbol{f}) \in \boldsymbol{V}^p(\Omega)$ is also analytic in an open interval I containing 0 (cf. Destuynder & Galbe [1978] for a St Venant–Kirchhoff material).

CHAPTER 7

EXISTENCE THEORY BASED ON THE MINIMIZATION OF THE ENERGY

INTRODUCTION

We have seen in Chapter 6 that the Riesz representation theorem in Hilbert spaces leads to a simple proof of the existence (and uniqueness) of a minimizer of a *coercive* and *quadratic* functional over a *Hilbert space*, and that this result had an application to linearized elasticity. In genuine three-dimensional hyperelasticity, these assumptions must be considerably weakened, since the problem now consists in minimizing *coercive*, but *nonquadratic*, functionals over *nonconvex* subsets of *more general reflexive Banach spaces*; besides, the integrand $\hat{W}(\boldsymbol{F})$ is nonconvex with respect to $\boldsymbol{F}$ and it becomes infinite when $\det \boldsymbol{F}$ approaches zero. For all these reasons, establishing the existence of minimizers of the energy in hyperelasticity is considerably more difficult.

A powerful and successful approach has been nevertheless devised by John Ball in 1977: His method, outlined in more details in Sect. 7.4, consists in considering an *infimizing sequence* $(\boldsymbol{\varphi}^k)$ *of the total energy* over a *set of admissible deformations* of the form (in the case of a displacement–traction problem, for definiteness):

$$\begin{aligned}\boldsymbol{\Phi} = \{&\boldsymbol{\psi} \in \boldsymbol{W}^{1,p}(\Omega);\ \mathbf{Cof}\,\boldsymbol{\nabla\psi} \in \boldsymbol{L}^q(\Omega),\ \det \boldsymbol{\nabla\psi} \in L^r(\Omega)\,,\\ &\boldsymbol{\psi} = \boldsymbol{\varphi}_0\ da\text{-a.e. on } \Gamma_0,\ \det \boldsymbol{\nabla\psi} > 0 \text{ a.e. in } \Omega\}\,.\end{aligned}$$

The numbers p, q, r, which are precisely those appearing in the coerciveness inequality:

$$\hat{W}(x, \boldsymbol{F}) \geq \alpha\{\|\boldsymbol{F}\|^p + \|\mathbf{Cof}\,\boldsymbol{F}\|^q + (\det \boldsymbol{F})^r\} + \beta \quad \text{for all } \boldsymbol{F} \in \mathbb{M}^3_+\,,$$

are assumed to be sufficiently large. By the *coerciveness inequality*, the sequence $(\boldsymbol{\varphi}^k, \mathbf{Cof}\,\boldsymbol{\nabla\varphi}^k, \det \boldsymbol{\nabla\varphi}^k)$ is bounded in the reflexive Banach space $\boldsymbol{W}^{1,p}(\Omega) \times \boldsymbol{L}^q(\Omega) \times L^r(\Omega)$. Thus there exists a subsequence $(\boldsymbol{\varphi}^l, \mathbf{Cof}\,\boldsymbol{\nabla\varphi}^l, \det \boldsymbol{\nabla\varphi}^l)$ that converges weakly in this space to an element $(\boldsymbol{\varphi}, \boldsymbol{H}, \delta)$, and remarkably, one precisely has $\boldsymbol{H} = \mathbf{Cof}\,\boldsymbol{\nabla\varphi}$ and $\delta = \det \boldsymbol{\nabla\varphi}$

(Sects. 7.5 and 7.6). The *behavior* $\hat{W}(x, F) \to +\infty$ *as* $\det F \to 0^+$ then implies that $\det \nabla\varphi > 0$ almost everywhere in Ω, and thus the weak limit of the infimizing sequence also belongs to the set Φ.

Another crucial observation of John Ball is that the assumption of *polyconvexity* of the stored energy functiom implies the *sequential weak lower semi-continuity of the total energy I*, viz.,

$$I(\varphi) \leq \liminf_{l \to \infty} I(\varphi^l),$$

from which it follows that $\varphi \in \Phi$ *is a minimizer of the total energy over the set* Φ. He obtains in this fashion *existence theorems in the space* $W^{1,p}(\Omega)$ for pure displacement and displacement–traction problems (Theorem 7.7-1) and also, with an additional condition on the admissible deformations, for the pure traction problem (Theorem 7.7-2).

We then extend John Ball's existence results in two directions. First, we allow (Theorems 7.8-1 and 7.8-2) a *unilateral boundary condition of place* $\varphi(x) \in C$, $x \in \Gamma_2 \subset \Gamma$, in the set Φ (this is a model of contact without friction; cf. Sect. 5.3). Secondly, we impose the *injectivity condition* $\int_\Omega \det \nabla\varphi \, dx \leq \operatorname{vol} \varphi(\Omega)$ in the set Φ (this is a model of self-contact without friction and of noninterpenetration of matter; cf. Sect. 5.6), and we show that this implies the existence of an *almost everywhere injective minimizer* of the total energy (Theorem 7.9-1).

We conclude this chapter by a brief review (Sect. 7.10) of various *open problems*, among which the question of finding reasonable conditions under which the minimizers are (even weak) solutions of the associated boundary value problem stands as a major unresolved issue.

*7.1. WEAK TOPOLOGY AND WEAK CONVERGENCE

The existence theory developed in this chapter makes an essential use of weak convergence and weakly lower semi-continuous functionals; for this reason, we briefly review these notions in the first two sections. For proofs and complements about weak topology and weak convergence, see Brezis [1983], Yosida [1966, Chapter V].

Let V be a normed vector space. We recall that its dual space V' consists of all the linear forms $L: V \to \mathbb{R}$ that are continuous with respect to the **strong topology** on V, that is, the topology induced by the norm of the space V. The **weak topology** on V is the weakest topology on V for which all the elements L of V' remain continuous; "weakest" means that any other topology on V with the same property contains more open sets. An open set for the weak topology is thus an open set for the strong topology, but the converse is not always true:

Theorem 7.1-1. (a) *If V is a finite-dimensional space, the strong and weak topologies coincide, and thus in this case the weak topology is normable, hence a fortiori metrizable.*

(b) *If V is an infinite-dimensional space, the weak topology is strictly contained in the strong topology, i.e., there are open sets for the strong topology that are not open for the weak topology. Moreover, the weak topology is not metrizable in this case.*

(c) *The weak topology is a Hausdorff topology.* ∎

A sequence (v_k) of elements of V is **strongly convergent** if it converges to an element $v \in V$ with respect to the strong topology; it is **weakly convergent** if it converges to an element $v \in V$ with respect to the weak topology. We shall respectively denote these convergences by:

$$v_k \rightarrow v \text{ (strong convergence)}; \; v_k \rightharpoonup v \text{ (weak convergence)}.$$

Note that *the limit of a weakly convergent sequence is unique* (Theorem 7.1-1(c)), and that *a strongly convergent sequence is weakly convergent*, by definition of the two topologies (the converse, however, does not necessarily hold in an infinite-dimensional space; cf. Theorem 7.1-1(a)). Finally, it is easy to prove that

$$\boxed{v_k \rightharpoonup v \text{ in } V \Leftrightarrow L(v_k) \rightarrow L(v) \quad \text{for all } L \in V'.}$$

Except in some "pathological" spaces (such as l^1), there exist sequences that weakly converge but do not strongly converge in an infinite dimensional normed vector space. For instance, the sequence (v_k) with $v_k(t) = \sin kt$, $0 \leq t \leq 2\pi$, weakly converges to the null function in the space $L^2(0, 2\pi)$, since the Fourier coefficients $\int_0^{2\pi} v(t) \sin kt \, dt$ of a function $v \in L^2(0, 2\pi)$ must converge to 0 as $k \rightarrow \infty$ by Parseval's identity. The same sequence does not converge strongly however for it could only converge to 0, and yet $\|v_k\|_{L^2(0, 2\pi)} = \sqrt{\pi}$. Nevertheless, there are several useful relations between weak and strong convergences in infinite-dimensional spaces, which are stated in the next theorems.

Theorem 7.1-2 (Mazur's theorem). *Let $v_k \rightharpoonup v$ in a normed vector space V. Then there exist convex combinations*

$$w_k := \sum_{m=k}^{N(k)} \lambda_m^k u_m , \quad \text{with} \sum_{m=k}^{N(k)} \lambda_m^k = 1,\ \lambda_m^k \geq 0,\ k \leq m \leq N(k)\,,$$

such that $w_k \to w$. ■

For a proof, see e.g. Ekeland & Temam [1974, p. 6]. Let C be a convex subset of V that is closed for the strong topology. Then the limit of any weakly convergent sequence of elements of C also belongs to C by Mazur's theorem, and thus the set C is "sequentially weakly closed". In fact, we have more (see e.g. Brezis [1983, Theorem III.7, p. 38]):

Theorem 7.1-3. *A convex set that is closed for the strong topology is also closed for the weak topology.* ■

Under the assumption of *completeness*, further useful results can be obtained:

Theorem 7.1-4. (a) *In a Banach space, a weakly convergent sequence (v_k) is bounded, and its limit v satisfies*

$$\|v\| \leq \liminf_{k \to \infty} \|v_k\|\,.$$

(b) *In a reflexive Banach space, a bounded sequence contains a weakly convergent subsequence.* ■

For a proof, see e.g. Brezis [1983, Proposition III.5, p. 35 and Theorem III.27, p. 50].

Theorem 7.1-5. (a) *Let V and W be Banach spaces, and let $A: V \to W$ be a compact linear operator. Then*

$$v_k \rightharpoonup v \text{ in } V \Rightarrow Av_k \to Av \text{ in } W\,.$$

(b) *Let V be a normed vector space, let W be a Banach space, and let $B: V \times W \to \mathbb{R}$ be a continuous bilinear mapping. Then*

$$v_k \to v \text{ and } w_k \rightharpoonup w \Rightarrow B(v_k, w_k) \to B(v, w)\,.$$

Proof. Both assertions are corollaries of Theorem 7.1-4(a): A linear operator that is continuous with respect to the strong topologies of V and W is also continuous with respect to their weak topologies (Brezis [1983, Theorem III.9, p. 39]) and thus $v_k \rightharpoonup v$ implies $Av_k \rightharpoonup Av$; since (v_k) is bounded and A is compact, the sequence (Av_k) contains a subsequence that strongly converges, but the limit Av of this sequence is unique, and therefore the whole sequence (Av_k) strongly converges to Av. To prove the second assertion, we note that, by the bilinearity of B,

$$B(v, w) - B(v_k, w_k) = B(v - v_k, w_k) + B(v, w - w_k),$$

and thus the implication follows by the continuity of B, the boundedness of the sequence (w_k), and the assumptions of weak and strong convergences. ∎

*7.2. LOWER SEMI-CONTINUITY

More details on lower semi-continuous functions may be found in Brezis [1983], Choquet [1964], Vainberg [1956, §8]. Let V be a topological space; a function $J: V \to \mathbb{R} \cup \{+\infty\}$ is **lower semi-continuous** on V if, for each $\alpha \in \mathbb{R}$, the inverse image

$$J^{-1}(]-\infty, \alpha]) = \{v \in V;\ J(v) \leq \alpha\}$$

is a closed subset of V. Clearly, a continuous function $J: V \to \mathbb{R}$ is lower semi-continuous; conversely, a lower semi-continuous function $J: V \to \mathbb{R}$ is continuous if and only if the inverse images $J^{-1}([\alpha, +\infty[)$ are also closed for all $\alpha \in \mathbb{R}$, that is, if and only if the function $-J$ is also lower semi-continuous.

The next theorem gives two useful characterizations of lower semi-continuity. The first one is illustrated in Fig. 7.2-1, and the second one is in a sense the "lower half" of the well known property of continuous functions:

$$\lim_{k\to\infty} u_k = u \text{ in } V \Rightarrow J(u) = \lim_{k\to\infty} J(u_k),$$

which in return implies continuity if the topology of the space V is metrizable. We recall that the *limit inferior* of a sequence (α_k) of extended real numbers is the extended real number

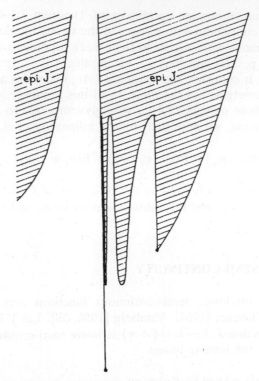

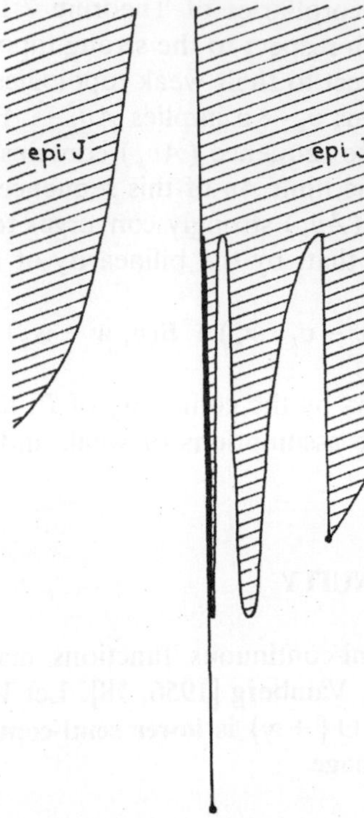

Fig. 7.2-1. The function $J:\mathbb{R}\to\mathbb{R}$ defined by $J(v)=-1/v$ for $v<0$, -1 for $v=0$, $1+\sin v^{-1}$ for $0<v<2/\pi$, and v^2 for $2/\pi\leq v$, is lower semi-continuous. Thus its epigraph epi J is a closed subset of $\mathbb{R}^2$.

$$\liminf_{k\to\infty}\alpha_k := \lim_{k\to\infty}\{\inf_{l\geq k}\alpha_l\},$$

which is well defined since a monotone sequence is always convergent in the set $\{-\infty\}\cup\mathbb{R}\cup\{+\infty\}$; equivalently, $\liminf_{k\to\infty}\alpha_k$ is the smallest limit of all convergent subsequences that can be extracted from (α_k). The epigraph of a function with values in the set $\mathbb{R}\cup\{+\infty\}$ has already been encountered in Theorem 4.7-10.

Theorem 7.2-1. (a) *Let V be a topological space. A function $J:V\to\mathbb{R}\cup\{+\infty\}$ is lower semi-continuous if and only if its epigraph*

$$\text{epi } J=\{(v,\alpha)\in V\times\mathbb{R};\ J(v)\leq\alpha\}$$

is a closed subset of the space $V\times\mathbb{R}$.

(b) *If a function* $J : V \to \mathbb{R} \cup \{+\infty\}$ *is lower semi-continuous, it is also* **sequentially lower semi-continuous**, i.e.,

$$\lim_{k\to\infty} u_k = u \text{ in } V \Rightarrow J(u) \leq \liminf_{k\to\infty} J(u_k)\,.$$

The converse property holds if the topology of the space V is metrizable. ∎

Remark. The function represented in Fig. 4.7-3 is not lower semicontinuous (its epigraph is not a closed subset of $\mathbb{R}^2$). ∎

We have seen (Theorem 7.1-1) that in an infinite-dimensional space V, the strong and the weak topologies are always *distinct*. Accordingly, a function $J : V \to \mathbb{R} \cup \{+\infty\}$ is said to be **strongly**, resp. **weakly, lower semi-continuous** if it is lower semi-continuous when V is endowed with the strong, resp. weak, topology. Likewise, a function $J : V \to \mathbb{R} \cup \{+\infty\}$ is said to be **strongly**, resp. **weakly, sequentially lower semi-continuous** if it is such that

$$\lim_{k\to\infty} u_k = u \text{ in } V \Rightarrow J(u) \leq \liminf_{k\to\infty} J(u_k)\,,$$

when V is equipped with the strong, resp. weak, topology.

Since the weak topology is *not* metrizable when V is infinite-dimensional (Theorem 7.1-1), the sequential weak lower semi-continuity is *not* equivalent to the weak lower semi-continuity in this case. Nevertheless, we shall see that *the weaker notion of sequential weak lower semi-continuity is sufficient for our purposes.*

Since a weakly closed set is strongly closed, the first characterization of Theorem 7.2-1 shows that *a weakly lower semi-continuous function is also strongly lower semi-continuous.* A sufficient condition for the converse implication is given in the following result.

Theorem 7.2-2. *Let V be a normed vector space. Then a convex and strongly lower semi-continuous function* $J : V \to \mathbb{R} \cup \{+\infty\}$ is *weakly lower semi-continuous.*

Proof. By Theorems 4.7-10 and 7.2-1, the set epi J is convex and closed for the strong topology of the space $V \times \mathbb{R}$. Hence epi J is weakly closed (Theorem 7.1-3), and thus the function J is weakly lower semi-continuous by Theorem 7.2-1. ∎

Remarks. (1) The *sequential* weak lower semi-continuity follows directly from Mazur's theorem (Theorem 7.1-2).

(2) If the function J is *real-valued* and *differentiable*, it is easy to prove that it is sequentially weakly lower semi-continuous, without resorting to Mazur's theorem: Given a sequence (u_k) that weakly converges to an element $u \in V$, the characterization of convexity for differentiable functions (Theorem 4.7-6) implies that

$$J(u) \leq J(u_k) - J'(u)(u_k - u) \quad \text{for all } k .$$

By definition of weak convergence, $\lim_{k\to\infty} J'(u)(u_k - u) = 0$ since $J'(u) \in V'$. Hence $J(u) \leq \liminf_{k\to\infty} J(u_k)$, and thus the function J is sequentially weakly lower semi-continuous. ■

*7.3. SEQUENTIALLY WEAKLY LOWER SEMI-CONTINUOUS FUNCTIONALS

We first prove a fundamental criterion for the sequential weak lower semi-continuity of functionals of the form

$$H : \zeta \in L^1(\Omega) \to H(\zeta) = \int_\Omega h(x, \zeta(x))\, dx ,$$

which is the basis for establishing the existence of minimizers for an important class of functionals (Theorem 7.3-2). The proof relies on two essential tools: *Fatou's lemma*, and *Mazur's theorem* (Theorem 7.1-2). We recall that $\rightharpoonup$ denotes weak convergence.

Theorem 7.3-1. *Let Ω be a bounded open subset of $\mathbb{R}^n$, let $\beta \in \mathbb{R}$, and let*

$$h : \Omega \times \mathbb{R}^\mu \to [\beta, +\infty]$$

be a function such that:

$$h(x, \cdot) : \zeta \in \mathbb{R}^\mu \to h(x, \zeta) \in [\beta, +\infty] \text{ is convex and continuous for almost all } x \in \Omega ,$$

$$h(\cdot, \zeta) : x \in \Omega \to h(x, \zeta) \in [\beta, +\infty] \text{ is measurable for all } \zeta \in \mathbb{R}^\mu .$$

Then

$$\zeta_k \rightharpoonup \zeta \text{ in } L^1(\Omega) = (L^1(\Omega))^\mu \Rightarrow$$
$$\int_\Omega h(x, \zeta(x))\,dx \le \liminf_{k\to\infty} \int_\Omega h(x, \zeta_k(x))\,dx\,.$$

Proof. Since the set Ω is bounded, constant functions are integrable over Ω, and consequently there is no loss of generality in assuming that $\beta = 0$ (if $\beta < 0$, replace the function h by the function $h - \beta$). The function h is a *Carathéodory function*, in the sense that the functions $h(x, \cdot)$ are continuous for almost $x \in \Omega$ and the functions $h(\cdot, \zeta)$ are measurable for all $\zeta \in \mathbb{R}^\mu$. Hence the function $x \in \Omega \to h(x, \zeta(x))$ is measurable whenever the function $\zeta : x \in \Omega \to \zeta(x) \in \mathbb{R}^\mu$ is itself measurable (Ekeland & Temam [1974, pp. 218 ff.]). Since the function h takes its values in the set $[0, +\infty]$, we conclude that for each measurable function $\zeta : \Omega \to \mathbb{R}^\mu$, hence in particular for each function $\zeta \in L^1(\Omega)$, *the integral* $\int_\Omega h(x, \zeta(x))\,dx$ *is a well defined extended real number in the interval* $[0, +\infty]$.

Let us next show that the functional

$$H : \zeta \in L^1(\Omega) \to H(\zeta) := \int_\Omega h(x, \zeta(x))\,dx \in [0, +\infty]$$

is lower semi-continuous with respect to the *strong* topology of the space $L^1(\Omega)$, i.e., that

$$\zeta_k \xrightarrow[k\to\infty]{} \zeta \text{ in } L^1(\Omega) \Rightarrow \int_\Omega h(x, \zeta(x))\,dx \le \liminf_{k\to\infty} \int_\Omega h(x, \zeta_k(x))\,dx$$

(for normed vector spaces, whose topology is metrizable, lower semi-continuity is equivalent to sequential lower semi-continuity; cf. Theorem 7.2-1). Let (ζ_k) be a sequence converging in the space $L^1(\Omega)$ to a limit ζ, and let (ζ_l) be any subsequence such that the sequence of extended real numbers $(\int_\Omega h(x, \zeta_l(x))\,dx)$ converges in the set $\mathbb{R} \cup \{+\infty\}$. By definition of the limit inferior, we must show that

$$\int_\Omega h(x, \zeta(x))\,dx \le \lim_{l\to\infty} \int_\Omega h(x, \zeta_l(x))\,dx\,.$$

Since the subsequence (ζ_l) converges to ζ in $L^1(\Omega)$, there exists a subsequence (ζ_m) of (ζ_l) such that $\zeta_m(x) \to \zeta(x)$ for almost all $x \in \Omega$, and so, by the assumed continuity of the functions $h(x, \cdot)$ for almost all $x \in \Omega$,

$$h(x, \zeta_m(x)) \to h(x, \zeta(x)) \quad \text{for almost all } x \in \Omega .$$

Thus, by *Fatou's lemma*,

$$\begin{aligned}\int_\Omega h(x, \zeta(x))\,dx &= \int_\Omega \lim_{m\to\infty} h(x, \zeta_m(x))\,dx \\ &\leq \liminf_{m\to\infty} \int_\Omega h(x, \zeta_m(x))\,dx = \lim_{l\to\infty} \int_\Omega h(x, \zeta_l(x))\,dx ,\end{aligned}$$

and therefore *the functional* $H : L^1(\Omega) \to [0, +\infty]$ *is strongly lower semi-continuous.*

On the other hand, *the functional* $H : L^1(\Omega) \to [0, +\infty]$ *is convex*: The assumed convexity of the function h with respect to its second argument implies that, for all $\lambda \in [0, 1]$ and all $\zeta, \eta \in L^1(\Omega)$:

$$\begin{aligned}H(\lambda\zeta + (1-\lambda)\eta) &= \int_\Omega h(x, \lambda\zeta(x) + (1-\lambda)\eta(x))\,dx \\ &\leq \int_\Omega \{\lambda h(x, \zeta(x)) + (1-\lambda) h(x, \eta(x))\}\,dx \\ &= \lambda H(\zeta) + (1-\lambda) H(\eta) .\end{aligned}$$

Since it is both convex and strongly lower semi-continuous, the functional H is weakly lower semi-continuous by Theorem 7.2-2. ∎

Remarks. (1) The continuity of the functions $h(x, \cdot)$ is not a superfluous assumption since the value $+\infty$ is allowed (convexity implies continuity only in the interior of the set $\{\zeta \in \mathbb{R}^\mu;\ h(x, \zeta) < +\infty\}$; cf. Theorem 4.7-10).

(2) If the function h is independent of $x \in \Omega$, the assumption of measurability is automatically satisfied.

(3) When Ω is bounded, weak convergence in any space $L^p(\Omega)$, $1 \leq p < \infty$, implies weak convergence in the space $L^1(\Omega)$.

(4) A converse to Theorem 7.3-1 holds if $n = 1$; cf. Exercise 7.1.

(5) The above result can be extended to more general functionals of the form $\int_\Omega h(x, v(x), \zeta(x))\,dx$, where the function $h : \Omega \times \mathbb{R}^\lambda \times \mathbb{R}^\mu \to [0, +\infty]$ has the following properties:

$h(x, \cdot, \cdot) : (\boldsymbol{v}, \boldsymbol{\zeta}) \in \mathbb{R}^\lambda \times \mathbb{R}^\mu \to [0, +\infty]$ is continuous for almost all $x \in \Omega$,

$h(\cdot, \boldsymbol{v}, \boldsymbol{\zeta}) : x \in \Omega \to [0, +\infty]$ is measurable for all $(\boldsymbol{v}, \boldsymbol{\zeta}) \in \mathbb{R}^\lambda \times \mathbb{R}^\mu$,

$h(x, \boldsymbol{v}, \cdot) : \boldsymbol{\zeta} \in \mathbb{R}^\mu \to [0, +\infty]$ is convex for all $\boldsymbol{v} \in \mathbb{R}^\lambda$ and almost all $x \in \Omega$.

Then these assumptions (which reduce to those of Theorem 7.3-1 if there is no dependence on $\boldsymbol{v} \in \mathbb{R}^\lambda$) imply that (Giaquinta [1983, p. 18]; see also Serrin [1961], Fichera [1967]):

$$(\boldsymbol{v}_k, \boldsymbol{\zeta}_k) \rightharpoonup (\boldsymbol{v}, \boldsymbol{\zeta}) \text{ in } \boldsymbol{L}^1(\Omega) \times \boldsymbol{L}^1(\Omega) \Rightarrow$$
$$\int_\Omega h(x, \boldsymbol{v}(x), \boldsymbol{\zeta}(x))\, dx \leq \liminf_{k\to\infty} \int_\Omega h(x, \boldsymbol{v}_k(x), \boldsymbol{\zeta}_k(x))\, dx .$$

In the same spirit, Ekeland & Temam [1974, p. 226] give sufficient conditions insuring that

$$\boldsymbol{v}_k(x) \to \boldsymbol{v}(x) \text{ for almost all } x \in \Omega,\ \boldsymbol{\zeta}_k \rightharpoonup \boldsymbol{\zeta} \text{ in } \boldsymbol{L}^1(\Omega)$$
$$\Rightarrow \int_\Omega h(x, \boldsymbol{v}(x), \boldsymbol{\zeta}(x))\, dx \leq \liminf_{k\to\infty} \int_\Omega h(x, \boldsymbol{v}_k(x), \boldsymbol{\zeta}_k(x))\, dx .$$

The proofs of these more general results are, however, substantially complicated by the new argument $\boldsymbol{v} \in \mathbb{R}^\lambda$ in the integrand. For recent results, see Dacorogna [1982a], Marcellini [1986a, 1986b], Marcellini & Sbordone [1983], Ambrosio [1987], and especially, the book of Dacorogna [1987]. ■

As an application of the criterion of weak lower semi-continuity, we prove the existence of *minimizers* in the Sobolev space $\boldsymbol{W}^{1,p}(\Omega)$, $p > 1$, for a special class of functionals with convex integrand. The set of all matrices with m rows and n columns is denoted $\mathbb{M}^{m\times n}$.

Theorem 7.3-2 (existence of minimizers). *Let Ω be a domain in $\mathbb{R}^n$ with boundary Γ, and let $h : \Omega \times \mathbb{M}^{m\times n} \to [\beta, +\infty]$, $\beta \in \mathbb{R}$, be a function with the following properties*:

(a) *Convexity*: *For almost all $x \in \Omega$, the function $h(x, \cdot) : \boldsymbol{F} \in \mathbb{M}^{m\times n} \to h(x, \boldsymbol{F})$ is convex*;

(b) *Continuity and measurability*: *For almost all* $x \in \Omega$, *the function* $h(x, \cdot): F \in \mathbb{M}^{m\times n} \to h(x, F)$ *is continuous, and the function* $h(\cdot, F): x \in \Omega \to h(x, F)$ *is measurable for all* $F \in \mathbb{M}^{m\times n}$;

(c) *Coerciveness*: *There exist constants* α *and* p *such that*

$$\alpha > 0,\ p > 1, \text{ and } h(x, F) \geq \alpha \|F\|^p + \beta \text{ for almost all } x \in \Omega, \text{ and for all } F \in \mathbb{M}^{m\times n}.$$

Let Γ_0 *be a measurable subset of* Γ *with* da-meas $\Gamma_0 > 0$, *let* $\varphi_0 : \Gamma_0 \to \mathbb{R}^m$ *be a measurable function such that the set*

$$\Phi = \{\psi \in W^{1,p}(\Omega);\ \psi = \varphi_0 \text{ da-a.e. on } \Gamma_0\},\ W^{1,p}(\Omega) := (W^{1,p}(\Omega))^m,$$

is nonempty, let L *be a continuous linear functional over the space* $W^{1,p}(\Omega)$. *Finally, let*

$$I(\psi) = \int_\Omega h(x, \nabla\psi(x))\,dx - L(\psi), \text{ with } \nabla\psi = (\partial_j \psi_i) \in \mathbb{M}^{m\times n},$$

and assume that $\inf_{\psi\in\Phi} I(\psi) < +\infty$.

Then there exists at least one function φ *such that*

$$\varphi \in \Phi \text{ and } I(\varphi) = \inf_{\psi\in\Phi} I(\psi).$$

Proof. By the assumed coerciveness of the function h and by the continuity of the linear form L,

$$I(\psi) \geq \alpha \int_\Omega \|\nabla\psi\|^p\,dx + \beta \operatorname{vol}\Omega - \|L\|\|\psi\|_{1,p,\Omega} \quad \text{for all } \psi \in W^{1,p}(\Omega).$$

By the generalized Poincaré inequality (Theorem 6.1-8(b)), there exists a constant $c_1 > 0$ such that

$$\int_\Omega |\psi|^p\,dx \leq c_1 \left\{ \int_\Omega |\mathbf{grad}\,\psi|^p\,dx + \left| \int_{\Gamma_0} \psi\,da \right|^p \right\} \quad \text{for all } \psi \in W^{1,p}(\Omega).$$

Hence there exist constants $c_2 > 0$ and c_3 such that

$$I(\psi) \geq c_2 \|\psi\|^p_{1,p,\Omega} - \|L\|\|\psi\|_{1,p,\Omega} + c_3 \quad \text{for all } \psi \in \Phi,$$

and since $p > 1$, there exist constants c and d such that

$$c > 0 \text{ and } I(\psi) \geq c\|\psi\|_{1,p,\Omega}^p + d \quad \text{for all } \psi \in \Phi .$$

Let (φ^k) be an *infimizing sequence* for the functional I, i.e., a sequence that satisfies

$$\varphi^k \in \Phi \quad \text{for all } k, \text{ and } \lim_{k\to\infty} I(\varphi^k) = \inf_{\psi\in\Phi} I(\psi) .$$

The assumption $\inf_{\psi\in\Phi} I(\psi) < +\infty$ and the relation $I(\psi) \to +\infty$ as $\|\psi\|_{1,p,\Omega} \to +\infty$ imply together that the infimizing sequence is bounded in the reflexive Banach space $W^{1,p}(\Omega)$ (Theorem 6.1-1). Hence the bounded sequence (φ^k) contains a subsequence that weakly converges to an element $\varphi \in W^{1,p}(\Omega)$ (Theorem 7.1-4). The closed convex set Φ is weakly closed (Theorem 7.1-3), and thus the weak limit φ belongs to the set Φ. Since

$$\varphi^l \rightharpoonup \varphi \text{ in } W^{1,p}(\Omega) \Rightarrow \nabla\varphi^l \rightharpoonup \nabla\varphi \text{ in } L^p(\Omega) \Rightarrow \nabla\varphi^l \rightharpoonup \nabla\varphi \text{ in } L^1(\Omega) ,$$

we conclude from Theorem 7.3-1 that

$$\int_\Omega h(x, \nabla\varphi(x))\, dx \leq \liminf_{l\to\infty} \int_\Omega h(x, \nabla\varphi^l(x))\, dx .$$

Also, as L is a continuous linear form on $W^{1,p}(\Omega)$, we infer that

$$L(\varphi) = \lim_{l\to\infty} L(\varphi^l)$$

by definition of weak convergence; consequently,

$$\inf_{\psi\in\Phi} I(\psi) \leq I(\varphi) \leq \liminf_{l\to\infty} I(\varphi^l) = \lim_{k\to\infty} I(\varphi^k) = \inf_{\psi\in\Phi} I(\psi) .$$

Therefore $I(\varphi) = \inf_{\psi\in\Phi} I(\psi)$ and the proof is complete. ■

Remarks. (1) If the function $h(x, \cdot)$ is *strictly convex on* $\mathbb{R}^{m\times n}$ for almost all $x \in \Omega$, the functional I is also strictly convex, and the minimizer is unique in this case (Theorem 4.7-8).

(2) Theorem 7.3-2 contains the existence result of linearized elasticity (Theorem 6.3-5) as a special case; see Exercise 7.3.

(3) More general existence results are proved in Ekeland & Temam

[1974, p. 232] and Morrey [1966, Theorem 1.9.1]. See also Berger [1977, p. 307], Aubin & Ekeland [1984], Ekeland & Turnbull [1984], Dacorogna [1987]. ■

When the integrand $h(x, F)$ is *not convex* with respect to the variable F, as is the case in hyperelasticity, considerable difficulties arise for proving weak lower semi-continuity and existence results. One means of circumventing these difficulties, and especially so in hyperelasticity, is the introduction of polyconvexity, as we shall see. Otherwise, "generalized" solutions may be obtained by the *duality* and *relaxation* techniques of the calculus of variations, as described in general by Ekeland & Temam [1974], and systematically applied by Dacorogna [1981, 1982a, 1982b, 1987] to hyperelasticity. A difficulty of this approach is that an actual knowledge of the "convexification" of the integrand is required, and there are few instances where this is possible (see however Kohn & Strang [1983, 1985], Gurtin & Temam [1981], for examples where the computations can be completely carried out). For analogous results with applications to elasticity, or more generally for references about nonconvex problems in the calculus of variations, see Ekeland [1979], Aubert & Tahraoui [1979, 1984], Atteia & Dedieu [1981], Atteia & Raissouli [1986], de Campos & Oden [1983], Marcellini [1986c], Mascolo & Schianchi [1983], Tahraoui [1986]. For the related question of weak closedness of the sets over which minimizers are sought, see Aubert & Tahraoui [1985, 1987].

7.4. OUTLINE OF JOHN BALL'S APPROACH TO EXISTENCE THEORY IN HYPERELASTICITY

Motivated by the proof of Theorem 7.3-2, our objectives are to consider an *infimizing sequence* $(\boldsymbol{\varphi}^k)$ of the total energy

$$I(\boldsymbol{\psi}) = \int_\Omega \hat{W}(x, \boldsymbol{\nabla}\boldsymbol{\psi}(x))\,\mathrm{d}x - L(\boldsymbol{\psi})$$

over an appropriate set $\boldsymbol{\Phi}$ of admissible deformations (to be defined later), show that this sequence is bounded as a consequence of the coerciveness inequality satisfied by the stored energy function, extract a subsequence $(\boldsymbol{\varphi}^l)$ that weakly converges to an element $\boldsymbol{\varphi}$, show that the weak limit belongs to the set $\boldsymbol{\Phi}$, and eventually, show that

$$\int_\Omega \hat{W}(x, \boldsymbol{\nabla}\boldsymbol{\varphi}(x))\,\mathrm{d}x \leqslant \liminf_{l\to\infty} \int_\Omega \hat{W}(x, \boldsymbol{\nabla}\boldsymbol{\varphi}^l(x))\,\mathrm{d}x$$

(since we shall restrict ourselves to dead loads, the remaining part of the total energy is a linear continuous functional that is thus simply handled, as in the proof of Theorem 7.3-2); establishing the sequential weak lower semi-continuity of the functional $\boldsymbol{\psi} \to \int_\Omega \hat{W}(x, \nabla\boldsymbol{\psi}(x))\, dx$ *will be, however, considerably more delicate than in Theorem 7.3-1, since the function* $\hat{W}(x, \boldsymbol{F})$ *is not convex in* $\boldsymbol{F}$, *and since it is not defined for* $\det \boldsymbol{F} \leqslant 0$. It will then follow that $\boldsymbol{\varphi} \in \boldsymbol{\Phi}$ is a **minimizer** of the energy, i.e., that $I(\boldsymbol{\varphi}) = \inf_{\boldsymbol{\varphi} \in \boldsymbol{\Phi}} I(\boldsymbol{\varphi})$.

A closer look at those steps allows us to list various observations and guidelines, which form the basis of John Ball's approach to existence theory in hyperelasticity.

(i) The "impossible convexity" of the stored energy function $\hat{W}$ with respect to its argument $\boldsymbol{F}$ will be replaced by the weaker assumption of *polyconvexity of the stored energy function* (Sect. 4.9): For almost all $x \in \Omega$, there exists a convex function $\mathbb{W}(x, \cdot): \mathbb{M}^3 \times \mathbb{M}^3 \times]0, +\infty[\to \mathbb{R}$ such that

$$\hat{W}(x, \boldsymbol{F}) = \mathbb{W}(x, \boldsymbol{F}, \mathbf{Cof}\, \boldsymbol{F}, \det \boldsymbol{F}) \quad \text{for all } \boldsymbol{F} \in \mathbb{M}^3_+ .$$

(ii) We have seen in Sect. 4.6 that the behavior of the stored energy function for large strains is reflected in part by a *coerciveness inequality* of the form

$$\hat{W}(x, \boldsymbol{F}) \geqslant \alpha\{\|\boldsymbol{F}\|^p + \|\mathbf{Cof}\, \boldsymbol{F}\|^q + (\det \boldsymbol{F})^r\} + \beta \quad \text{for all } \boldsymbol{F} \in \mathbb{M}^3_+ ,$$

with $\alpha > 0$, $\beta \in \mathbb{R}$, and "sufficiently large" exponents p, q, r. This inequality in turn implies that

$$\int_\Omega \hat{W}(x, \nabla\boldsymbol{\psi}(x))\, dx \geqslant \alpha\{|\nabla\boldsymbol{\psi}|^p_{0,p,\Omega} + |\mathbf{Cof}\, \nabla\boldsymbol{\psi}|^q_{0,q,\Omega} + |\det \nabla\boldsymbol{\psi}|^r_{0,r,\Omega}\} + \beta \operatorname{vol} \Omega .$$

Hence *any function* $\boldsymbol{\psi}$ *satisfying* $\int_\Omega \hat{W}(x, \nabla\boldsymbol{\psi}(x))\, dx < +\infty$ (as will be the case of the terms of an infimizing sequence of the total energy) *must be such that*

$$\nabla\boldsymbol{\psi} \in \boldsymbol{L}^p(\Omega),\ \mathbf{Cof}\, \nabla\boldsymbol{\psi} \in \boldsymbol{L}^q(\Omega),\ \det \nabla\boldsymbol{\psi} \in L^r(\Omega) .$$

How large must then be the exponents p, q, r in the coerciveness inequality? A first observation is that they must all be >1 in order that the spaces $L^p(\Omega)$, $L^q(\Omega)$, $L^r(\Omega)$ be *reflexive*, so that we may extract weakly convergent subsequences from bounded sequences. If we restrict ourselves to functions $\psi \in W^{1,p}(\Omega)$ satisfying a *boundary condition of place* $\psi = \varphi_0$ on $\Gamma_0 \subset \Gamma$, with area $\Gamma_0 > 0$, the generalized Poincaré inequality implies, as in Theorem 7.3-2, that *the semi-norm* $|\nabla\psi|_{0,p,\Omega}$ *can be replaced by the norm* $\|\psi\|_{1,p,\Omega}$ *in the lower bound of the integral* $\int_\Omega \hat{W}(x, \nabla\psi(x))\,dx$. If the remaining part of the total energy, which takes into account the applied forces, is assumed to be a continuous linear form $L: W^{1,p}(\Omega) \to \mathbb{R}$ (this corresponds to dead loads, but more general applied forces can be as well considered), we obtain the following *lower bound for the total energy*: There exist constants $a > 0$ and $b \in \mathbb{R}$ such that, for all functions $\psi \in W^{1,p}(\Omega)$ satisfying $\psi = \varphi_0$ on Γ_0,

$$I(\psi) = \int_\Omega \hat{W}(x, \nabla\psi)\,dx - L(\psi)$$

$$\geq a\{\|\psi\|^p_{1,p,\Omega} + |\mathbf{Cof}\,\nabla\psi|^q_{0,q,\Omega} + |\det \nabla\psi|^r_{0,r,\Omega}\} + b\,.$$

(iii) The definition of the *set of admissible deformations* is thus gradually imposing itself in a natural way: From (ii), we first infer that it should consist of functions $\psi \in W^{1,p}(\Omega)$ satisfying the boundary conditions of place $\psi = \varphi_0$ on Γ_0, and which are such that $\mathbf{Cof}\,\nabla\psi \in L^q(\Omega)$ and $\det \nabla\psi \in L^r(\Omega)$. From the definition of a deformation, we next infer that the functions $\psi \in W^{1,p}(\Omega)$ should also be *orientation-preserving*. If, following John Ball, we only take these requirements into account, we conclude that *the set of admissible deformations is of the form* ("a.e." means almost everywhere):

$$\Phi = \{\psi \in W^{1,p}(\Omega);\ \mathbf{Cof}\,\nabla\psi \in L^q(\Omega),\ \det \nabla\psi \in L^r(\Omega)\,,$$

$$\psi = \varphi_0\ da\text{-a.e. on } \Gamma_0,\ \det \nabla\psi > 0 \text{ a.e. in } \Omega\}\,,$$

where the exponents p, q, r *are thus governed by the coerciveness inequality satisfied by the stored energy function*. Notice that the *orientation-preserving condition* $\det \nabla\psi > 0$ *can be only asked to hold almost everywhere in* Ω, since $\det \nabla\psi$ is only in $L^r(\Omega)$. Thus, even though we shall continue for convenience to call **deformations** arbitrary elements in the set Φ, they are not in general deformations according to the general definition given in Sect. 1.4 (besides, the functions in $W^{1,p}(\Omega)$ are not continuous in general, unless $p > 3$; cf. Exercise 6.1).

We will also generalize John Ball's approach in allowing *more general boundary conditions* in the set Φ (Sect 7.8), and in including in Φ a condition of *injectivity of deformations* (Sect. 7.9).

(iv) As we shall show, the other facet of the behavior of the stored energy function for large strains, viz., $\hat{W}(x, F) \rightarrow +\infty$ as $\det F \rightarrow 0^+$ (Sect. 4.6), implies that the weak limit of the infimizing sequence also satisfies the orientation-preserving condition. In other words, the behavior as $\det F \rightarrow 0^+$ compensates the restriction that the stored energy function $\hat{W}(x, F)$ be only defined for matrices F with $\det F > 0$.

(v) As expected, *the set* Φ *is not convex* (Exercise 7.6); this observation indicates that difficulties will arise when taking weak limits, since Theorem 7.1-3 cannot be applied in the present situation. Accordingly, we will have to give sufficient conditions insuring that

$$\left.\begin{array}{rl} \varphi^l \rightharpoonup \varphi & \text{in } W^{1,p}(\Omega) \\ \mathbf{Cof}\, \nabla\varphi^l \rightharpoonup H & \text{in } L^q(\Omega) \\ \det \nabla\varphi^l \rightharpoonup \delta & \text{in } L^r(\Omega) \end{array}\right\} \Rightarrow \left\{\begin{array}{l} H = \mathbf{Cof}\, \nabla\varphi\,, \\ \delta = \det \nabla\varphi\,. \end{array}\right.$$

We will show in the next two sections that this is the case if $p \geq 2$, $q \geq p/(p-1)$. Hence this imposes further restrictions on the exponents p and q, which so far were only required, as r, to be >1.

7.5. THE MAPPING $\psi \in W^{1,p}(\Omega) \rightarrow \mathbf{Cof}\, \nabla\psi$

Keeping the above program in mind, we first take a closer look at the properties of the mappings $\psi \in W^{1,p}(\Omega) \rightarrow \mathbf{Cof}\, \nabla\psi$ and $\psi \in W^{1,p}(\Omega) \rightarrow \det \nabla\psi$, notably with respect to weak convergence (which is as usual denoted by $\rightharpoonup$). We follow here Ball [1977, Lemma 6.1 and Theorem 6.2].

Theorem 7.5-1. *Let Ω be a domain in $\mathbb{R}^3$. For each $p \geq 2$, the mapping*

$$\psi \in W^{1,p}(\Omega) \rightarrow \mathbf{Cof}\, \nabla\psi = (\partial_{i+1}\psi_{j+1}\partial_{i+2}\psi_{j+2} - \partial_{i+2}\psi_{j+1}\partial_{i+1}\psi_{j+2}) \in L^{p/2}(\Omega)$$

is well defined and continuous. Furthermore,

$$\left.\begin{array}{rl} \varphi^l \rightharpoonup \varphi & \text{in } W^{1,p}(\Omega),\ p \geq 2\,, \\ \mathbf{Cof}\, \nabla\varphi^l \rightharpoonup H & \text{in } L^q(\Omega),\ q \geq 1 \end{array}\right\} \Rightarrow H = \mathbf{Cof}\, \nabla\varphi\,.$$

Proof. (i) By Hölder's inequality, the bilinear mapping

$$(\xi, \eta) \in (L^p(\Omega))^2 \rightarrow \xi\eta \in L^{p/2}(\Omega)$$

is well defined and continuous for $p \geqslant 2$. Consequently, the mapping $\boldsymbol{\psi} \in \boldsymbol{W}^{1,p}(\Omega) \to \mathbf{Cof}\, \boldsymbol{\nabla}\boldsymbol{\psi} \in \boldsymbol{L}^{p/2}(\Omega)$ is well defined and continuous for $p \geqslant 2$.

(ii) For sufficiently smooth functions $\boldsymbol{\psi}$, for instance in the space $\mathscr{C}^2(\bar{\Omega})$, we can also write

$$(\mathbf{Cof}\, \boldsymbol{\nabla}\boldsymbol{\psi})_{ij} = \partial_{i+2}(\psi_{j+2}\partial_{i+1}\psi_{j+1}) - \partial_{i+1}(\psi_{j+2}\partial_{i+2}\psi_{j+1})$$
$$\text{(no summation)},$$

and consequently an application of Green's formula shows that, for all functions $\boldsymbol{\psi} \in \mathscr{C}^2(\bar{\Omega})$ and all functions $\theta \in \mathscr{D}(\Omega)$,

$$\int_\Omega (\mathbf{Cof}\, \boldsymbol{\nabla}\boldsymbol{\psi})_{ij}\theta \,\mathrm{d}x = -\int_\Omega \psi_{j+2}\partial_{i+1}\psi_{j+1}\partial_{i+2}\theta \,\mathrm{d}x$$
$$+ \int_\Omega \psi_{j+2}\partial_{i+2}\psi_{j+1}\partial_{i+1}\theta \,\mathrm{d}x \quad \text{(no summation)}.$$

For a fixed function $\theta \in \mathscr{D}(\Omega)$, the two sides of this relation are continuous if the space $\mathscr{C}^2(\bar{\Omega})$ is equipped with the norm $\|\cdot\|_{1,\Omega}$, since

$$\left|\int_\Omega (\mathbf{Cof}\, \boldsymbol{\nabla}\boldsymbol{\psi})_{ij}\theta \,\mathrm{d}x\right| \leqslant |(\mathbf{Cof}\, \boldsymbol{\nabla}\boldsymbol{\psi})_{ij}|_{0,1,\Omega}|\theta|_{0,\infty,\Omega} \leqslant c_1(\theta)\|\boldsymbol{\psi}\|_{1,\Omega}^2 ,$$
$$\left|\int_\Omega \psi_i\partial_j\psi_k\partial_l\theta \,\mathrm{d}x\right| \leqslant |\psi_i|_{0,\Omega}|\psi_k|_{1,\Omega}|\theta|_{1,\infty,\Omega} \leqslant c_2(\theta)\|\boldsymbol{\psi}\|_{1,\Omega}^2 .$$

Therefore this relation remains valid for functions $\boldsymbol{\psi}$ in the space $\boldsymbol{H}^1(\Omega)$, whence in any space $\boldsymbol{W}^{1,p}(\Omega)$, $p \geqslant 2$, since the space $\mathscr{C}^2(\bar{\Omega})$ is dense in the space $H^1(\Omega)$ when Ω is a domain (Theorem 6.1-6).

(iii) Let $p \geqslant 2$. Given an arbitrary function $\theta \in \mathscr{D}(\Omega)$, we show next that

$$\boldsymbol{\varphi}^l \to \boldsymbol{\varphi} \text{ in } \boldsymbol{W}^{1,p}(\Omega) \Rightarrow \int_\Omega \varphi_i^l\partial_j\varphi_k^l\partial_m\theta \,\mathrm{d}x \xrightarrow[l\to\infty]{} \int_\Omega \varphi_i\partial_j\varphi_k\partial_m\theta \,\mathrm{d}x ,$$

which, by step (ii), will in turn imply that

$$\boldsymbol{\varphi}^l \to \boldsymbol{\varphi} \text{ in } \boldsymbol{W}^{1,p}(\Omega) \Rightarrow \int_\Omega (\mathbf{Cof}\, \boldsymbol{\nabla}\boldsymbol{\varphi}^l)_{ij}\theta \,\mathrm{d}x \xrightarrow[l\to\infty]{} \int_\Omega (\mathbf{Cof}\, \boldsymbol{\nabla}\boldsymbol{\varphi})_{ij}\theta \,\mathrm{d}x .$$

By Hölder's inequality, the bilinear mapping

$$(\xi, \chi) \in L^r(\Omega) \times W^{1,p}(\Omega) \to \int_\Omega \xi \partial_j \chi \partial_m \theta \, dx$$

is continuous if $p^{-1} + r^{-1} \leqslant 1$ (the function $\theta \in \mathcal{D}(\Omega)$ is held fixed in the argument). Hence by Theorem 7.1-5, the implication

$$\left.\begin{array}{l} \xi^l \to \xi \text{ in } L^r(\Omega) \\ \chi^l \rightharpoonup \chi \text{ in } W^{1,p}(\Omega) \end{array}\right\} \Rightarrow \int_\Omega \xi^l \partial_j \chi^l \partial_m \theta \, dx \xrightarrow[l\to\infty]{} \int_\Omega \xi \partial_j \chi \partial_m \theta \, dx$$

holds. From the compact imbedding (Theorem 6.1-5)

$$W^{1,p}(\Omega) \Subset L^r(\Omega) \quad \text{for all } 1 \leqslant r < p^* = \begin{cases} \dfrac{3p}{3-p} & \text{if } p < 3, \\ +\infty & \text{if } p \geqslant 3, \end{cases}$$

we infer that

$$\varphi^l \rightharpoonup \varphi \text{ in } W^{1,p}(\Omega) \Rightarrow \varphi^l \to \varphi \text{ in } L^r(\Omega) \quad \text{for all } 1 \leqslant r < p^* .$$

Thus our assertion is proved since for any $p \geqslant 2$ (in fact for any $p > \frac{3}{2}$), we can find a number r that simultaneously satisfies

$$\frac{1}{p} + \frac{1}{r} \leqslant 1 \text{ and } r < p^* .$$

(iv) Let (φ^l) be a sequence in the space $W^{1,p}(\Omega)$, $p \geqslant 2$, such that $\mathbf{Cof}\,\nabla\varphi^l \in L^q(\Omega)$, $q \geqslant 1$, and such that

$$\varphi^l \rightharpoonup \varphi \text{ in } W^{1,p}(\Omega), \text{ and } \quad \mathbf{Cof}\,\nabla\varphi^l \rightharpoonup \boldsymbol{H} \text{ in } L^q(\Omega) .$$

Therefore

$$\int_\Omega (\mathbf{Cof}\,\nabla\varphi^l)_{ij} \theta \, dx \xrightarrow[l\to\infty]{} \int_\Omega (\mathbf{Cof}\,\nabla\varphi)_{ij} \theta \, dx \quad \text{for all } \theta \in \mathcal{D}(\Omega),$$

by step (iii), and

$$\int_\Omega (\mathbf{Cof}\,\nabla\varphi^l)_{ij} \theta \, dx \xrightarrow[l\to\infty]{} \int_\Omega H_{ij} \theta \, dx$$

by assumption. We conclude that each function $(\mathbf{Cof}\,\nabla\varphi - \boldsymbol{H})_{ij} \in L^1(\Omega)$ satisfies

$$\int_\Omega (\mathbf{Cof}\,\boldsymbol{\nabla}\boldsymbol{\varphi} - \boldsymbol{H})_{ij}\theta \,\mathrm{d}x = 0 \quad \text{for all } \theta \in \mathscr{D}(\Omega).$$

By a classical result in integration theory (see e.g. Vo-Khac [1972, p. 166]), this implies that $(\mathbf{Cof}\,\boldsymbol{\nabla}\boldsymbol{\varphi} - \boldsymbol{H})_{ij} = 0$ a.e. in Ω, and the proof is complete. ∎

Remarks. (1) We infer from Theorem 7.5-1 that *the nonconvex set*

$$\{(\boldsymbol{\psi}, \boldsymbol{K}) \in \boldsymbol{W}^{1,p}(\Omega) \times \boldsymbol{L}^q(\Omega);\ \boldsymbol{K} = \mathbf{Cof}\,\boldsymbol{\nabla}\boldsymbol{\psi}\},\ p \geq 2,\ q \geq 1,$$

is weakly closed in the space $\boldsymbol{W}^{1,p}(\Omega) \times \boldsymbol{L}^q(\Omega)$. This does *not* mean, however, that the set

$$\{\boldsymbol{\psi} \in \boldsymbol{W}^{1,p}(\Omega);\ \mathbf{Cof}\,\boldsymbol{\nabla}\boldsymbol{\psi} \in \boldsymbol{L}^q(\Omega)\},\ p \geq 2,\ q \geq 1,$$

is weakly closed in the space $\boldsymbol{W}^{1,p}(\Omega)$, and indeed this is not always the case (Exercise 7.4).

(2) In step (ii), we showed that the functions $\boldsymbol{\psi} \in \boldsymbol{W}^{1,p}(\Omega)$, $p \geq 2$, satisfy

$$\int_\Omega (\mathbf{Cof}\,\boldsymbol{\nabla}\boldsymbol{\psi})_{ij}\theta \,\mathrm{d}x = -\int_\Omega \psi_{j+2}\partial_{i+1}\psi_{j+1}\partial_{i+2}\theta \,\mathrm{d}x + \int_\Omega \psi_{j+2}\partial_{i+2}\psi_{j+1}\partial_{i+1}\theta \,\mathrm{d}x \quad \text{for all } \theta \in \mathscr{D}(\Omega).$$

Hence, for $p \geq 2$, we also have

$$\mathbf{Cof}\,\boldsymbol{\nabla}\boldsymbol{\psi} = \mathbf{Cof}^{\#}\,\boldsymbol{\nabla}\boldsymbol{\psi} \text{ in } \mathscr{D}'(\Omega),$$

with

$$(\mathbf{Cof}^{\#}\,\boldsymbol{\nabla}\boldsymbol{\psi})_{ij} := \partial_{i+2}(\psi_{j+2}\partial_{i+1}\psi_{j+1}) - \partial_{i+1}(\psi_{j+2}\partial_{i+2}\psi_{j+1}).$$

In addition, this alternative expression can be used for extending the definition of $\mathbf{Cof}\,\boldsymbol{\nabla}\boldsymbol{\psi}$ to functions $\boldsymbol{\psi} \in \boldsymbol{W}^{1,p}(\Omega)$ with $\frac{3}{2} \leq p < 2$ (in which case it is no longer an integrable function; cf. Exercise 7.5). ∎

7.6. THE MAPPING $\psi \in W^{1,p}(\Omega) \to \det \nabla\psi$

Since by Hölder's inequality, the trilinear mapping

$$(\xi, \eta, \zeta) \in (L^p(\Omega))^3 \to \xi\eta\zeta \in L^{p/3}(\Omega)$$

is well defined and continuous, and since

$$\det \nabla\psi = \tfrac{1}{6}\varepsilon_{ijk}\varepsilon_{lmn}\partial_l\psi_i\partial_m\psi_j\partial_n\psi_k ,$$

it seems that we need $p \geqslant 3$ in order that the mapping $\psi \in W^{1,p}(\Omega) \to \det \nabla\psi \in L^1(\Omega)$ be well defined and continuous. With *additional* information on the function $\mathbf{Cof}\,\nabla\psi$ however, we can weaken this requirement by taking advantage of the expansion of $\det \nabla\psi$ as

$$\det \nabla\psi = \partial_j\psi_1(\mathbf{Cof}\,\nabla\psi)_{1j} .$$

The choice of the first row is arbitrary; we could likewise consider the expansion of $\det \nabla\psi$ along any other row or any one of the columns of the matrix $\nabla\psi$. Then another application of Hölder's inequality shows that $\det \nabla\psi$ is well determined as an element of the space $L^s(\Omega)$ if $\psi \in W^{1,p}(\Omega)$ with $p \geqslant 2$ and $\mathbf{Cof}\,\nabla\psi \in L^q(\Omega)$ with

$$\frac{1}{s} := \frac{1}{p} + \frac{1}{q} \leqslant 1$$

(the inequality $p \geqslant 2$ insures that $\mathbf{Cof}\,\nabla\psi$ is in the space $L^1(\Omega)$; cf. Theorem 7.5-1). If $p \geqslant 3$, there is no need to assume that $\mathbf{Cof}\,\nabla\psi \in L^q(\Omega)$ with $p^{-1} + q^{-1} \leqslant 1$, since we then have $\mathbf{Cof}\,\nabla\psi \in L^{p/2}(\Omega)$ and $p^{-1} + 2p^{-1} \leqslant 1$.

Following again Ball [1977, Lemma 6.1 and Theorem 6.2], we now extend Theorem 7.5-1.

Theorem 7.6-1. *Let Ω be a domain in $\mathbb{R}^3$. For each number $p \geqslant 2$ and each number q such that $s^{-1} := p^{-1} + q^{-1} \leqslant 1$, the mapping*

$$(\psi, \mathbf{Cof}\,\nabla\psi) \in W^{1,p}(\Omega) \times L^q(\Omega) \to \det \nabla\psi := \partial_j\psi_1(\mathbf{Cof}\,\nabla\psi)_{1j} \in L^s(\Omega)$$

is well defined and continuous. Furthermore,

$$\left.\begin{array}{l} \boldsymbol{\varphi}^l \rightharpoonup \boldsymbol{\varphi} \text{ in } \boldsymbol{W}^{1,p}(\Omega),\ p \geq 2\,, \\ \mathbf{Cof}\,\boldsymbol{\nabla}\boldsymbol{\varphi}^l \rightharpoonup \boldsymbol{H} \text{ in } \boldsymbol{L}^q(\Omega)\,,\ \dfrac{1}{p}+\dfrac{1}{q} \leq 1\,, \\ \det \boldsymbol{\nabla}\boldsymbol{\varphi}^l \rightharpoonup \delta \text{ in } L^r(\Omega)\,,\ r \geq 1 \end{array}\right\} \Rightarrow \left\{\begin{array}{l} \boldsymbol{H} = \mathbf{Cof}\,\boldsymbol{\nabla}\boldsymbol{\varphi}\,, \\ \delta = \det \boldsymbol{\nabla}\boldsymbol{\varphi}\,. \end{array}\right.$$

Proof. (i) The bilinear mapping

$$(\boldsymbol{\psi}, \mathbf{Cof}\,\boldsymbol{\nabla}\boldsymbol{\psi}) \in \boldsymbol{W}^{1,p}(\Omega) \times \boldsymbol{L}^q(\Omega) \rightarrow \partial_j\psi_1(\mathbf{Cof}\,\boldsymbol{\nabla}\boldsymbol{\psi})_{1j} \in L^s(\Omega)$$

is well defined and continuous by Hölder's inequality.

(ii) For sufficiently smooth functions $\boldsymbol{\psi}$, for instance in the space $\mathscr{C}^2(\bar{\Omega})$,

$$\partial_j(\mathbf{Cof}\,\boldsymbol{\nabla}\boldsymbol{\psi})_{1j} = 0\,,$$

as a consequence of the *Piola identity* $\mathbf{div}\,\mathbf{Cof}\,\boldsymbol{\nabla}\boldsymbol{\psi} = \boldsymbol{o}$ (proved in Theorem 1.7-1), and hence

$$\partial_j\psi_1(\mathbf{Cof}\,\boldsymbol{\nabla}\boldsymbol{\psi})_{1j} = \partial_j\{\psi_1(\mathbf{Cof}\,\boldsymbol{\nabla}\boldsymbol{\psi})_{1j}\} = \det \boldsymbol{\nabla}\boldsymbol{\psi}$$

for such smooth functions $\boldsymbol{\psi}$. An application of Green's formula then shows that, for all functions $\boldsymbol{\psi} \in \mathscr{C}^2(\bar{\Omega})$ and all functions $\theta \in \mathscr{D}(\Omega)$,

$$\int_\Omega \partial_j\psi_1(\mathbf{Cof}\,\boldsymbol{\nabla}\boldsymbol{\psi})_{1j}\theta \,\mathrm{d}x = -\int_\Omega \psi_1(\mathbf{Cof}\,\boldsymbol{\nabla}\boldsymbol{\psi})_{1j}\partial_j\theta \,\mathrm{d}x\,.$$

Our aim is to show that *this relation still holds for all functions* $\boldsymbol{\psi} \in \boldsymbol{W}^{1,p}(\Omega)$, $p \geq 2$, *such that* $\mathbf{Cof}\,\boldsymbol{\nabla}\boldsymbol{\psi} \in \boldsymbol{L}^{p'}(\Omega)$, $p^{-1} + p'^{-1} = 1$, hence *a fortiori* such that $\mathbf{Cof}\,\boldsymbol{\nabla}\boldsymbol{\psi} \in \boldsymbol{L}^q(\Omega)$, $p^{-1} + q^{-1} \leq 1$. There is, however, a difficulty in applying a straightforward density argument as in part (ii) of the proof of Theorem 7.5-1, since the function

$$\boldsymbol{\psi} \rightarrow \int_\Omega \partial_j\psi_1(\mathbf{Cof}\,\boldsymbol{\nabla}\boldsymbol{\psi})_{1j}\theta \,\mathrm{d}x$$

is not continuous with respect to the norm $\|\cdot\|_{1,p,\Omega}$, unless $p \geq 3$. On the other hand, the bilinear form

$$(\boldsymbol{\psi}, \boldsymbol{H}) \in \boldsymbol{W}^{1,p}(\Omega) \times \boldsymbol{L}^{p'}(\Omega) \rightarrow \int_\Omega \partial_j\psi_1 H_{1j}\theta \,\mathrm{d}x$$

is clearly continuous if $p^{-1} + p'^{-1} = 1$, but then the relation

$$\int_\Omega \partial_j \psi_1 H_{1j} \theta \, dx = -\int_\Omega \psi_1 H_{1j} \partial_j \theta \, dx$$

does not hold for smooth functions ψ and H_{1j} in general, unless we consider functions H_{1j} that satisfy $\partial_j H_{1j} = 0$, as is the case of the function $(\mathbf{Cof}\, \nabla\boldsymbol{\psi})_{1j}$ when $\boldsymbol{\psi}$ is smooth. We therefore resort to a more refined argument, which takes this last property into account.

The relation $\partial_j(\mathbf{Cof}\, \nabla\boldsymbol{\psi})_{1j} = 0$ for all $\boldsymbol{\psi} \in \boldsymbol{\mathscr{C}}^2(\bar{\Omega})$ implies that

$$\int_\Omega (\mathbf{Cof}\, \nabla\boldsymbol{\psi})_{1j} \partial_j \chi \, dx = 0 \quad \text{for all } \chi \in \mathscr{D}(\Omega)\,.$$

For each $\chi \in \mathscr{D}(\Omega)$, the mapping

$$\boldsymbol{\psi} \in \boldsymbol{\mathscr{C}}^2(\bar{\Omega}) \rightarrow \int_\Omega (\mathbf{Cof}\, \nabla\boldsymbol{\psi})_{1j} \partial_j \chi \, dx$$

is continuous if the space $\boldsymbol{\mathscr{C}}^2(\bar{\Omega})$ is equipped with the norm $\|\cdot\|_{1,p,\Omega}$, $p \geq 2$, since

$$\left| \int_\Omega (\mathbf{Cof}\, \nabla\boldsymbol{\psi})_{1j} \partial_j \chi \, dx \right| \leq |\mathbf{Cof}\, \nabla\boldsymbol{\psi}|_{0,1,\Omega} |\chi|_{1,\infty,\Omega} \leq c_1(\chi) \|\boldsymbol{\psi}\|_{1,\Omega}^2 \,.$$

From the density of $\mathscr{C}^2(\bar{\Omega})$ in $W^{1,p}(\Omega)$, we thus infer that

$$\int_\Omega (\mathbf{Cof}\, \nabla\boldsymbol{\psi})_{1j} \partial_j \chi \, dx = 0 \quad \text{for all } \boldsymbol{\psi} \in \boldsymbol{W}^{1,p}(\Omega),\ p \geq 2,$$
$$\text{and all } \chi \in \mathscr{D}(\Omega)\,.$$

We now show that, given any function $\psi \in W^{1,p}(\Omega)$ and any function $\boldsymbol{w} = (w_j) \in \boldsymbol{L}^{p'}(\Omega)$, $p^{-1} + p'^{-1} = 1$, that satisfies $\int_\Omega w_j \partial_j \chi \, dx = 0$ for all $\chi \in \mathscr{D}(\Omega)$, we have

$$-\int_\Omega \psi w_j \partial_j \theta \, dx = \int_\Omega (\partial_j \psi) w_j \theta \, dx \quad \text{for all } \theta \in \mathscr{D}(\Omega)\,,$$

so that our assertion will follow by letting $\psi = \psi_1$ and $w_j = (\mathbf{Cof}\, \nabla\boldsymbol{\psi})_{1j}$. When the functions $\boldsymbol{w}$ and θ are held fixed, both sides of this relation are continuous linear forms with respect to $\psi \in W^{1,p}(\Omega)$. Hence it suffices to consider the case where $\psi \in \mathscr{C}^\infty(\bar{\Omega})$ since $\{\mathscr{C}^\infty(\bar{\Omega})\}^- = W^{1,p}(\Omega)$, but then $\psi\theta \in \mathcal{D}(\Omega)$, and thus by assumption,

$$0 = \int_\Omega w_j \partial_j(\psi\theta)\, dx = \int_\Omega \psi w_j \partial_j \theta\, dx + \int_\Omega (\partial_j \psi) w_j \theta\, dx\,.$$

(iii) We next show that, given an arbitrary function $\theta \in \mathcal{D}(\Omega)$,

$$\left.\begin{array}{l} \varphi^l \rightharpoonup \varphi \text{ in } W^{1,p}(\Omega),\ p \leqslant 2\,, \\ \operatorname{Cof} \nabla\varphi^l \rightharpoonup \operatorname{Cof} \nabla\varphi \text{ in } L^{p'}(\Omega)\,,\ \dfrac{1}{p} + \dfrac{1}{p'} = 1 \end{array}\right\} \Rightarrow$$

$$\Rightarrow \int_\Omega (\det \nabla\varphi^l)\theta\, dx \to \int_\Omega (\det \nabla\varphi)\theta\, dx\,.$$

By definition of $\det \nabla\varphi$ and by the result of step (ii), it suffices to show that

$$\int_\Omega \varphi_1^l (\operatorname{Cof} \nabla\varphi^l)_{1j} \partial_j \theta\, dx \xrightarrow[l\to\infty]{} \int_\Omega \varphi_1 (\operatorname{Cof} \nabla\varphi)_{1j} \partial_j \theta\, dx\,.$$

Arguing as in step (iii) of the proof of Theorem 7.5-1, we infer that this will be the case if

$$\varphi^l \rightharpoonup \varphi \text{ in } W^{1,p}(\Omega) \Rightarrow \varphi^l \to \varphi \text{ in } L^r(\Omega) \text{ with } \frac{1}{r} - \frac{1}{p'} \leqslant 1\,,$$

i.e., if the compact inclusion $W^{1,p}(\Omega) \Subset L^r(\Omega)$ holds. If $2 \leqslant p < 3$ (the only case that needs a proof), this inclusion holds provided $r < p^* = 3p/(3-p)$, and since

$$\frac{1}{p^*} + \frac{1}{p'} = \left(\frac{1}{p} - \frac{1}{3}\right) + \left(1 - \frac{1}{p}\right) = \frac{2}{3}\,,$$

we conclude that there exist numbers $r < p^*$ such that $r^{-1} + p'^{-1} \leqslant 1$.

(iv) The implications announced in the theorem are proved as in part (iv) of the proof of Theorem 7.5-1. ∎

Remarks. (1) We infer from Theorem 7.6-1 that *the nonconvex set*

$$\{(\psi, K, \varepsilon) \in W^{1,p}(\Omega) \times L^q(\Omega) \times L^r(\Omega);\ K = \operatorname{Cof} \nabla\psi,\ \varepsilon = \det \nabla\psi\}\,,$$

$$p \geqslant 2,\ \frac{1}{p} + \frac{1}{q} \leqslant 1,\ r \geqslant 1\,,$$

is weakly closed in the space $W^{1,p}(\Omega) \times L^q(\Omega) \times L^r(\Omega)$. This does *not* mean that the set

$$\{\psi \in W^{1,p}(\Omega);\ \mathbf{Cof}\,\nabla\psi \in L^q(\Omega),\ \det \nabla\psi \in L^r(\Omega)\},$$

$$p \geq 2, \frac{1}{p} + \frac{1}{q} \leq 1, r \geq 1,$$

is weakly closed in the space $W^{1,p}(\Omega)$, and indeed this is not always the case (Exercise 7.6).

(2) We can interpret the results of part (ii) of the proof *in the sense of distributions*. First, the relation

$$\int_\Omega (\mathbf{Cof}\,\nabla\psi)_{1j}\,\partial_j\chi\,\mathrm{d}x = 0 \quad \text{for all } \chi \in \mathcal{D}(\Omega)$$

means that

$$\partial_j(\mathbf{Cof}\,\nabla\psi)_{1j} = 0 \text{ in } \mathcal{D}'(\Omega).$$

Hence this relation, which holds for smooth functions ψ by Piola's identity, also holds in the sense of distributions for functions $\psi \in W^{1,p}(\Omega)$, $p \geq 2$. Likewise, the main result of part (ii) can be equivalently stated as

$$\left.\begin{array}{l} \psi \in W^{1,p}(\Omega),\ p \geq 2, \\ \mathbf{Cof}\,\nabla\psi \in L^{p'}(\Omega),\ \dfrac{1}{p} + \dfrac{1}{p'} = 1 \end{array}\right\} \Rightarrow$$

$$\partial_j\psi_1(\mathbf{Cof}\,\nabla\psi)_{1j} = \partial_j(\psi_1(\mathbf{Cof}\,\nabla\psi)_{1j}) \text{ in } \mathcal{D}'(\Omega).$$

This relation can be used for extending the definition of $\det \nabla\psi$ as a distribution, but not as an integrable function (Exercise 7.7). ∎

The results of Theorems 7.5-1 and 7.6-1 can be put in a more general perspective: Let (φ^k) be a sequence such that

$$\varphi^k \rightharpoonup \varphi \text{ in } W^{1,p}(\Omega),\ p \geq 2,$$

and assume in addition that the sequence $(\mathbf{Cof}\,\nabla\varphi^k)$ is bounded in the space $L^q(\Omega)$, $p^{-1} + q^{-1} = 1$. Since $q > 1$, the space $L^q(\Omega)$ is reflexive and so we can extract a subsequence (φ^l) such that $\mathbf{Cof}\,\nabla\varphi^l \rightharpoonup \boldsymbol{H}$ in $L^q(\Omega)$

(Theorem 7.1-4). Besides, $\boldsymbol{H} = \mathbf{Cof}\,\boldsymbol{\nabla\varphi}$ by Theorem 7.5-1, so that the limit $\boldsymbol{H}$ is unique, and therefore the whole sequence converges:

$$\mathbf{Cof}\,\boldsymbol{\nabla\varphi}^k \rightharpoonup \mathbf{Cof}\,\boldsymbol{\nabla\varphi} \text{ in } \boldsymbol{L}^q(\Omega)\,.$$

By step (iii) of the proof of Theorem 7.6-1, it then follows that

$$\int_\Omega (\det \boldsymbol{\nabla\varphi}^k)\theta \,\mathrm{d}x \to \int_\Omega (\det \boldsymbol{\nabla\varphi})\theta \,\mathrm{d}x \quad \text{for all } \theta \in \mathcal{D}(\Omega)\,,$$

or equivalently, in the sense of distributions,

$$\det \boldsymbol{\nabla\varphi}^k \to \det \boldsymbol{\nabla\varphi} \text{ in } \mathcal{D}'(\Omega)\,.$$

In other words, *if some appropriate combinations of partial derivatives* (the components of the matrix $\mathbf{Cof}\,\boldsymbol{\nabla\varphi}^k$) *remain bounded* in $\boldsymbol{L}^q(\Omega)$, *a nonlinear function* (the function $\det \boldsymbol{\nabla\varphi} := \partial_j\varphi_1(\det \boldsymbol{\nabla\varphi})_{1j}$) *is continuous with respect to sequential weak convergence*, in the sense that

$$\boldsymbol{\varphi}^k \rightharpoonup \boldsymbol{\varphi} \text{ in } \boldsymbol{W}^{1,p}(\Omega) \Rightarrow \det \boldsymbol{\nabla\varphi}^k \to \det \boldsymbol{\nabla\varphi} \text{ in } \mathcal{D}'(\Omega)\,.$$

This is a special case of the general phenomenon of **compensated compactness**, introduced, and extensively studied, by François Murat and Luc Tartar. Their starting point was the following result, which plays an important rôle in homogenization theory:

Theorem 7.6-2 (**div–curl lemma**). *Let Ω be a bounded open subset of $\mathbb{R}^n$, and let there be given two sequences $(\boldsymbol{u}^k)$ and $(\boldsymbol{v}^k)$ that satisfy*:

$$\begin{cases} (\boldsymbol{u}^k, \boldsymbol{v}^k) \rightharpoonup (\boldsymbol{u}, \boldsymbol{v}) \quad \text{in} \quad \boldsymbol{L}^2(\Omega) \times \boldsymbol{L}^2(\Omega)\,, \\ (\operatorname{div} \boldsymbol{u}^k, \mathbf{curl}\, \boldsymbol{v}^k) \text{ is a bounded sequence in } L^2(\Omega) \times \boldsymbol{L}^2(\Omega)\,, \end{cases}$$

where $\mathbf{curl}\, \boldsymbol{v} = (\partial_j v_i - \partial_i v_j)$. *Then*

$$\boldsymbol{u}^k \cdot \boldsymbol{v}^k \to \boldsymbol{u} \cdot \boldsymbol{v} \text{ in } \mathcal{D}'(\Omega). \qquad \blacksquare$$

The essence of this result is that the Euclidean inner product $(\boldsymbol{u}, \boldsymbol{v}) \to \boldsymbol{u} \cdot \boldsymbol{v}$ remains continuous with respect to weak convergence even though neither sequence is assumed to be relatively compact in $\boldsymbol{L}^2(\Omega)$ (if one of the sequences were bounded in $\boldsymbol{H}^1(\Omega)$, the conclusion would follow from the Rellich–Kondrašov theorem combined with Theorem 7.1-5(b)). The *lack of compactness* is thus *compensated* here by the *boundedness of some combinations of partial derivatives* (here: div · and

curl $\cdot$), themselves adapted to the mapping under consideration (here: the mapping $(\boldsymbol{u}, \boldsymbol{v}) \to \boldsymbol{u} \cdot \boldsymbol{v}$).

Remark. F. Murat and L. Tartar also showed that the mapping $(\boldsymbol{u}, \boldsymbol{v}) \to \boldsymbol{u} \cdot \boldsymbol{v}$ is the only one, apart from a linear one, that remains continuous with respect to weak convergence in $\boldsymbol{L}^2(\Omega) \times \boldsymbol{L}^2(\Omega)$ if the sequence $(\operatorname{div} \boldsymbol{u}^k, \mathbf{curl}\, \boldsymbol{v}^k)$ is the only one that remains bounded in $L^2(\Omega) \times \boldsymbol{L}^2(\Omega)$. ■

Compensated compactness has proved to be a very efficient tool in the study of nonlinear partial differential equations. See Murat [1978, 1979, 1981, 1987], Tartar [1979, 1983a], DiPerna [1985b].

7.7. JOHN BALL'S EXISTENCE RESULTS IN THE SPACES $W^{1,p}(\Omega)$, $p \geq 2$

All prerequisite ground has now been laid for establishing the existence of minimizers in hyperelasticity. Following Ball [1977, Theorems 7.3 and 7.6], we begin by considering a class of *displacement–traction problems*, which include *pure displacement problems* as special cases, with *dead loads* (for the extension to more general applied forces, see Exercises 7.8 and 7.9). Notice that the statement and proof of this existence result are both reminiscent of the statement and proof of Theorem 7.3-2; the proof, however, is significantly more delicate in the present case. We recall that a domain is a bounded open connected subset with a Lipschitz-continuous boundary (Sect. 1.6).

Theorem 7.7-1 (existence for pure displacement and displacement–traction problems). *Let Ω be a domain in $\mathbb{R}^3$, and let $\hat{W} : \Omega \times \mathbb{M}^3_+ \to \mathbb{R}$ be a stored energy function with the following properties*:

(a) *Polyconvexity*: *For almost all $x \in \Omega$, there exists a convex function $\mathbb{W}(x, \cdot) : \mathbb{M}^3 \times \mathbb{M}^3 \times]0, +\infty[\to \mathbb{R}$ such that*

$$\boxed{\mathbb{W}(x, \boldsymbol{F}, \mathbf{Cof}\, \boldsymbol{F}, \det \boldsymbol{F}) = \hat{W}(x, \boldsymbol{F}) \quad \text{for all } \boldsymbol{F} \in \mathbb{M}^3_+ \,;}$$

the function $\mathbb{W}(\cdot, \boldsymbol{F}, \boldsymbol{H}, \delta) : \Omega \to \mathbb{R}$ is measurable for all $(\boldsymbol{F}, \boldsymbol{H}, \delta) \in \mathbb{M}^3 \times \mathbb{M}^3 \times]0, +\infty[$.

(b) *Behavior as* $\det \boldsymbol{F} \to 0^+$: *For almost all $x \in \Omega$,*

$$\boxed{\lim_{\det \boldsymbol{F} \to 0^+} \hat{W}(x, \boldsymbol{F}) = +\infty \,.}$$

(c) *Coerciveness*: *There exist constants* α, β, p, q, r *such that*

$$\alpha > 0,\ p \geq 2,\ q \geq \frac{p}{p-1},\ r > 1,$$
$$\hat{W}(x, \boldsymbol{F}) \geq \alpha(\|\boldsymbol{F}\|^p + \|\mathbf{Cof}\,\boldsymbol{F}\|^q + (\det \boldsymbol{F})^r) + \beta$$
$$\text{for almost all } x \in \Omega \text{ and for all } \boldsymbol{F} \in \mathbb{M}^3_+ .$$

Let $\Gamma = \Gamma_0 \cup \Gamma_1$ *be a* da-*measurable partition of the boundary* Γ *of* Ω *with* area $\Gamma_0 > 0$, *and let* $\varphi_0 : \Gamma_0 \to \mathbb{R}^3$ *be a measurable function such that the set*

$$\Phi := \{\psi \in W^{1,p}(\Omega);\ \mathbf{Cof}\,\nabla\psi \in L^q(\Omega),\ \det \nabla\psi \in L^r(\Omega),\ \psi = \varphi_0 \text{ da-a.e. on } \Gamma_0,\ \det \nabla\psi > 0 \text{ a.e. in } \Omega\}$$

is nonempty. Let $f \in L^\rho(\Omega)$ and $g \in L^\sigma(\Gamma_1)$ *be such that the linear form*

$$L : \psi \in W^{1,p}(\Omega) \to L(\psi) := \int_\Omega f \cdot \psi \,dx + \int_{\Gamma_1} g \cdot \psi \,da$$

is continuous, let

$$I(\psi) = \int_\Omega \hat{W}(x, \nabla\psi(x))\,dx - L(\psi),$$

and assume that $\inf_{\psi \in \Phi} I(\psi) < +\infty$.

Then there exists at least one function φ *such that*

$$\varphi \in \Phi \text{ and } I(\varphi) = \inf_{\psi \in \Phi} I(\psi).$$

Proof. (i) *The integrals* $\int_\Omega \hat{W}(x, \nabla\psi(x))\,dx$ *are well defined for all* $\psi \in \Phi$. We first note the following consequences of assumptions (a): For almost all $x \in \Omega$, the function $\mathbb{W}(x, \cdot) : \mathbb{M}^3 \times \mathbb{M}^3 \times]0, +\infty[\to \mathbb{R}$ is continuous (it is convex and real-valued on an open subset of a finite-dimensional space; cf. Theorem 4.7-10); for all $(\boldsymbol{F}, \boldsymbol{H}, \delta) \in \mathbb{M}^3 \times \mathbb{M}^3 \times]0, +\infty[$, the function $\mathbb{W}(\cdot, \boldsymbol{F}, \boldsymbol{H}, \delta) : \Omega \to \mathbb{R}$ is measurable, and $\mathbb{M}^3 \times \mathbb{M}^3 \times]0, +\infty[$ is a Borel set. Therefore the function $\mathbb{W} : \Omega \times \mathbb{M}^3 \times \mathbb{M}^3 \times]0, +\infty[\to \mathbb{R}$ is a

Carathéodory function, and consequently (Ekeland & Temam [1974, pp. 218 ff.]) the function

$$x \in \Omega \to \mathbb{W}(x, \nabla\psi(x), \mathbf{Cof}\,\nabla\psi(x), \det \nabla\psi(x)) \in \mathbb{R}$$

is measurable for each $\psi \in \Phi$ (recall that $\det \nabla\psi(x) \in\,]0, +\infty[$ for almost all $x \in \Omega$). Since the function $\hat{W}$ is in addition bounded below (by the coerciveness inequality), we conclude that the integral

$$\int_\Omega \hat{W}(x, \nabla\psi(x))\,dx = \int_\Omega \mathbb{W}(x, \nabla\psi(x), \mathbf{Cof}\,\nabla\psi(x), \det \nabla\psi(x))\,dx$$

is a well defined extended real number in the interval $[\beta \operatorname{vol} \Omega, +\infty]$ for each $\psi \in \Phi$.

(ii) We next find a *lower bound for* $I(\psi)$, $\psi \in \Phi$. By the assumed coerciveness of the function $\hat{W}$ and by the assumed continuity of the linear form L,

$$I(\psi) \geq \alpha \int_\Omega \{\|\nabla\psi\|^p + \|\mathbf{Cof}\,\nabla\psi\|^q + (\det \nabla\psi)^r\}\,dx + \beta \operatorname{vol} \Omega$$
$$- \|L\| \|\psi\|_{1,p,\Omega} \quad \text{for all } \psi \in \Phi\,.$$

Combining the boundary condition $\psi = \varphi_0$ on Γ_0 with the generalized Poincaré inequality (Theorem 6.1-8(b)), we conclude that there exist constants c and d such that (a similar argument was used in the proof of Theorem 7.3-2):

$$c > 0 \text{ and } I(\psi) \geq c\{\|\psi\|^p_{1,p,\Omega} + |\mathbf{Cof}\,\nabla\psi|^q_{0,q,\Omega} + |\det \nabla\psi|^r_{0,r,\Omega}\} + d$$
$$\text{for all } \psi \in \Phi\,.$$

(iii) Let (φ^k) be an *infimizing sequence* for the functional I, i.e., a sequence that satisfies

$$\varphi^k \in \Phi \quad \text{for all } k, \text{ and } \lim_{k\to\infty} I(\varphi^k) = \inf_{\psi\in\Phi} I(\psi)\,.$$

By assumption, $\inf_{\psi\in\Phi} I(\psi) < +\infty$, and thus by (ii) the sequence $(\varphi^k, \mathbf{Cof}\,\nabla\varphi^k, \det \nabla\varphi^k)$ is bounded in the reflexive Banach space $W^{1,p}(\Omega) \times L^q(\Omega) \times L^r(\Omega)$ (each number p, q, r is >1). Hence there exists a subsequence $(\varphi^l, \mathbf{Cof}\,\nabla\varphi^l, \det \nabla\varphi^l)$ that converges weakly to an element $(\boldsymbol{\varphi}, \boldsymbol{H}, \delta)$ in the space $\boldsymbol{W}^{1,p}(\Omega) \times \boldsymbol{L}^q(\Omega) \times L^r(\Omega)$, and by Theorem 7.6-1, $\boldsymbol{H} = \mathbf{Cof}\, \nabla\boldsymbol{\varphi}$ and $\delta = \det \nabla\boldsymbol{\varphi}$. To sum up, *there exists a subsequence of the infimizing sequence that satisfies*:

$$\begin{cases} \boldsymbol{\varphi}^l \rightharpoonup \boldsymbol{\varphi} \text{ in } \boldsymbol{W}^{1,p}(\Omega)\,, \\ \mathbf{Cof}\, \nabla\boldsymbol{\varphi}^l \rightharpoonup \mathbf{Cof}\, \nabla\boldsymbol{\varphi} \text{ in } \boldsymbol{L}^q(\Omega)\,, \\ \det \nabla\boldsymbol{\varphi}^l \rightharpoonup \det \nabla\boldsymbol{\varphi} \text{ in } L^r(\Omega)\,. \end{cases}$$

(iv) *We now show that* $\boldsymbol{\varphi} \in \boldsymbol{\Phi}$. To prove this, it remains to establish that $\det \nabla\boldsymbol{\varphi} > 0$ almost everywhere in Ω and that $\boldsymbol{\varphi} = \boldsymbol{\varphi}_0$ on Γ_0. Since $\det \nabla\boldsymbol{\varphi}^l \rightharpoonup \det \nabla\boldsymbol{\varphi}$ in $L^r(\Omega)$, there exist for each l by Mazur's theorem (Theorem 7.1-2) integers $i(l) \geq l$ and numbers λ_s^l, $l \leq s \leq i(l)$, such that

$$\begin{cases} \lambda_s^l \geq 0, \ \sum_{s=l}^{i(l)} \lambda_s^l = 1\,, \\ d^l := \sum_{s=l}^{i(l)} \lambda_s^l \det \nabla\boldsymbol{\varphi}^s \xrightarrow[l \to \infty]{} \det \nabla\boldsymbol{\varphi} \text{ in } L^r(\Omega)\,; \end{cases}$$

hence there exists a subsequence (d^m) of (d^l) that converges almost everywhere to $\det \nabla\boldsymbol{\varphi}$. Since the functions d^l are >0 almost everywhere (≥ 0 would suffice), we conclude that $\det \nabla\boldsymbol{\varphi} \geq 0$ almost everywhere in Ω.

Assume that $\det \nabla\boldsymbol{\varphi} = 0$ on a subset A of Ω with $\operatorname{vol} A > 0$. Since $\det \nabla\boldsymbol{\varphi}^l > 0$ almost everywhere on A (again, ≥ 0 would suffice) and $\det \nabla\boldsymbol{\varphi}^l \rightharpoonup \det \nabla\boldsymbol{\varphi}$, we have

$$\int_A |\det \nabla\boldsymbol{\varphi}^l|\, dx = \int_A \det \nabla\boldsymbol{\varphi}^l\, dx \to \int_A \det \nabla\boldsymbol{\varphi}\, dx = 0\,,$$

by definition of weak convergence (the characteristic function of the set A belongs to the dual space of $L^r(\Omega)$), and hence $\det \nabla\boldsymbol{\varphi}^l \to 0$ in $L^1(A)$. Therefore there exists a subsequence $(\boldsymbol{\varphi}^m)$ of $(\boldsymbol{\varphi}^l)$ such that

$$\det \nabla\boldsymbol{\varphi}^m(x) \to 0 \quad \text{for almost all } x \in A\,.$$

Consider the sequence of measurable functions (f_m) defined by

$$f^m : x \in A \to f^m(x) := \hat{W}(x, \nabla\boldsymbol{\varphi}^m(x))\,.$$

Since $f^m \geq \beta$ for all m, we can apply Fatou's lemma:

$$\int_A \liminf_{m\to\infty} f^m(x)\,\mathrm{d}x \leq \liminf_{m\to\infty} \int_A f^m(x)\,\mathrm{d}x\,.$$

By assumption (b),

$$\liminf_{m\to\infty} f^m(x) = \lim_{m\to\infty} \hat{W}(x, \nabla\varphi^m(x)) = \lim_{\det F\to 0^+} \hat{W}(x, F) = +\infty$$
$$\text{for almost all } x \in A\,,$$

and thus

$$\lim_{m\to\infty} \int_A f^m(x)\,\mathrm{d}x = \lim_{m\to\infty} \int_A \hat{W}(x; \nabla\varphi^m(x))\,\mathrm{d}x = +\infty\,.$$

But this last relation contradicts the relation $\lim_{m\to\infty} I(\varphi^m) = \inf_{\psi\in\Phi} I(\psi) < +\infty$ and the inequalities

$$I(\varphi^m) \geq \int_A \hat{W}(x, \nabla\varphi^m(x))\,\mathrm{d}x + \beta \operatorname{vol}(\Omega - A) - \|L\| \|\varphi^m\|_{1,p,\Omega}$$

(a weakly convergent sequence is bounded in a Banach space; cf. Theorem 7.1-4). Hence *we must have* $\det \nabla\varphi > 0$ *almost everywhere in* Ω.

To show that $\varphi = \varphi_0$ on Γ_0, we observe that the trace operator $\mathrm{tr} \in \mathscr{L}(W^{1,p}(\Omega); L^p(\Gamma))$ is compact (Theorem 6.1-7(b)). Therefore,

$$\varphi^l \to \varphi \text{ in } W^{1,p}(\Omega) \Rightarrow \mathrm{tr}\,\varphi^l \to \mathrm{tr}\,\varphi \text{ in } L^p(\Gamma)\,,$$

and by extracting a subsequence that pointwise converges da-almost everywhere on Γ, we conclude that $\varphi = \varphi_0$ on Γ_0.

(v) We now show that

$$\int_\Omega \hat{W}(x, \nabla\varphi(x))\,\mathrm{d}x \leq \liminf_{l\to\infty} \int_\Omega \hat{W}(x, \nabla\varphi^l(x))\,\mathrm{d}x\,.$$

By definition of the limit inferior, we must show that, given any subsequence (φ^m) of (φ^l) such that the sequence $(\int_\Omega \hat{W}(x, \nabla\varphi^m(x))\,\mathrm{d}x)$ converges, then

$$\int_\Omega \hat{W}(x, \nabla\varphi(x))\,\mathrm{d}x \leq \lim_{m\to\infty} \int_\Omega \hat{W}(x, \nabla\varphi^m(x))\,\mathrm{d}x\,.$$

Let us consider such a subsequence. Using the result of step (iii) and

Mazur's theorem, we conclude that for each m, there exist integers $j(m) \geq m$ and numbers μ_t^m, $m \leq t \leq j(m)$, such that

$$\mu_t^m \geq 0, \sum_{t=m}^{j(m)} \mu_t^m = 1 ,$$

$$D^m := \sum_{t=m}^{j(m)} \mu_t^m (\nabla \varphi^t, \operatorname{Cof} \nabla \varphi^t, \det \nabla \varphi^t) \xrightarrow[m \to \infty]{} (\nabla \varphi, \operatorname{Cof} \nabla \varphi, \det \nabla \varphi)$$

$$\text{in } L^p(\Omega) \times L^q(\Omega) \times L^r(\Omega) ;$$

hence there exists a subsequence (D^n) of (D^m) such that

$$\sum_{t=n}^{j(n)} \mu_t^n (\nabla \varphi^t(x), \operatorname{Cof} \nabla \varphi^t(x), \det \nabla \varphi^t(x)) \xrightarrow[n \to \infty]{}$$

$$(\nabla \varphi(x), \operatorname{Cof} \nabla \varphi(x), \det \nabla \varphi(x)) \quad \text{for almost all } x \in \Omega .$$

Since the function $\mathbb{W}(x, \cdot)$ is continuous on the set $\mathbb{M}^3 \times \mathbb{M}^3 \times]0, +\infty[$ for almost all $x \in \Omega$ as a consequence of assumption (a), and since we have proved in step (iv) that $\det \nabla \varphi(x) > 0$ for almost all $x \in \Omega$, we have

$$\begin{aligned} \hat{W}(x, \nabla \varphi(x)) &= \mathbb{W}(x, (\nabla \varphi(x), \operatorname{Cof} \nabla \varphi(x), \det \nabla \varphi(x))) \\ &= \lim_{n \to \infty} \mathbb{W}\Big(x, \sum_{t=n}^{j(n)} \mu_t^n (\nabla \varphi^t(x), \operatorname{Cof} \nabla \varphi^t(x), \det \nabla \varphi^t(x))\Big) \end{aligned}$$

for almost all $x \in \Omega$. Using this relation, Fatou's lemma, and the assumed convexity of the function $\mathbb{W}(x, \cdot)$ for almost all $x \in \Omega$, we next obtain

$$\begin{aligned} &\int_\Omega \hat{W}(x, \nabla \varphi(x)) \, dx \\ &\leq \liminf_{n \to \infty} \int_\Omega \mathbb{W}\Big(x, \sum_{t=n}^{j(n)} \mu_t^n (\nabla \varphi^t(x), \operatorname{Cof} \nabla \varphi^t(x), \det \nabla \varphi^t(x))\Big) \, dx \\ &\leq \liminf_{n \to \infty} \sum_{t=n}^{j(n)} \mu_t^n \int_\Omega \hat{W}(x, \nabla \varphi^t(x)) \, dx \\ &= \lim_{n \to \infty} \int_\Omega \hat{W}(x, \nabla \varphi^n(x)) \, dx \\ &= \lim_{m \to \infty} \int_\Omega \hat{W}(x, \nabla \varphi^m(x)) \, dx . \end{aligned}$$

We have used here a simple observation: Let (α^n) be a convergent sequence, and let

$$\beta^n := \sum_{l=n}^{j(n)} \mu_l^n \alpha^l, \quad \text{with } \mu_l^n \geq 0 \text{ and } \sum_{l=n}^{j(n)} \mu_l^n = 1 ;$$

then the sequence (β_n) is also convergent, and $\lim_{n\to\infty} \beta^n = \lim_{n\to\infty} \alpha^n$.

Since on the other hand $L(\varphi) = \lim_{l\to\infty} L(\varphi^l)$ by definition of weak convergence, we have proved that

$$I(\varphi) \leq \liminf_{l\to\infty} I(\varphi^l) .$$

(vi) The function φ is thus a solution of the minimization problem since $\varphi \in \Phi$ by step (iv), and since

$$I(\varphi) \leq \liminf_{l\to\infty} I(\varphi^l) = \inf_{\psi\in\Phi} I(\psi) \Rightarrow I(\varphi) = \inf_{\psi\in\Phi} I(\psi) . \quad \blacksquare$$

Remarks. (1) That the hypotheses on the stored energy function need only hold for almost all $x \in \Omega$ is not a gratuitous assumption. For instance, if an elastic body consists of several different materials glued together, the stored energy function is not defined along the separation surfaces, whose union has zero volume. In the same spirit, notice that the existence result applies to hyperelastic materials that are not necessarily isotropic.

(2) In Ball [1977], the behavior of the stored energy function as $\det \boldsymbol{F} \to 0^+$ was replaced by the stronger assumption that

$$\lim_{k\to\infty} \mathbb{W}(x, \boldsymbol{F}_k, \boldsymbol{H}_k, \delta_k) = +\infty$$

as $\boldsymbol{F}_k \to \boldsymbol{F}$ in $\mathbb{M}^3_+$, $\boldsymbol{H}_k \to \boldsymbol{H}$ in $\mathbb{M}^3_+$, $\delta_k \to 0^+$, which allows for a simpler proof of the existence of a minimizer (the proof then follows more closely that of Theorem 7.3-2; for more details, see Exercise 7.11). The replacement by the weaker assumption that $\lim \hat{W}(x, \boldsymbol{F}) = +\infty$ as $\det \boldsymbol{F} \to 0^+$ (as considered here) is due to Ball, Currie & Olver [1981, Theorem 6.2].

(3) There are situations where the coerciveness inequality takes a slightly different form and yet, an analogous existence theorem, holds; see Ball [1981c], Ball, Currie & Olver [1981], Ball & Murat [1984].

(4) If the material is isotropic, the stored energy function is of the form $\hat{W}(x, \boldsymbol{F}) = \dot{W}(x, \boldsymbol{\iota}_C)$, where $\boldsymbol{\iota}_C = (\iota_1, \iota_2, \iota_3)$ denotes the triple formed by the three principal invariants $\iota_k = \iota_k(\boldsymbol{C})$ of the matrix $\boldsymbol{C} = \boldsymbol{F}^{\mathrm{T}}\boldsymbol{F}$ (Theorem

4.4-1). There are practical instances, however, where the stored energy function is more naturally expressed in terms of the "modified invariants $\iota_1^* = \iota_1 \iota_3^{-1/3}$, $\iota_2^* = \iota_2 \iota_3^{-2/3}$, and ι_3. The present existence theory can be extended to this case, as shown by Charrier, Dacorogna, Hanouzet & Laborde [1985].

(5) Theorem 7.7-1 does not provide an existence result for *St Venant–Kirchhoff materials*, since their stored energy functions, though coercive (Exercise 4.10), are not polyconvex (Raoult [1986]; cf. Theorem 4.10-1). It is, however, always possible to construct a polyconvex stored energy function that agrees with that of a St Venant–Kirchhoff material for small values of $\|E\|$ (Ciarlet & Geymonat [1982]; cf. Theorem 4.10-2), and then Theorem 7.7-1 can be applied with this more general stored energy function (which, incidentally, is also well suited for numerical computations; cf. Le Tallec & Vidrascu [1984]). Another approach consists in looking for "generalized solutions" of an appropriately modified problem, as in Atteia & Dedieu [1981] or Atteia & Raissouli [1986].

(6) The assumption that the set $\boldsymbol{\Phi}$ be nonempty (a necessary condition for the existence of a minimizer!) is essentially an assumption on the given function $\boldsymbol{\varphi}_0$.

(7) By Theorem 6.1-3, the linear form $\boldsymbol{\psi} \in \boldsymbol{W}^{1,p}(\Omega) \rightarrow \int_\Omega \boldsymbol{f} \cdot \boldsymbol{\psi}\, dx$ is well defined and continuous if $\boldsymbol{f} \in \boldsymbol{L}^\rho(\Omega)$, with

$$\rho = (p^*)' = \frac{3p}{4p-3} \text{ if } p < 3,\ \rho > 1 \text{ if } p = 3,\ \rho = 1 \text{ if } p > 3\,,$$

and, by Theorem 6.1-7(a), the linear form $\boldsymbol{\psi} \in \boldsymbol{W}^{1,p}(\Omega) \rightarrow \int_{\Gamma_1} \boldsymbol{g} \cdot \boldsymbol{\psi}\, da$ is well defined and continuous if $\boldsymbol{g} \in \boldsymbol{L}^\sigma(\Gamma_1)$, with

$$\sigma = \frac{2}{3}\,\frac{p}{(p-1)} \text{ if } p < 3,\ \sigma > 1 \text{ if } p = 3,\ \sigma = 1 \text{ if } p > 3\,.$$

(8) As noted earlier (cf. Sect. 7.6; see also Exercise 7.6), the set

$$\{\boldsymbol{\psi} \in \boldsymbol{W}^{1,p}(\Omega);\ \mathbf{Cof}\,\nabla\boldsymbol{\psi} \in \boldsymbol{L}^q(\Omega),\ \det \nabla\boldsymbol{\psi} \in L^r(\Omega),\ \boldsymbol{\psi} = \boldsymbol{\varphi}_0 \text{ on } \Gamma_0\}\,,$$

let alone the set $\boldsymbol{\Phi}$ of the theorem (i.e., with the additional orientation-preserving condition), is not in general weakly closed in the space $\boldsymbol{W}^{1,p}(\Omega)$. Instead, it is the coerciveness of the stored energy function (combined with the result of Theorem 7.6-1) that guarantees that the weak limit of the infimizing sequence is still in the same set (step (iii) of the proof). In the same vein, it is the behavior of the stored energy function as $\det \boldsymbol{F} \to 0^+$ that guarantees that the weak limit of the infimizing sequence still satisfies the orientation-preserving condition (step (iv) of the proof).

(9) Applied forces that are not dead loads (such as a pressure load) can be also considered. However, it must be checked in each instance whether the corresponding part L (which is then nonlinear) of the total energy still satisfies $L(\boldsymbol{\varphi}) = \lim_{l\to\infty} L(\boldsymbol{\varphi}^l)$ (Exercises 7.8 and 7.9).

(10) As shown by Ball [1977, Sect. 9], his existence theory is compatible with the *nonuniqueness of solutions* observed in practice (Sect. 5.8). ■

In the case of a *pure traction problem*, and additional condition has to be included in the set $\boldsymbol{\Phi}$ of admissible deformations in order that an infimizing sequence $(\boldsymbol{\varphi}^k)$ of the total energy be bounded with respect to the *norm* $\|\cdot\|_{1,p,\Omega}$. If otherwise the set $\boldsymbol{\Phi}$ were as in Theorem 7.7-1, we could only conclude from the assumed coerciveness of the stored energy function that the *semi-norms* $|\boldsymbol{\varphi}^k|_{1,p,\Omega}$ are bounded. We follow here Ball [1977, Theorem 7.10].

Theorem 7.7-2 (existence for pure traction problems). *Let Ω be a domain in $\mathbb{R}^3$, and let $\hat{W}: \Omega \times \mathbb{M}^3_+ \to \mathbb{R}$ be a stored energy function that satisfies assumptions* (a), (b), (c) *of Theorem 7.7-1 (polyconvexity, behavior as* $\det \boldsymbol{F} \to 0^+$, *coerciveness). Let $\boldsymbol{e}$ be a vector in $\mathbb{R}^3$, and let*

$$\boldsymbol{\Phi} := \Big\{ \boldsymbol{\psi} \in \boldsymbol{W}^{1,p}(\Omega);\ \mathbf{Cof}\,\nabla\boldsymbol{\psi} \in \boldsymbol{L}^q(\Omega),\ \det \nabla\boldsymbol{\psi} \in L^r(\Omega),\ \det \nabla\boldsymbol{\psi} > 0 \text{ a.e. in } \Omega,\ \int_\Omega \boldsymbol{\psi}(x)\,dx = \boldsymbol{e} \Big\}.$$

Let the linear form L defined by $L(\boldsymbol{\psi}) = \int_\Omega \boldsymbol{f}\cdot\boldsymbol{\psi}\,dx + \int_\Gamma \boldsymbol{g}\cdot\boldsymbol{\psi}\,da$ be continuous on $\boldsymbol{W}^{1,p}(\Omega)$, let $I(\boldsymbol{\psi}) = \int_\Omega \hat{W}(x, \nabla\boldsymbol{\psi}(x))\,dx - L(\boldsymbol{\psi})$, and finally, assume that $\inf_{\boldsymbol{\psi}\in\boldsymbol{\Phi}} I(\boldsymbol{\psi}) < +\infty$. Then there exists at least one function $\boldsymbol{\varphi} \in \boldsymbol{\Phi}$ such that

$$\boldsymbol{\varphi} \in \boldsymbol{\Phi} \text{ and } I(\boldsymbol{\varphi}) = \operatorname*{Inf}_{\boldsymbol{\psi}\in\boldsymbol{\Phi}} I(\boldsymbol{\psi}).$$

Proof. We only need to slightly modify the proof of Theorem 7.7-1. In step (ii), we now use the generalized Poincaré inequality in the following form (Theorem 6.1-8(a)): There exists a constant c_0 such that

$$\int_\Omega |\boldsymbol{\psi}|^p \,\mathrm{d}x \leq c_0\left\{\int_\Omega |\mathbf{grad}\,\boldsymbol{\psi}|^p \,\mathrm{d}x + \left|\int_\Omega \boldsymbol{\psi}\,\mathrm{d}x\right|^p\right\}$$

$$\text{for all } \boldsymbol{\psi} \in \boldsymbol{W}^{1,p}(\Omega)\,.$$

We then infer from this inequality and from the relation $\int_\Omega \boldsymbol{\psi}\,\mathrm{d}x = \boldsymbol{e}$ that there exist constants c and d such that

$$c > 0 \text{ and } I(\boldsymbol{\psi}) \geq c\{\|\boldsymbol{\psi}\|^p_{1,p,\Omega} + |\mathbf{Cof}\,\boldsymbol{\nabla\psi}|^q_{0,q,\Omega} + |\det \boldsymbol{\nabla\psi}|^r_{0,r,\Omega}\} + d$$

$$\text{for all } \boldsymbol{\psi} \in \boldsymbol{\Phi}\,.$$

In step (iii), we must check that $\int_\Omega \boldsymbol{\varphi}\,\mathrm{d}x = \boldsymbol{e}$, but this follows from the implication

$$\boldsymbol{\varphi}' \rightharpoonup \boldsymbol{\varphi} \text{ in } \boldsymbol{W}^{1,p}(\Omega) \Rightarrow \int_\Omega \boldsymbol{\varphi}'\,\mathrm{d}x \to \int_\Omega \boldsymbol{\varphi}\,\mathrm{d}x\,. \quad ■$$

Remarks. (1) Whereas it was observed in Sect. 5.1 that the pure traction problem has a smooth solution only if the applied forces satisfy the compatibility condition $\int_\Omega \boldsymbol{f}\,\mathrm{d}x + \int_\Gamma \boldsymbol{g}\,\mathrm{d}a = \boldsymbol{o}$, this condition is not needed here for proving the existence of a minimizer of the energy in the set $\boldsymbol{\Phi}$ of Theorem 7.7-2. If this condition is not satisfied, however, a smooth minimizer solves a boundary value problem that is *not* the associated pure traction problem (Exercise 5.2).

(2) The set $\boldsymbol{\Phi}$ as defined in Theorem 7.7-2 is nonempty since it is always possible to find a vector field of the form $\boldsymbol{\psi} = \boldsymbol{id} + \boldsymbol{a}$, where $\boldsymbol{a}$ is a constant vector, that belongs to $\boldsymbol{\Phi}$. ■

John Ball's approach can be likewise applied to *incompressible materials*. The *incompressibility condition* (Sect. 5.7) is then included in the form "$\det \boldsymbol{\nabla\psi} = 1$ a.e. in Ω" in the definition of the set $\boldsymbol{\Phi}$; cf. Ball [1977] (see Exercise 7.12). In this respect, we also mention the interesting results of Rostamian [1978] and Le Dret [1986a], who consider compressible materials with a stored energy function of the form introduced by Ogden [1972b]:

$$\hat{W}_\varepsilon : \boldsymbol{F} \in \mathbb{M}^3_+ \to \hat{W}_\varepsilon(\boldsymbol{F}) = W(\boldsymbol{F}) + \frac{1}{\varepsilon} h(\det \boldsymbol{F}),\ \varepsilon > 0\,.$$

They then establish that, as $\varepsilon \to 0$, the corresponding "compressible" minimizers $\boldsymbol{\varphi}^\varepsilon$ weakly (even strongly in some cases) converge in the space $\boldsymbol{W}^{1,p}(\Omega)$ to an "incompressible" minimizer. Assuming that $\boldsymbol{\varphi}^\varepsilon \in \boldsymbol{W}^{2,p}(\Omega)$ for some $p > 3$, Le Dret [1986] shows in addition that $\boldsymbol{\varphi} = \boldsymbol{\varphi}^0 + \varepsilon\boldsymbol{\varphi}^1 + o(\varepsilon)$ in $\boldsymbol{W}^{2,p}(\Omega)$, where $\boldsymbol{\varphi}^0$ is the solution of the "limit" boundary value problem of incompressible elasticity. Of particular interest are also the results of Le Tallec & Oden [1980, 1981] who established the existence of a hydrostatic pressure (the Lagrange multiplier associated with the incompressibility condition $\det \nabla\boldsymbol{\psi} = 1$; cf. Exercise 5.9) when the minimizer of the energy is smooth enough (for related ideas, see also Glowinski & Le Tallec [1982], Fosdick & MacSithigh [1986]).

7.8. PROBLEMS WITH UNILATERAL CONSTRAINTS

Following Ciarlet & Nečas [1985], we extend the existence result of Theorem 7.7-1, by allowing in the set $\boldsymbol{\Phi}$ of admissible deformations a *unilateral boundary condition of place* of the form introduced in Sect. 5.3; we recall that this condition is a mathematical model of *contact without friction with an obstacle* for a smooth minimizer of the energy (Theorem 5.3-1). We simultaneously allow in the set $\boldsymbol{\Phi}$ a *locking constraint* of the form introduced in Sect. 5.7.

We first consider the case where a boundary condition of place $\boldsymbol{\varphi} = \boldsymbol{\varphi}_0$ is imposed on a portion Γ_0 of the boundary. The boundary condition "$\boldsymbol{\psi} \in C$ da-a.e. on Γ_2" means that $\boldsymbol{\psi}(x) \in C$ for da-almost all $x \in \Gamma_2$.

Theorem 7.8-1 (existence for displacement–traction problems with a unilateral boundary condition of place). *Let Ω be a domain in $\mathbb{R}^3$, and let $\hat{W}: \Omega \times \mathbb{M}^3_+ \to \mathbb{R}$ be a stored energy function that satisfies assumptions* (a), (b), (c) *of Theorem 7.7-1 (polyconvexity, behavior as* $\det \boldsymbol{F} \to 0^+$, *coerciveness). Let $\hat{L}: \mathbb{M}^3_+ \to \mathbb{R}$ be a polyconvex locking function, i.e., there exists a convex function $\mathbb{L}: \mathbb{M}^3 \times \mathbb{M}^3 \times]0, +\infty[\to \mathbb{R}$ such that*

$$\boxed{\hat{L}(\boldsymbol{F}) = \mathbb{L}(\boldsymbol{F}, \mathbf{Cof}\,\boldsymbol{F}, \det \boldsymbol{F}) \quad \text{for all } \boldsymbol{F} \in \mathbb{M}^3_+ .}$$

Let Γ_0, Γ_1, Γ_2 be disjoint relatively open subsets of $\Gamma = \partial\Omega$ with area $\Gamma_0 > 0$ *and* area $\{\Gamma - (\Gamma_0 \cup \Gamma_1 \cup \Gamma_2)\} = 0$, *let C be a closed subset of $\mathbb{R}^3$, and let $\boldsymbol{\varphi}_0: \Gamma_0 \to \mathbb{R}^3$ be a measurable function such that the set*

$$\boxed{\begin{aligned}\boldsymbol{\Phi} := \{&\boldsymbol{\psi} \in \mathbf{W}^{1,p}(\Omega);\ \mathbf{Cof}\,\boldsymbol{\nabla}\boldsymbol{\psi} \in \mathbf{L}^q(\Omega),\ \det \boldsymbol{\nabla}\boldsymbol{\psi} \in L^r(\Omega),\\ &\boldsymbol{\psi} = \boldsymbol{\varphi}_0\ da\text{-a.e. on } \Gamma_0,\\ &\boldsymbol{\psi} \in C\ da\text{-a.e. on } \Gamma_2,\ \det \boldsymbol{\nabla}\boldsymbol{\psi} > 0 \text{ a.e. in } \Omega,\\ &L(\boldsymbol{\nabla}\boldsymbol{\psi}) \leq 0 \text{ a.e. in } \Omega\}\end{aligned}}$$

is nonempty. Let the linear form L defined by $L(\boldsymbol{\psi}) = \int_\Omega \boldsymbol{f} \cdot \boldsymbol{\Psi}\, dx + \int_{\Gamma_1} \boldsymbol{g} \cdot \boldsymbol{\psi}\, da$ *be continuous on* $\mathbf{W}^{1,p}(\Omega)$, *let* $I(\boldsymbol{\psi}) = \int_\Omega \hat{W}(x, \boldsymbol{\nabla}\boldsymbol{\psi}(x))\, dx - L(\boldsymbol{\psi})$, *and finally, assume that* $\inf_{\boldsymbol{\psi} \in \boldsymbol{\Phi}} I(\boldsymbol{\psi}) < +\infty$.

Then there exists at least one function $\boldsymbol{\varphi}$ *such that*

$$\boxed{\boldsymbol{\varphi} \in \boldsymbol{\Phi} \text{ and } I(\boldsymbol{\varphi}) = \inf_{\boldsymbol{\psi} \in \boldsymbol{\Phi}} I(\boldsymbol{\psi})\,.}$$

Proof. It suffices to show that the function $\boldsymbol{\varphi}$ found in step (iv) of the proof of Theorem 7.7-1 satisfies the two additional conditions that are now imposed; otherwise, the proof is identical. First, the compactness of the trace operator $\mathrm{tr} \in \mathcal{L}(\mathbf{W}^{1,p}(\Omega); \mathbf{L}^p(\Gamma))$ allows us to extract a subsequence of the infimizing sequence $(\boldsymbol{\varphi}^l)$ that converges pointwise da-almost everywhere on Γ; hence $\boldsymbol{\varphi} \in C$ da-almost everywhere on Γ_2 since the set C is closed.

Secondly, we can find as in step (v) of the proof of Theorem 7.7-1 a subsequence indexed by n such that

$$\sum_{l=n}^{j(n)} \mu_l^n(\boldsymbol{\nabla}\boldsymbol{\varphi}^l(x), \mathbf{Cof}\,\boldsymbol{\nabla}\boldsymbol{\varphi}^l(x), \det \boldsymbol{\nabla}\boldsymbol{\varphi}^l(x)) \xrightarrow[n\to\infty]{} (\boldsymbol{\nabla}\boldsymbol{\varphi}(x), \mathbf{Cof}\,\boldsymbol{\nabla}\boldsymbol{\varphi}(x), \det \boldsymbol{\nabla}\boldsymbol{\varphi}(x)) \quad \text{for almost all } x \in \Omega\,.$$

The function $\mathbb{L} : \mathbb{M}^3 \times \mathbb{M}^3 \times]0, +\infty[$ is convex by the assumed polyconvexity of the locking function $\hat{L}$; hence it is continuous on the open set $\mathbb{M}^3 \times \mathbb{M}^3 \times]0, +\infty[$ (Theorem 4.7-10(c)), and thus

$$\begin{aligned}\hat{L}(\boldsymbol{\nabla}\boldsymbol{\varphi}(x)) &= \mathbb{L}(\boldsymbol{\nabla}\boldsymbol{\varphi}(x), \mathbf{Cof}\,\boldsymbol{\nabla}\boldsymbol{\varphi}(x), \det \boldsymbol{\nabla}\boldsymbol{\varphi}(x))\\ &= \lim_{n\to\infty} \mathbb{L}\Big(\sum_{l=n}^{j(n)} \mu_l^n(\boldsymbol{\nabla}\boldsymbol{\varphi}^l(x), \mathbf{Cof}\,\boldsymbol{\nabla}\boldsymbol{\varphi}^l(x), \det \boldsymbol{\nabla}\boldsymbol{\varphi}^l(x))\Big)\\ &\leq \liminf_{n\to\infty} \Big\{\sum_{l=n}^{j(n)} \mu_l^n \mathbb{L}(\boldsymbol{\nabla}\boldsymbol{\varphi}^l(x), \mathbf{Cof}\,\boldsymbol{\nabla}\boldsymbol{\varphi}^l(x), \det \boldsymbol{\nabla}\boldsymbol{\varphi}^l(x))\Big\} \leq 0\end{aligned}$$

for almost all $x \in \Omega$, and the proof is complete. ■

The case where $\Gamma_0 = \Gamma_2 = \emptyset$ has been treated in Theorem 7.7-2. We now turn to the more interesting case where $\Gamma_0 = \emptyset$ and $\Gamma_2 \neq \emptyset$. We observed in Sect. 5.3 (cf. notably Fig. 5.3-2) that a necessary condition for the existence of smooth minimizers of the energy in this case is that

$$\left\{\int_\Omega \boldsymbol{f}\,\mathrm{d}x + \int_{\Gamma_1} \boldsymbol{g}\,\mathrm{d}a\right\}\cdot \boldsymbol{d} \leq 0$$

for all vectors $\boldsymbol{d}$ in a set D of "directions where the body can escape". Following Ciarlet & Nečas [1985, Theorem 4.2], we now give a precise definition of the set D, and we prove that these inequalities are very close to forming also a sufficient condition for the existence of a minimizer (we now restrict them to be strict). A noticeable feature of the present approach is that *we no longer need to impose an additional condition in the set* $\boldsymbol{\Phi}$ *such as* $\int_\Omega \boldsymbol{\psi}\,\mathrm{d}x = \boldsymbol{e}$, *as in Theorem 7.7-2*. For simplicity, we do not impose a locking constraint, which would otherwise be dealt with exactly as in Theorem 7.8-1.

Theorem 7.8-2 (existence for pure traction problems with a unilateral boundary condition of place). *Let* Ω *be a domain in* $\mathbb{R}^3$, *and let* $\hat{W}: \Omega \times \mathbb{M}^3_+ \to \mathbb{R}$ *be a stored energy function that satisfies assumptions* (a), (b), (c) *of Theorem 7.7-1 (polyconvexity, behavior as* $\det \boldsymbol{F} \to 0^+$, *coerciveness). Let* Γ_1, Γ_2 *be disjoint relatively open subset of* Γ *with* area $\Gamma_2 > 0$ *and* area $\{\Gamma - (\Gamma_1 \cup \Gamma_2)\} = 0$, *and let* C *be a closed subset of* $\mathbb{R}^3$ *such that the set*

$$\boldsymbol{\Phi} := \{\boldsymbol{\psi} \in \boldsymbol{W}^{1,p}(\Omega);\ \mathbf{Cof}\,\nabla\boldsymbol{\psi} \in \boldsymbol{L}^q(\Omega),\ \det \nabla\boldsymbol{\psi} \in L^r(\Omega)\,,\ \boldsymbol{\psi} \in C \text{ a.e. on } \Gamma_2,\ \det \nabla\boldsymbol{\psi} > 0 \text{ a.e. in } \Omega\}$$

is nonempty. Let the linear form L *defined by* $L(\boldsymbol{\psi}) = \int_\Omega \boldsymbol{f}\cdot\boldsymbol{\psi}\,\mathrm{d}x + \int_{\Gamma_1} \boldsymbol{g}\cdot\boldsymbol{\psi}\,\mathrm{d}a$ *be continuous on* $\boldsymbol{W}^{1,p}(\Omega)$, *and assume that*

$$L(\boldsymbol{d}) = \left\{\int_\Omega \boldsymbol{f}\,\mathrm{d}x + \int_{\Gamma_1} \boldsymbol{g}\,\mathrm{d}a\right\}\cdot \boldsymbol{d} > 0 \quad \text{for all } \boldsymbol{d} \in D\,,$$

where

$$D := cl\left\{ d \in \mathbb{R}^3;\ d = \lim_{k\to\infty} \frac{\hat{\varphi}^k}{|\hat{\varphi}^k|},\ \text{where } \hat{\varphi}^k := \frac{\int_{\Gamma_2} \varphi^k \,\mathrm{d}a}{\int_{\Gamma_2} \mathrm{d}a},\ \varphi^k \in \Phi,\ \lim_{k\to\infty} |\hat{\varphi}^k| = +\infty \right\}.$$

Finally, let $I(\psi) = \int_\Omega \hat{W}(x, \nabla\psi(x))\,\mathrm{d}x - L(\psi)$, *and assume that* $\inf_{\psi\in\Phi} I(\psi) < +\infty$. *Then there exists at least one function* φ *such that*

$$\varphi \in \Phi \text{ and } I(\varphi) = \inf_{\psi\in\Phi} I(\psi).$$

Proof. (i) By the assumed coerciveness of the stored energy function, we deduce that

$$I(\psi) \geq \alpha\{|\psi|^p_{1,p,\Omega} + |\mathbf{Cof}\,\nabla\psi|^q_{0,q,\Omega} + |\det \nabla\psi|^r_{0,r,\Omega}\} + \beta \operatorname{vol} \Omega - L(\psi)$$

for all $\psi \in \Phi$. The difficulty here is that, since we have neither a boundary condition of place $\psi = \varphi_0$ on Γ_0 (as in Theorems 7.7-1 and 7.8-1) nor an additional constraint such as $\int_\Omega \psi(x)\,\mathrm{d}x = e$ (as in Theorem 7.7-2) in the set Φ, we cannot use the generalized Poincaré inequality to replace the semi-norm $|\cdot|_{1,p,\Omega}$ by the norm $\|\cdot\|_{1,p,\Omega}$ in the lower bound for $I(\psi)$, $\psi \in \Phi$. We shall thus resort to another argument, by showing that

$$\varphi^k \in \Phi \text{ and } \lim_{k\to\infty} \|\varphi^k\|_{1,p,\Omega} = +\infty \Rightarrow \gamma := \liminf_{k\to\infty} \frac{I(\varphi^k)}{\|\varphi^k\|_{1,p,\Omega}} > 0.$$

(ii) Assume that this implication is not true. Then there exists a sequence (φ^k) with

$$\varphi^k \in \Phi,\ \lim_{k\to\infty} \|\varphi^k\|_{1,p,\Omega} = +\infty,\ I(\varphi^k) = \gamma^k \|\varphi^k\|_{1,p,\Omega},$$
$$\text{and } \lim_{k\to\infty} \gamma^k \leq 0.$$

Using the lower bound on the energy, we deduce

$$\gamma^k \|\boldsymbol{\varphi}^k\|_{1,p,\Omega} \geq \alpha |\boldsymbol{\varphi}^k|^p_{1,p,\Omega} + \beta \operatorname{vol} \Omega - \left\{ \int_\Omega \boldsymbol{f} \cdot \boldsymbol{\varphi}^k \, dx + \int_{\Gamma_1} \boldsymbol{g} \cdot \boldsymbol{\varphi}^k \, da \right\},$$

and this inequality may be rewritten as

$$\gamma^k \geq \alpha \|\boldsymbol{\varphi}^k\|^{p-1}_{1,p,\Omega} |\tilde{\boldsymbol{\varphi}}^k|^p_{1,p,\Omega} + \frac{\beta \operatorname{vol} \Omega}{\|\boldsymbol{\varphi}^k\|_{1,p,\Omega}} - \left\{ \int_\Omega \boldsymbol{f} \cdot \tilde{\boldsymbol{\varphi}}^k \, dx + \int_{\Gamma_1} \boldsymbol{g} \cdot \tilde{\boldsymbol{\varphi}}^k \, da \right\},$$

with

$$\tilde{\boldsymbol{\varphi}}^k = \frac{\boldsymbol{\varphi}^k}{\|\boldsymbol{\varphi}^k\|_{1,p,\Omega}}.$$

From the relations $\alpha > 0$, $\lim_{k\to\infty} \gamma^k \leq 0$, $\lim_{k\to\infty} \|\boldsymbol{\varphi}^k\|_{1,p,\Omega} = +\infty$, $p > 1$, and $\|\tilde{\boldsymbol{\varphi}}^k\|_{1,p,\Omega} = 1$, we deduce that

$$\lim_{k\to\infty} |\tilde{\boldsymbol{\varphi}}^k|_{1,p,\Omega} = 0.$$

(iii) By the generalized Poincaré inequality, we also have

$$\|\boldsymbol{\psi}\|^p_{1,p,\Omega} \leq c_1 \left\{ |\boldsymbol{\psi}|^p_{1,p,\Omega} + \left| \int_{\Gamma_2} \boldsymbol{\psi} \, da \right|^p \right\} \quad \text{for all } \boldsymbol{\psi} \in \boldsymbol{W}^{1,p}(\Omega).$$

Applying this inequality to the particular functions $\|\boldsymbol{\varphi}^k\|^{-1}_{1,p,\Omega}(\boldsymbol{\varphi}^k - \hat{\boldsymbol{\varphi}}^k)$, with

$$\hat{\boldsymbol{\varphi}}^k = \frac{1}{\int_{\Gamma_2} da} \int_{\Gamma_2} \boldsymbol{\varphi}^k \, da,$$

we obtain

$$\frac{\|\boldsymbol{\varphi}^k - \hat{\boldsymbol{\varphi}}^k\|_{1,p,\Omega}}{\|\boldsymbol{\varphi}^k\|_{1,p,\Omega}} \leq c_2^{1/p} |\tilde{\boldsymbol{\varphi}}^k|_{1,p,\Omega},$$

and thus $\lim_{k\to\infty} |\tilde{\boldsymbol{\varphi}}^k|_{1,p,\Omega} = 0$ implies

$$\lim_{k\to\infty} \frac{\|\varphi^k - \hat{\varphi}^k\|_{1,p,\Omega}}{\|\varphi^k\|_{1,p,\Omega}} = 0 .$$

Since $\lim_{k\to\infty} \|\varphi^k\|_{1,p,\Omega} = +\infty$ and

$$\left| \frac{\|\hat{\varphi}^k\|_{1,p,\Omega}}{\|\varphi^k\|_{1,p,\Omega}} - 1 \right| \leq \frac{\|\varphi^k - \hat{\varphi}^k\|_{1,p,\Omega}}{\|\varphi^k\|_{1,p,\Omega}} ,$$

we thus conclude that

$$\lim_{k\to\infty} \|\hat{\varphi}^k\|_{1,p,\Omega} = +\infty .$$

(iv) Each $\hat{\varphi}^k \in W^{1,p}(\Omega)$ is a constant function, so that it can be identified with a vector of $\mathbb{R}^3$. Since the relations $d = \lim_{k\to\infty} \hat{\varphi}^k/|\hat{\varphi}^k|$ and $\lim_{k\to\infty} |\hat{\varphi}^k| = +\infty$ may thus be also understood as convergences in $W^{1,p}(\Omega)$, and since there exists a constant c_3 such that

$$c_3 > 0 \text{ and } c_3^{-1}|d| \leq \|d\|_{1,p,\Omega} \leq c_3|d| \quad \text{for all } d \in \mathbb{R}^3 ,$$

the sequence (φ^k) is a "candidate" for defining an element of the set D, as defined in the theorem. Since the constant functions

$$\delta^k = \frac{\hat{\varphi}^k}{\|\varphi^k\|_{1,p,\Omega}}$$

are bounded in $W^{1,p}(\Omega)$ independently of k by step (iii), there exists a subsequence (δ^l) such that

$$\lim_{k\to\infty} \delta^l = \lambda d \text{ in both } \mathbb{R}^3 \text{ and } W^{1,p}(\Omega) \text{ with } d \in D \text{ and } \lambda > 0 ,$$

and thus we also have

$$\lim_{l\to\infty} \tilde{\varphi}^l = \lambda d \text{ in } W^{1,p}(\Omega) ,$$

by (iii). From step (ii), we deduce that

$$\begin{aligned}\lambda L(d) &= \lambda\left\{\int_\Omega f \cdot d \,\mathrm{d}x + \int_{\Gamma_1} g \cdot d \,\mathrm{d}a\right\} \\ &= \lim_{l\to\infty}\left\{\int_\Omega f \cdot \varphi^l \,\mathrm{d}x + \int_{\Gamma_1} g \cdot \varphi^l \,\mathrm{d}a\right\}\end{aligned}$$

$$\geqslant \limsup_{l\to\infty} \{\alpha \|\boldsymbol{\varphi}^l\|_{1,p,\Omega}^{p-1} |\tilde{\boldsymbol{\varphi}}^l|_{1,p,\Omega}^p\} + \lim_{l\to\infty} \left\{ \frac{\beta \operatorname{vol} \Omega}{\|\boldsymbol{\varphi}^l\|_{1,p,\Omega}} - \gamma^l \right\} \geqslant 0 ,$$

but this inequality contradicts the assumption that $L(\boldsymbol{d}) < 0$ for all $\boldsymbol{d} \in D$, and thus the implication announced in (i) is true.

(v) Any infimizing sequence $(\boldsymbol{\varphi}^k)$ of the total energy with $\boldsymbol{\varphi}^k \in \boldsymbol{\Phi}$ is thus necessarily bounded in the space $\boldsymbol{W}^{1,p}(\Omega)$, and the proof then proceeds as the proof of Theorem 7.7-1. ■

Remark. In linearized elasticity, Signorini [1933, 1959] first posed the problem, since then called the *Signorini problem*, of finding equilibrium positions of a body resting without friction on a horizontal plane. Then Fichera [1964] initiated a long series of mathematical studies of analogous problems. In this respect, the inequalities $\{\int_\Omega \boldsymbol{f}\,dx + \int_{\Gamma_1} \boldsymbol{g}\,da\} \cdot \boldsymbol{d} < 0$ for all $\boldsymbol{d} \in D$ of Theorem 7.8-2 are the natural extensions to nonlinear elasticity of analogous inequalities used by Duvaut & Lions [1972, p. 149] and Fichera [1972, p. 413] for proving existence theorems in linearized elasticity. An historical perspective is given in Fichera [1977]. ■

Motivated by the discussion given in Sect. 5.3, we call **set of directions of escape** the set D defined in Theorem 7.8-2. In the light of this definition, let us review some examples: Suppose first that *the set C is bounded*. Then *the set D of directions of escape is empty in this case*, since we cannot have $\lim \|\hat{\boldsymbol{\varphi}}^k\|_{1,p,\Omega} = +\infty$ with $\boldsymbol{\varphi}^k \in \boldsymbol{\Phi}$: For if $\boldsymbol{\psi} \in \boldsymbol{\Phi}$, then $\boldsymbol{\psi} \in C$ on Γ_2 by definition and thus the Euclidean norms $|\hat{\boldsymbol{\psi}}|$, $\hat{\boldsymbol{\psi}} \in \boldsymbol{\Phi}$, are uniformly bounded since

$$\sup_{\boldsymbol{\psi}\in\boldsymbol{\Phi}} |\hat{\boldsymbol{\psi}}| \leqslant \frac{\sqrt{3}}{\int_{\Gamma_2} da} \max_i \int_{\Gamma_2} |\psi_i|\, da < +\infty .$$

Hence the assumption that $L(\boldsymbol{d}) < 0$ for all $\boldsymbol{d} \in D$ is superfluous if C is bounded, and accordingly, there always exists a minimizer of the energy in this case (provided of course all the other assumptions of Theorem 7.8-2 are satisfied).

Suppose next that the set C is unbounded and that there is a closed convex cone C_1 with vertex $\boldsymbol{a}_1$ such that $C \subset C_1$ and that C_1 is not a half space. If $\boldsymbol{\psi} \in \boldsymbol{\Phi}$, then $\boldsymbol{\psi} \in C$ on Γ_2 by definition, and so $\boldsymbol{\psi} \in C_1$ on Γ_2, whence $\boldsymbol{\psi} \in \boldsymbol{\Phi}$ implies $\hat{\boldsymbol{\psi}} \in C_1$. The relations $\boldsymbol{d} = \lim_{k\to\infty} \hat{\boldsymbol{\varphi}}^k / |\hat{\boldsymbol{\varphi}}^k|$ with $|\boldsymbol{d}| = 1$ and $\lim_{k\to\infty} |\boldsymbol{\varphi}^k| = +\infty$ with $\boldsymbol{\varphi}^k \in \boldsymbol{\Phi}$ force the vector $\boldsymbol{d}$ to belong to the cone $\{C_1 - \boldsymbol{a}_1\}$, i.e., the cone C_1 translated by the vector $-\boldsymbol{a}_1$.

Therefore we have in this case

$$D \subset S_1 \cap \{C_1 - a_1\},$$

where S_1 denotes the unit sphere in $\mathbb{R}^3$.

Suppose finally that C is unbounded and that there exists a closed convex cone C_2 with a nonempty interior such that $C_2 \subset C$. Denoting by a_2 the vertex of the cone C_2, we have in this case

$$S_1 \cap \{C_2 - a_2\} \subset D.$$

To see this, let $d \in \text{int } C_2$ with $|d| = 1$. Then the particular functions $\varphi^k(x) = x + kd$ for $k \geq 0$ large enough, can be used in the definition of the set D.

Observe that *if there exists a closed convex cone* C_0 *with a nonempty interior, and which is not a half-space, such that* (Fig. 7.8-1)

$$C_0 = C_1 - a_1 = C_2 - a_2,$$

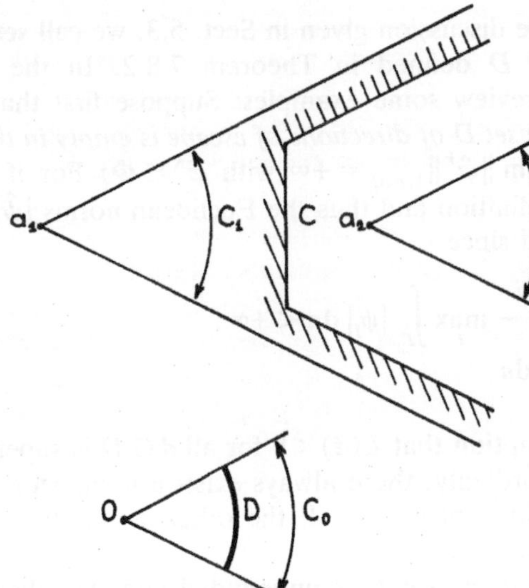

Fig. 7.8-1. A situation in which the set D of directions of escape can be easily identified: There exists a closed convex cone C_0 with a nonempty interior, and which is not a half-space, such that $C_1 = a_1 + C_0 \supset C \supset C_2 = C_0 + a_2$. Then D is intersection of C_0 and the unit sphere of $\mathbb{R}^3$.

Otherwise for a certain $\varepsilon > 0$, there are sequences (x^m), (y^m), (δ^m) satisfying

$$x^m \in \varphi(\bar{\Omega}),\ y^m \notin O_\varepsilon,\ |x^m - y^m| < \delta^m \text{ and } \lim_{m\to\infty} \delta^m = 0\,.$$

Since the set $\varphi(\bar{\Omega})$ is compact, there exists a subsequence (x^n) that converges to an element $x \in \varphi(\bar{\Omega})$, and therefore the subsequence (y^n) also converges to the same element x since $\lim_{n\to\infty}|y^n - x^n| = 0$. But the limit of the sequence (y^n) should also belong to the closed set $\mathbb{R}^3 - O_\varepsilon$, which is impossible by virtue of the inclusion $\varphi(\bar{\Omega}) \subset O_\varepsilon$. Hence $\bigcup_{x\in\varphi(\bar{\Omega})} B_x(\delta(\varepsilon)) \subset O_\varepsilon$ for some $\delta(\varepsilon) > 0$ and thus there exists an integer $l_0 = l_0(\delta) = l_0(\varepsilon)$ such that

$$\varphi^l(\bar{\Omega}) \subset O_\varepsilon \quad \text{for all } l \geq l_0\,,$$

since the sequence (φ^l) converges uniformly to φ. Since $\varphi^l \in \Phi$, we thus have

$$\int_\Omega \det \nabla\varphi^l \,dx \leq \operatorname{vol} \varphi^l(\bar{\Omega}) \leq \operatorname{vol} O_\varepsilon \quad \text{for all } l \geq l_0\,.$$

Since the sequence $(\det \nabla\varphi^l)$ weakly converges to $\det \nabla\varphi$ in the space $L^r(\Omega)$ (cf. step (iii) of the proof of Theorem 7.7-1), we conclude that

$$\int_\Omega \det \nabla\varphi \,dx = \lim_{l\to\infty} \int_\Omega \det \nabla\varphi^l \,dx \leq \operatorname{vol} O_\varepsilon\,,$$

and since $\operatorname{vol} O_\varepsilon = \operatorname{vol} \varphi(\bar{\Omega}) + \operatorname{vol}(O_\varepsilon - \varphi(\bar{\Omega}))$, with $\operatorname{vol}(O_\varepsilon - \varphi(\bar{\Omega})) < \varepsilon$, we infer from the arbitrariness of ε that

$$\int_\Omega \det \nabla\varphi \,dx \leq \operatorname{vol} \varphi(\bar{\Omega}) = \operatorname{vol} \varphi(\Omega)\,,$$

and thus we have established that $\varphi \in \Phi$.

Finally, let us show that any $\varphi \in \Phi$, hence in particular any minimizer of the total energy, is almost everywhere injective. By a result of Marcus & Mizel [1973] (see also Vodopyanov, Goldshtein & Reshetnyak [1979], Bojarski & Iwaniec [1983]), if Ω is a bounded open subset of $\mathbb{R}^3$ and $\varphi \in W^{1,p}(\Omega)$ with $p > 3$,

$$\int_\Omega |\det \nabla\varphi| \,dx = \int_{\varphi(\Omega)} \operatorname{card}\{\varphi^{-1}(x')\} \,dx'\,,$$

whenever one of the two sides is meaningful (this is thus the extension to mappings with values in Sobolev spaces of the same relation for smooth maps, already used in (Theorem 5.6-1). Combining this relation with the definition of the set $\boldsymbol{\Phi}$, we obtain

$$\operatorname{vol} \boldsymbol{\varphi}(\Omega) = \int_{\boldsymbol{\varphi}(\Omega)} \mathrm{d}x' \leqslant \int_{\boldsymbol{\varphi}(\Omega)} \operatorname{card}\{\boldsymbol{\varphi}^{-1}(x')\}\, \mathrm{d}x' = \int_{\Omega} \det \boldsymbol{\nabla}\boldsymbol{\varphi}\, \mathrm{d}x \leqslant \operatorname{vol} \boldsymbol{\varphi}(\Omega)\,,$$

which shows that

$$\operatorname{card}\{\boldsymbol{\varphi}^{-1}(x')\} = 1 \quad \text{for almost all } x' \in \boldsymbol{\varphi}(\Omega)\,,$$

and the proof is complete. ∎

Remarks. (1) There is no need to impose the injectivity condition $\int_{\Omega} \det \boldsymbol{\nabla}\boldsymbol{\psi}\, \mathrm{d}x \leqslant \operatorname{vol} \boldsymbol{\psi}(\Omega)$ to the admissible deformation $\boldsymbol{\psi}$ in the case of the *pure displacement problem*, provided the boundary function $\boldsymbol{\varphi}_0 : \Gamma \to \mathbb{R}^3$ and the stored energy satisfy suitable hypotheses, since the injectivity of the minimizers of the energy is automatic in this case (Ball [1981b, Theorem 3]; cf. Exercise 7.13).

(2) The link between the invertibility of equilibrium solutions in nonlinear elasticity (hence in particular of the minimizers of the energy) and the behavior of the stored energy function as $\det \boldsymbol{F} \to 0^+$ has been studied by Antman [1976], Antman & Brezis [1978], Ball [1981b], in the case of one-dimensional problems.

(3) We could likewise treat the problem where *several noninterpenetrating bodies* are simultaneously considered. In this case, the set Ω has a finite number of connected components, and the constituting material may vary according to which connected component of the set Ω is considered, i.e., the stored energy functions may differ from one connected component to another. ∎

Using delicate regularity results of Šverák [1987], Tang [1987] has extended the existence and almost everywhere injectivity results of Theorem 7.9-1 to the situation where one only has $p > 2$. The image $\boldsymbol{\varphi}(\Omega)$ must then be properly redefined, since $\boldsymbol{\varphi}$ need not be continuous in this case.

7.10. EPILOGUE: SOME OPEN PROBLEMS

A *major open problem* consists in stating sufficient conditions that would imply additional *regularity of the minimizers* found in John Ball's approach. For instance, under strict growth assumptions on the integrand $\hat{W}$ and its partial derivatives (Exercise 7.14), or under the *a priori* assumption that the minimizer is in the space $W^{1,\infty}(\Omega)$ (Exercise 7.15), one can show that a minimizer $\varphi \in W^{1,p}(\Omega)$ of the integral

$$I(\psi) = \int_\Omega \hat{W}(x, \nabla\psi(x))\, dx - \left\{\int_\Omega f \cdot \psi \, dx + \int_{\Gamma_1} g \cdot \psi \, da\right\}$$

(in the case of a displacement–traction problem, for definiteness) is a *weak solution of the Euler–Lagrange equations* (Sect. 4.1), in the sense that

$$\int_\Omega \frac{\partial \hat{W}}{\partial F}(x, \nabla\varphi(x)) : \nabla\theta(x)\, dx = \int_\Omega f \cdot \theta \, dx + \int_{\Gamma_1} g \cdot \theta \, da$$

for all $\theta \in \mathscr{C}^\infty(\bar{\Omega})$ that vanish on Γ_0. However, these assumptions, whose raison d'être is to insure that the the integral I is differentiable at φ, so that the above equations simply express that $I'(\varphi)\theta = 0$, are not realistic: The growth assumptions are too severe for actual stored energy functions; no acceptable set of assumptions is as yet known that would guarantee the $W^{1,\infty}(\Omega)$-regularity of the minimizers.

As regards the smoothness of the minimizers, a key rôle is played by the *ellipticity conditions* mentioned in Sect. 5.10, although this rôle is not yet fully understood. In this direction, Ball [1980] has shown that the strong ellipticity condition is necessary for the regularity of a minimizer; see also Ball [1981a]. But it is not sufficient: even in dimension one, there are examples of integrals of the form

$$J(v) = \int_I h(x, v(x), v'(x))\, dx \,,$$

with a smooth enough integrand h satisfying $h(x, v, p) \geq 0$, and $(\partial^2 h/\partial p^2)(x, v, p) > 0$ (the one-dimensional analog of the strong ellipticity condition) for all $(x, v, p) \in I \times \mathbb{R} \times \mathbb{R}$, such that the minimizers in the space $W^{1,1}(I)$ of absolutely continuous functions over $I =]0, 1[$ do *not* satisfy the associated Euler–Lagrange equation; see Ball & Mizel [1984, 1985]. General references about the relation between ellipticity conditions and regularity of solutions have been given at the end of Sect. 5.10.

These questions are also related to the surprising *Lavrentiev phenomenon*, discovered by Lavrentiev [1926]: Again within the class of one-dimensional problems of the calculus of variations, it is possible to construct examples such that

$$\inf_{v \in \Phi_\infty} J(v) > \inf_{v \in \Phi_1} J(v),$$

where

$$\Phi_p = \{v \in W^{1,p}(I);\ v(0) = -\alpha,\ v(1) = \alpha\},\ p = 1, \infty;$$

in addition, there exists a minimizer in the set Φ_1. In this direction, see Ball [1984], Ball & Mizel [1985], Ball & Knowles [1987], Sivaloganathan [1986b], Podio-Guidugli, Vergara-Caffarelli & Virga [1987], Dacorogna [1987].

On the other hand, unqualified regularity is not to be expected in all cases, since there are physical situations, such as *cavitation*, *fracture*, or *crystal twinning*, where the minimizers $\varphi \in W^{1,p}(\Omega)$ should not lie in smoother spaces, such as $W^{1,\infty}(\Omega)$ or $\mathscr{C}^1(\bar{\Omega})$. In this respect, Ball [1982] has shown by means of striking three-dimensional examples that indeed a minimizer is not always smooth; in particular, there may be points of discontinuity, corresponding to *cavitation* or *fracture*. Hence in this case, the exponent p that appears in the coerciveness inequality cannot be >3, since $W^{1,p}(\Omega) \hookrightarrow C^0(\bar{\Omega})$ for $p > 3$ (the modeling of cavitation also rules out the "limit" exponent $p = 3$; cf. Ball [1978]). For these and related questions, see also Noll & Podio-Guidugli [1986], Sivaloganathan [1986a, 1986b], Podio-Guidugli, Vergara-Caffarelli & Virga [1986], Stuart [1985], Šverák [1987].

Discontinuities in the *gradients* of the solutions at material surfaces occur in the phenomenon of *crystal twinning*. The associated mathematical model is substantially complicated by the unusual features that the stored energy functions $\hat{W}$ of solid crystals must exhibit, because of special geometrical invariances related to the lattice structure of the crystal: In particular the usual properties of weak lower semi-continuity of the integral $\int_\Omega \hat{W}(\nabla\psi(x))\,dx$ and of coerciveness of the stored energy function (which must remain bounded in some directions) are lost. For these questions see Ericksen [1979, 1986, 1987], James [1981], Kinderlehrer [1987a, 1987b], Pitteri [1984, 1985], Fonseca [1987a, 1987b, 1987c], Chipot & Kinderlehrer [1987].

Discontinuities in the gradients of the solutions also occur in some solid–solid phase transformations known as *martensitic transformations*, in which plane interfaces separate a homogeneous phase, called *austenite*, from a fine mixture of twins of the other phase, called *martensite*. As shown by Ball & James [1987], a fascinating feature of this phenomenon is that the associated deformation can no longer be interpreted as a minimizer of the energy, but rather as a weak limit of a minimizing sequence of the energy, which converges to a deformation which is not a minimizer of the energy! Such solutions are reminiscent of the "generalized curves" introduced by Young [1937, 1942] (cf. also Young [1980] for a recent account) in order to handle integrals that do not have minimizers in the usual sense. Similar ideas are found in the theory of homogenization, that we next briefly introduce.

Of considerable interest is the analysis of *periodic* or *cellular elastic materials*, whose stored energy function is of the form

$$\hat{W}^{\varepsilon}:(x, \boldsymbol{F}) \in \bar{\Omega} \times \mathbb{M}^3_+ \to \hat{W}^{\varepsilon}(x, \boldsymbol{F}) = \hat{W}\left(\frac{x}{\varepsilon}, \boldsymbol{F}\right) \in \mathbb{R},$$

where the function $\hat{W}$ is periodic on $[0, 1]^3$ with respect to the first variable, and ε is a "small" parameter. Typical questions are then to examine whether, and in which sense, the associated functionals

$$\int_{\Omega} \hat{W}\left(\frac{x}{\varepsilon}, \nabla\varphi(x)\right) \mathrm{d}x$$

converge as $\varepsilon \to 0$ to a simpler *homogenized functional* of the form $\int_{\Omega} \hat{W}^0(\nabla\varphi(x))\, \mathrm{d}x$, where $\hat{W}^0$ is an appropriate *homogenized stored energy function*; to examine whether, and in which sense, the minimizers of the energy converge as $\varepsilon \to 0$ to a minimizer of the homogenized energy functional; to obtain an explicit expression for $\hat{W}^0$, or equivalently, for the associated response function. This last question, which is fundamental from a practical viewpoint, is especially challenging, as recognized in the pioneering work of Gent & Thomas [1959], or more recently, by Gibson & Ashby [1982].

The study of these problems, or of analogous problems originating from other fields, is the object of *homogenization theory*, a field which has received considerable attention during the past decade. See notably Babuška [1974, 1975], De Giorgi [1975, 1983], Spagnolo [1976], Tartar [1977, 1978, 1983], Bensoussan, Lions & Papanicolaou [1978], Kozlov, Oleinik, Zhikov & Kha Ten Ngoan [1979], Sanchez-Palencia [1980], Attouch [1984], for general references. Interesting applications to three-dimensional linearized elasticity have been obtained by Duvaut [1978], Cioranescu & Saint Jean Paulin [1979, 1986], Oleinik [1984], Caillerie [1987], Francfort & Murat [1986] (references to homogenization of two-dimensional structures are given in Vol. II). The extension of these results to nonlinear elasticity, which presents special difficulties, is only beginning; we mention in this respect the work of Müller [1987].

A problem closely related to the occurrence of singularities, as in the onset of fracture or cavitation, can be stated as follows: Physical evidence shows that actual problems have *solutions that can be continuously followed*, at least in some "intervals of moderate variations" of the applied loads; in particular, solutions close to those found in linearized elasticity should be expected in most cases for small enough applied forces. Therefore the question arises as to whether "generic" results of this kind can be obtained, in particular within the existence theory described in this chapter; "generic" means that such results can be only expected to hold in "most" situations, in order to be also compatible with phenomena such as bifurcations, snap-through, etc. We recall that the local existence theory of smooth solutions developed in Chapter 6 does provide a positive answer to these questions, but that it is unfortunately limited to very special situations, such as pure displacement problems.

Substantial progress has been recently made in the mathematical analysis of *plasticity* when the "elastic" part of the model is linear, i.e., when only the linearized strain tensor $\boldsymbol{e}(\boldsymbol{u}) = \frac{1}{2}(\boldsymbol{\nabla u}^{\mathrm{T}} + \boldsymbol{\nabla u})$ is considered. Serious mathematical difficulties already arise, for the energy integrals are of the form $\int_{\Omega} f(\mu)$, where μ is a *bounded measure* on Ω, and f is a convex function which has at most a *linear growth at infinity*. Motivated by this problem, Demengel [1985b] and Demengel & Temam [1984, 1986] have thoroughly analyzed the notion of "convex function of a measure", which has proved to be a powerful tool for establishing existence results in this case. For related ideas and results, see Suquet [1979, 1981], Temam & Strang [1980a, 1980b], Kohn & Temam [1983], Anzellotti [1983a, 1984, 1985a, 1985b], Anzellotti & Giaquinta [1980, 1982], Del Piero [1985a], Hadhri [1985, 1986]. Thorough introductions to plasticity are given in Duvaut & Lions [1972], Nečas & Hlaváček [1981], Temam [1983], for the mathematical aspects, and Salençon [1983], for the experimental and mechanical aspects.

The extension to plasticity models where the elastic part is that of genuine nonlinear elasticity would be of considerable interest, since elasto-plastic models are undoubtedly more accurate: In a metal such as steel for instance, plasticity effects coexist with nonlinearly elastic effects even for moderate ("of the order of 3%") deformations. By contrast, materials such as polymers, foam, rubber, are accurately modeled by elasticity alone in a much wider range of deformations ("of the order of 100%").

Another open problem of importance consists in modeling and mathematically analyzing *contact with friction*. Introductions to this question may be found in Duvaut & Lions [1972, Chapter 3], who give various existence theorems in linearized elasticity, Moreau [1974], who models friction via "pseudo-potentials", Oden & Martins [1985], who thoroughly describe the physics and the computational methods for friction phenomena. See also Panagiotopoulos [1975, 1985], Nečas, Jarušek & Haslinger [1980], Villagio [1980], Cocu [1984], Del Piero [1985b], Frémond [1985], Kikuchi & Oden [1987]. We recall that the modeling and existence theory for contact *without* friction can be satisfactorily treated for hyperelastic materials (Sects. 5.3 and 7.8).

Pour finir en beauté, we mention the fundamental problem of finding existence results for the time-dependent problem of *nonlinear elastodynamics*. For a pure displacement problem and a homogeneous elastic material, for definiteness, this problem consists in solving the following initial value problem:

$$\begin{cases} \dfrac{\partial^2 \varphi_i}{\partial x}(x,t) - a_{ijkl}(\nabla\varphi(x,t))\dfrac{\partial^2 \varphi_k}{\partial x_j \partial x_l}(x,t) = f_i(x,t), \\ \qquad\qquad\qquad\qquad\qquad (x,t) \in \Omega \times \,]0,+\infty[\,, \\ \varphi(x,0) = \chi_0(x),\ \dfrac{\partial \varphi}{\partial t}(x,0) = \chi_1(x),\ x \in \bar{\Omega}\,, \\ \varphi(x,t) = \varphi_0(x,t),\ x \in \partial\Omega,\ t \geqslant 0\,, \end{cases}$$

where $\varphi(x,t) = (\varphi_i(x,t))$ is the unknown deformation at $x \in \bar{\Omega}$ and time $t \geqslant 0$, the tensor $(a_{ijkl}(F))$ is the elasticity tensor (Sect. 5.9), $f(x,t) = (f_i(x,t))$ is the density of the applied body force, $\chi_0 : \bar{\Omega} \to \mathbb{R}^3$, $\chi_1 : \bar{\Omega} \to \mathbb{R}^3$, and $\varphi_0 : \partial\Omega \times [0,+\infty] \to \mathbb{R}^3$ are prescribed initial and boundary values. "Local in time" existence results, i.e., for small enough $t > 0$, of smooth solutions have been obtained by Hughes, Kato & Marsden [1976] when $\Omega = \mathbb{R}^3$ and by Kato [1979], Chen & von Wahl [1982], Dafermos & Hrusa [1985], Sablé-Tougeron [1987], when Ω is a domain in $\mathbb{R}^3$. These results have been recently extended to the incompressible case, where the constraint $\det \nabla\varphi(x,t) = 1$, $x \in \bar{\Omega}$, $t \geqslant 0$, is added, by Schochet [1985], Ebin & Saxton [1986] when $\Omega = \mathbb{R}^3$, and by Hrusa & Renardy [1986] when Ω is a domain in $\mathbb{R}^3$.

Characteristic feature of nonlinear hyperbolic equations, such as those of nonlinear elastodynamics, is that whatever the smoothness of the initial data may be, shock waves and other discontinuities rapidly develop with time (Lax [1964]; see also the discussion given in Dafermos [1986]) and thus the existence of smooth solutions can only be proved for small times. The occurrence of these singularities render the mathematical analysis very difficult. Deep existence results have nevertheless been obtained "for large times", but only for one space variable, by Glimm [1965], Dafermos [1973], Diperna [1983, 1985a, 1985b].

EXERCISES

7.1. Let $h:\mathbb{M}^n \to [\beta, +\infty[$ be a continuous function. Then, for each $1 \leqslant p < \infty$, the functional

$$I: \boldsymbol{\varphi} \in \boldsymbol{W}^{1,p}(\Omega) = (W^{1,p}(\Omega))^n \to I(\boldsymbol{\varphi}) = \int_\Omega h(\boldsymbol{\nabla}\boldsymbol{\varphi}(x))\,\mathrm{d}x$$

is well defined (since the integrand is a bounded from below Carathéodory function, cf. the proof of Theorem 7.3-1); note that $I(\boldsymbol{\varphi})$ may be equal to $+\infty$.

(1) By Theorem 7.3-1, the functional I is sequentially weakly lower semi-continuous if the integrand is convex. Show that the converse holds if $n = 1$; this is a special case of *Tonelli's theorem* (Tonelli [1920]).

Hint. Let $\Omega =]0, 1[$; show that, for $\lambda \in]0, 1[$, $a \in \mathbb{R}$, $b \in \mathbb{R}$, the sequence $(\zeta_k)_{k \geqslant 1}$ defined by

$$\zeta_k(x) = \begin{cases} a \text{ for } \dfrac{j}{k} \leqslant x < (j+\lambda)\dfrac{1}{k}\,, \\[2ex] b \text{ for } (j+\lambda)\dfrac{1}{k} \leqslant x < \dfrac{(j+1)}{k}\,,\ 0 \leqslant j \leqslant k-1\,, \end{cases}$$

weakly converges in $L^1(\Omega)$ to the function ζ defined by

$$\zeta(x) = \lambda a + (1-\lambda) b,\ 0 \leqslant x \leqslant 1\,.$$

(2) Show that if the functional I is sequentially weakly lower semi-continuous, the integrand h is *quasiconvex* (Morrey [1952]): For all $\boldsymbol{F} \in \mathbb{M}^n$ and all functions $\boldsymbol{\theta} \in \mathcal{D}(\Omega) = (\mathcal{D}(\Omega))^n$,

$$\frac{1}{\mathrm{d}x\text{-meas}\,\Omega} \int h(\boldsymbol{F} + \boldsymbol{\nabla}\boldsymbol{\theta}(x))\,\mathrm{d}x \geqslant h(\boldsymbol{F})\,.$$

Remarks. See Serrin [1961, Theorem 12], Marcellini & Sbordone [1980], for generalizations of Tonnelli's theorem; Meyers [1965], Ball [1977, Theorem 6.1 and Corollary 6.1.1], Acerbi & Fusco [1984], for various extensions of (2). Quasiconvexity was already introduced in Exercise 5.14 for integrands defined on the set $\mathbb{M}^3_+$. ■

7.2. (1) Show that the functional

$$J: v \in W^{1,4}(0,1) \to J(v) = \int_0^1 \{v'(x) + \tfrac{1}{2}(v'(x))^2\}^2\,\mathrm{d}x$$

is not weakly lower semi-continuous (see Nečas [1976] for a similar example).

Hint. Consider the sequence $(v_k)_{k\geq 1}$, where

$$v_k(x) = \begin{cases} -2x + \dfrac{j}{k} \text{ for } \dfrac{j}{k} \leq x < (j + \frac{1}{2})\dfrac{1}{k}, \\ -\dfrac{(j+1)}{k} \text{ for } (j+\frac{1}{2})\dfrac{1}{k} \leq x < \dfrac{(j+1)}{k}, 0 \leq j \leq k-1. \end{cases}$$

(2) Let $U = \{v \in W^{1,4}(0,1);\ v(0) = 0,\ v(1) = -1\}$. Find all functions $u \in U$ that satisfy $J(u) = \inf_{v\in U} J(v)$.

7.3. With the same notation and assumptions as in Theorem 6.3-5, let

$$J(\boldsymbol{v}) = \tfrac{1}{2}B(\boldsymbol{v}, \boldsymbol{v}) - L(\boldsymbol{v}),$$

and

$$\boldsymbol{U} = \{\boldsymbol{v} \in \boldsymbol{H}^1(\Omega);\ \boldsymbol{v} = \boldsymbol{0}\ da\text{-a.e. on } \Gamma_0,\ v_3 \geq 0\ da\text{-a.e. on } \Gamma_2\},$$

where $\Gamma = \Gamma_0 \cup \Gamma_1 \cup \Gamma_2$ is a measurable partition of the boundary Γ with area $\Gamma_0 > 0$. Show that there exists one and only one function $\boldsymbol{u}$ such that

$$\boldsymbol{u} \in \boldsymbol{U} \text{ and } J(\boldsymbol{u}) = \inf_{v\in U} J(\boldsymbol{v}).$$

Hint. Adapt Theorem 7.3-2 to this situation

7.4. This exercise is a complement to Theorem 7.5-1. (1) Show that the set

$$\{(\boldsymbol{\psi}, \boldsymbol{K}) \in \boldsymbol{W}^{1,p}(\Omega) \times \boldsymbol{L}^q(\Omega);\ \boldsymbol{K} = \mathbf{Cof}\,\boldsymbol{\nabla\psi}\},\ p \geq 2,\ q \geq 1,$$

which is weakly closed by Theorem 7.5-1, is a nonconvex subset of the space $\boldsymbol{W}^{1,p}(\Omega) \times \boldsymbol{L}^q(\Omega)$. Is it strongly closed?

(2) Let X and Y be normed vector spaces. A (possibly nonlinear) mapping $f: X \to Y$ is said to be *sequentially weakly continuous* if

$$x^k \rightharpoonup x \text{ in } X \Rightarrow f(x^k) \rightharpoonup f(x) \text{ in } Y.$$

Show that (Ball [1977, Corollary 6.2.2]) the mapping

$$\boldsymbol{\psi} \in \boldsymbol{W}^{1,p}(\Omega) \to \mathbf{Cof}\,\boldsymbol{\nabla\psi} \in \boldsymbol{L}^{p/2}(\Omega)$$

is sequentially weakly continuous if $p > 2$.

(3) For what values of p and q is the set

$$\{\psi \in W^{1,p}(\Omega);\ \mathbf{Cof}\,\nabla\psi \in L^q(\Omega)\}$$

weakly closed in the space $W^{1,p}(\Omega)$?

7.5. This exercise is a complement to Theorem 7.5-1. (1) Show that the expression

$$(\mathbf{Cof}^{\#}\,\nabla\psi)_{ij} := \partial_{i+2}(\psi_{j+2}\partial_{i+1}\psi_{j+1}) - \partial_{i+1}(\psi_{j+2}\partial_{i+2}\psi_{j+1})$$

defines a distribution when $\psi \in W^{1,p}(\Omega)$, $\frac{3}{2} \leqslant p$. Note that $\mathbf{Cof}^{\#}\,\nabla\psi = \mathbf{Cof}\,\nabla\psi$ if $p \geqslant 2$.

(2) Show that (Ball [1977, Theorem 6.2])

$$\varphi^l \rightharpoonup \varphi \text{ in } W^{1,p}(\Omega),\ p > \tfrac{3}{2} \Rightarrow (\mathbf{Cof}^{\#}\,\nabla\varphi^l)_{ij}(\theta) \to (\mathbf{Cof}^{\#}\,\nabla\varphi)_{ij}(\theta)$$

for all $\theta \in \mathcal{D}(\Omega)$; observe that the inequality $p \geqslant \frac{3}{2}$ of (1) has to be replaced by the corresponding strict inequality in (2).

7.6. This exercise is a complement to Theorem 7.6-1. (1) Show that the set

$$\{(\psi, K, \varepsilon) \in W^{1,p}(\Omega) \times L^q(\Omega) \times L^r(\Omega);\ K = \mathbf{Cof}\,\nabla\psi,\ \varepsilon = \det \nabla\psi\},$$

$$p \geqslant 2,\ \frac{1}{p} + \frac{1}{q} \leqslant 1,\ r \geqslant 1\,,$$

which is weakly closed by Theorem 7.6-1, is a nonconvex subset of the space $W^{1,p}(\Omega) \times L^q(\Omega) \times L^r(\Omega)$. It is strongly closed?

(2) Show that (Ball [1977, Corollary 6.2.2]) the mapping

$$\psi \in W^{1,p}(\Omega) \to \det \nabla\psi \in L^{p/3}(\Omega)$$

is sequentially weakly continuous if $p > 3$ (according to the definition given in Exercise 7.2).

(3) For what values of p, q, r is the set

$$\{\psi \in W^{1,p}(\Omega);\ \mathbf{Cof}\,\nabla\psi \in L^q(\Omega),\ \det \nabla\psi \in L^r(\Omega)\}$$

weakly closed in the space $W^{1,p}(\Omega)$?

(4) Let $p \geqslant 2$, $p^{-1} + q^{-1} \leqslant 1$, $r \geqslant 1$. Show that the set

$$\{\psi \in W^{1,p}(\Omega);\ \mathbf{Cof}\,\nabla\psi \in L^q(\Omega),\ \det \nabla\psi \in L^r(\Omega),\ \det \nabla\psi > 0 \text{ a.e. in } \Omega\}$$

is not convex.

7.7. This exercise is a complement to Theorem 7.6-1. (1) Show that the expression

$$\det{}^{\#} \nabla \psi := \partial_j(\psi_1 (\mathbf{Cof}^{\#} \nabla \psi)_{ij})$$

defines a distribution when $\psi \in W^{1,p}(\Omega)$, $\frac{3}{2} \leqslant p$, and $\mathbf{Cof}^{\#} \nabla \psi \in L^q(\Omega)$, $p^{-1} + q^{-1} \leqslant \frac{4}{3}$; the distribution $\mathbf{Cof}^{\#} \nabla \psi$ is defined as in Exercise 7.5. Note that $\det{}^{\#} \nabla \psi = \det \nabla \psi$ if $p \geqslant 2$ and $p^{-1} + q^{-1} \leqslant 1$.

(2) Show that (Ball [1977, Theorem 6.2]):

$$\left.\begin{aligned} &\varphi^l \rightharpoonup \varphi \text{ in } W^{1,p}(\Omega),\ p \geqslant \tfrac{3}{2} \\ &\mathbf{Cof}^{\#} \nabla \varphi^l \rightharpoonup \mathbf{Cof}^{\#} \nabla \varphi \text{ in } L^q(\Omega),\ \frac{1}{p} + \frac{1}{q} < \tfrac{4}{3}, \end{aligned}\right\} \Rightarrow$$

$$(\det{}^{\#} \nabla \varphi^l)(\theta) \to (\det{}^{\#} \nabla \varphi)(\theta)$$

for all $\theta \in \mathcal{D}(\Omega)$; observe that the inequality $p^{-1} + q^{-1} \leqslant \frac{4}{3}$ of (1) has to be replaced by the corresponding strict inequality in (2).

7.8. Consider the pure displacement problem (for definiteness) where the applied body force is the *centrifugal force* described in Sect. 2.7; we recall that such an applied body force is conservative (Exercise 2.7). Can the existence result of Theorem 7.7-1, with the same assumptions on the stored energy function, be extended to this situation?

7.9. Consider a *displacement–pressure problem*, where a boundary condition of place $\varphi = \varphi_0$ is imposed on a da-measurable subset Γ_0 of Γ with area $\Gamma_0 > 0$, and a boundary condition of pressure $\boldsymbol{Tn} = -\pi(\mathbf{Cof}\, \nabla \varphi)\boldsymbol{n}$ is imposed on $\Gamma_1 = \Gamma - \Gamma_0$, where the constant $\pi \in \mathbb{R}$ is given; we recall that such an applied surface force is conservative (Theorem 2.7-1). Show that the existence result of Theorem 7.7-1, with the same assumptions on the stored energy functions and on the applied body force, can be extended to this situation (Ball [1977, Theorem 7.11]).

7.10. In Theorem 7.3-1, the strong lower semi-continuity of the functional is established first, and its weak lower semi-continuity is then a consequence of Theorem 7.2-2. Show that an argument similar to that used in step (v) of the proof of Theorem 7.7-1 provides a more direct proof of the sequential weak lower semi-continuity.

7.11. Assume that hypothesis (b) in Theorem 7.7-1 is replaced by hypothesis (b′):

$$\left.\begin{array}{l} \boldsymbol{F}_k \to \boldsymbol{F} \text{ in } \mathbb{M}^3_+ \\ \boldsymbol{H}_k \to \boldsymbol{H} \text{ in } \mathbb{M}^3_+ \\ \delta_k \to 0^+ \end{array}\right\} \Rightarrow \mathbb{W}(x, \boldsymbol{F}_k, \boldsymbol{H}_k, \delta_k) \to +\infty$$

for almost all $x \in \Omega$. The other assumptions of Theorem 7.7-1 are unmodified.

(1) Show that hypothesis (b′) implies hypothesis (b), but that the converse implication does not necessarily hold.

(2) Let $\mathbb{W}(x, \boldsymbol{F}, \boldsymbol{H}, \delta) = +\infty$ when $\delta \leq 0$. Show that the extended function $\mathbb{W}(x, \cdot) : \mathbb{M}^3 \times \mathbb{M}^3 \times \mathbb{R} \to [\beta, +\infty]$ defined in this fashion is convex and continuous.

(3) Infer directly from Theorem 7.3-1 that

$$(\boldsymbol{\nabla}\boldsymbol{\varphi}^l, \mathbf{Cof}\, \boldsymbol{\nabla}\boldsymbol{\varphi}^l, \det \boldsymbol{\nabla}\boldsymbol{\varphi}^l) \rightharpoonup (\boldsymbol{\nabla}\boldsymbol{\varphi}, \mathbf{Cof}\, \boldsymbol{\nabla}\boldsymbol{\varphi}, \det \boldsymbol{\nabla}\boldsymbol{\varphi}) \text{ in } L^1(\Omega)$$

$$\Rightarrow \int_\Omega \hat{W}(x, \boldsymbol{\nabla}\boldsymbol{\varphi}(x))\, dx < \liminf_{l \to \infty} \int_\Omega \hat{W}(x, \boldsymbol{\nabla}\boldsymbol{\varphi}^l(x))\, dx\,,$$

so that the proof of Theorem 7.7-1 becomes closer to the proof of Theorem 7.3-2 in this case.

Remark. The stored energy function of an Ogden's material (Sect. 4.9) satisfies the stronger assumption (b′).

7.12. The object of this problem is the extension of the existence result of Theorem 7.7-1 to *incompressible materials* (Ball [1977, Theorem 7.8]).

(1) Let $\Gamma = \Gamma_0 \cup \Gamma_1$ be a da-measurable partition of Γ with area $\Gamma_0 > 0$, and let $\boldsymbol{\varphi}_0 : \Gamma_0 \to \mathbb{R}^3$ be a measurable function such that the set

$$\boldsymbol{\Phi} := \{(\boldsymbol{\psi}, \boldsymbol{K}) \in \boldsymbol{H}^1(\Omega) \times \boldsymbol{L}^2(\Omega);\ \boldsymbol{K} = \mathbf{Cof}\, \boldsymbol{\nabla}\boldsymbol{\psi}\,,\ \boldsymbol{\psi} = \boldsymbol{\varphi}_0 \ da\text{-a.e. on } \Gamma_0,\ \det \boldsymbol{\nabla}\boldsymbol{\psi} = 1 \text{ a.e. in } \Omega\}$$

is nonempty. Show that the set $\boldsymbol{\Phi}$ is sequentially weakly closed in the space $\boldsymbol{H}^1(\Omega) \times \boldsymbol{L}^2(\Omega)$.

(2) Let

$$\mathbb{M}^3_1 = \{\boldsymbol{F} \in \mathbb{M}^3;\ \det \boldsymbol{F} = 1\}\,,$$

and let

$$\hat{W}: F \in \mathbb{M}_1^3 \to \hat{W}(F) = a\|F\|^2 + b\|\mathbf{Cof}\, F\|^2 ,$$

with $a > 0$, $b > 0$, denote the stored energy function of an *incompressible Mooney–Rivlin material*. Let $f \in L^p(\Omega)$ and $g \in L^\sigma(\Gamma_1)$ be such that the linear form $L: H^1(\Omega) \to \mathbb{R}$ is continuous, and assume that $\inf_{\psi \in \Phi} I(\psi) < +\infty$ (the linear form L and the total energy I are as in Theorem 7.7-1). Show that there exists at least one function φ such that

$$\varphi \in \Phi \text{ and } I(\varphi) = \inf_{\psi \in \Phi} I(\psi) .$$

7.13. As shown by Ball [1981b, Theorem 3], the existence of *everywhere injective minimizers* for *pure displacement problems* can be established directly (i.e., without an injectivity condition in the set of admissible deformations) if the stored energy function $\hat{W}$ grows sufficiently rapidly to $+\infty$ as $\det F \to 0^+$. More specifically, let Ω be a domain in $\mathbb{R}^3$, let $\hat{W}: \Omega \times \mathbb{M}_+^3 \to \mathbb{R}$ be a polyconvex stored energy function, and assume that there exist constants α, β, p, q, r, s such that (compare with assumptions (b) and (c) in Theorem 7.7-1):

$$\alpha > 0,\ p > 3,\ q > 3,\ r > 1,\ s > \frac{2q}{q-3} ,$$

$$\hat{W}(x, F) \geq \alpha(\|F\|^p + \|\mathbf{Cof}\, F\|^q + (\det F)^r + (\det F)^{-s}) + \beta$$

for almost all $x \in \Omega$ and all $F \in \mathbb{M}_+^3$. Let $\varphi_0: \bar{\Omega} \to \mathbb{R}^3$ be a mapping that satisfies:

$$\begin{cases} \varphi_0 \in W^{1,p}(\Omega) , \\ \varphi_0 \text{ is injective on } \Omega , \\ \det \nabla\varphi_0 > 0 \text{ a.e. in } \Omega , \\ \varphi_0(\Omega) \text{ is a domain in } \mathbb{R}^3 , \end{cases}$$

and let φ denote a minimizer of the total energy over the set

$$\Phi := \{\psi \in W^{1,p}(\Omega);\ \mathbf{Cof}\, \nabla\psi \in L^q(\Omega),\ \det \nabla\psi \in L^r(\Omega) ,$$
$$\psi = \varphi_0\ da\text{-a.e. on } \Gamma,\ \det \nabla\psi > 0 \text{ a.e. in } \Omega\} .$$

Show that $\varphi: \bar{\Omega} \to \varphi(\bar{\Omega})$ is a homemorphism.

Hint. Use Exercise 5.7.

7.14. (1) Let Ω be a domain in $\mathbb{R}^3$, and let $\hat{W}:\Omega \times \mathbb{M}^3 \to \mathbb{R}$ be a function with the following properties: For almost all $x \in \Omega$, there exists a convex function $\mathbb{W}(x,\cdot):\mathbb{M}^3 \times \mathbb{M}^3 \times \mathbb{R} \to \mathbb{R}$ such that

$$\mathbb{W}(x, \boldsymbol{F}, \mathbf{Cof}\,\boldsymbol{F}, \det \boldsymbol{F}) = \hat{W}(x, \boldsymbol{F}) \quad \text{for all } \boldsymbol{F} \in \mathbb{M}^3 ,$$

and the function $\mathbb{W}(\cdot, \boldsymbol{F}, \boldsymbol{H}, \delta):\Omega \to \mathbb{R}$ is measurable for all $(\boldsymbol{F}, \boldsymbol{H}, \delta) \in \mathbb{M}^3 \times \mathbb{M}^3 \times \mathbb{R}$ (the function $\hat{W}(x,\cdot):\mathbb{M}^3 \to \mathbb{R}$ is thus polyconvex for almost all $x \in \Omega$); there exist constants

$$\alpha > 0,\ \beta \in \mathbb{R},\ p \geq 2,\ q \geq \frac{p}{p-1},\ r > 1 ,$$

such that the coerciveness inequality

$$\mathbb{W}(x, \boldsymbol{F}, \boldsymbol{H}, \delta) \geq \alpha(\|\boldsymbol{F}\|^p + \|\boldsymbol{H}\|^q + |\delta|^r) + \beta ,$$

holds for almost all $x \in \Omega$ and for all $\boldsymbol{F} \in \mathbb{M}^3$. Let $\Gamma = \Gamma_0 \cup \Gamma_1$ be a da-measurable partition of $\Gamma = \partial\Omega$ with area $\Gamma_0 > 0$, and let $\boldsymbol{\varphi}_0 : \Gamma_0 \to \mathbb{R}^3$ be a measurable function such that the set

$$\boldsymbol{\Phi} := \{\boldsymbol{\psi} \in \boldsymbol{W}^{1,p}(\Omega);\ \mathbf{Cof}\,\boldsymbol{\nabla\psi} \in \boldsymbol{L}^q(\Omega),\ \det \boldsymbol{\nabla\psi} \in L^r(\Omega),\ \boldsymbol{\psi} = \boldsymbol{\varphi}_0\ da\text{-a.e. on } \Gamma_0\}$$

is nonempty. Let L be a continuous linear functional over the space $\boldsymbol{W}^{1,p}(\Omega)$. Finally, let

$$I(\boldsymbol{\psi}) = \int_\Omega \hat{W}(x, \boldsymbol{\nabla\psi}(x))\, dx - L(\boldsymbol{\psi}) ,$$

and assume that $\inf_{\boldsymbol{\psi} \in \boldsymbol{\Phi}} I(\boldsymbol{\psi}) < +\infty$. Show that these exists at least one function $\boldsymbol{\varphi} \in \boldsymbol{\Phi}$ such that $I(\boldsymbol{\varphi}) = \inf_{\boldsymbol{\psi} \in \boldsymbol{\Phi}} I(\boldsymbol{\psi})$.

(2) Assume further that the function $\mathbb{W}(x,\cdot):\mathbb{M}^3 \times \mathbb{M}^3 \times \mathbb{R} \to \mathbb{R}$ is of class $\mathscr{C}^1$ for almost all $x \in \Omega$, and that there exist constants A and B such that ($p'^{-1} = (p-1)/p$ and $q'^{-1} = (q-1)/q$):

$$\left|\frac{\partial \mathbb{W}}{\partial \boldsymbol{F}}(x, \boldsymbol{F}, \boldsymbol{H}, \delta)\right| \leq A(\|\boldsymbol{F}\|^p + \|\boldsymbol{H}\|^q + |\delta|^r) + B ,$$

$$\left|\frac{\partial \mathsf{W}}{\partial \boldsymbol{H}}(x, \boldsymbol{F}, \boldsymbol{H}, \delta)\right| \leq A(\|\boldsymbol{F}\|^{p-1} + \|\boldsymbol{H}\|^{q/p'} + |\delta|^{r/p'}) + B\,,$$

$$\left|\frac{\partial \mathsf{W}}{\partial \delta}(x, \boldsymbol{F}, \boldsymbol{H}, \delta)\right| \leq A(\|\boldsymbol{F}\|^{p/q'} + \|\boldsymbol{H}\|^{q-1} + |\delta|^{r/q'}) + B\,,$$

for almost all $x \in \Omega$ and for all $(\boldsymbol{F}, \boldsymbol{H}, \delta) \in \mathbb{M}^3 \times \mathbb{M}^3 \times \mathbb{R}$. Show that any minimizer $\boldsymbol{\varphi} \in \boldsymbol{\Phi}$ of the functional I over the set $\boldsymbol{\Phi}$ is a *weak solution of the Euler–Lagrange equations*, in the sense that

$$\int_\Omega \frac{\partial \hat{W}}{\partial \boldsymbol{F}}(x, \nabla\boldsymbol{\varphi}(x)) : \nabla\boldsymbol{\theta}(x)\, dx = L(\boldsymbol{\theta})$$

for all $\boldsymbol{\theta} \in \mathscr{C}^\infty(\bar{\Omega})$ that vanish on Γ_0 (Ball [1977, Theorem 7.12]).

Remark. The result of (2) holds at the expense of *strict growth conditions* on the partial derivatives of the function W; besides the integrand $\hat{W}(x, \boldsymbol{F})$ is here defined *for all* $\boldsymbol{F} \in \mathbb{M}^3$, and consequently it cannot approach $+\infty$ as $\det \boldsymbol{F} \to 0^+$; this in turn precludes the orientation-preserving condition. Hence these assumptions are not realistic for actual hyperelastic materials.

7.15. Let all the assumptions of Theorem 7.7-1 be satisfied. In addition, assume that the function W is independent of $x \in \Omega$ and that it is in the space $\mathscr{C}^1(\mathbb{M}^3 \times \mathbb{M}^3 \times]0, +\infty[)$. Let $\boldsymbol{\varphi} \in \boldsymbol{\Phi}$ be a minimizer of the energy in the space $\boldsymbol{W}^{1,\infty}(\Omega)$, with the additional property that there exists $\delta > 0$ such that $\det \nabla\boldsymbol{\varphi}(x) \geq \delta$ for almost all $x \in \Omega$.

Show that $\boldsymbol{\varphi}$ is a *weak solution of the Euler–Lagrange equations*.

Remark. The growth assumptions on the partial derivatives of Exercise 7.14 are replaced here by an assumption of *regularity on the minimizer*.

7.16. Consider the *pure displacement problem for a St Venant–Kirchhoff material*:

$$(*)\quad \begin{cases} -\operatorname{div}\{(\boldsymbol{I} + \nabla\boldsymbol{u})(\lambda(\operatorname{tr}\boldsymbol{E}(\boldsymbol{u}))\boldsymbol{I} + 2\mu\boldsymbol{E}(\boldsymbol{u}))\} = \boldsymbol{f} \text{ in } \Omega\,, \\ \boldsymbol{u} = \boldsymbol{o} \text{ on } \Gamma\,, \end{cases}$$

and its associated total energy (expressed in terms of the displacement field):

$$J(\boldsymbol{v}) = \tfrac{1}{2}\int_\Omega \left\{\frac{\lambda}{2}(\operatorname{tr}\boldsymbol{E}(\boldsymbol{v}))^2 + \mu \operatorname{tr}(\boldsymbol{E}(\boldsymbol{v}))^2\right\} dx - \int_\Omega \boldsymbol{f}\cdot\boldsymbol{v}\, dx\,,$$

which is well defined and differentiable over the space $\boldsymbol{W}^{1,4}(\Omega)$, whence on any space continuously inbedded in $\boldsymbol{W}^{1,4}(\Omega)$. Assume that $\boldsymbol{f} \in \boldsymbol{L}^p(\Omega)$ for some $p > 3$ and that $|\boldsymbol{f}|_{0,p,\Omega}$ is small enough, so that $(*)$ has a unique solution $\boldsymbol{u}$ in a neighborhood of the origin in the space $\boldsymbol{V}^p(\Omega) = \{\boldsymbol{v} \in \boldsymbol{W}^{2,p}(\Omega);\ \boldsymbol{v} = \boldsymbol{o} \text{ on } \Gamma\}$ (Theorem 6.4-1).

(1) Show that $J'(\boldsymbol{u}) = 0$.

(2) Show that there exists a constant $\alpha > 0$ such that

$$J''(\boldsymbol{o})(\boldsymbol{v}, \boldsymbol{v}) \geq \alpha |\boldsymbol{v}|_{1,\Omega}^2 \quad \text{for all } \boldsymbol{v} \in \boldsymbol{W}_0^{1,4}(\Omega)\,.$$

Hence if $\boldsymbol{f} = \boldsymbol{o}$ (in which case $J'(\boldsymbol{o}) = 0$ by (1)) we cannot conclude that $\boldsymbol{u} = \boldsymbol{o}$ is a strict local minimum in the space $\boldsymbol{W}_0^{1,4}(\Omega)$ (Theorem 1.3-1(a) would require that the semi-norm $|\cdot|_{1,4,\Omega}$, instead of the semi-norm $|\cdot|_{1,\Omega}$, appear in the right-hand side of the inequality), and indeed it is possible to construct examples where a minimizer in $\boldsymbol{W}_0^{1,p}(\Omega)$, $1 \leq p < \infty$, is not a strict local minimum in this space (see Exercise 7.17).

(3) Show that there exist constants $\varepsilon > 0$ and $\beta > 0$ such that

$$|\boldsymbol{u}|_{1,\infty,\Omega} < \varepsilon \Rightarrow J''(\boldsymbol{u})(\boldsymbol{v}, \boldsymbol{v}) \geq \beta |\boldsymbol{v}|_{1,\Omega}^2 \quad \text{for all } \boldsymbol{v} \in \boldsymbol{W}_0^{1,4}(\Omega)\,.$$

Hence we can conclude from Theorem 1.3-1(b) that $\boldsymbol{u}$ is a strict local minimum in the space $\{\boldsymbol{v} \in \boldsymbol{W}^{1,\infty}(\Omega);\ \boldsymbol{v} = \boldsymbol{o} \text{ on } \Gamma\}$.

Remark. For related results, see notably John [1972a], Ball & Marsden [1984], Simpson & Spector [1987], Quintela-Estevez [1986].

7.17. This problem is adapted from Ball, Knops & Marsden [1978]. Let $h(p) = \{p(p^2 - 1)\}^2$, $p \in \mathbb{R}$, and $I =]0, 1[$. Then the functional

$$J : v \in W^{1,6}(I) \to J(v) = \int_I h(v'(x))\,\mathrm{d}x$$

is well defined, and the problem: Find u such that

$$u \in U_p := \{v \in W^{1,p}(I);\ v(0) = v(1) = 0\}, \text{ and } J(u) = \inf_{v \in U_p} J(v)\,,$$

has at least $u_0 := 0$ as a solution, for all $6 \leq p \leq \infty$.

(1) Show that J is strictly convex on a neighborhood of u_0 in U_∞, and thus that u_0 is a strict local minimum of J in U_∞.

(2) Show that $J''(u_0)(v, v) > 0$ for all $v \in U_p - \{0\}$, $p \geq 6$.

(3) Show that if $6 \leq p < \infty$, for each $\varepsilon > 0$, there exists u such that

$$u \in U_p,\ u \neq u_0,\ \|u - u_0\|_{1,p,I} < \varepsilon,\ J(u) = J(u_0) = \inf_{v \in U_p} J(v)\,.$$

Hence there is no neighborhood of u_0 in U_p, $6 \leq p < \infty$, on which J is strictly convex (compare with (1)).

Remarks. This problem shows that the "natural" solution of a minimization problem (in particular, that found by the implicit function theorem applied to the associated Euler–Lagrange equations, as in Exercise 7.16) need not be isolated in any Sobolev space $W^{1,p}(\Omega)$, $p < \infty$. For other instances of the "pivotal" rôle played by the space $W^{1,\infty}(\Omega)$, see Exercises 7.15 and 7.16.

7.18. Let Ω be a domain in $\mathbb{R}^n$. Show that a function $\varphi \in W^{1,n}(\Omega)$ that satisfies $\det \nabla\varphi > 0$ a.e. in Ω can be identified with a continuous function (cf. Vodopyanov & Goldshtein [1977]; see also Šverák [1987]).

BIBLIOGRAPHY

ABRAHAM, R.; MARSDEN, J.E.; RATIU, T. [1983]: *Manifolds, Tensor Analysis, and Applications*, Global Analysis Series, No. 2, Addison-Wesley, Reading.

ACERBI, E.; FUSCO, N. [1984]: Semicontinuity problems in the calculus of variations, *Arch. Rational Mech. Anal.* **86**, 125–145.

ACERBI, E.; FUSCO, N. [1987]: A regularity theorem for minimizers of quasiconvex integrals, *Arch. Rational Mech. Anal.* **99**, 261–281.

ADAMS, R.A. [1975]: *Sobolev Spaces*, Academic Press, New York.

ADELEKE, S.A. [1983]: On the problem of eversion for incompressible elastic materials, *J. Elasticity* **13**, 63–69.

AGMON, S.; DOUGLIS, A.; NIRENBERG, L. [1964]: Estimates near the boundary for solutions of elliptic partial differential equations satisfying general boundary condition II, *Comm. Pure Appl. Math.* **17**, 35–92.

ALEXANDER, J.C.; ANTMAN, S.S. [1982]: The ambiguous twist of Love, *Quart. Appl. Math.* **49**, 83–92.

AMBROSIO, L. [1987]: New lower semicontinuity results for integral functionals, *Rend. Accad. Naz. Sci. XL Mem. Mat. Sci. Fis. Natur.*, to appear.

AMBROSIO, L.; BUTTAZZO, G.; LEACI, A. [1987]: Continuous operators of the form $T_f = f(x, u, Du)$, to appear.

ANSELONE, P.M.; MOORE [1966]: An extension of the Newton-Kantorovic method for solving nonlinear equations with an application to elasticity, *J. Math. Anal. Appl.* **13**, 476–501.

ANTMAN, S.S. [1970]: Existence of solutions of the equilibrium equations for nonlinearly elastic rings and arches, *Indiana University Mathematics Journal* **20**, 281–302.

ANTMAN, S.S. [1976a]: Ordinary differential equations of non-linear elasticity I: Foundations of the theories of nonlinearly elastic rods and shells, *Arch. Rational Mech. Anal.* **61**, 307–351.

ANTMAN, S.S. [1976b]: Ordinary differential equations of non-linear elasticity II: Existence and regularity for conservative boundary value problems, *Arch. Rational Mech. Anal.* **61**, 352–393.

ANTMAN, S.S. [1978a]: Buckled states of nonlinearly elastic plates, *Arch. Rational Mech. Anal.* **67**, 111–149.

ANTMAN, S.S. [1978b]: A family of semi-inverse problems of non-linear elasticity, in *Contemporary Developments in Continuum Mechanics and Partial Differential Equations* (G.M. DE LA PENHA & L.A.J. MEDEIROS, Editors), pp. 1–24, North-Holland, Amsterdam.

ANTMAN, S.S. [1979]: The eversion of thick spherical shells, *Arch. Rational Mech. Anal.* **70**, 113–123.

ANTMAN, S.S. [1983a]: Regular and singular problems for large elastic deformations of tubes, wedges, and cylinders, *Arch. Rational Mech. Anal.* **83**, 1–52.

ANTMAN, S.S. [1983b]: The influence of elasticity on analysis: Modern developments, *Bull. Amer. Mach. Soc.* **9**, 267–291.

ANTMAN, S.S. [1984]: Geometrical and analytical questions in nonlinear elasticity, in *Seminar on Nonlinear Partial Differential Equations* (S.S. CHERN, Editor), Springer-Verlag, New York.

ANTMAN, S.S. [1988]: *Nonlinear Problems of Elasticity*, to appear.

ANTMAN, S.S.; BREZIS, H. [1978]: The existence of orientation-preserving deformations in nonlinear elasticity, in *Nonlinear Analysis and Mechanics: Heriot-Watt Symposium, Vol. II* (R.J. KNOPS, Editor), pp. 1–29, Pitman, London.

ANTMAN, S.S.; OSBORN, J.E. [1979]: The principle of virtual work and integral laws of motion, *Arch. Rational Mech. Anal.* **69**, 231–262.

ANZELLOTI, G. [1983]: On the existence of the rates of stress and displacements for Prandtl-Reuss plasticity, *Quart. Appl. Math.* **41**, 181–208.

Anzellotti,, G. [1984]: On the extremal stress and displacement in Hencky plasticity, *Duke Math. J.* **51**, 133–147.

ANZELLOTTI, G. [1985a]: The Euler equation for functionals with linear growth, *Trans. Amer. Math. Soc.* **290**, 483–501.

ANZELLOTTI, G. [1985b]: A class of convex non-coercive functionals and masonry-like materials, *Ann. Inst. Henri Poincaré* **2**, 261–307.

ANZELLOTTI, G.; GIAQUINTA, M. [1980]: Existence of the displacements field for an elasto-plastic body subject to Hencky's law and von Mises yield condition, *Manuscripta Math.* **32**, 101–136.

ANZELLOTTI, G.; GIAQUINTA, M. [1982]: On the existence of the fields of stresses and displacements for an elasto-perfectly plastic body in elastic equilibrium, *J. Math. Pures Appl.* **61**, 219–244.

ARGYRIS, J.H.; KLEIBER, M. [1977]: Incremental formulation in nonlinear mechanics and large strain elasto-plasticity; Natural approach, Part I, *Comput. Methods Appl. Mech. Engrg.* **11**, 215–248.

ARNOLD, V.I. [1973]: *Ordinary Differential Equations*, MIT Press, Cambridge.

ARON, M. [1978]: On the physical meaning of a uniqueness condition in finite elasticity, *J. Elasticity* **8**, 111–115.

ARON, M. [1979]: A continuous dependence result for a mixed boundary value problem in finite elastostatics, *Acta Mechanica* **32**, 205–208.

ARON, M.; ROSEMAN, J.J. [1977]: Integral estimates for the displacement and strain energy in nonlinear elasticity in terms of the body force, *Int. J. Engrg. Sc.* **15**, 317.

ATTEIA, M.; DEDIEU, J.P. [1981]: Minimization of energy in nonlinear elasticity, in *Nonlinear Problems of Analysis in Geometry and Mechanics* (M. ATTEIA, D. BANCEL, I. GUMOWSKI, Editors), pp. 73–79, Pitman, Boston.

ATTEIA, M.; RAISSOULI, M. [1986]: On the Saint-Venant problem with mixed condition in nonlinear three dimensional elasticity, to appear.

ATTOUCH, H. [1984]: *Variational convergence for Functions and Operators*, Pitman, London.

AUBERT, G. [1987]: Quelques remarques sur les notions de polyconvexité et de 1-rang convexité en dimensions 2 et 3, *Modél. Math. et Anal. Numér.*, to appear.

AUBERT, G.; TAHRAOUI, R. [1979]: Théorèmes d'existence en calcul des variations, *J. Differential Equations* **33**, 1–15.

AUBERT, G.; TAHRAOUI, R. [1984], Théorèmes d'existence en optimisation non convexe, *Applicable Analysis* **18**, 75–100.

AUBERT, G.; TAHRAOUI, R. [1985]: Conditions nécessaires de faible fermeture et de 1-rang convexité en dimension 3, *Rendiconti del Circolo Matematico di Palermo*, *Ser. 2*, **34**, 460–488.

AUBERT, G.; TAHRAOUI, R. [1987]: Sur la faible fermeture de certains ensembles de contraintes en élasticité non-linéaire plane, *Arch. Rational Mech. Anal.* **97**, 33–58.

AUBIN, J.P.; EKELAND, I. [1984]: *Applied Nonlinear Analysis*, J. Wiley, New York.

AVEZ, A. [1983]: *Calcul Différentiel*, Masson, Paris (English translation: *Differential Calculus*, J. Wiley, New York, 1986).

BABUŠKA, I. [1974]: Solution of problems with interfaces and singularities, in *Mathematical Aspects of Finite Elements in Partial Differential Equations*, pp. 213–277, Academic Press, New York.

BABUŠKA, I. [1975]: Homogenization approach in engineering, in *Lecture Notes in Economics and Mathematical Systems* (M. BECKMAN & H.P. KINZI, Editors), pp. 137–153, Springer-Verlag, Berlin.

BAIOCCHI, C.; BUTTAZZO, G.; GASTALDI, F.; TOMARELLI, F. [1986]: General existence results for unilateral problems in continuum mechanics, report, Università di Pavia.

BIAOCCHI, C.; CAPELO, A. [1978]: *Disequazioni Variazionali e Quasi Variazionali. Applicazioni a Problemi di Frontiera Libera* (Two Volumes), Pitagora, Bologna (English translation: *Variational and Quasivariational Inequalities. Applications to Free Boundary-Problems*, Wiley-Interscience, New York, 1984).

BALL, J.M. [1977]: Convexity conditions and existence theorems in nonlinear elasticity, *Arch. Rational Mech. Anal.* **63**, 337–403.

BALL, J.M. [1978]: Finite time blow-up in nonlinear problems, in *Nonlinear Evolution Equations* (M.G. CRANDALL, Editor), pp. 189–205, Academic Press, New York.

BALL, J.M. [1980]: Strict convexity, strong ellipticity and regularity in the calculus of variations, *Math. Proc. Camb. Phil. Soc.* **87**, 501–513.

BALL, J.M. [1981a]: Remarques sur l'existence et la régularité des solutions d'élastostatique non linéaire, in *Recent Contributions to Nonlinear Partial Differential Equations*, pp. 50–62, Research Notes in Math. 50, Pitman, Boston.

BALL, J.M. [1981b]: Global invertibility of Sobolev functions and the interpenetration of matter, *Proc. Roy. Soc. Edinburgh* **88A**, 315–328.

BALL, J.M. [1982]: Discontinuous equilibrium solutions and cavitation in nonlinear elasticity, *Phil. Trans. Roy. Soc. London* **A 306**, 557–611.

BALL, J.M. [1984]: Minimizers and the Euler-Lagrange equations, in *Trends and Applications of Pure Mathematics to Mechanics* (P.G. CIARLET & M. ROSEAU, Editors), pp. 1–4, Springer-Verlag, Berlin.

BALL, J.M.; CURRIE, J.C.; OLVER, P.J. [1981]: Null Lagrangians, weak continuity, and variational problems of arbitrary order, *J. Functional Anal.* **41**, 135–174.

BALL, J.M.; JAMES, R.D. [1987]: Fine phase mixtures as minimizers of energy, *Arch. Rational Mech. Anal.*, to appear.

BALL, J.M.; KNOPS, R.J.; MARSDEN, J.E. [1978]: Two examples in nonlinear elasticity, in *Proceedings Conference on Nonlinear Analysis, Besançon*, pp. 41–49, Lecture Notes in Mathematics, Vol. 466, Springer-Verlag, Berlin.

BALL, J.M.; KNOWLES, G. [1987]: A numerical method for detecting singular minimizers, *Numer. Math.* **51**, 181–197.

BALL, J.M.; MARSDEN, J.E. [1984]: Quasiconvexity at the boundary, positivity of the second variation and elastic stability, *Arch. Rational Mech. Anal.* **86**, 251–277.

BALL, J.M.; MIZEL, V.J. [1984]: Singular minimizers for regular one-dimensional problems in the calculus of variations, *Bull. Amer. Math. Soc.* **11**, 143–146.

BALL, J.M.; MIZEL, V.J. [1985]: One-dimensional variational problems whose minimizers do not satisfy the Euler-Lagrange equation, *Arch. Rational Mech. Anal.* **90**, 325–388.

BALL, J.M.; MURAT, F. [1984]: $W^{1,p}$-quasiconvexity and weak convergence in the calculus of variations, *J. Functional Analysis* **58**, 225–253.

BAMBERGER, Y. [1981]: *Mécanique de l'Ingénieur II. Milieux Déformables*, Hermann, Paris.

BAMPI, F.; MORRO, A. [1982]: Objective constitution relations in elasticity and viscoelasticity, *Meccanica* **17**, 138–142.

BATRA, R.C. [1972]: On non-classical boundary conditions, *Arch. Rational Mech. Anal.* **48**, 163–191.

BEATTY, M.F. [1970]: Stability of hyperelastic bodies subject to hydrostatic loads, *Int. J. Non-linear Mech.* **5**, 367–383.

BEATTY, M.F. [1987]: A class of universal relations in isotropic elasticity theory, *J. Elasticity* **17**, 113–121.

BECKMAN, F.S.; QUARLES, JR., D.A. [1953]: On isometries of Euclidean spaces, *Proc. Amer. Math. Soc.* **4**, 810–815.

BELL, J.F. [1973]: The experimental foundations of solid mechanics, *Handbuch der Physik*, Vol. VIa/1. [C. TRUESDELL, Editor], Springer-Verlag, Berlin.

BENSOUSSAN, A.; LIONS, J.L.; PAPANICOLAOU, G.C. [1978]: *Asymptotic Analysis for Periodic Structures*, North-Holland, Amsterdam.

BERGER, M.S. [1967]: On von Kármán equations and the buckling of a thin elastic plate I, *Comm. Pure Appl. Math.* **20**, 687–718.

BERGER, M.S. [1977]: *Nonlinearity and Functional Analysis*, Academic Press, New York.

BERGER, M.S.; FIFE, P. [1968]: On von Kármán equations and the buckling of a thin elastic plate, II, Plate with general boundary conditions, *Comm. Pure Appl. Math.* **22**, 227–247.

BERNADOU, M.; CIARLET, P.G.; HU, JIAN-WEI [1982]: On the convergence of incremental methods in finite elasticity, in *Proceedings of the Chinese-French Conference on Finite Element Methods* (Beijing, 19–23 April, 1982).

BERNADOU. M.; CIARLET, P.G.; HU, JIAN-WEI [1984]: On the convergence of the semi-discrete incremental method in nonlinear, three-dimensional, elasticity, *J. Elasticity* **14**, 425–440.

BHARATHA, S.; LEVINSON, M. [1977]: Signorini's perturbation scheme for a general reference configuration in finite elastostatics, *Arch. Rational Mech. Anal.* **67**, 365–394.

BIELSKI, W.R.; TELEGA, J.J. [1986]: The complementary energy principle in finite elastostatics as a dual problem, to appear.

BLANCHARD, D.; CIARLET, P.G. [1983]: A remark on the von Kármán equations, *Comput. Methods Appl. Mech. Engrg.* **37**, 79–92.

BLATZ, P.J. [1971]: On the thermostatic behavior of elastomers: in *Polymer Networks, Structure and Mechanical Properties*, pp. 23–45, Plenum Press, New York.

BLATZ, P.J.; KO, W.L. [1962]: Application of finite elasticity theory to the deformation of rubbery materials, *Trans. Soc. Rheology* **6**, 223–251.

BOEHLER, J.-P. [1978]: Lois de comportement anisotrope des milieux continus, *J. Mécanique* **17**, 153–190.

BOJARSKI, B.; IWANIEC, T. [1983]: Analytical foundations of the theory of quasiconformal mappings in R^n, *Annales Academiae Scientiarum, Ser. A.I. Mathematica*, **8**, 257–324.

BOURBAKI, N. [1966]: *General Topology*, Part 1, Hermann, Paris.

BOURBAKI, N. [1970]: *Théorie des Ensembles*, Hermann, Paris.

BREUER, S.; ARON, M. [1982]: Sufficient conditions for continuous dependence in nonlinear hyperelasticity, *J. Elasticity* **12**, 161–166.

BREUER, S.; ROSEMAN, J.J. [1978]: Integral bounds on the strain energy for the traction problem in finite elasticity, *Arch. Rational Mech. Analysis* **68**, 333–342.

BREUER, S.; ROSEMAN, J.J. [1979]: An integral bound for the strain energy in nonlinear elasticity in terms of the boundary displacements, *J. Elasticity* **9**, 21–27.

BREUER, S.; ROSEMAN, J.J. [1980]: A bound on the strain energy for the traction problem in finite elasticity with localized non-zero surface data, *J. Elasticity* **10**, 11–22.

BREZIS, H. [1983]: *Analyse Fonctionnelle, Théorie et Applications*, Masson, Paris; (*English translation*: Springer-Verlag, Heidelberg, 1987).

BROUWER, L.E.J. [1912]: Uber Abbildung der Mannigfaltigkeiten, *Math. Ann.* **71**, 97–115.

BUFLER, H. [1983]: On the work theorem for finite and incremental elastic deformations with discontinuous fields: a unified treatment of different versions, *Comput. Meth. Appl. Mech. Engrg.* **36**, 95–124.

BUFLER, H. [1984]: Pressure loaded structures under large deformations, ZAMM, *Z. Angew. Math. u. Mech.* **64**, 287–295.

VAN BUREN, M. [1968]: *On the Existence and Uniqueness of Solutions to Boundary Value Problems in Finite Elasticity*, Thesis, Carnegie-Mellon University.

BURGESS, I.W.; LEVINSON, M. [1972]: The instability of slightly compressible rectangular solids under biaxial loadings, *Int. J. Solids Structures* **8**, 133–148.

BUSEMANN, H., EWALD, G.; SHEPHARD, G.C. [1963]: Convex bodies and convexity on Grassman cones, Parts I–IV, *Math. Ann.* **151**, 1–41.

CABANE, R. [1981]: Une caractérisation simple des isométries, *Revue de l'Association des Professeurs de Mathématiques de l'Enseignement Pubic*, 821–829.

CAILLERIE, D. [1987]: Non homogeneous plate theory and conduction in fibered composites, in *Homogenization Techniques for Composite Media* (E. SANCHEZ-PALENCIA, A. ZAOUI, Editors), pp. 1–62, Springer-Verlag, Berlin.

CALDERÓN, A.P. [1961]: Lebesgue spaces of differentiable functions and distributions, *Proc. Symp. in Pure Maths IV*, American Mathematical Society, Providence.

DE CAMPOS, L.T.; ODEN, J.T. [1983]: Non-quasi-convex problems in nonlinear elastostatics, *Advances Appl. Math.* **4**, 380–401.

DE CAMPOS, L.T.; ODEN, J.T. [1985]: On the principle of stationary complementary energy in finite elastostatics, *Intern. J. Engrg. Sci.* **23**, 57–63.

CAPRIZ, G.; PODIO-GUIDUGLI, P. [1974]: On Signorini's perturbation method in finite elasticity, *Arch. Rational Mech. Anal.* **57**, 1–30.

CAPRIZ, G.; PODIO-GUIDUGLI, P. [1979]: The role of Fredholm conditions in Signorini's perturation method, *Arch. Rational Mech. Anal.* **70**, 261–288.

CAPRIZ, G.; PODIO-GUIDUGLI, P. [1981]: Questions of uniqueness in finite elasticity, in *Trends in Applications of Pure Mathematics to Mechanics, Vol. III*, pp. 76–95, Pitman, Boston.

CAPRIZ, G.; PODIO-GUIDUGLI, P. [1982]: A generalization of Signorini's perturbation method suggested by two problems of Grioli, *Rend. Sem. Mat. Padova* **68**, 149–162.

CARLSON, D.E.; HOGER, A. [1986]: On the derivatives of the principle invariants of a second-order tensor, *J. Elasticity* **16**, 221–224.

CARTAN, H. [1967]: *Calcul Différentiel*, Hermann, Paris.

CAUCHY, A.-L. [1823]: Recherches sur l'équilibre et le mouvement intérieur des corps solides ou fluides, élastiques ou non élastiques, *Bulletin de la Société Philomatique*, 9-13 (Oeuvres (2) 2, pp. 300–304, Gauthier-Villars, Paris, 1889).

CAUCHY, A.-L. [1827a]: De la pression ou tension dans un corps solide, *Exercices de Mathématiques* **2**, 42-56 (Oeuvres (2) 7, pp. 60–93, Gauthier-Villars, Paris, 1889).

CAUCHY, A.-L. [1827b]: Sur les relations qui existent dans l'état d'équilibre d'un corps solide ou fluide (Oeuvres (2) 7, pp. 141–145, Gauthier-Villars, Paris, 1889).

CAUCHY, A.-L. [1828]: Sur les équations qui expriment les conditions d'équilibre ou les lois du mouvement intérieur d'un corps solide, élastique ou non élastique (Oeuvres (2) 8, pp. 195–226, Gauthier-Villars, Paris, 1890).

CESCOTTO, S.; FREY, F.; FONDER, G. [1979]: Total and updated Lagrangian descriptions in nonlinear structural analysis: a unified approach, in *Energy Methods in Finite Element Analysis* (R. GLOWINSKI, E.Y. RODIN, O.C. ZIENKIEWICZ, Editors), pp. 283–296, John Wiley, New York.

CHADWICK, P. [1972]: The existence and uniqueness of solutions to two problems in the Mooney-Rivlin theory for rubber, *J. Elasticity* **2**, 123–128.

CHADWICK, P.; HADDON, E.W. [1972]: Inflation-extension and eversion of a tube of incompressible isotropic elastic material, *J. Inst. Math. Appl.* **10**, 258–278.

CHARRIER, P.; DACOROGNA, B.; HANOUZET, B.; LABORDE, P. [1986]: An existence theorem for slightly compressible material in non linear elasticity, to appear.

CHEN, C.; VON WAHL, W. [1982]: Das Rand-Anfangswertproblem für quasilineare Wellengleichungen in Sobolevräumen niedriger Ordnung, *J. reine angew. Math.* **337**, 77–112.

CHEN, W.F.; SALEEB, A.F. [1982]: *Constitutive Equations for Engineering Materials, Vol. 1 – Elasticity and modeling*, Wiley, New York.

CHENEY, E.W. [1966]: *Introduction to Approximation Theory*, McGraw-Hill, New York.

CHILLINGWORTH, D.R.J.; MARSDEN, J.E.; WAN, Y.H. [1982]: Symmetry and bifurcation in three dimensional elasticity, Part I, *Arch. Rational Mech. Anal.* **80**, 295–331.

CHILLINGWORTH, D.R.J.; MARSDEN, J.E.; WAN, Y.H. [1983]: Symmetry and bifurcation in three-dimensional elasticity, Part II, *Arch. Rational Mech. Anal.* **83**, 363–395.

CHIPOT, M.; KINDERLEHRER, D. [1987]: Equilibrium configurations of crystals, Preprint 326, Institute for Mathematics and its Applications, University of Minnesota, Minneapolis.

CHOQUET, G. [1964]: *Cours d'Analyse: Topologie*, Masson, Paris.

CHOQUET-BRUHAT, Y. [1973]: *Distributions, Théorie et Problèmes*, Masson, Paris.

CHOQUET-BRUHAT, Y.; DEWITT-MORETTE, C.; DILLARD-BLEICK, M. [1977]: *Analysis, Manifolds and Physics*, North-Holland, Amsterdam.

CIARLET, P.G. [1978]: *The Finite Element Method for Elliptic Problems*, North-Holland, Amsterdam.

CIARLET, P.G. [1980]: A justification of the von Kármán equations, *Arch. Rational Mech. Anal.* **73**, 349–389.

CIARLET, P.G. [1982]: Two-dimensional approximations of three-dimensional models in nonlinear plate theory, in *Proceedings of the IUTAM Symposium on Finite Elasticity* (SHIELD, R.T. & CARLSON, D.E., Editors), pp. 123–141, Martinus Nijhoff, The Hague.

CIARLET, P.G. [1983]: *Introduction à l'Analyse Numérique Matricielle et à l'Optimisation*, Masson, Paris (*English Translation: Introduction to Numerical Linear Algebra and Optimization*, Cambridge University Press, 1987).

CIARLET, P.G. [1985]: *Elasticité Tridimensionnelle*, Masson, Paris.

CIARLET, P.G.; DESTUYNDER, P. [1979]: A justification of a nonlinear model in plate theory, *Comput. Methods Appl. Mech. Engrg.* **17/18**, 227–258.

CIARLET, P.G.; GEYMONAT, G. [1982]: Sur les lois de comportement en élasticité non-linéaire compressible, *C.R. Acad. Sci. Paris Sér. II* **295**, 423–426.

CIARLET, P.G.; LAURENT, F. [1987]: Global existence of displacement fields whose Cauchy-Green strain tensor is known, to appear.

CIARLET, P.G.; NEČAS, J. [1985]: Unilateral problems in nonlinear, three-dimensional elasticity, *Arch. Rational Mech. Anal.* **87**, 319–338.

CIARLET, P.G.; NEČAS, J. [1987]: Injectivity and self-contact in nonlinear elasticity, *Arch. Rational Mech. Anal.* **19**, 171–188.

CIARLET, P.G.; PAUMIER, J.C. [1986]: A justification of the Marguerre-von Kármán equations, *Computational Mechanics*, **1**, 177–202.

CIARLET, P.G.; QUINTELA-ESTEVEZ, P. [1987]: Polyconvexity and the von Kármán equations, to appear.

CIARLET, P.G.; RABIER, P. [1980]: *Les Equations de von Kármán*, Lecture Notes in Mathematics, Vol. 826, Springer-Verlag, Berlin.

CIORANESCU, D.; SAINT JEAN PAULIN, J. [1979]: Homogenization in open sets with holes, *J. Math. Anal. Appl.* **71**, 590–607.

CIORANESCU, D.; SAINT JEAN PAULIN, J. [1986]: Reinforced and honeycomb structures, *J. Math. Pures et Appliquées*, **65**, 403–422.

COCU, M. [1984]: Existence of solutions of Signorini problems with friction, *Internat. J. Engrg. Sci.* **22**, 567.

CODDINGTON, E.A.; LEVINSON, N. [1955]: *Theory of Ordinary Differential Equations*, McGraw Hill, New York.

COHEN, H.; WANG, C.-C. [1984]: A note on hyperelasticity, *Arch. Rational Mech. Anal.* **85**, 213–236.

COHEN, H.; WANG, C.-C. [1987]: On the response and symmetry of elastic materials with internal constraints, *Arch. Rational Mech. Anal.* **99**, 1–36.

COLEMAN, B.D.; NOLL, W. [1959]: On the thermostatics of continuous media, *Arch. Rational Mech. Anal.* **4**, 97–128.

COLEMAN, B.D., NOLL, W. [1963]: The thermodynamics of elastic materials with heat conduction and viscosity, *Arch. Rational Mech. Anal.* **13**, 167–178.

COLEMAN, B.D.; NOLL, W. [1964]: Material symmetry and thermostatic inequalities in finite elastic deformations, *Arch. Rational Mech. Anal.* **15**, 87–111.

CROUZEIX, M.; MIGNOT, A. [1984]: *Analyse Numérique des Equations Différentielles*, Masson, Paris.

DACOROGNA, B. [1981]: A relaxation theorem and its application to the equilibrium of gases, *Arch. Rational Mech. Anal.* **77**, 359–386.

DACOROGNA, B. [1982a]: *Weak Continuity and Weak Lower Semicontinuity of Non-Linear Functionals*, Lecture Notes in Mathematics, Vol. 922, Springer-Verlag, Berlin.

DACOROGNA, B. [1982b]: Quasiconvexity and relaxation of non convex problems in the calculus of variations, *J. Functional Analysis* **46**, 102–118.

DACOROGNA, B. [1985]: Remarques sur les notions de polyconvexité, quasiconvexité et convexité de rang 1, *J. Math. Pures Appl.* **64**, 403–438.

DACOROGNA, B. [1987]: *Direct Methods in the Calculus of Variations*, to appear.

DAFERMOS, C.M. [1973]: Solutions of the Riemann problem for a class of hyperbolic systems of conservation laws by the viscosity method, *Arch. Rational Mech. Anal.* **52**, 1–9.

DAFERMOS, C.M. [1986]: Quasilinear hyperbolic systems with involutions, *Arch. Rational Mech. Anal.* **94**, 373–389.

DAFERMOS, C.M.; HRUSA, W.J. [1985]: Energy methods for quasilinear hyperbolic initial-boundary value problems. Applications to elastodynamics, *Arch. Rational Mech. Anal.* **87**, 267–292.

DAHLBERG, B.E.J. [1979]: A note on Sobolev spaces, in *Proceedings of Symposia in Pure Mathematics, Vol. 35, Part 1*, pp. 183–185, American Mathematical Society, Providence.

DAUTRAY, R.; LIONS, J.-L. [1984]: *Analyse Mathématique et Calcul Numérique pour les Sciences et les Techniques*, Collection du Commissariat à l'Energie Atomique, Masson, Paris.

Davet, J.L. [1985]: Sur les densités d'énergie en élasticité non linéaire: confrontation de modèles et de travaux expérimentaux, *Annales des Ponts et Chaussées*, 3ème trimestre, 2–33.

de Boor, C. [1985]: A naive proof of the representation theorem for isotropic, linear asymmetric stress-strain relations, *J. Elasticity* **15**, 225–227.

De Giorgi, E. [1975]: Sulla convergenza di alcune successioni d'integrali del tipo dell'area, *Rend. Matematica* **8**, 277–294.

De Giorgi, E. [1983]: G-operators and Γ-convergence, in *Proceedings International Congress of Mathematicians, Warsaw, 1983*, pp. 1175–1191.

Deimling, K. [1985]: *Nonlinear Functional Analysis*, Springer-Verlag, Berlin.

Del Piero, G. [1985a]: A mathematical characterization of plastic phenomena in the context of Noll's new theory of simple materials, *Engrg. Fracture Mech.* **21**, 633–639.

Del Piero, G. [1985b]: *Unilateral Problems in Structural Analysis*, Atti dell'Istituto di Meccanica Teorica ed Applicata, Università degli Studi, Udine.

Demengel, F. [1985a] Relaxation et existence pour le problème des matériaux à blocage, *Math. Modelling & Numer. Anal.* **19**, 351–395.

Demengel, F. [1985b]: Déplacements à déformations bornées et champs de contrainte mesure, *Ann. Scuola Norm. Sup Pisa, Classe di Scienze, Ser. IV*, **12**, 243–318.

Demengel, F.; Suquet, P. [1986]: On locking materials, *Acta Applicandae Matematicae* **6**, 185–211.

Demengel, F.; Temam, R. [1984]: Convex functions of a measure and applications, *Indiana Univ. Math. J.* **33**, 673–709.

Demengel, F.; Temam, R. [1986]: Convex function of a measure: The unbounded case, in *Fermat Days 1985: Mathematics for Optimization* (J.-B. Hirriart-Urruty, Editor), pp. 103–134, North-Holland, Amsterdam.

Destuynder, P.; Galbe, G. [1978]: Analyticité de la solution d'un problème hyperélastique non linéaire, *C.R. Acad. Sci. Paris Sér. A* **287**, 365–368.

De Turck, D.M.; Yang, D. [1983]: Existence of elastic deformations with prescribed principal strains and triply orthogonal systems.

Dieudonné, J. [1968]: *Eléments d'Analyse, Tome 1: Fondements de l'Analyse Moderne*, Gauthier-Villars, Paris.

Dieudonné, J. [1982]: *Eléments d'Analyse, Tome 9*, Gauthier-Villars, Paris.

DiPerna, R. [1983]: Convergence of approximate solutions to conservation laws, *Arch. Rational Mech. Anal.* **82**, 27–70.

DiPerna, R.J. [1985a]: Measure-valued solutions to conservation laws, *Arch. Rational Mech. Anal.* **88**, 223–270.

DiPerna, R.J. [1985b]: Compensated compactness and general systems of conservation laws, *Trans. Amer. Math. Soc.* **292**, 383–420.

Doubrovine, B.; Novikov, S.; Fomenko, A. [1982a]: *Géométrie Contemporaine, Méthodes et Applications. Première Partie: Géométrie des Surfaces, des Groupes de Transformations, et des Champs*, Editions Mir, Moscou (original Russian edition: 1979; English translation: *Modern Geometry, Vol. I*, Springer-Verlag, New York, 1984).

Doubrovine, B.; Novikov, S.; Fomenko, A. [1982b]: *Géométrie Contemporaine, Méthodes et Applications. Deuxième Partie: Géométrie et Topologie des Variétés*, Editions Mir, Moscou (original Russian edition: 1979; English translation: *Modern geometry, Vol II*, Springer-Verlag, New York, 1985).

Doyle, T.C.; Ericksen, J.L. [1956]: Nonlinear elasticity, in *Advances in Applied Mechanics* (H.L. Dryden & T. von Kármán, Editors), pp. 53–115, Academic Press, New York.

Dunn, J.E. [1981]: Possibilities and impossibilities in affine constitutive theories for stress in solids, report.

Duvaut, G. [1978]: Analyse fonctionnelle et mécanique des milieux continus, applications à l'étude des matériaux composites élastiques à structure périodique, homogénéisation, in *Theoretical and Applied Mechanics* (W.T. Koiter, Editor), pp. 119–131, North-Holland, Amsterdam.

Duvaut, G.; Lions, J.-L. [1972]: *Les Inéquations en Mécanique et en Physique*, Dunod, Paris (English translation: Springer-Verlag, Berlin, 1976).

Ebin, D.G.; Saxton, R.A. [1986]: The initial value problem for incompressible elastodynamics, *Arch. Rational Mech. Anal.* **94**, 15–38.

Edelen, D.G.B. [1969a]: Protoelastic bodies with large deformation, *Arch. Rational Mech. Anal.* **34**, 283–300.

Edelen, D.G.B. [1969b]: Nonlocal variational mechanics, *Internat. J. Engrg. Sci.* **7**, Part I: 269–285, Part II: 287–293, Part III: 373–389, Part IV: 391–399, Part V: 401–415, Part VI: 677–688, Part VII: 843–847.

Edelen, D.G.B. [1970]: Operation separable problems in the nonlocal calculus of variations, *J. Math. Anal. Appl.* **32**, 445–452.

Ekeland, I. [1979]: Nonconvex minimization problems, *Bull. Amer. Math. Soc.* **1**, 443–474.

Ekeland, I.; Temam, R. [1974]: *Analyse Convexe et Problèmes Variationnels*, Dunod, Paris (English translation: *Convex Analysis and Variational Problems*, North-Holland, Amsterdam, 1975).

Ekeland, I.; Turnbull, T. [1984]: *Infinite-Dimensional Optimization and Convexity*, The University of Chicago Press, Chicago.

Ericksen, J.E. [1954]: Deformations possible in every isotropic, incompressible perfectly elastic body, *Z. angew. Math. Phys.* **5**, 466–486.

Ericksen, J.L. [1955a]: Deformations possible in every compressible, isotropic, perfectly elastic material, *J. Math. Phys.* **34**, 126–128.

Ericksen, J.L. [1955b]: Inversion of a perfectly elastic spherical shell, *Z. angew. Math. Mech.* **35**, 382–385.

Ericksen, J.L. [1963]: Nonexistence theorems in linear elasticity theory, *Arch. Rational Mech. Anal.* **14**, 180–183.

Ericksen, J.L. [1965]: Nonexistence theorems in linearized elastostatics, *J. Differential Equ.* **1**, 446–451.

Ericksen, J.L. [1979]: On the symmetry of deformable crystals, *Arch. Rational Mech. Anal.* **72**, 1–13.

Ericksen, J.L. [1983]: Ill-posed problems in thermoelasticity theory, in *Systems of Nonlinear Partial Differential Equations* (J.M. Ball, Editor), pp. 71–93, Reidel, Dordrecht.

Ericksen, J.L. [1986]: Stable equilibrium configurations of elastic crystals, *Arch. Rational Mech. Anal.* **94**, 1–14.

Ericksen, J.L. [1987]: Twinning of crystals I, in *Metastability and Incompletely Posed Problems, IMA Vol. Math. Appl. 3* (S.S. Antman, J.L. Ericksen, D. Kinderlehrer, I. Müller, Editors), pp. 77–96, Springer-Verlag, New York.

Ericksen, J.L.; Toupin, R.A. [1956]: Implications of Hadamard's condition for elastic stability with respect to uniqueness theorems, *Canad. J. Math.* **8**, 432–436.

Eringen, A.C. [1962]: *Nonlinear Theory of Continuous Media*, McGraw-Hill, New York.

Eringen, A.C. [1966]: A unified theory of thermomechanical materials, *Internat. J. Engrg. Sci.* **4**, 179–202.

Eringen, A.C. [1978]: Nonlocal continuum mechanics and some applications, in *Nonlinear Equations in Physics and Mathematics* (A.O. Barut, Editor), pp. 271–318, Reidel.

Eringen, A.C.; Edelen, D.G.B. [1972]: On nonlocal elasticity, *Internat. J. Engrg. Sci.* **10**, 233–248.

EULER, L. [1757]: Continuation des recherches sur la théorie du mouvement des fluides, *Hist. Acad. Berlin*, 316–361.

EULER, L. [1758]: Du mouvement de rotation des corps solides autour d'un axe variable, *Mem. Acad. Roy. Sci. et Belles-Lettres de Berlin* **14**, 154–193.

EULER, L. [1771]: Sectio tertia de motu fluidorum lineari potissimum aquae, *Novi Comm. Petrop.* **15**, 219–360.

EVANS, L.C. [1986]: Quasiconvexity and partial regularity in the calculus of variations, *Arch. Rational Mech. Anal.* **95**, 227–252.

FEDERER, H. [1969]: *Geometric Measure Theory*, Springer-Verlag, New York.

FICHERA, G. [1964]: Problemi elastostatici con vincoli unilaterali: il problema di Signorini con ambigue condizioni al contorno, *Mem. Accad. Naz. Lincei Ser. VIII* **7**, 91–140.

FICHERA, G. [1967]: Semicontinuità ed esistenza del minimo per una classe di integrali multipli, *Rev. Roum. Math. Pures et Appl.* **12**, 1217–1220.

FICHERA, G. [1972a]: Existence theorems in elasticity, *Handbuch der Physik*, VIa/2, pp. 347–389, Springer-Verlag, Berlin.

FICHERA, G. [1972b]: Boundary value problems of elasticity with unilateral constraints, *Handbuch der Physik*, VIa/2, 391–424, Springer-Verlag, Berlin.

FICHERA, G. [1972c]: *Linear Elliptic Differential Systems and Eigenvalue Problems*, Lecture Notes Series, Vol. 8, Springer-Verlag, Berlin.

FICHERA, G. [1977]: Problemi unilaterali nella statica dei sistemi continui, in *Problemi Attuali di Meccanica Teorica e Applicata*, pp. 169–178, Torino.

FLORY, P.J.; TATARA, Y. [1975]: The elastic free energy and the elastic equation of state: elongation and swelling of polydimethylsiloxane networks; *J. Polymer Science* **13**, 683–702.

FONG, J.T.; PENN, W. [1975]: Construction of a strain-energy function for an isotropic elastic material, *Trans. Soc. Rheology* **19**, 99–113.

FONSECA, I. [1987a]: Variational methods for elastic crystals, *Arch. Rational Mech. Anal.* **19**, 189–220.

FONSECA, I. [1987b]: The lower quasiconvex envelope of the stored energy function for an elastic crystal, to appear.

FONSECA, I. [1987c]: Stability of elastic crystals, to appear.

FOSDICK, R.L.; MACSITHIGH, G.P. [1986]: Minimization in incompressible nonlinear elasticity theory, *J. Elasticity* **16**, 267–301.

FOSDICK, R.L.; SERRIN, J. [1979]: On the impossibility of linear Cauchy and Piola-Kirchhoff constitutive theories for stress in solids, *J. Elasticity* **9**, 83–89.

FRAEIJS DE VEUBEKE, B.M. [1979]: *A Course in Elasticity*, Springer-Verlag, New York.

FRANCFORT, G.A.; MURAT, F. [1986]: Homogenization and optimal bounds in linear elasticity, *Arch. Rational Mech. Anal.* **94**, 307–334.

FRÉMOND, M. [1985]: Contact unilatéral avec adhérence, in *Unilateral Problems in Structural Analysis* (G. DEL PIERO, F. MACERI, Editors), Springer-Verlag, Heidelberg.

FRIEDRICHS, K.O. [1947]: On the boundary-value problems of the theory of elasticity and Korn's inequality, *Annals of Math.* **48**, 441–471.

GALKA, A.; TELEGA, J.J. [1982]: A variational method for finite elasticity in the case of non-potential loadings, I. First Piola-Kirchhoff stress tensor; II. Symmetric stress tensor and some comments, *Bull. Acad. Polonaise des Sc., Sér. Sci. Tech.* **30**, 471–485.

GEAR, C.W. [1971]: *Numerical Initial Value Problems in Ordinary Differential Equations*, Prentice Hall, Englewood Cliffs.

GENT, A.N.; THOMAS, A.G. [1959]: The deformation of foamed elastic materials, *J. Applied Polymer Science* **1**, 107–113.

GERMAIN, P. [1972]: *Mécanique des Milieux Continus*, Tome I, Masson, Paris.

GERMAIN, P. [1973]: La méthode des puissance virtuelles en mécanique des milieux continus, *J. Mécanique* **12**, 236–274.

GEYMONAT, G. [1965]: Sui problemi ai limiti per i sistemi lineari ellitici, *Ann. Mat. Pura Appl.* **69**, 207–284.

GIAQUINTA, M. [1983a]: The regularity of extremals of variational integrals, in *Systems of Nonlinear Partial differential Equations* (J.M. BALL, Editor), pp. 115–145, Reidel, Dordrecht.

GIAQUINTA, M. [1983b]: *Multiple Integrals in the Calculus of Variations and Nonlinear Elliptic Systems*, Princeton University Press, Princeton.

GIAQUINTA, M.; MODICA, G. [1982]: Non linear systems of the type of the stationary Navier-Stokes system, *J. reine angew. Math.* **330**, 173–214.

GIBSON, L.J.; ASHBY, M.F. [1982]: The mechanics of three-dimensional cellular materials, *Proc. Roy. Soc. London* **A 382**, 43–59.

GIUSTI, E. [1983]: Some aspects of the regularity theory for nonlinear elliptic systems, in *Systems of Nonlinear Partial Differential Equations* (J.M. BALL, Editor), pp. 147–172, Reidel, Dordrecht.

GLIMM, J. [1965]: Solutions in the large for nonlinear hyperbolic systems of equations, *Comm. Pure Appl. Math.* **18**, 697–715.

GLOWINSKI, R. [1984]: *Numerical Methods for Nonlinear Variational Problems*, Springer-Verlag, New York.

GLOWINSKI, R.; LE TALLEC, P. [1982]: Numerical solution of problems in incompressible finite elasticity by augmented Lagrangian methods. I. Two-dimensional and axisymmetric problems, *SIAM J. Appl. Math.* **42**, 400–429.

GLOWINSKI, R.; LIONS, J.-L.; TRÉMOLIÈRES, R. [1976]: *Analyse Numérique des Inéquations Variationnelles, Vol. I: Théorie Générale et Premières Applications*; *Vol. II: Applications aux Phénomènes Stationnaires et d'Evolution*, Dunod, Paris (English translation: *Numerical Analysis of Variational Inequalities*, North-Holland, Amsterdam, 1981).

GOBERT, J. [1962]: Une inégalité fondamentale de la théorie de l'élasticité, *Bull. Soc. Royale Sciences Liège* **3–4**, 182–191.

GOHBERG, I.C.; KREJN, M.G. [1971]: *Introduction à la Théorie des Opérateurs Linéaires Non Auto-Adjoints dans un Espace Hilbertien*, Dunod, Paris, 1971 (original Russian edition: Nauka, Moscow, 1965).

GREEN, A.E.; ADKINS, J.E. [1970]: *Large Elastic Deformations*, Second Edition, Clarendon Press, Oxford.

GREEN, A.E.; RIVLIN, R.S. [1964]: Multipolar continuum mechanics, *Arch. Rational Mech. Anal.* **17**, 113–147.

GREEN, A.E.; ZERNA, W. [1968]: *Theoretical Elasticity*, University Press, Oxford.

GRIOLI, G. [1962]: *Mathematical Theory of Elastic Equilibrium*, Ergebnisse der Angew. Math. Vol. 67, Springer, Berlin.

GRIOLI, G. [1983]: Mathematical problems in elastic equilibrium with finite deformations, *Applicable Analysis* **15**, 171–186.

GRISVARD, P. [1985]: *Elliptic Problems in Nonsmooth Domains*, Pitman, Boston.

GRISVARD, P. [1987]: Singularities in elasticity thoery, in *Applications of Multiple Scaling in Mechanics* (P.G. CIARLET & M. ROSEAU, Editors), Masson, Paris.

GUO, ZHONG-HENG [1963]: Homographic representation of the theory of finite thermoelastic deformations, *Arch. Mech. Stos.* **15**, 475–505.

GUO, ZHONG-HENG [1980]: Unified theory of variation principles in non-linear theory of elasticity, *Applied Mathematics and Mechanics* **1**, 1–22.

GUO, ZHONG-HENG [1981]: Representation of orthogonal tensors, *SM Archives* **6**, 451–466.

GUO, ZHONG-HENG [1983a]: An alternative proof of the representation theorem for isotropic, linear asymmetric stress-strain relations, *Quart. Appl. Math.* **41**, 119–123.

GUO, ZHONG-HENG [1983b]: The representation theorem for isotropic, linear asymmetric stress-strain relations, *J. Elasticity* **13**, 121–124.

GUO, ZHONG-HENG [1984]: Rates of stretch tensors, *J. Elasticity* **14**, 263–267.

GURTIN, M.E. [1972]: The linear theory of elasticity, in *Handbuch der Physik*, pp. 1–295, Vol. VIa/2 (S. FLÜGGE & C. TRUESDELL, Editors), Springer-Verlag, Berlin.

GURTIN, M.E. [1973]: Thermodynamics and the potential energy of an elastic solid, *J. Elasticity* **3**, 23–26.

GURTIN, M.E. [1974]: A short proof of the representation theorem for isotropic, linear stress-strain relations, *J. Elasticity* **4**, 243–245.

GURTIN, M.E. [1978]: On the nonlinear theory of elasticity, in *Contemporary Developments in Continuum Mechanics and Partial Differential Equations* (G.M. de LA PENHA & L.A.J. MEDEIROS, Editors), pp. 237–253, North-Holland, Amsterdam.

GURTIN, M.E. [1981a]: *Topics in Finite Elasticity*, CBMS-NSF Regional Conference Series in Applied Mathematics, SIAM, Philadelphia.

GURTIN, M.E. [1981b]: *An Introduction to Continuum Mechanics*, Academic Press, New York.

GURTIN, M.E. [1982]: On uniqueness in finite elasticity, in *Finite Elasticity* (D.E. CARLSON & R.T. SHIELD, Editors), pp. 191–199, Martinus Nijhoff Publishers, The Hague.

GURTIN, M.E.; MARTINS, L.C. [1976]: Cauchy's theorem in classical physics, *Arch. Rational Mech. Anal.* **60**, 305–324, 325–328.

GURTIN, M.E.; MIZEL, V.J; WILLIAMS, W.O. [1968]: A note on Cauchy's stress theorem, *J. Math. Anal. Appl.* **22**, 398–401.

GURTIN, M.E.; SPECTOR, S.J. [1979]: On stability and uniqueness in finite elasticity, *Arch. Rational Mech. Anal.* **70**, 153–165.

GURTIN, M.E.; TEMAM, R. [1981]: On the anti-plane shear problem in finite elasticity, *J. Elasticity* **11**, 197–206.

GURTIN, M.E.; WILLIAMS, W.O. [1967]: An axiomatic foundation for continuum thermodynamics, *Arch. Rational Mech. Anal.* **26**, 83–117.

GWINNER, J. [1986]: A penalty approximation for unilateral contact problems in nonlinear elasticity, Preprint Nr. 1024, Fachbereich Mathematik, Technische Hochschule Darmstadt.

HADHRI, T. [1985]: Fonction convexe d'une measure, *C.R. Acad. Sci. Paris, Sér. I* **301**, 687–690.

HADHRI, T. [1986]: Convex function of a measure and application to a problem of nonhomogeneous elastoplastic material, to appear.

HANYGA, A. [1985]: *Mathematical Theory of Non-Linear Elasticity*, PWN-Polish Scientific Publishers, Warsaw, and Ellis Horwood, Chichester.

HARTMAN, P. [1964]: *Ordinary Differential Equations*, John Wiley, London.

HENRICI, P. [1962]: *Discrete Variable Methods in Ordinary Differential Equations*, John Wiley, New York.

HILDEBRANDT, S. [1977]: Regularity results for solutions of quasilinear elliptic systems, *Convegno su: Sistemi ellitici nonlineari ed applicazioni*, Ferrara.

HILL, R. [1957]: On uniqueness and stability in the theory of finite elastic strain, *J. Mech. Phys. Solids* **5**, 229–241.

HILL, R. [1961]: Bifurcation and uniqueness in non-linear mechanics of continua, in *Problems of Continuum Mechanics*, SIAM, Philadelphia.

HILL, R. [1968]: On constitutive inequalities for simple materials, I, *J. Mech. Phys. Solids* **16**, 229–242.

HILL, R. [1970]: Constitutive inequalities for isotropic elastic solids under finite strain, *Proc. Roy. Soc. London* **A314**, 457–472.

HLAVÁČEK, I., NEČAS, J. [1970a]: On inequalities of Korn's type, I. Boundary value problems for elliptic systems of partial differential equations, *Arch. Rational Mech. Anal.* **36**, 305–311.

HLAVÁČEK, I.; NEČAS, J. [1970b]: On inequalities of Korn's type, II. Applications to linear elasticity, *Arch. Rational Mech. Anal.* **36**, 312–334.

HOGER, A. [1985]: On the residual stress possible in an elastic body with material symmetry, *Arch. Rational Mech. Anal.* **88**, 271–289.

HOGER, A. [1986]: On the determination of residual stress in an elastic body, *J. Elasticity* **16**, 303–324.

HOGER, A.; CARLSON, D.E. [1984a]: On the derivative of the square root of a tensor and Guo's rate theorems, *J. Elasticity* **14**, 329–336.

HOGER, A.; CARLSON, D.E. [1984b]: Determination of the stretch and rotation in the polar decomposition of the deformation gradient, *Quart. Appl. Math.* **42**, 113–117.

HÖRMANDER, L. [1983]: *The Analysis of Partial Differential Equations*; *I*; Grundlehren der mathematischen Wissenschaften, Vol. 256, Springer-Verlag, Berlin.

HRUSA, W.J.; RENARDY, M. [1986]: An existence theorem for the Dirichlet initial-boundary value problem in incompressible nonlinear elasticity, to appear.

HUGHES, T.J.R.; KATO, T.; MARSDEN, J.E. [1976]: Well-posed quasi linear second-order hyperbolic systems with applications to non-linear elastodynamics and general relativity, *Arch. Rational Mech. Anal.* **63**, 273–294.

HUREWICZ, W.; WALLMAN, H. [1948]: *Dimension Theory*, Princeton.

IOFFE, A.D.; TIKHOMIROV, V.M. [1974]: *Theory of Extremal Problems* (in Russian), Nauka, Moscow (English translation: North-Holland, Amsterdam, 1979).

JAMES, R.D. [1981]: Finite deformations by mechanical twinning, *Arch. Rational Mech. Anal.* **77**, 143–176.

JOHN, F. [1961]: Rotation and strain, *Comm. Pure Appl. Math.* **14**, 391–413.

JOHN, F. [1964]: Remarks on the non-linear theory of elasticity, *Seminari Ist. Naz. Alta Matem. 1962/1963*, 474–482.

JOHN, F. [1972a]: Uniqueness of non-linear elastic equilibrium for prescribed boundary displacements and sufficiently small strains, *Comm. Pure Appl. Math.* **25**, 617–634.

JOHN, F. [1972b]: Bounds for deformations in terms of average strains, in *Inequalities, III* (O. SHISHA, Editor), pp. 129–144, Academic Press, New York.

JOHN, F. [1975]: A priori estimates, geometric effects and asymptotic behavior, *Bull. Amer. Math. Soc.* **81**, 1013.

VON KÁRMÁN, T.; BIOT, M.A. [1940]: *Mathematical Methods in Engineering*, McGraw-Hill, New York.

KATO, T. [1979]: *Linear and Quasi-Linear Equations of Evolution of Hyperbolic Type*, CIME Lectures, Cortona.

KEELING, S.L. [1987]: On Lipschitz continuity of nonlinear differential operators, *ICASE Report No. 87-12*, NASA Langley Research Center.

KIKUCHI, N.; ODEN, J.T. [1987]: *Contact Problems in Elasticity*, SIAM Publications, Philadelphia.

KINDERLEHRER, D. [1987a]: Twinning in crystals II, in *Metastability and Incompletely Posed Problems, IMA Vol. Math. Appl. 3.* (S.S. ANTMAN, J.L. ERICKSEN, D. KINDERLEHRER, I. MÜLLER, Editors), pp. 185–211, Springer-Verlag, New York.

KINDERLEHRER, D. [1987b]: Remarks about the equilibrium configurations of crystals, in *Proceedings, Symposium on Material Instabilities in Continuum Mechanics, Heriot-Watt University* (J. BALL, Editor), Oxford.

KINDERLEHRER, D.; STAMPACCHIA, G. [1980]: *An Introduction to Variational Inequalities and their Applications*, Academic Press.

KIRCHHOFF, G. [1852]: Über die Gleichungen des Gleichgewichts eines elastischen Körpers bei nicht unendlich kleinen Verschiebungen seiner Theile, *Sitzgsber. Akad. Wiss. Wien* **9**, 762–773.

KNOPS, R.J.; PAYNE, L.E. [1971]: *Uniqueness Theorems in Linear Elasticity*, Springer Tracts in Natural Philosophy, Vol. 19.

KNOPS, R.J.; STUART, C.A. [1984]: Quasiconvexity and uniqueness of equilibrium solutions in nonlinear elasticity, *Arch. Rational Mech. Anal.* **86**, 233–249.

KNOWLES, J.K.; STERNBERG, E. [1973]: An asymptotic finite-deformation analysis of the elastostatic field near the tip of a crack, *J. Elasticity* **3**, 67–107.

KNOWLES, J.K.; STERNBERG, E. [1975]: On the ellipticity of the equations of non-linear elastostatics for special material, *J. Elasticity* **5**, 341–361.

KNOWLES, J.K.; STERNBERG, E. [1977]: On the failure of ellipticity of the equations for finite elastic plane strain, *Arch. Rational Mech. Anal.* **63**, 321–336.

KNOWLES, J.K.; STERNBERG, E. [1978]: On the failure of ellipticity and the emergence of discontinuous deformation gradients in plane finite elastostatics, *J. Elasticity* **8**, 329–379.

KOHN, R.V. [1982]: New integral estimates for deformations in terms of their nonlinear strains, *Arch. Rational Mech. Anal.* **78**, 131–172.

KOHN, R.V.; STRANG, G. [1983]: Explicit relaxation of a variational problem in optimal design, *Bull. Amer. Math. Soc.* **9**, 211–214.

KOHN, R.V.; STRANG, G. [1986]: Optimal design and relaxation of variational problems, *Comm. Pure Appl. Math.* **39**, Part I: 113–137, Part II: 139–182, Part III: 353–377.

KOHN, R.; TEMAM, R. [1983]: Dual spaces of stresses and strains with applications to Hencky plasticity, *Appl. Math. Optimization* **10**, 1–35.

KORN, A. [1907]: Sur un problème fondamental dans la théorie de l'élasticité, *C.R. Acad. Sci. Paris* **145**, 165–169.

KORN, A. [1908]: Solution générale du problème d'équilibre dans la théorie de l'élasticité dans le cas où les efforts sont donnés à la surface, *Ann. Fac. Sci. Univ. Toulouse* **10**, 165–269.

KORN, A. [1914]: Über die Lösungen der Grundprobleme der Elastizitatstheorie, *Math. Annal.* **75**, 497.

KOSELEV, A.I. [1958]: A priori estimates in L_p and generalized solutions of elliptic equations and systems; *Uspehi Mat. Nauk (N.S.)* **13**, n° 4 (82), 29–88 (English translation: *American Mathematical Society Translations, Series 2*, **20** (1962), 105–171).

KOSLOV, S.M.; OLEINIK, O.A.; ZHIKOV, V.V.; KHA TEN NGOAN [1979]: Averaging and G-convergence of differential operators, *Russian Math. Surveys* **34**, 69–147.

KRASNOSELSKII, M.A.; ZABREYKO, P.P.; PUSTYLNIK, E.I.; SBOLEVSKII, P.E. [1976]: *Integral Operators in Spaces of Summable Functions*, Noordhoff, Leyden.

LABISCH, F.K. [1982]: On the dual formulation of boundary value problems in nonlinear elastostatics, *Internat. J. Engrg. Sci.* **20**, 413–431.

LANDAU, L.; LIFCHITZ, E. [1967]: *Théorie de l'Elasticité*, Mir, Moscou.

LANG, S. [1962]: *Introduction to Differential Manifolds*, John Wiley, New York.

LANZA DE CRISTOFORIS, M.; VALENT, T. [1982]: On Neumann's problem for a quasilinear differential system of the finite elastostatics type. Local theorems of existence and uniqueness, *Rend. Sem. Mat. Univ. Padova* **68**, 183–206.

LAVRENTIEV, M. [1926]: Sur quelques problèmes du calcul des variations, *Ann. Mat. Pura Appl.* **4**, 7–28.

LAX, P.D. [1964]: Development of singularities of solutions of nonlinear hyperbolic differential equations, *J. Math. Phys.* **5**, 611–613.

LAY, S.R. [1982]: *Convex Sets and their Applications*, J. Wiley, New York.

LE DRET, H. [1985]: Constitutive laws and existence questions in incompressible nonlinear elasticity, *J. Elasticity* **15**, 369–387.

LE DRET, H. [1986]: Incompressible limit behaviour of slightly compressible nonlinear elastic materials, *Modélisation Mathématique et Analyse Numérique* **20**, 315–340.

LE DRET, H. [1987]: Structure of the set of equilibrated loads in nonlinear elasticity and applications to existence and nonexistence, *J. Elasticity* **17**, 123–141.

LELONG-FERRAND, J. [1963]: *Géométrie Différentielle*, Masson, Paris.

LERAY, J.; SCHAUDER, J. [1934]: Topologie et équations fonctionnelles, *Ann. Sci. Ecole Normale Sup.* **51**, 45–78.

LE TALLEC, P.; ODEN, J.T. [1980]: On the existence of hydrostatic pressure in regular finite deformations of incompressible hyperelastic solids, in *Nonlinear Partial Differential Equations in Engineering and Applied Science* (R.L. STERNBERG; A.J. KALINOWSKI; J.S. PAPADAKIS, Editors), pp. 1–13, Marcel Dekker, New York.

LE TALLEC, P.; ODEN, J.T. [1981]: Existence and characterization of hydrostatic pressure in finite deformations of incompressible elastic bodies, *J. Elasticity* **11**, 341–358.

LE TALLEC, P.; VIDRASCU, M. [1984]: Une méthode numérique pour les problèmes d'équilibre de corps hyperélastiques compressibles en grandes déformations, *Numer. Math.* **43**, 199–224.

LIONS, J.-L. [1965]: *Problèmes aux Limites dans les Equations aux Dérivées Partielles*, Presses de l'Université de Montréal, Montréal.

LIONS, J.-L. [1969]: *Quelques Méthodes de Résolution des Problèmes aux Limites Non Linéaires*, Dunod, Paris.

LIONS, J.-L.; MAGENES, E. [1968]: *Problèmes aux Limites Non Homogènes et Applications*, Tome 1, Dunod, Paris.

LIONS, J.-L.; STAMPACCHIA, G. [1967]: Variational inequalities, *Comm. Pure Applied Math.* **20**, 493–519.

LLOYD, N.G. [1978]: *Degree Theory*, Cambridge University Press, Cambridge.

LOVE, A.E.H. [1927]: *A Treatise on the Mathematical Theory of Elasticity*, Cambridge University Press, Cambridge (reprinted by Dover Publications, New York, 1944).

MALLIAVIN, P. [1972]: *Géométrie Différentielle Intrinsèque*, Hermann, Paris.

MARCELLINI, P. [1984]: Quasiconvex quadratic forms in two dimensions, *Appl. Math. Optimization* **11**, 183–189.

MARCELLINI, P. [1986a]: On the definition and the lower semicontinuity of certain quasiconvex integrals, *Ann. Inst. Henri Poincaré* **3**, 391–409.

MARCELLINI, P. [1986b]: Approximation of quasiconvex functions, and lower semicontinuity of multiple integrals, to appear.

MARCELLINI, P. [1986c]: A relation between existence of minima for non convex integrals and uniqueness for non strictly convex integrals of the calculus of variations, to appear.

MARCELLINI, P.; SBORDONE, C. [1980]: Semicontinuity problems in the calculus of variations, *Nonlinear Anal.* **4**, 241–257.

MARCELLINI, P.; SBORDONE, C. [1983]: On the existence of minima of multiple integrals of the calculus of variations, *J. Math. Pures Appl.* **62**, 1–9.

MARCUS, M.; MIZEL, V.J. [1973]: Transformations by functions in Sobolev spaces and lower semicontinuity for parametric variational problems, *Bull. Amer. Math. Soc.* **79**, 790–795.

MARQUES, M.D.P.M. [1984]: Hyperélasticité et existence de fonctionnelle d'énergie, *J. Mécanique Théorique et Appliquée* **3**, 339–347.

MARQUES, M.D.P.M.; MOREAU, J.J. [1982]: Isotropie et convexité dans l'espace des tenseurs symétriques, *Séminaire d'Analyse Convexe*, Exposé n° 6, Montpellier.

MARSDEN, J.E.; HUGHES, T.J.R. [1978]: Topics in the mathematical foundations of elasticity, in *Nonlinear Analysis and Mechanics: Heriot-Watt Symposium*, Vol. 2, pp. 30–285, Pitman, London.

MARSDEN, J.E.; HUGHES, T.J.R. [1983]: *Mathematical Foundations of Elasticity*, Prentice-Hall, Englewood Cliffs.

MARSDEN, J.E.; WAN, Y.H. [1983]: Linearization stability and Signorini series for the traction problem in elastostatics, *Proc. Roy. Soc. Edinburgh* **A95**, 171–180.

MARTINI, R. [1976]: On the Fréchet differentiability of certain energy functionals, *Proc. Kon. Ned. Ak. Wet.* **A79**, 326–330.

MARTINS, L.C. [1976]: On Cauchy's theorem in classical physics: some counterexamples, *Arch. Rational Mech. Anal.* **60**, 325–328.

MARTINS, L.C.; OLIVEIRA, R.F.; PODIO-GUIDUGLI, P. [1987]: On the vanishing of the additive measures of strain and rotation for finite deformations, *J. Elasticity* **17**, 189–193.

MARTINS, L.C.; PODIO-GUIDUGLI, P. [1978]: A new proof of the representation theorem for isotropic, linear constitutive relations, *J. Elasticity* **8**, 319–322.

MARTINS, L.C.; PODIO-GIUDUGLI, P. [1979]: A variational approach to the polar decomposition theorem, *Rend. Accad. Naz. Lincei.* **66**, 487–493.

MARTINS, L.C.; PODIO-GUIDUGLI, P. [1980]: An elementary proof of the polar decomposition theorem, *American Math. Monthly* **87**, 288–290.

MASCOLO, E.; SCHIANCHI, R. [1983]: Existence for non convex problems, *J. Math. Pures Appl.* **62**, 349–359.

MASON, J. [1980]: *Variational, Incremental and Energy Methods in Solid Mechanics and Shell Theory*, Elsevier, Amsterdam.

MEISTERS, G.H.; OLECH, C. [1963]: Locally one-to-one mappings and a classical theorem on Schlicht functions, *Duke Math. J.* **30**, 63–80.

MEYERS, N.G. [1965]: Quasi-convexity and lower semicontinuity of multiple variational integrals of any order, *Trans. Amer. Math. Soc.* **119**, 125–149.

MINDLIN, R.D. [1964]: Micro-structure in linear elasticity, *Arch. Rational Mech. Anal.* **16**, 51–78.

MINDLIN, R.D. [1965]: Second gradient of strain and surface tension in linear elasticity, *Int. J. Solids Structures* **1**, 417–438.

MIRSKY, L. [1959]: On the trace of matrix products, *Math. Nach.* **20**, 171–174.

MIRSKY, L. [1975]: A trace inequality of John von Neumann, *Monat. für Math.* **79**, 303–306.

MIYOSHI, T. [1985]: *Foundations of the Numerical Analysis of Plasticity*, North-Holland, Amsterdam.

MOREAU, J.J. [1974]: On unilateral constraints, friction and plasticity, in *New Variational Techniques in Mathematical Physics* (G. CAPRIZ & G. STAMPACCHIA, Editors), pp. 173–322, C.I.M.E., Edizioni Cremonese, Rome.

MOREAU, J.J. [1979]: Lois d'élasticité en grande déformation, *Séminaire d'Analyse Convexe*, Exposé n° 12, Montpellier.

MORREY, JR., C.B. [1952]: Quasi-convexity and the lower semicontinuity of multiple integrals, *Pacific J. Math.* **2**, 25–53.

MORREY, JR., C.B. [1966]: *Multiple Integrals in the Calculus of Variations*, Springer, Berlin.

MÜLLER, S. [1987]: Homogenization of nonconvex integral functionals and cellular elastic materials, *Arch. Rational Mech. Anal.* **99**, 189–212.

MURAT, F. [1978]: Compacité par compensation, *Annali Scu. Norm. Sup. Pisa Ser. IV*, **5**, 489–507.

MURAT, F. [1979]: Compacité par compensation II, in *Proceedings International Conference on Recent Methods in Non-Linear Analysis, Rome, 1978* (E. DE GIORGI, E. MAGENES & U. Mosco, Editors), pp. 245–256, Pitagora, Bologna.

MURAT, F. [1981]: Compacité par compensation: condition nécessaire et suffisante de continuité faible sous une hypothèse de rang constant, *Annali Sc. Norm. Sup. Pisa* **8**, 69–102.

MURAT, F. [1987]: A survey on compensated compactness, in *Contributions to Modern Calculus of Variations* (L. CESARI, Editor), pp. 145–183, Longman, Harlow.

MURDOCH, A.I. [1979]: Symmetry considerations for materials of second grade, *J. Elasticity* **9**, 43–50.

MURNAGHAN, F.D. [1937]: Finite deformations of an elastic solid, *American Journal of Mathematics* **59**, 235–260.

MURNAGHAN, F.D. [1951]: *Finite Deformations of an Elastic Solid*, John Wiley, New York.

NEČAS, J. [1967]: *Les Méthodes Directes en Théorie des Equations Elliptiques*, Masson, Paris.

NEČAS, J. [1976]: Theory of locally monotone operators modeled on the finite displacement theory for hyperelasticity, *Beiträge zur Analysis* **8**, 103–114.

NEČAS, J. [1981]: *Régularité des Solutions Faibles d'Equations Elliptiques Non Linéaires; Applications à l'Elasticité*, Lecture Notes, Laboratoire d'Analyse Numérique, Université Pierre et Marie Curie, Paris.

NEČAS, J. [1983a]: On regular solutions to the displacement boundary value problem in finite elasticity, in *Trends in Applications of Pure Mathematics to Mechanics, Vol. IV* (J. BRILLA, Editor), Pitman, Boston.

NEČAS, J. [1983b]: *Introduction to the Theory of Nonlinear Elliptic Equations*, Teubner–Texte zur Mathematik, Band 52, Leipzig.

NEČAS, J. HLAVÁČEK, I. [1981]: *Mathematical Theory of Elastic and Elasto-Plastic Bodies: An Introduction*, Elsevier, Amsterdam.

NEČAS, J.; JARUŠEK, J.; HASLINGER, J. [1980]: On the solution of the variational inequality to the Signorini problem with small friction, Boll. UMI **17**, 796.

VON NEUMANN, J. [1937]: Some matrix-inequalities and metrization of matric-space, *Tomsk Univ. Rev.* 1, 286–300.

NIRENBERG, L. [1974]: *Topics in Nonlinear Functional Analysis*, Lecture Notes, Courant Institute, New York.

NITSCHE, J.A. [1981]: On Korn's second inequality, *RAIRO Analyse Numérique* **15**, 237–248.

NOLL, W. [1955]: On the continuity of the solid and fluid states, *J. Rational Mech. Anal.* **4**, 3–81.

NOLL, W. [1958]: A mathematical theory of the mechanical behavior of continuous media, *Arch. Rational Mech. Anal.* **2**, 197–226.

NOLL, W. [1959]: The Foundations of classical mechanics in the light of recent advances in continuum mechanics, in *The axiomatic method, with Special Reference to Geometry and Physics*, pp. 266–281, North-Holland, Amsterdam.

NOLL, W. [1966]: The foundations of mechanics, in *C.I.M.E., Non Linear Continuum Theories* [G. GRIOLI & C. TRUESDELL, Editors], pp. 159–200, Cremonese, Rome.

NOLL, W. [1972]: A new mathematical theory of simple materials, *Arch. Rational Mech. Anal.* **48**, 1–50.

Noll, W. [1973]: Lectures on the foundations of continuum mechanics and thermodynamics, *Arch. Rational Mech. Anal.* **52**, 62–92.

Noll, W. [1978]: A general framework for problems in the statics of finite elasticity, in *Contemporary Developments in Continuum Mechanics and Partial Differential Equations* (G.M. de La Penha & L.A.J. Medeiros, Editors), pp. 363–387, North-Holland, Amsterdam.

Noll, W.; Podio-Guidugli, P. [1986]: Discontinuous displacements in elasticity, in *Contributi del Centro Linceo Interdisciplinare di Scienze Matematiche e loro Applicazioni, n. 76.*, pp. 151–163, Accademia dei Lincei, Roma.

Novozhilov, V.V. [1953]: *Foundations of the Nonlinear Theory of Elasticity*, Graylock Press, Rochester.

Novozhilov, V.V. [1961]: *Theory of Elasticity*, Pergamon Press, Oxford, 1961.

Nzengwa, R. [1987]: Incremental methods in nonlinear, three-dimensional, incompressible elasticity, *Modélisation Mathématique et Analyse Numérique*, to appear.

Oden, J.T. [1972]: *Finite Elements of Nonlinear Continua*, McGraw-Hill, New York.

Oden, J.T. [1979]: Existences theorems for a class of problems in nonlinear elasticity, *J. Math. Anal. Appl.* **69**, 51–83.

Oden, J.T. [1986]: *Qualitative Methods in Nonlinear Mechanics*, Prentice-Hall.

Oden, J.T.; Martins, J.A.C. [1985]: Models and computational methods for dynamic friction phenomena, *Comput. Meth. Applied Mech. Engrg.* **52**, 527–634.

Oden, J.T.; Reddy, J.N. [1983]: *Variational Methods in Theoretical Mechanics*, Universitext series, Springer-Verlag, Berlin (second edition).

Ogden, R.W. [1970]: Compressible isotropic elastic solids under finite strain-constitutive inequalities, *Quart. J. Mech. Appl. Math.* **23**, 457–468.

Ogden, R.W. [1972a]: Large deformation isotropic elasticity: on the correlation of theory and experiment for incompressible rubber-like solids, *Proc. Roy. Soc. London* **A326**, 565–584.

Ogden, R.W. [1972b]: Large deformation isotropic elasticity: on the correlation of theory and experiment for compressible rubber-like solids, *Proc. Roy. Soc. London* **A328**, 567–583.

Ogden, R.W. [1976]: Volume changes associated with the deformation of rubber-like solids, *J. Mech. Phys. Solids* **24**, 323–338.

Ogden, R.W. [1977]: Inequalities associated with the inversion of elastic stress-deformation relations and their implications, *Math. Proc. Camb. Phil. Soc.* **81**, 313–324.

Ogden, R.W. [1984]: *Non-Linear Elastic Deformations*, Ellis Horwood, Chichester, and John Wiley.

Oleinik, O.A. [1984]: On homogenization problems, in *Trends and Applications of Pure Mathematics to Mechanics* (P.G. Ciarlet & M. Roseau, Editors), pp. 248–272, Springer-Verlag, Berlin.

Ostrowski, A.M. [1966]: *Solution of Equations and Systems of Equations*, (second edition), Academic Press, New York.

Owen, N. [1986]: On a variational problem from one-dimensional nonlinear elasticity, *Lefchetz Center for Dynamical Systems Report No. 86-30*, Brown University, Providence.

Owen, N. [1987]: Existence and stability of necking deformations for nonlinearly elastic rods, *Arch. Rational Mech. Anal.* **98**, 357–383.

Panagiotopoulos, P.D. [1975]: A nonlinear programming approach to the unilateral contact and friction boundary value problem in the theory of elasticity, *Ing. Archiv.* **44**, 421.

PANAGIOTOPOULOS, P.D. [1985]: *Inequality Problems in Mechanics and Applications*, Birkhäuser, Boston.

PAUMIER, J.C. [1985]: *Analyse de Certains Problèmes Non Linéaires: Modèles de Plaques et de Coques*, Thesis, Université Pierre et Marie Curie, Paris.

PEARSON, C.E. [1956]: General theory of elastic stability, *Quart. Appl. Math.* **14**, 133–144.

PENG, S.T.J.; LANDEL, R.F. [1975]: Stored energy function and compressibility of compressible rubber-like materials under large strain, *J. Applied Phys.* **46**, 2599–2604.

PETROVSKII, I.G. [1950], *Lectures on Partial Differential Equations*; Gosudarstv. Izdat. Techn. – Teor. Lit., Moscow.

PIAN, T.H.H. [1976]: Variational principles for incremental finite element methods, *Journal of the Franklin Institute* **302**, 473–488.

PIERCE, J.F.; WHITMAN, A.P. [1980]: Topological properties of the manifolds of configurations for several simple deformable bodies, *Arch. Rational Mech. Anal.* **74**, 101–113.

PIETRASZKIEWICZ, W. [1982]: Determination of displacements from given strains in the non-linear Continuum Mechanics, *ZAMM* **62**, 154–156.

PIETRASZKIEWICZ, W.; BADUR, J. [1983a]: Finite rotations in the description of continuum deformation, *Int. J. Engrg. Sci.* **21**, 1097–1115.

PIETRASZKIEWICZ, W.; BADUR, J. [1983b]: On non-classical forms of compatibility conditions in continuum mechanics; in *Trends in Application of Pure Mathematics to Mechanics, Vol. IV*, pp. 197–227, Pitman, Boston.

PITTERI, M. [1984]: Reconciliation of local and global symmetries of crystals, *J. Elasticity* **14**, 175–190.

PITTERI, M. [1985]: On the kinematics of mechanical twinning in crystals, *Arch. Rational Mech. Anal.* **88**, 25–57.

PODIO-GUIDUGLI, P. [1987a]: The Piola-Kirchhoff stress may depend linearly on the deformation gradient, *J. Elasticity* **17**, 123–141.

PODIO-GUIDUGLI, P. [1987b]: A variational approach to live loadings in finite elasticity, *J. Elasticity*, to appear.

PODIO-GUIDUGLI, P.; VERGARA-CAFFARELLI, G. [1984]: On a class of live traction problems in elasticity, in *Trends and Applications of Pure Mathematics to Mechanics* (P.G. CIARLET & M. ROSEAU, Editors), pp. 291–304, Springer-Verlag, Berlin.

PODIO-GIUDUGLI, P.; VERGARA-CAFFARELLI, G.; VIRGA, E.G. [1986]: Discontinuous energy minimizers in nonlinear elastostatics: an example of J. BALL revisited, *J. Elasticity* **16**, 75–96.

PODIO-GUIDUGLI, P.; VERGARA-CAFFARELLI, G.; VIRGA, E.G. [1987]: The role of ellipticity and normality assumptions in formulating live-boundary conditions in elasticity, *Quarterly of Applied Mathematics* **44**, 659–664.

POTIER-FERRY, M. [1982]: On the mathematical foundations of elastic stability theory. I, *Arch. Rational Mech. Anal.* **78**, 55–72.

POURCIAU, B.H. [1983]: Univalence and degree for Lipschitz continuous maps, *Arch. Rational Mech. Anal.* **81**, 289–299.

PRAGER, W. [1957]: On ideal locking materials, *Transactions of the Society of Rheology* **1**, 169–175.

PRAGER, W. [1958]: Elastic solids of limited compressibility, in *Proceedings 9th International Congress of Applied Mechanics, Brussels*, 5, pp. 205–211.

PRAGER, W. [1964]: Unilateral constraints in Mechanics of Continua, in *Estratto dagli Atti del Simposio Lagrangiano*, Accademia delle Scienze di Torino, pp. 1–11.

QUINTELA-ESTEVEZ, P. [1986]: *Elasticidad No Lineal. Modelo de von Kármán*, Thesis, Universidad Autonoma, Madrid.

RABINOWITZ, P.H. [1975]: *Théorie du Degré Topologique et Application à des Problèmes aux Limites Non Linéaires* (rédigé par H. BERESTYCKI), publication 75010 du Laboratoire d'Analyse Numérique, Université de Paris VI, Paris.

RADO, T.; REICHELDERFER, P.V. [1955]: *Continuous transformations in Analysis*, Springer-Verlag, Berlin.

RAJAGOPAL, K.R.; WINEMAN, A.S. [1987]: New universal relations for nonlinear isotropic elastic materials, *J. Elasticity* **17**, 75–83.

RAOULT, A. [1985]: Construction d'un modèle d'évolution de plaques avec terme d'inertie de rotation, *Annali di Matematica Pura ed Applicata* **139**, 361–400.

RAOULT, A. [1986]: Non-polyconvexity of the stored energy function of a Saint Venant-Kirchhoff material, *Aplikace Matematiky*, **6**, 417–419.

REISSNER, E. [1984]: Formulation of variational theorems in geometrically nonlinear elasticity, *J. Engrg. Mech.* **110**, 1377–1390.

REISSNER, E. [1986]: Some aspects of the variational principles problem in elasticity, *Computational Mechanics* **1**, 3–9.

RHEINBOLDT, W.C. [1974]: *Methods for Solving Systems of Nonlinear Equations*, CBMS Series, N° 14, SIAM, Philadelphia.

RHEINBOLDT, W.C. [1981]: Numerical analysis of continuation methods for nonlinear structural problems, *Computers & Structures* **13**, 103–113.

RHEINBOLDT, W.C. [1986]: *Numerical Analysis of Parametrized Nonlinear Equations*, John Wiley, New York.

RIVLIN, R.S. [1948]: Large elastic deformations of isotropic materials. II. Some uniqueness theorems for pure homogeneous deformation, *Philos. Trans. Roy. Soc. London Ser. A* **240**, 491–508.

RIVLIN, R.S. [1973]: Some restrictions on constitutive equations, in *Proceedings, International Symposium on the Foundations of Continuum Thermodynamics*, Bussaco.

RIVLIN, R.S. [1974]: Stability of pure homogeneous deformations of an elastic cube under dead loading, *Quart. Applied Math.* **32**, 265–271.

RIVLIN, R.S.; ERICKSEN, J.L. [1955]: Stress-deformation relations for isotropic materials, *J. Rational Mech. Anal.* **4**, 323–425.

ROBERTS, A.W.; VARBERG, D.E. [1973]: *Convex Functions*, Academic Press, New York.

ROCKAFELLAR, R.T. [1970]: *Convex Analysis*, Princeton University Press, Princeton.

RODRIGUES, J.F. [1987]: *Obstacle Problems in Mathematical Physics*, North-Holland, Amsterdam.

ROMANO, G. [1972]: Potential operators and conservative systems, *Meccanica* **7**, 141–146.

ROSEAU, M. [1976]: *Equations Différentielles*, Masson, Paris.

ROSEMAN, J.J. [1981]: An integral bound on the strain energy for the traction problem in non-linear elasticity with sufficiently small strains, *Internat. J. Non-Linear Mech.* **16**, 317–325.

ROSTAMIAN, R. [1978]: Internal constraints in boundary value problems of continuum mechanics, *Indiana Univ. Mathematics J.* **27**, 637–656.

SABLÉ-TOUGERON, M. [1987]: Existence pour un problème de l'élastodynamique Neumann non linéaire en dimension 2, to appear.

SALENÇON, J. [1983]: *Calcul à la Rupture et Analyse Limite*, Presses de l'Ecole Nationale des Ponts et Chaussées, Paris.

SANCHEZ-PALENCIA, E. [1980]: *Nonhomogeneous Media and Vibration Theory*, Lecture Notes in Physics, Vol. 127, Springer-Verlag, Berlin.

SAWYERS, K.N. [1976]: Stability of an elastic cube under dead loading, *Internat. J. Non-Linear Mech.* **11**, 11–23.

SCHOCHET, S. [1985]: The incompressible limit in nonlinear elasticity, *Comm. Math. Phys.* **102**, 207–215.

SCHWARTZ, L. [1966]: *Théorie des Distributions*, Hermann, Paris.

SCHWARTZ, L. [1967]: *Cours d'Analyse*, Hermann, Paris.

SCHWARTZ, L. [1970]: *Analyse, Deuxième Partie: Topologie Générale et Analyse Fonctionnelle*, Hermann, Paris.

SCOTT, N.H. [1986]: The slowness surfaces of incompressible and nearly incompressible elastic materials, *J. Elasticity* **16**, 239–250.

SERRE, D. [1981a]: Relations d'ordre entre formes quadratiques en compacité par compensation, *C.R. Acad. Sci. Paris Sér. I*, **292**, 785–787.

SERRE, D. [1981b]: Condition de Legendre-Hadamard; espaces de matrices de rang $\neq 1$, *C.R. Acad. Sci. Paris Sér. I* **293**, 23–26.

SERRE, D. [1983]: Formes quadratiques et calcul des variations, *J. Math. Pures Appl.* **62**, 177–196.

SERRIN, J. [1961]: On the definition and properties of certain variational integrals, *Trans. Amer. Math. Soc.* **101**, 139–167.

SEWELL, M.J. [1967]: On configuration-dependent loading, *Arch. Rational Mech. Anal.* **23**, 327–351.

SHAMINA, V.A. [1974]: Determination of displacement vector from the components of strain tensor in nonlinear continuum mechanics (in Russian), *Izv. AN SSR, Mech. of Solids*, **1**, 14–22.

SHIELD, R. [1971]: Deformations possible in every compressible, isotropic, perfectly elastic material, *J. Elasticity* **1**, 91–92.

SIDOROFF, R. [1974]: Sur les restrictions à imposer à l'énergie de déformation d'un matériau hyperélastique, *C.R. Acad. Sci. Paris* **279**, 379–382.

SIGNORINI, A. [1930]: Sulle deformazioni termoelastiche finite, in *Proc. 3rd Intern. Congress Applied Mechanics* **2**, pp. 80–89.

SIGNORINI, A. [1933]: Sopra alcune questioni di elastostatica, *Atti Soc. Ital. per il Progresso delle Scienze*.

SIGNORINI, A. [1943]: Trasformazioni termoelastiche finite, Memoria 1ª, *Annali Mat. Pura Appl.* **22**, 33–143.

SIGNORINI, A. [1949]: Trasformazioni termoelastiche finite, Memoria 2ª, *Annali Mat. Pura Appl.* **30**, 1–72.

SIGNORINI, A. [1959]: Questioni di elasticità nonlinearizzata o semilinearizzata, *Rend. di Matem. e delle sue Appl.* **18**, 1–45.

SIMO, J.C.; MARSDEN, J.E. [1984a]: On the rotated stress tensor and the material version of the Doyle-Ericksen formula, *Arch. Rational Mech. Anal.* **86**, 213–231.

SIMO, J.C.; MARSDEN, J.E. [1984b]: Stress tensors, Riemannian metrics and the alternative descriptions in elasticity, in *Trends and Applications of Pure Mathematics to Mechanics* (P.G. CIARLET & M. ROSEAU, Editors), pp. 369–383, Springer-Verlag, Berlin.

SIMPSON, H.C.; SPECTOR, S.J. [1983]: On compositive matrices and strong ellipticity for isotropic elastic materials, *Arch. Rational Mech. Anal.* **84**, 55–68.

SIMPSON, H.C.; SPECTOR, S.J. [1984a]: On barrelling instabilities in finite elasticity, *J. Elasticity* **14**, 103–125.

SIMPSON, H.C.; SPECTOR, S.J. [1984b]: On barrelling for a special material in finite elasticity, *Quart. Applied Math.* **42**, 99–111.

SIMPSON, H.C.; SPECTOR, S.J. [1987]: On the positivity of the second variation in finite elasticity, *Arch. Rational Mech. Anal.* **98**, 1–30.

SIVALOGANATHAN, J. [1986a]: Uniqueness of regular and singular equilibria for spherically symmetric problems of nonlinear elasticity, *Arch. Rational Mech. Anal.* **96**, 97–136.

SIVALOGANATHAN, J. [1986b]: A field theory approach to stability of radial equilibria in nonlinear elasticity, to appear.

SMITH, K.T. [1983]: *Primer of Modern Analysis*, Second Edition, Springer-Verlag, New York.

SOBOLEV, S.L. [1938]: On a theorem of functional analysis, *Mat. Sb.* **46**, 471–496.

SOBOLEV, S.L. [1950]: *Applications of Functional Analysis in Mathematical Physics*, Leningrad (English translation: *Transl. Amer. Math. Soc.*, Mathematical Monograph 7, 1963).

SOKOLNIKOFF, I.S. [1956]: *Mathematical Theory of Elasticity*, McGraw-Hill.

SPAGNOLO, S. [1976]: Convergence in energy for elliptic operators, in *Numerical Solutions of Partial Differential Equations*, SYNSPADE III (B.E. HUBBARD, editor), pp. 469–498, Academic Press.

SPECTOR, S.J. [1980]: On uniqueness in finite elasticity with general loading, *J. Elasticity* **10**, 145–161.

SPECTOR, S.J. [1982]: On uniqueness for the traction problem in finite elasticity, *J. Elasticity* **12**, 367–383.

STEPHENSON, R.A. [1980]: On the uniqueness of the square-root of a symmetric, positive-definite tensor, *J. Elasticity* **10**, 213–214.

STETTER, H.J. [1973]: *Analysis of Discretization Methods for Ordinary Differential Equations*, Springer-Verlag, New York.

STOKER, J.J. [1968]: *Nonlinear Elasticity*, Gordon and Breach, New York.

STOPPELLI, F. [1954]: Un teorema di esistenza e di unicità relativo alle equazioni dell'elastostatica isoterma per deformazioni finite, *Ricerche di Matematica* **3**, 247–267.

STOPPELLI, F. [1955]: Sulla svilluppabilità in serie di potenze di un parametro delle soluzioni delle equazioni dell'elastostatic isoterma, *Ricerche di Matematica* **4**, 58–73.

STORÅKERS, B. [1979]: An explicit method to determine response coefficients in finite elasticity, *J. Elasticity* **9**, 207–214.

STRANG, G. [1976]: *Linear Algebra and its Applications*, Academic Press, New York.

STUART, C.A. [1985]: Radially symmetric cavitation for hyperelastic materials, *Ann. Institut Henri Poincaré* **2**, 33–66.

STUART, C.A. [1986]: Special problems involving uniqueness and multiplicity in hyperelasticity, in *Nonlinear Functional Analysis and its Applications* (S.P. SINGH, editor), pp. 131–145, D. Reidel.

DE ST VENANT, A.-J.-C.B. [1844]: Sur les pressions qui se développent à l'intérieur des corps solides lorsque les déplacements de leurs points, sans altérer l'élasticité, ne peuvent cependant pas être considérés comme très petits, *Bull. Soc. Philomath.* **5**, 26–28.

SUQUET, P. [1979]: Sur les équations de la plasticité, *Ann. Fac. Sci. Toulouse* **1**, 77–87.

SUQUET, P. [1981]: Sur les équations de la plasticité, existence et régularité des solutions, *J. Mécanique* **20**, 3–39.

ŠVERÁK, V. [1987]: Regularity properties of deformations with finite energy, *Arch. Rational Mech. Anal.*, to appear.

SYNGE, J.L. [1960]: Classical dynamics, in *Handbuch der Physic* (S. FLÜGGE, Editor), Vol. III/1, pp. 1–225.

TAHRAOUI, R. [1986]: Théorèmes d'existence en calcul des variations et applications à l'élasticité non linéaire, *C.R. Acad. Sci. Paris Sér. I*, **14**, 495–498.

TANG Qi [1987]: Almost everywhere injectivity in nonlinear elasticity, *Proc. Royal Soc. Edinburgh*, to appear.

Tartar, L. [1977]: *Problèmes d'Homogénéisation dans les Equations aux Dérivées Partielles*, Cours Peccot, Collège de France, Paris.

Tartar, L. [1978]: Quelques remarques sur l'homogénéisation, in *Functional Analysis and Numerical Analysis* (H. Fujita, Editor), pp. 469–481, Japan Society for the Promotion of Science, Tokyo.

Tartar, L. [1979]: Compensated compactness and partial differential equations, in *Nonlinear Analysis and Mechanics, Heriot-Watt Symposium, Vol. IV* (R.J. Knops, Editor), pp. 136–212, Pitman.

Tartar, L. [1983a]: The compensated compactness method applied to systems of conservation laws, in *Systems of Nonlinear Partial Differential Equations* (J.M. Ball, Editor), pp. 263–285, Reidel, Dordrecht.

Tartar, L. [1983b]: Etude des oscillations dans les équations aux dérivées partielles non linéaires, in *Trends and Applications of Pure Mathematics to Mechanics* (P.G. Ciarlet & M. Roseau, Editors), pp. 384–412.

Taylor, A.E. [1965]: *General Theory of Functions and Integration*, Blaisdell, Waltham.

Temam, R. [1983]: *Problèmes Mathématiques en Plasticité*, Gauthier-Villars, Paris (English translation: *Mathematical Problems in Plasticity*, Gauthier-Villars, Paris, 1985).

Temam, R.; Strang, G. [1980a]: Duality and relaxation in the variational problems of plasticity, *J. de Mécanique* **19**, 493–527.

Temam, R.; Strang, G. [1980b]: Functions of bounded deformation, *Arch. Rational Mech. Anal.* **75**, 7–21.

Thompson, J.L. [1969]: Some existence theorems for the traction boundary value problem of linearized elastostatics, *Arch. Rational Mech. Anal.* **32**, 369–399.

Thompson, R.C.; Freede, L.J. [1971]: On the eigenvalues of sums of Hermitian matrices, *Linear Algebra and Appl.* **4**, 369–376.

van Tiel, J. [1984]: *Convex Analysis: An Introductory Text*, J. Wiley, New York.

Timoshenko, S. [1951]: *Theory of Elasticity*, McGraw-Hill.

Ting, T.C.T. [1985]: Détermination of $C^{1/2}$, $C^{-1/2}$ and more general isotropic tensor functions of C, *J. Elasticity* **15**, 319–323.

Tonelli, L. [1920]: La semicontinuità nel calcolo delle variazioni, *Rend. Circ. Matem. Palermo* **44**, 167–249.

Toupin, R.A. [1962]: Elastic materials with couple-stresses, *Arch. Rational Mech. Anal.* **11**, 385–414.

Toupin, R.A. [1964]: Theories of elasticity with couple-stress, *Arch. Rational Mech. Anal.* **17**, 85–112.

Treloar, L.R.G. [1975]: *The Physics of Rubber Elasticity*, Oxford University Press, Oxford.

Triantafyllidis, N.; Aifantis, E.C. [1986]: A gradient approach to localization of deformation. I. Hyperelastic materials, *J. Elasticity* **16**, 225–237.

Troianiello, G.M. [1987]: *Elliptic Differential Equations and Obstacle Problems*, Plenum, New York.

Truesdell, C. [1977]: *A First Course in Rational Continuum Mechanics*, Academic Press, New York.

Truesdell, C. [1978]: Some challenges offered to analysis by rational thermomechanics, in *Contemporary Developments in Continuum Mechanics and Partial Differential Equations* (G.M. de La Penha & L.A.J. Medeiros, Editors), pp. 495–603, North-Holland, Amsterdam.

Truesdell, C. [1983]: The influence of elasticity on analysis: The classic heritage, *Bull. Amer. Math. Soc.* **9**, 293–310.

TRUESDELL, C.; NOLL, W. [1965]: *The Non-Linear Field Theories of Mechanics, Handbuch der Physik*, Vol. III/3, Springer, Berlin.

TRUESDELL, C.; TOUPIN, R.A. [1960]: *The Classical Field Theories, Handbuch der Physik*, Vol. III/1, Springer, Berlin.

TRUESDELL, C.; TOUPIN, R.A. [1963]: Static grounds for inequalities in finite elastic strain, *Arch. Rational Mech. Anal.* **12**, 1–33.

VAINBERG, M.M. [1952]: Some questions of differential calculus in linear spaces, *Usp. Math. Nauk* **7**, 55.

VAINBERG, M.M. [1956]: *Variational Methods for the Study of Nonlinear Operators* (in Russian), G.I.T.T.L., Moscow (English translation, Holden-Day, San Francisco, 1964).

VALENT, T. [1978a]: Sulla differenziabilità dell'operatore di Nemytsky, *Rend. Acc. Naz. Lincei* **65**, 1–12.

VALENT, T. [1978b]: Osservazioni sulla linearizzazione di un operatore differenziale, *Rend. Acc. Naz. Lincei* **65**, 1–11.

VALENT, T. [1978c]: Teoremi di esistenza e unicità in elastostatica finita, *Rend. Sem. Mat. Univ. Padova* **60**, 165–181.

VALENT, T. [1982]: Local theorems of existence and uniqueness in finite elastostatics, in *Finite Elasticity* (D.E. CARLSON & R.T. SHIELD, Editors), pp. 401–421, Nijhoff, The Hague.

VALENT, T. [1985]: A property of multiplication in Sobolev spaces. Some applications, *Rend. Sem. Mat. Univ. Padova* **74**, 63–73.

VALENT, T.; ZAMPIERI, G. [1977]: Sulla differenziabilità di un operatore legato a una classe di sistemi differenziali quasi-lineari, *Rend. Sem. Mat. Univ. Padova* **57**, 311–322.

VALID, R. [1977]: *La Mécanique des Milieux Continus et le Calcul des Structures*, Eyrolles, Paris (English translation: *Mechanics of Continuous Media and Analysis of Structures*, North-Holland, Amsterdam, 1981).

VARGA, O.H.: *Stress-strain Behaviour of Elastic Materials*, Interscience, Wiley, New York, 1966.

VILLAGGIO, P. [1972]: Energetic bounds in finite elasticity, *Arch. Rational Mech. Anal.* **45**, 282–293.

VILLAGGIO, P. [1977]: *Qualitative Methods in Elasticity*, Noordhoff, Leyden.

VILLAGGIO, P. [1980]: A unilateral contact problem in linear elasticity, *J. Elasticity* **10**, 113–119.

VIŠIK, M.I. [1976]: Sobolev-Slobodeckii spaces of variable order with weighted norms and their applications to mixed elliptic boundary problems, *Amer. Math. Sol. Transl. Ser. 2*, **105**, 104–110.

VODOPYANOV, S.K.; GOLDSHTEIN, V.M. [1977]: Quasiconformal mappings and spaces of functions with generalized first derivatives, *Siberian Math. J.* **17**, 515–531.

VODOPYANOV, S.K.; GOLDSHTEIN, V.M.; RESHETNYAK, YU. G. [1979]: On geometric properties of functions with generalized first derivatives, *Russian Math. Surveys* **34**, 19–74.

VOIGT, W. [1893–1894]: Ueber eine anscheinend notwendige Erweiterung der Theorie der Elasticität, *Nachrichten von der Königlichen Gesellochaft der Wissenschaften zu Göttingen*, pp. 534–552 (1893); pp. 33–42 (1894).

VO-KHAC, K. [1972]: *Distributions, Analyse de Fourier, Opérateurs aux Dérivées Partielles*, Vol. 1, Vuibert, Paris.

WAN, Y.H. [1986]: The traction problem for incompressible materials, to appear.

WAN, Y.H.; MARSDEN, J.E. [1984]: Symmetry and bifurcation in three-dimensional elasticity, Part III: Stressed reference configuration, *Arch. Rational Mech. Anal.* **84**, 203–233.

WANG, C.-C.; TRUESDELL, C. [1973]: *Introduction to Rational Elasticity*, Noordhoff, Groningen.

WASHIZU, K. [1975]: *Variational Methods in Elasticity and Plasticity*, Second Edition, Pergamon, Oxford.

WEINSTEIN, A. [1985]: A global invertibility theorem for manifolds with boundary, *Proc. Royal Soc. of Edinburgh* **99**, 283–284.

WHEELER, L. [1977]: A uniqueness theorem for the displacement problem in finite elastodynamics, *Arch Rational Mech. Anal.* **63**, 183–189.

YALE, P.B. [1968]: *Geometry and Symmetry*, Holden-Day, San Francisco.

YANG, W.H. [1980]: A generalized von Mises criterion for yield and fracture, *J. Applied Mechanics* **47**, 297–300.

YOSIDA, K. [1966]: *Functional Analysis*, Springer-Verlag, Berlin.

YOUNG, L.C. [1937]: Generalized curves and the existence of an attained absolute minimum in the calculus of variations, *C.R. Soc. Sci. Lett. Varsovie Classe III* **30**, 212–234.

YOUNG, L.C. [1942]: Generalized surfaces in the calculus of variations I and II, *Ann. Math.* **43**, 84–103 and 530–544.

YOUNG, L.C. [1980]: *Lectures on the Calculus of Variations and Optimal Control Theory*, Chelsea.

ZEE, L.; STRENBERG, E. [1983]: Ordinary and strong ellipticity in the equilibrium theory of incompressible hyperelastic solids, *Arch. Rational Mech. Anal.* **83**, 53–90.

ZEIDLER, E. [1986]: *Nonlinear Functional Analysis and its Applications, Vol. I: Fixed-Point Theorems*, Springer-Verlag, New York.

ZIEMER, W.P. [1983]: Cauchy flux and sets of finite perimeters, *Arch. Rational Mech. Anal.* **84**, 189–201.

ZIENKIEWICZ, O.C. [1977]: *The Finite Element Method in Engineering Science*, McGraw-Hill, New York (third edition).

INDEX

Printed and bound by CPI Group (UK) Ltd, Croydon, CR0 4YY
03/10/2024
01040432-0020